D1256389

Springer Series in Computational Physics

Editors
H. Cabannes M. Holt
H. B. Keller J. Killeen S. A. Orszag

Springer Series in Computational Physics

Editors: H. Cabannes, M. Holt, H. B. Keller, J. Killeen, S. A. Orszag

F. Bauer/ O. Betancourt/ P. Garabedian: A Computational Method in Plasma Physics
1978. vi, 144 pages. 22 figures. ISBN 08833-4

D. Book (ed.): Finite-Difference Techniques for Vectorized Fluid Dynamics Calculations
1981. viii, 240 pages. 60 figures. ISBN 10482-8

C.A.J. Fletcher: Computational Galerkin Methods
1984 xi, 309 pages. 107 figures. ISBN 12633-3

R. Glowinski: Numerical Methods for Nonlinear Variational Problems
1984. xv, 493 pages. 80 figures. ISBN 12434-9

M. Holt: Numerical Methods in Fluid Dynamics, 2nd ed.
1984. xi, 273 pages. 114 figures. ISBN 12799-2

Kubicek/ Marek: Computational Methods in Bifurcation Theory and Dissipative Structures
1983. xi, 243 pages. 91 figures. ISBN 12070-X

Peyret/Taylor: Computational Methods for Fluid Flow
1982. x, 358 pages. 129 figures. ISBN 11147-6

O. Pironneau: Optimal Shape Design for Elliptic Systems
1983. xiii, 192 pages. 57 figures. ISBN 12069-6

Yu. I. Shokin: The Method of Differential Approximation
1983. xiii, 296 pages. ISBN 12225-7

D. Telionis: Unsteady Viscous Flows
1981. xxiii, 406 pages. 127 figures. ISBN 10481-8

F. Thomasset: Implementation of Finite Element Methods for Navier-Stokes Equations
1981. xii, 161 pages. 86 figures. ISBN 10771-1

GLOWINSKI, R.

Roland Glowinski

Numerical Methods for Nonlinear Variational Problems

With 82 Illustrations

Springer-Verlag
New York Berlin Heidelberg Tokyo

Roland Glowinski
Institut de Recherche d'Informatique et d'Automatique (IRIA)
Domaine de Voluceau, Rocquencourt, B.P. 105
F-78150 Le Chesnay, France

Editors

Henri Cabannes
Mécanique Théorique
Université Pierre et Marie Curie
Tour 66. 4, Place Jussieu
F-75005 Paris
France

M. Holt
Department of Mechanical Engineering
College of Engineering
University of California
Berkeley, CA 94720
U.S.A.

H. B. Keller
Applied Mathematics 101-50
Firestone Laboratory
California Institute of Technology
Pasadena, CA 91125
U.S.A.

John Killeen
Lawrence Livermore Laboratory
P.O. Box 808
Livermore, CA 94551
U.S.A.

Stephen A. Orszag
Department of Mathematics
Massachusetts Institute of Technology
Cambridge, MA 02139
U.S.A

Library of Congress Cataloging in Publication Data
Glowinski, R.
Numerical methods for nonlinear variational problems.
(Springer series in computational physics)
Bibliography: p.
Includes indexes.
1. Variational inequalities (Mathematics)
2. Numerical analysis. I. Title. II. Series.
QA316.G56 1983 515'.26 83-6732

A preliminary version of this book was originally published as part of a set of monographs on numerical analysis in the series Lectures on Mathematics and Physics, by the Tata Institute of Fundamental Research.

© 1984 by Springer-Verlag New York Inc.
All rights reserved. No part of this book may be translated or reproduced in any form without written permission from Springer-Verlag, 175 Fifth Avenue, New York, New York 10010, U.S.A.

Typeset by Composition House Ltd., Salisbury, England.
Printed and bound by R. R. Donnelley & Sons, Harrisonburg, Virginia.
Printed in the United States of America.

9 8 7 6 5 4 3 2 1

ISBN 0-387-12434-9 Springer-Verlag New York Berlin Heidelberg Tokyo
ISBN 3-540-12434-9 Springer-Verlag Berlin Heidelberg New York Tokyo

. . . To my wife Angela and to Mrs. Madeleine Botineau . . .

Preface

When Herb Keller suggested, more than two years ago, that we update our lectures held at the Tata Institute of Fundamental Research in 1977, and then have it published in the collection Springer Series in Computational Physics, we thought, at first, that it would be an easy task. Actually, we realized very quickly that it would be more complicated than what it seemed at first glance, for several reasons:

1. The first version of *Numerical Methods for Nonlinear Variational Problems* was, in fact, part of a set of monographs on numerical mathematics published, in a short span of time, by the Tata Institute of Fundamental Research in its well-known series Lectures on Mathematics and Physics; as might be expected, the first version systematically used the material of the above monographs, this being particularly true for *Lectures on the Finite Element Method* by P. G. Ciarlet and *Lectures on Optimization—Theory and Algorithms* by J. Cea. This second version had to be more self-contained. This necessity led to some minor additions in Chapters I–IV of the original version, and to the introduction of a chapter (namely, Chapter V of this book) on relaxation methods, since these methods play an important role in various parts of this book. For the same reasons we decided to add an appendix (Appendix I) introducing linear variational problems and their approximation, since many of the methods discussed in this book try to reduce the solution of a nonlinear problem to a succession of linear ones (this is true for Newton's method, but also for the augmented Lagrangian, preconditioned conjugate gradient, alternating-direction methods, etc., discussed in several parts of this book).
2. Significant progress has been achieved these last years in computational fluid dynamics, using finite element methods. It was clear to us that this second version had to include some of the methods and results whose efficiency has been proved in the above important applied field. This led to Chapter VII, which completes and updates Chapter VI of the original version, and in which approximation and solution methods for some important problems in fluid dynamics are discussed, such as transonic flows for compressible inviscid fluids and the Navier–Stokes equations

for incompressible viscous fluids. Like the original version, the main goal of this book is to serve as an introduction to the study of nonlinear variational problems, and also to provide tools which may be used for their numerical solution. We sincerely believe that many of the methods discussed in this book will be helpful to those physicists, engineers, and applied mathematicians who are concerned with the solution of nonlinear problems involving differential operators. Actually this belief is supported by the fact that some of the methods discussed in this book are currently used for the solution of nonlinear problems of industrial interest in France and elsewhere (the last illustrations of the book represent a typical example of such situations).

The numerical integration of nonlinear hyperbolic problems has not been considered in this book; a good justification for this omission is that this subject is in the midst of an important evolution at the moment, with many talented people concentrating on it, and we think that several more years will be needed in order to obtain a clear view of the situation and to see which methods take a definitive lead, particularly for the solution of multidimensional problems.

Let us now briefly describe the content of the book.

Chapters I and II are concerned with elliptic variational inequalities (EVI), more precisely with their approximation (mostly by finite element methods) and their iterative solution. Several examples, originating from continuum mechanics, illustrate the methods which are described in these two chapters.

Chapter III is an introduction to the approximation of parabolic variational inequalities (PVI); in addition, we discuss in some detail a particular PVI related to the unsteady flow of some viscous plastic media (Bingham fluids) in a cylindrical pipe.

In Chapter IV we show how variational inequality concepts and methods may be useful in studying some nonlinear boundary-value problems which can be reduced to nonlinear variational equations.

In Chapters V and VI we discuss the iterative solution of some variational problems whose very specific structure allows their solution by relaxation methods (Chapter V) and by decomposition-coordination methods via augmented Lagrangians (Chapter VI); several iterative methods are described and illustrated with examples taken mostly from mechanics.

Chapter VII is mainly concerned with the numerical solution of the full potential equation governing transonic potential flows of compressible inviscid fluids, and of the Navier–Stokes equations for incompressible viscous fluids. We discuss the approximation of the above nonlinear fluid flow problems by finite element methods, and also iterative methods of solution of the approximate problems by nonlinear least-squares and preconditioned conjugate gradient algorithms. In Chapter VII we also emphasize the solution of the Stokes problem by either direct or iterative methods. The results of numerical experiments illustrate the possibilities of the solution methods discussed in Chapter VII, which also contains an introduction to arc-length-continuation methods (H. B. Keller) for solving nonlinear boundary-value problems with *multiple* solutions.

As already mentioned, Appendix I is an introduction to the theory and numerical analysis of linear variational problems, and one may find in it details (some being practical) about the finite element solution of such important boundary-value problems, like those of Dirichlet, Neumann, Fourier, and others.

In Appendix II we describe a finite element method with *upwinding* which may be helpful for solving elliptic boundary-value problems with *large first-order terms.*

Finally, Appendix III, which contains various information and results useful for the practical solution of the Navier–Stokes equations, is a complement to Chapter VII, Sec. 5. (Actually the reader interested in computational fluid mechanics will find much useful theoretical and practical information about the numerical solution of fluid flow problems—Navier–Stokes equations, in particular—in the following books: *Implementation of Finite Element Methods for Navier–Stokes Equations* by F. Thomasset, and *Computational Methods for Fluid Flow* by R. Peyret and T. D. Taylor, both published in the Springer Series in Computational Physics.)

Exercises (without answers) have been scattered throughout the text; they are of varying degrees of difficulty, and while some of them are direct applications of the material in this book, many of them give the interested reader or student the opportunity to prove by him- or herself either some technical results used elsewhere in the text, or results which complete those explicitly proved in the book.

Concerning references, we have tried to include all those available to us and which we consider relevant to the topics treated in this book. It is clear, however, that many significant references have been omitted (due to lack of knowledge and/or organization of the author). Also we apologize in advance to those authors whose contributions have not been mentioned or have not received the attention they deserve.

Large portions of this book were written while the author was visiting the following institutions: the Tata Institute of Fundamental Research (Bombay and Bangalore), Stanford University, the University of Texas at Austin, the Mathematical Research Center of the University of Wisconsin at Madison, and the California Institute of Technology. We would like to express special thanks to K. G. Ramanathan, G. H. Golub, J. Oliger, J. T. Oden, J. H. Nohel, and H. B. Keller, for their kind hospitality and the facilities provided for us during our visits.

We would also like to thank C. Baiocchi, P. Belayche, J. P. Benque, M. Bercovier, H. Beresticky, J. M. Boisserie, H. Brezis, F. Brezzi, J. Cea, T. F. Chan, P. G. Ciarlet, G. Duvaut, M. Fortin, D. Gabay, A. Jameson, G.

Labadie, C. Lemarechal, P. Le Tallec, P. L. Lions, B. Mercier, F. Mignot, C. S. Moravetz, F. Murat, J. C. Nedelec, J. T. Oden, S. Osher, R. Peyret, J. P. Puel, P. A. Raviart, G. Strang, L. Tartar, R. Temam, R. Tremolieres, V. Girault, and O. Widlund, whose collaboration and/or comments and suggestions were essential for many of the results presented here.

We also thank F. Angrand, D. Begis, M. Bernadou, J. F. Bourgat, M. O. Bristeau, A. Dervieux, M. Goursat, F. Hetch, A. Marrocco, O. Pironneau, L. Reinhart, and F. Thomasset, whose permanent and friendly collaboration with the author at INRIA produced a large number of the methods and results discussed in this book.

Thanks are due to P. Bohn, B. Dimoyat, Q. V. Dinh, B. Mantel, J. Periaux, P. Perrier, and G. Poirier from Avions Marcel Dassault/Bréguet Aviation, whose faith, enthusiasm, and friendship made (and still make) our collaboration so exciting, who showed us the essence of a real-life problem, and who inspired us (and still do) to improve the existing solution methods or to discover new ones.

We are grateful to the Direction des Recherches et Etudes Techniques (D.R.E.T.), whose support was essential to our researches on computational fluid dynamics.

We thank Mrs. Françoise Weber, from INRIA, for her beautiful typing of the manuscript, and for the preparation of some of the figures in this book, and Mrs. Frederika Parlett for proofreading portions of the manuscript.

Finally, we would like to express our gratitude to Professors W. Beiglböck and H. B. Keller, who accepted this book for publication in the Springer Series in Computational Physics, and to Professor J. L. Lions who introduced us to variational methods in applied mathematics and who constantly supported our research in this field.

Chevreuse
September 1982

ROLAND GLOWINSKI

Contents

Some Preliminary Comments

To those who might think our approach is too mathematical for a book published in a collection oriented towards computational physics, we would like to say that many of the methods discussed here are used by engineers in industry for solving practical problems, and that, in our opinion, mastery of most of the tools of functional analysis used here is not too difficult for anyone with a reasonable background in applied mathematics. In fact, most of the time the choice of the functional spaces used for the formulation and the solution of a given problem is not at all artificial, but is based on well-known physical principles, such as energy conservation, the virtual work principle, and others.

From a *computational* point of view, a proper choice of the functional spaces used to formulate a problem will suggest, for example, what would be the "good" finite element spaces to approximate it and also the good preconditioning techniques for the iterative solution of the corresponding approximate problem.

"*The Buddha, the Godhead, resides quite as comfortably in the circuits of a digital computer or the gears of a cycle transmission as he does at the top of a mountain or in the petals of a flower.*"

Robert M. Pirsig
Zen and the Art of Motorcycle Maintenance,
William Morrow and Company Inc., New York, 1974

"*En tennis comme en science, certains écarts minimes à la source d'un phénomène peuvent parfois provoquer d'énormes différences dans les effets qu'ils provoquent.*" *

Phillipe Bouin,
L'Equipe, Paris, 2-26-1981

* "In tennis, as in science, certain tiny gaps at the very beginning of a phenomenon can occasionally produce enormous differences in the ensuing results."

CHAPTER I

Generalities on Elliptic Variational Inequalities and on Their Approximation

1. Introduction

An important and very useful class of nonlinear problems arising from mechanics, physics, etc. consists of the so-called *variational inequalities*. We consider mainly the following two types of variational inequalities, namely:

1. elliptic variational inequalities (EVI),
2. parabolic variational inequalities (PVI).

In this chapter (following Lions and Stampacchia [1]), we shall restrict our attention to the study of the *existence*, *uniqueness*, and *approximation* of the solution of EVI (PVI will be considered in Chapter III).

2. Functional Context

In this section we consider two classes of EVI, namely EVI of the *first kind* and EVI of the *second kind*.

2.1. Notation

- V: real Hilbert space with scalar product $(\cdot, \cdot)$ and associated norm $\|\cdot\|$,
- V^*: the dual space of V,
- $a(\cdot, \cdot)$: $V \times V \to \mathbb{R}$ is a bilinear, continuous and *V-elliptic* form on $V \times V$.

A bilinear form $a(\cdot, \cdot)$ is said to be *V-elliptic* if there exists a positive constant α such that $a(v, v) \geq \alpha\|v\|^2$, $\forall\, v \in V$.

In general we do not assume $a(\cdot, \cdot)$ to be symmetric, since in some applications nonsymmetric bilinear forms may occur naturally (see, for instance, Comincioli [1]).

- L: $V \to \mathbb{R}$ continuous, linear functional,
- K is a closed convex nonempty subset of V,
- $j(\cdot)$: $V \to \overline{\mathbb{R}} = \mathbb{R} \cup \{\infty\}$ is a convex lower semicontinuous (l.s.c.) and proper functional ($j(\cdot)$ is proper if $j(v) > -\infty$, $\forall\, v \in V$ and $j \not\equiv +\infty$).

2.2. EVI of the first kind

Find $u \in V$ such that u is a solution of the problem

$$a(u, v - u) \geq L(v - u), \qquad \forall\, v \in K, \quad u \in K. \tag{P_1}$$

2.3. EVI of the second kind

Find $u \in V$ such that u is a solution of the problem

$$a(u, v - u) + j(v) - j(u) \geq L(v - u), \qquad \forall\, v \in V, \quad u \in V. \tag{P_2}$$

2.4. Remarks

Remark 2.1. The cases considered above are the simplest and most important. In Bensoussan and Lions [1] some generalization of problem (P_1) called *quasivariational inequalities* (QVI) are considered, which arises, for instance, from decision sciences. A typical problem of QVI is:

Find $u \in V$ such that

$$a(u, v - u) \geq L(v - u), \qquad \forall\, v \in K(u), \quad u \in K(u),$$

where $v \to K(v)$ is a family of closed convex nonempty subsets of V.

Remark 2.2. If $K = V$ and $j \equiv 0$, then problems (P_1) and (P_2) reduce to the classical variational equation

$$a(u, v) = L(v), \qquad \forall\, v \in V, \quad u \in V.$$

Remark 2.3. The distinction between (P_1) and (P_2) is artificial, since (P_1) can be considered to be a particular case of (P_2) by replacing $j(\cdot)$ in (P_2) by the *indicator functional* I_K of K defined by

$$I_K(v) = \begin{cases} 0 & \text{if } v \in K, \\ +\infty & \text{if } v \notin K. \end{cases}$$

Even though (P_1) is a particular case of (P_2), it is worthwhile to consider (P_1) directly because in most cases it arises naturally, and doing so we will obtain geometrical insight into the problem.

EXERCISE 2.1. Prove that I_K is a convex l.s.c. and proper functional.

EXERCISE 2.2. Show that (P_1) is equivalent to the problem of finding $u \in V$ such that $a(u, v - u) + I_K(v) - I_K(u) \geq L(v - u)$, $\forall\, v \in V$.

3. Existence and Uniqueness Results for EVI of the First Kind

3.1. A theorem of existence and uniqueness

Theorem 3.1 (Lions and Stampacchia [1]). *The problem* (P_1) *has a unique solution.*

PROOF. We first prove the uniqueness and then the existence.

(1) *Uniqueness.* Let u_1 and u_2 be solutions of (P_1). We then have

$$a(u_1, v - u_1) \geq L(v - u_1), \qquad \forall\, v \in K, \quad u_1 \in K, \tag{3.1}$$

$$a(u_2, v - u_2) \geq L(v - u_2), \qquad \forall\, v \in K, \quad u_2 \in K. \tag{3.2}$$

Taking $v = u_2$ in (3.1), $v = u_1$ in (3.2) and adding, we obtain, by using the V-ellipticity of $a(\cdot, \cdot)$,

$$\alpha \|u_2 - u_1\|^2 \leq a(u_2 - u_1, u_2 - u_1) \leq 0,$$

which proves that $u_1 = u_2$ since $\alpha > 0$.

(2) *Existence.* We use a generalization of the proof used by Ciarlet [1]–[3], for example, for proving the Lax–Milgram lemma, i.e., we will reduce the problem (P_1) to a *fixed-point* problem.

By the *Riesz representation theorem* for Hilbert spaces, there exist $A \in \mathscr{L}(V, V)$ ($A = A^t$ if $a(\cdot, \cdot)$ is symmetric) and $l \in V$ such that

$$(Au, v) = a(u, v), \qquad \forall\, u, v \in V \quad \text{and} \quad L(v) = (l, v), \qquad \forall\, v \in V. \tag{3.3}$$

Then the problem (P_1) is equivalent to finding $u \in V$ such that

$$(u - \rho(Au - l) - u, v - u) \leq 0, \qquad \forall\, v \in K, \quad u \in K, \quad \rho > 0. \tag{3.4}$$

This is equivalent to finding u such that

$$u = P_K(u - \rho(Au - l)) \quad \text{for some } \rho > 0, \tag{3.5}$$

where P_K denotes the *projection operator* from V to K in the $\|\cdot\|$ norm. Consider the mapping $W_\rho \colon V \to V$ defined by

$$W_\rho(v) = P_K(v - \rho(Av - l)). \tag{3.6}$$

Let $v_1, v_2 \in V$. Then since P_K is a *contraction* we have

$$\|W_\rho(v_1) - W_\rho(v_2)\|^2 \leq \|v_2 - v_1\|^2 + \rho^2 \|A(v_2 - v_1)\|^2 - 2\rho a(v_2 - v_1, v_2 - v_1).$$

Hence we have

$$\|W_\rho(v_1) - W_\rho(v_2)\|^2 \leq (1 - 2\rho\alpha + \rho^2 \|A\|^2) \|v_2 - v_1\|^2. \tag{3.7}$$

Thus W_ρ is a strict contraction mapping if $0 < \rho < 2\alpha/\|A\|^2$. By taking ρ in this range, we have a unique solution for the fixed-point problem which implies the existence of a solution for (P_1). □

3.2. Remarks

Remark 3.1. If $K = V$, Theorem 3.1 reduces to Lax–Milgram lemma (see Ciarlet [1]–[3]).

Remark 3.2. If $a(\cdot, \cdot)$ is symmetric, then Theorem 3.1 can be proved using optimization methods (see Cea [1], [2]); such a proof is sketched below.

Let $J: V \to \mathbb{R}$ be defined by

$$J(v) = \tfrac{1}{2}a(v, v) - L(v). \tag{3.8}$$

Then

(i) $\lim_{\|v\| \to +\infty} J(v) = +\infty$

since $J(v) = \frac{1}{2}a(v, v) - L(v) \geq (\alpha/2)\|v\|^2 - \|L\| \, \|v\|$.

(ii) J is strictly convex.

Since L is linear, to prove the strict convexity of J it suffices to prove that the functional

$$v \to a(v, v)$$

is strictly convex. Let $0 < t < 1$ and $u, v \in V$ with $u \neq v$; then $0 < a(v - u, v - u) = a(u, u) + a(v, v) - 2a(u, v)$. Hence we have

$$2a(u, v) < a(u, u) + a(v, v). \tag{3.9}$$

Using (3.9), we have

$$\begin{aligned} a(tu + (1 - t)v, tu + (1 - t)v) &= t^2 a(u, u) \\ &\quad + 2t(1 - t)a(u, v) + (1 - t)^2 a(v, v) \\ &< ta(u, u) + (1 - t)a(v, v). \end{aligned} \tag{3.10}$$

Therefore $v \to a(v, v)$ is strictly convex.

(iii) Since $a(\cdot, \cdot)$ and L are continuous, J is continuous.

From these properties of J and standard results of optimization theory (cf. Cea [1], [2], Lions [4], Ekeland and Temam [1]), it follows that the minimization problem of finding u such that

$$J(u) \leq J(v), \qquad \forall\, v \in K, \quad u \in K \tag{π}$$

has a unique solution. Therefore (π) is equivalent to the problem of finding u such that

$$(J'(u), v - u) \geq 0, \qquad \forall\, v \in K, \quad u \in K, \tag{3.11}$$

where $J'(u)$ is the *Gateaux derivative* of J at u. Since $(J'(u), v) = a(u, v) - L(v)$, we see that (P_1) and (π) are equivalent if $a(\cdot, \cdot)$ is symmetric.

EXERCISE 3.1. Prove that $(J'(u), v) = a(u, v) - L(v)$, $\forall\, u,\ v \in V$ and hence deduce that $J'(u) = Au - l, \forall\, u \in V$.

Remark 3.3. The proof of Theorem 3.1 gives a natural algorithm for solving (P_1) since $v \to P_K(v - \rho(Av - l))$ is a *contraction* mapping for $0 < \rho < 2\alpha/\|A\|^2$. Hence we can use the following algorithm to find u:

$$\text{Let } u^0 \in V, \text{ arbitrarily given,} \tag{3.12}$$

then for $n \geq 0$, assuming that u^n is known, define u^{n+1} by

$$u^{n+1} = P_K(u^n - \rho(Au^n - l)). \tag{3.13}$$

Then $u^n \to u$ strongly in V, where u is the solution of (P_1). In practice it is not easy to calculate l and A unless $V = V^*$. To project over K may be as difficult as solving (P_1). In general this method cannot be used for computing the solution of (P_1) if $K \neq V$ (at least not so directly).

We observe that if $a(\cdot, \cdot)$ is symmetric then $J'(u) = Au - l$ and hence (3.13) becomes

$$u^{n+1} = P_K(u^n - \rho(J'(u^n))). \tag{3.13$'$}$$

This method is known as the *gradient-projection* method (with constant step ρ).

4. Existence and Uniqueness Results for EVI of the Second Kind

Theorem 4.1 (Lions and Stampacchia [1]). *Problem* (P_2) *has a unique solution.*

PROOF. As in Theorem 3.1, we shall first prove uniqueness and then existence.

(1) *Uniqueness*. Let u_1 and u_2 be two solutions of (P_2); we then have

$$a(u_1, v - u_1) + j(v) - j(u_1) \geq L(v - u_1), \qquad \forall\, v \in V, \quad u_1 \in V, \tag{4.1}$$

$$a(u_2, v - u_2) + j(v) - j(u_2) \geq L(v - u_2), \qquad \forall\, v \in V, \quad u_2 \in V. \tag{4.2}$$

Since $j(\cdot)$ is a proper functional, there exists $v_0 \in V$ such that $-\infty < j(v_0) < \infty$. Hence, for $i = 1, 2$,

$$-\infty < j(u_i) \leq j(v_0) - L(v_0 - u_i) + a(u_i, v_0 - u_i). \tag{4.3}$$

This shows that $j(u_i)$ is finite for $i = 1, 2$. Hence, by taking $v = u_2$ in (4.1), $v = u_1$ in (4.2), and adding, we obtain

$$\alpha\|u_1 - u_2\|^2 \leq a(u_1 - u_2, u_1 - u_2) \leq 0. \tag{4.4}$$

Hence $u_1 = u_2$.

(2) *Existence.* For each $u \in V$ and $\rho > 0$ we associate a problem (π_ρ^u) of type (P_2) defined as follows.

Find $w \in V$ such that

$$(w, v - w) + \rho j(v) - \rho j(w) \geq (u, v - w)$$
$$+ \rho L(v - w) - \rho a(u, v - w), \qquad \forall v \in V, \quad w \in V. \tag{4.5) (π_ρ^u)}$$

The advantage of considering this problem instead of problem (P_2) is that the bilinear form associated with (π_ρ^u) is the inner product of V which is symmetric.

Let us first assume that (π_ρ^u) has a unique solution for all $u \in V$ and $\rho > 0$. For each ρ define the mapping $f_\rho: V \to V$ by $f_\rho(u) = w$, where w is the unique solution of (π_ρ^u).

We shall show that f_ρ is a uniformly strict contraction mapping for suitably chosen ρ.

Let $u_1, u_2 \in V$ and $w_i = f_\rho(u_i)$, $i = 1, 2$. Since $j(\cdot)$ is proper we have $j(u_i)$ finite which can be proved as in (4.3). Therefore we have

$$(w_1, w_2 - w_1) + \rho j(w_2) - \rho j(w_1) \geq (u_1, w_2 - w_1)$$
$$+ \rho L(w_2 - w_1) - \rho a(u_1, w_2 - w_1), \tag{4.6}$$
$$(w_2, w_1 - w_2) + \rho j(w_1) - \rho j(w_2) \geq (u_2, w_1 - w_2)$$
$$+ \rho L(w_1 - w_2) - \rho a(u_2, w_1 - w_2). \tag{4.7}$$

Adding these inequalities, we obtain

$$\begin{aligned} \|f_\rho(u_1) - f_\rho(u_2)\|^2 &= \|w_2 - w_1\|^2 \\ &\leq ((I - \rho A)(u_2 - u_1), w_2 - w_1) \\ &\leq \|I - \rho A\| \, \|u_2 - u_1\| \, \|w_2 - w_1\|. \end{aligned} \tag{4.8}$$

Hence

$$\|f_\rho(u_1) - f_\rho(u_2)\| \leq \|I - \rho A\| \, \|u_2 - u_1\|.$$

It is easy to show that $\|I - \rho A\| < 1$ if $0 < \rho < 2\alpha/\|A\|^2$. This proves that f_ρ is uniformly a strict contracting mapping and hence has a unique fixed point u. This u turns out to be the solution of (P_2) since $f_\rho(u) = u$ implies $(u, v - u) + \rho j(v) - \rho j(u) \geq (u, v - u) + \rho L(v - u) - \rho a(u, v - u)$, $\forall v \in V$. Therefore

$$a(u, v - u) + j(v) - j(u) \geq L(v - u), \qquad \forall v \in V. \tag{4.9}$$

Hence (P_2) has a unique solution. □

The existence and uniqueness of the problem (π_ρ^u) follows from the following lemma.

Lemma 4.1. *Let $b: V \times V \to \mathbb{R}$ be a symmetric continuous bilinear V-elliptic form with V-ellipticity constant β. Let $L \in V^*$ and $j: V \to \bar{\mathbb{R}}$ be a convex, l.s.c. proper functional. Let $J(v) = \frac{1}{2}b(v, v) + j(v) - L(v)$. Then the minimization problem (π):*

Find u such that

$$J(u) \leq J(v), \qquad \forall v \in V, \quad u \in V \tag{π}$$

has a unique solution which is characterized by

$$b(u, v - u) + j(v) - j(u) \geq L(v - u), \qquad \forall\, v \in V, \quad u \in V. \tag{4.10}$$

Proof. (1) *Existence and uniqueness of* u: Since $b(v, v)$ is strictly convex, j is convex, and L is linear, we have J strictly convex; J is l.s.c. because $b(\cdot, \cdot)$ and L are continuous and j is l.s.c..

Since j is convex, l.s.c., and proper, there exists $\lambda \in V^*$ and $\mu \in \mathbb{R}$ such that

$$j(v) \geq \lambda(v) + \mu$$

(cf. Ekeland and Temam [1]), then

$$\begin{aligned} J(v) &\geq \frac{\beta}{2}\|v\|^2 - \|\lambda\|\,\|v\| - \|L\|\,\|v\| + \mu \\ &= \left(\sqrt{\frac{\beta}{2}}\,\|v\| - \frac{(\|\lambda\| + \|L\|)}{2}\sqrt{\frac{2}{\beta}}\right)^2 + \mu - \frac{(\|\lambda\| + \|L\|)^2}{2\beta}. \end{aligned} \tag{4.11}$$

Hence

$$J(v) \to +\infty \quad \text{as } \|v\| \to +\infty. \tag{4.12}$$

Hence (cf. Cea [1], [2])[1] there exists a unique solution for the optimization problem (π).

Characterization of u: We show that the problem (π) is equivalent to (4.10) and thus get a characterization of u.

(2) *Necessity of* (4.10): Let $0 < t \leq 1$. Let u be the solution of (π). Then for all $v \in V$ we have

$$J(u) \leq J(u + t(v - u)). \tag{4.13}$$

Set $J_0(v) = \frac{1}{2}b(v, v) - L(v)$, then (4.13) becomes

$$\begin{aligned} 0 &\leq J_0(u + t(v - u)) - J_0(u) + j(u + t(v - u)) - j(u) \\ &\leq J_0(u + t(v - u)) - J_0(u) + t[j(v) - j(u)], \qquad \forall\, v \in V \end{aligned} \tag{4.14}$$

obtained by using the convexity of j. Dividing by t in (4.14) and taking the limit as $t \to 0$, we get

$$0 \leq (J_0'(u), v - u) + j(v) - j(u), \qquad \forall\, v \in V. \tag{4.15}$$

Since $b(\cdot, \cdot)$ is symmetric, we have

$$(J_0'(v), w) = b(v, w) - L(w), \qquad \forall\, v, w \in V. \tag{4.16}$$

From (4.15) and (4.16) we obtain

$$b(u, v - u) + j(v) - j(u) \geq L(v - u), \qquad \forall\, v \in V.$$

This proves the necessity.

(3) *Sufficiency of* (4.10): Let u be a solution of (4.10); for $v \in V$,

$$J(v) - J(u) = \tfrac{1}{2}[b(v, v) - b(u, u)] + j(v) - j(u) - L(v - u). \tag{4.17}$$

---

[1] See also Ekeland and Temam [1].

But

$$b(v, v) = b(u + v - u, u + v - u)$$
$$= b(u, u) + 2b(u, v - u) + b(u - v, u - v).$$

Therefore

$$J(v) - J(u) = b(u, v - u) + j(v) - j(u) - L(v - u) + \tfrac{1}{2}b(v - u, v - u). \quad (4.18)$$

Since u is a solution of (4.10) and $b(v - u, v - u) \geq 0$, we obtain

$$J(v) - J(u) \geq 0. \quad (4.19)$$

Hence u is a solution of (π).

By taking $b(\cdot, \cdot)$ to be the inner product in V and replacing $j(v)$ and $L(v)$ in Lemma 4.1 by $\rho j(v)$ and $(u, v) + \rho L(v) - \rho a(u, v)$, respectively, we get the solution for (π_ρ^u). □

Remark 4.1. From the proof of Theorem 4.1 we obtain an algorithm for solving (P_2). This algorithm is given by

$$u^0 \in V, \text{ arbitrarily given,} \quad (4.20)$$

then for $n \geq 0$, u^n known, we define u^{n+1} from u^n as the solution of

$$(u^{n+1}, v - u^{n+1}) + \rho j(v) - \rho j(u^{n+1}) \geq (u^n, v - u^{n+1}) + \rho L(v - u^{n+1})$$
$$- \rho a(u^n, v - u^{n+1}), \qquad \forall\, v \in V, \quad u^{n+1} \in V. \quad (4.21)$$

If ρ is chosen such that

$$0 < \rho < \frac{2\alpha}{\|A\|^2},$$

we can easily see that $u^n \to u$ strongly in V, where u is the solution of (P_2). Actually, practical difficulties may arise since the problem that we have to solve at each iteration is then a problem of the same order of difficulty as that of the original problem (actually, conditionning can be better provided that ρ has been conveniently chosen). If $a(\cdot, \cdot)$ is not symmetrical the fact that $(\cdot, \cdot)$ is symmetric can also provide some simplification.

5. Internal Approximation of EVI of the First Kind

5.1. Introduction

In this section we shall study the approximation of EVI of the first kind from an abstract axiomatic point of view.

5.2. The continuous problem

The assumptions on V, K, L, and $a(\cdot,\cdot)$ are as in Sec. 2. We are interested in the approximation of

$$a(u, v - u) \geq L(v - u), \qquad \forall\, v \in K, \quad u \in K, \tag{P$_1$}$$

which has a unique solution by Theorem 3.1.

5.3. The approximate problem

5.3.1. *Approximation of V and K*

We suppose that we are given a parameter h converging to 0 and a family $\{V_h\}_h$ of closed subspaces of V. (In practice, the V_h are finite dimensional and the parameter h varies over a sequence). We are also given a family $\{K_h\}_h$ of closed convex nonempty subsets of V with $K_h \subset V_h$, $\forall\, h$ (in general, we do not assume $K_h \subset K$) such that $\{K_h\}_h$ satisfies the following two conditions:

(i) If $\{v_h\}_h$ is such that $v_h \in K_h$, $\forall\, h$ and $\{v_h\}_h$ is bounded in V, then the weak cluster points of $\{v_h\}_h$ belong to K.
(ii) There exists $\chi \subset V$, $\bar{\chi} = K$ and $r_h\colon \chi \to K_h$ such that $\lim_{h\to 0} r_h v = v$ strongly in V, $\forall\, v \in \chi$.

Remark 5.1. If $K_h \subset K$, $\forall\, h$, then (i) is trivially satisfied because K is weakly closed.

Remark 5.2. $\bigcap_h K_h \subset K$.

Remark 5.3. A useful variant of condition (ii) for r_h is

(ii)′ There exist a subset $\chi \subset V$ such that $\bar{\chi} = K$ and $r_h\colon \chi \to V_h$ with the property that for each $v \in \chi$, there exists $h_0 = h_0(v)$ with $r_h v \in K_h$ for all $h \leq h_0(v)$ and $\lim_{h\to 0} r_h v = v$ strongly in V.

5.3.2. *Approximation of* (P$_1$)

The problem (P$_1$) is approximated by

$$a(u_h, v_h - u_h) \geq L(v_h - u_h), \qquad \forall\, v_h \in K_h, \quad u_h \in K_h. \tag{P$_{1h}$}$$

Theorem 5.1. (P$_{1h}$) *has a unique solution.*

PROOF. In Theorem 3.1, taking V to be V_h and K to be K_h, we have the result.

Remark 5.4. In most cases it will be necessary to replace $a(\cdot,\cdot)$ and L by $a_h(\cdot,\cdot)$ and L_h (usually defined, in practical cases, from $a(\cdot,\cdot)$ and L by a *numerical integration procedure*). Since there is nothing very new on that matter compared to the classical linear case, we shall say nothing about this problem for which we refer to Ciarlet [1, Chapter 8], [2], [3].

5.4. Convergence results

Theorem 5.2. *With the above assumptions on* K *and* $\{K_h\}_h$, *we have* $\lim_{h \to 0} \|u_h - u\|_V = 0$ *with* u_h *the solution of* (P_{1h}) *and* u *the solution of* (P_1).

PROOF. For proving this kind of convergence result, we usually divide the proof into three parts. First we obtain *a priori* estimates for $\{u_h\}_h$, then the weak convergence of $\{u_h\}_h$, and finally with the help of the weak convergence, we will prove strong convergence.

(1) *Estimates for* u_h. We will now show that there exist two constants C_1 and C_2 independent of h such that

$$\|u_h\|^2 \le C_1 \|u_h\| + C_2, \qquad \forall\, h. \tag{5.1}$$

Since u_h is the solution of (P_{1h}), we have

$$a(u_h, v_h - u_h) \ge L(v_h - u_h), \qquad \forall\, v_h \in K_h \tag{5.2}$$

i.e.,

$$a(u_h, u_h) \le a(u_h, v_h) - L(v_h - u_h).$$

By V-ellipticity, we get

$$\alpha \|u_h\|^2 \le \|A\|\, \|u_h\|\, \|v_h\| + \|L\|(\|v_h\| + \|u_h\|), \qquad \forall\, v_h \in K_h. \tag{5.3}$$

Let $v_0 \in \chi$ and $v_h = r_h v_0 \in K_h$. By condition (ii) on K_h we have $r_h v_0 \to v_0$ strongly in V and hence $\|v_h\|$ is uniformly bounded by a constant m. Hence (5.3) can be written as

$$\|u_h\|^2 \le \frac{1}{\alpha}\{(m\|A\| + \|L\|)\|u_h\| + \|L\| m\} = C_1 \|u_h\| + C_2,$$

where $C_1 = (1/\alpha)(m\|A\| + \|L\|)$ and $C_2 = (m/\alpha)\|L\|$; then (5.1) implies $\|u_h\| \le C, \forall\, h$.

(2) *Weak convergence of* $\{u_h\}_h$. Relation (5.1) implies that u_h is uniformly bounded. Hence there exists a subsequence, say $\{u_{h_i}\}$, such that u_{h_i} converges to u^* weakly in V. By condition (i) on $\{K_h\}_h$, we have $u^* \in K$. We will prove that u^* is a solution of (P_1). We have

$$a(u_{h_i}, u_{h_i}) \le a(u_{h_i}, v_{h_i}) - L(v_{h_i} - u_{h_i}), \qquad \forall\, v_{h_i} \in K_{h_i}. \tag{5.4}$$

Let $v \in \chi$ and $v_{h_i} = r_{h_i} v$. Then (5.4) becomes

$$a(u_{h_i}, u_{h_i}) \le a(u_{h_i}, r_{h_i} v) - L(r_{h_i} v - u_{h_i}). \tag{5.5}$$

Since $r_{h_i} v$ converges strongly to v and u_{h_i} converges to u^* weakly as $h_i \to 0$, taking the limit in (5.5), we obtain

$$\liminf_{h_i \to 0} a(u_{h_i}, u_{h_i}) \le a(u^*, v) - L(v - u^*), \qquad \forall\, v \in \chi. \tag{5.6}$$

Also we have

$$0 \leq a(u_{h_i} - u^*, u_{h_i} - u^*) \leq a(u_{h_i}, u_{h_i}) - a(u_{h_i}, u^*) - a(u^*, u_{h_i}) + a(u^*, u^*)$$

i.e.,

$$a(u_{h_i}, u^*) + a(u^*, u_{h_i}) - a(u^*, u^*) \leq a(u_{h_i}, u_{h_i}).$$

By taking the limit, we obtain

$$a(u^*, u^*) \leq \liminf_{h_i \to 0} a(u_{h_i}, u_{h_i}). \tag{5.7}$$

From (5.6) and (5.7), we obtain

$$a(u^*, u^*) \leq \liminf_{h_i \to 0} a(u_{h_i}, u_{h_i}) \leq a(u^*, v) - L(v - u^*), \qquad \forall\, v \in \chi.$$

Therefore we have

$$a(u^*, v - u^*) \geq L(v - u^*), \qquad \forall\, v \in \chi, \quad u^* \in K. \tag{5.8}$$

Since χ is dense in K and $a(\cdot, \cdot)$, L are continuous, from (5.8) we obtain

$$a(u^*, v - u^*) \geq L(v - u^*), \qquad \forall\, v \in K, \quad u^* \in K. \tag{5.9}$$

Hence u^* is a solution of (P_1). By Theorem 3.1, the solution of (P_1) is unique and hence $u^* = u$ is the unique solution. Hence u is the only cluster point of $\{u_h\}_h$ in the weak topology of V. Hence the whole $\{u_h\}_h$ converges to u weakly.

(3) *Strong convergence*. By V-ellipticity of $a(\cdot, \cdot)$, we have

$$0 \leq \alpha \|u_h - u\|^2 \leq a(u_h - u, u_h - u) = a(u_h, u_h) - a(u_h, u) - a(u, u_n) + a(u, u), \tag{5.10}$$

where u_h is the solution of (P_{1h}) and u is the solution of (P_1). Since u_h is the solution of (P_{1h}) and $r_h v \in K_h$ for any $v \in \chi$, from (P_{1h}) we obtain

$$a(u_h, u_h) \leq a(u_h, r_h v) - L(r_h v - u_h), \qquad \forall\, v \in \chi. \tag{5.11}$$

Since $\lim_{h \to 0} u_h = u$ *weakly* in V and $\lim_{h \to 0} r_h v = v$ *strongly* in V [by condition (ii)], we obtain (5.11) from (5.10), and after taking the limit, $\forall\, v \in \chi$, we have

$$0 \leq \alpha \liminf \|u_h - u\|^2 \leq \alpha \limsup \|u_h - u\|^2 \leq a(u, v - u) - L(v - u). \tag{5.12}$$

By *density* and *continuity*, (5.12) also holds for $\forall\, v \in K$; then taking $v = u$ in (5.12), we obtain

$$\lim_{h \to 0} \|u_h - u\|^2 = 0,$$

i.e., the strong convergence. □

Remark 5.5. Error estimates for the EVI of the first kind can be found in Falk [1], [2], [3], Mosco and Strang, [1], Strang [1], Glowinski, Lions, and Tremolieres (G.L.T.) [1], [2], [3], Ciarlet [1], [2], [3], Falk and Mercier [1], Glowinski [1], and Brezzi, Hager, and Raviart [1], [2]. But as in many nonlinear problems, the methods used to obtain these estimates are specific to the particular problem under consideration (as we shall see in the following sections). This remark also holds for the approximation of the EVI of the second kind which is the subject of Sec. 6.

Remark 5.6. If for a given problem, several approximations are available, and if numerical results are needed, the choice of the approximation to be used is not obvious. We have to take into account not only the convergence properties of the method, but also the computations involved in that method. Some iterative methods are well suited only to specific problems. For example, some methods are easier to code than others.

6. Internal Approximation of EVI of the Second Kind

6.1. The continuous problem

With the assumptions on V, $a(\cdot, \cdot)$, L, and $j(\cdot)$ as in Sec. 2.1, we shall consider the approximation of

$$a(u, v - u) + j(v) - j(u) \geq L(v - u), \qquad \forall\, v \in V, \quad u \in V \tag{P$_2$}$$

which has a unique solution by Theorem 4.1.

6.2. Definition of the approximate problem

Preliminary remark: We assume in the sequel that $j: V \to \mathbb{R}$ is continuous. However, we can prove the same sort of results as in this section under less restrictive hypotheses (see Chapter 4, Sec. 2).

6.2.1. *Approximation of V*

Given a real parameter h converging to 0 and a family $\{V_h\}_h$ of closed subspaces of V (in practice, we will take V_h to be finite dimensional and h to vary over a sequence), we suppose that $\{V_h\}_h$ satisfies:

(i) There exists $U \subset V$ such that $\overline{U} = V$, and for each h, a mapping $r_h: U \to V_h$ such that $\lim_{h \to 0} r_h v = v$ strongly in V, $\forall\, v \in U$.

6.2.2. *Approximation of j(·)*

We approximate the functional $j(\cdot)$ by $\{j_h\}_h$ where for each h, j_h satisfies

$$j_h: V_h \to \overline{\mathbb{R}}, \quad j_h \text{ is convex, l.s.c., and uniformly proper in } h. \tag{6.1}$$

The family $\{j_h\}_h$ is said to be *uniformly proper in h* if there exist $\lambda \in V^*$ and $\mu \in \mathbb{R}$ such that

$$j_h(v_h) \geq \lambda(v_h) + \mu, \qquad \forall\, v_h \in V_h, \forall\, h. \tag{6.2}$$

Furthermore we assume that $\{j_h\}_h$ satisfies:

(ii) If $v_h \to v$ weakly in V, then

$$\liminf_{h \to 0} j_h(v_h) \geq j(v),$$

(iii) $\lim_{h \to 0} j_h(r_h v) = j(v), \forall v \in U$.

Remark 6.1. In all the applications that we know, if $j(\cdot)$ is a continuous functional, then it is always possible to construct continuous j_h satisfying (ii) and (iii).

Remark 6.2. In some cases we are fortunate enough to have $j_h(v_h) = j(v_h)$, $\forall v_h, \forall h$, and then (ii) and (iii) are trivially satisfied.

6.2.3. *Approximation of* (P_2)

We approximate (P_2) by

$$a(u_h, v_h - u_h) + j_h(v_h) - j_h(u_h) \geq L(v_h - u_h), \quad \forall v_h \in V_h, \quad u_h \in V_h. \tag{P_{2h}}$$

Theorem 6.1. *Problem* (P_{2h}) *has a unique solution.*

PROOF. In Theorem 4.1, taking V to be V_h, $j(\cdot)$ to be $j_h(\cdot)$, we get the result.

Remark 6.3. Remark 5.4 of Sec. 5 still holds for (P_2) and (P_{2h}).

6.3. Convergence results

Theorem 6.2. *Under the above assumptions on* $\{V_h\}_h$ *and* $\{j_h\}_h$, *we have*

$$\lim_{h \to 0} \|u_h - u\| = 0, \qquad \lim_{h \to 0} j_h(u_h) = j(u). \tag{6.3}$$

PROOF. As in the proof of Theorem 5.2, we divide the proof into three parts.

(1) *Estimates for* u_h. We will show that there exist positive constants C_1 and C_2 independent of h such that

$$\|u_h\|^2 \leq C_1\|u_h\| + C_2, \qquad \forall h. \tag{6.4}$$

Since u_h is the solution of (P_{2h}), we have

$$a(u_h, u_h) + j_h(u_h) \leq a(u_h, v_h) + j_h(v_h) - L(v_h - u_h), \qquad \forall v_h \in V_h. \tag{6.5}$$

By using relation (6.2), we obtain

$$\alpha\|u_h\|^2 \leq \|\lambda\| \|u_h\| + |\mu| + \|A\| \|u_h\| \|v_h\| + |j_h(v_h)| + \|L\|(\|v_h\| + \|u_h\|). \tag{6.6}$$

Let $v_0 \in U$ and $v_h = r_h v_0$. By using conditions (i) and (iii), there exists a constant m, independent of h, such that $\|v_h\| \le m$ and $|j_h(v_h)| \le m$. Therefore (6.6) can be written as

$$\|u_h\|^2 \le \frac{1}{\alpha}(\|\lambda\| + \|A\|m + \|L\|)\|u_h\| + \frac{m}{\alpha}(1 + \|L\|) + \frac{|\mu|}{\alpha}$$

$$= C_1\|u_h\| + C_2$$

where

$$C_1 = \frac{1}{\alpha}(\|\lambda\| + \|A\|m + \|L\|) \quad \text{and} \quad C_2 = \frac{m}{\alpha}(1 + \|L\|) + \frac{|\mu|}{\alpha}$$

and (6.4) implies

$$\|u_h\| \le C, \qquad \forall\, h,$$

where C is a constant.

(2) *Weak convergence of* $\{u_h\}_h$. Relation (6.4) implies that u_h is uniformly bounded. Therefore there exists a subsequence $\{u_{h_i}\}_{h_i}$ such that $u_{h_i} \to u_h$ weakly in V.

Since u_h is the solution of (P_{1h}) and $r_h v \in V_h$, $\forall\, h$ and $\forall\, v \in U$, we have

$$a(u_{h_i}, u_{h_i}) + j_{h_i}(u_{h_i}) \le a(u_{h_i}, r_{h_i}v) + j_{h_i}(r_{h_i}v) - L(r_{h_i}v - u_{h_i}). \tag{6.7}$$

By condition (iii) and from the weak convergence of $\{u_{h_i}\}$, we have

$$\liminf_{h_i \to 0}\,[a(u_{h_i}, u_{h_i}) + j_{h_i}(u_{h_i})] \le a(u^*, v) + j(v) - L(v - u^*), \qquad \forall\, v \in U. \tag{6.8}$$

As in (5.7), and using condition (ii), we obtain

$$a(u^*, u^*) + j(u^*) \le \liminf_{h_i \to 0}\,[a(u_{h_i}, u_{h_i}) + j_{h_i}(u_{h_i})]. \tag{6.9}$$

From (6.8) and (6.9), and using the density of U, we have

$$a(u^*, v - u^*) + j(v) - j(u^*) \ge L(v - u^*), \qquad \forall\, v \in V, \quad u^* \in V.$$

This implies that u^* is a solution of (P_2). Hence $u^* = u$ is the unique solution of (P_2), and this shows that $\{u_h\}_h$ converges to u *weakly*.

(3) *Strong convergence of* $\{u_h\}_h$. From the V-ellipticity of $a(\cdot, \cdot)$ and from (P_{2h}) we have

$$\begin{aligned}
\alpha\|u_h - u\|^2 + j_h(u_h) &\le a(u_h - u, u_h - u) + j_h(u_h) \\
&= a(u_h, u_h) - a(u, u_h) - a(u_h, u) + a(u, u) + j_h(u_h) \\
&\le a(u_h, r_h v) + j_h(r_h v) - L(r_h v - u_h) - a(u, u_h) \\
&\quad - a(u_h, u) + a(u, u), \qquad \forall\, v \in U.
\end{aligned} \tag{6.10}$$

The right-hand side of inequality (6.10) converges to $a(u, v - u) + j(v) - L(v - u)$ as $h \to 0$, $\forall\, v \in U$. Therefore we have

$$\begin{aligned}
\liminf_{h \to 0} j_h(u_h) &\le \liminf_{h \to 0}\,[\alpha\|u_h - u\|^2 + j_h(u_h)] \\
&\le \limsup_{h \to 0}\,[\alpha\|u_h - u\|^2 + j_h(u_h)] \\
&\le a(u, v - u) + j(v) - L(v - u), \qquad \forall\, v \in U.
\end{aligned} \tag{6.11}$$

By the density of U, (6.11) holds, $\forall\, v \in V$. Replacing v by u in (6.11) and using condition (ii), we obtain

$$j(u) \leq \liminf_{h \to 0} j_h(u_h) \leq \limsup_{h \to 0} [\alpha \|u - u_h\|^2 + j_h(u_h)] \leq j(u).$$

This implies that

$$\lim_{h \to 0} j_h(u_h) = j(u)$$

and

$$\lim_{h \to 0} \|u_h - u\| = 0.$$

This proves the theorem. □

7. Penalty Solution of Elliptic Variational Inequalities of the First Kind

7.1. Synopsis

In this section we would like to discuss the approximation of elliptic variational inequalities of the first kind by *penalty methods*. In fact these penalty techniques can be applied to more complicated problems as shown in Lions [1], [4] (see also Chapter VII, Sec. 4, where a penalty method is applied to the solution of transonic flow problems).

7.2. Formulation of the penalized problem

Consider the EVI problem.

Find $u \in K$ ($\subset V$) such that

$$a(u, v - u) \geq L(v - u), \qquad \forall\, v \in K, \tag{7.1}$$

where the properties of V, $a(\cdot, \cdot)$, $L(\cdot)$, and K are those given in Sec. 2.1. Now suppose that a functional $j\colon V \to \overline{\mathbb{R}}$ has the following properties:

$$j \text{ is convex, proper, l.s.c.,} \tag{7.2}$$

$$j(v) = 0 \Leftrightarrow v \in K, \tag{7.3}$$

$$j(v) \geq 0, \qquad \forall\, v \in V. \tag{7.4}$$

Let $\varepsilon > 0$; we define $j_\varepsilon\colon V \to \overline{\mathbb{R}}$ by

$$j_\varepsilon = \frac{1}{\varepsilon} j. \tag{7.5}$$

The *penalized problem* associated to $j(\cdot)$ is defined by:

Find $u_\varepsilon \in V$ such that

$$a(u_\varepsilon, v - u_\varepsilon) + j_\varepsilon(v) - j_\varepsilon(u_\varepsilon) \geq L(v - u_\varepsilon), \qquad \forall\, v \in V. \tag{7.6}$$

(7.6) is definitely an EVI of the *second kind*, and from the properties of V, $a(\cdot, \cdot)$, $L(\cdot)$, and $j(\cdot)$, it has (see Sec. 4) a unique solution according to Theorem 4.1.

Remark 7.1. Suppose that j_ε is differentiable; the solution u_ε of (7.6) is then characterized by the fact that it is the unique solution of the following non-linear variational equation:

$$a(u_\varepsilon, v) + \langle j'_\varepsilon(u_\varepsilon), v\rangle = L(v), \qquad \forall\, v \in V, \quad u_\varepsilon \in V, \tag{7.7}$$

where $j'_\varepsilon(v)$ ($\in V^*$, the topological dual space of V) denotes the differential of j_ε at v, and where $\langle \cdot, \cdot \rangle$ is the duality pairing between V^* and V. That differentiability property (if it exists) can be helpful for solving (7.6), (7.7) by efficient iterative methods like Newton's method or the conjugate gradient method (see Chapter IV, Sec. 2.6 and Chapter VII for references and also some applications of these methods).

7.3. Convergence of $\{u_\varepsilon\}_\varepsilon$

Concerning the behavior of $\{u_\varepsilon\}_\varepsilon$ as $\varepsilon \to 0$, we have the following:

Theorem 7.1. *If the hypotheses on V, K, $a(\cdot, \cdot)$, $L(\cdot)$, $j(\cdot)$ are those of Secs.* 2.1 *and* 7.2, *we have*

$$\lim_{\varepsilon\to 0} \|u_\varepsilon - u\| = 0, \tag{7.8}$$

$$\lim_{\varepsilon\to 0} j_\varepsilon(u_\varepsilon) = 0, \tag{7.9}$$

where u (resp., u_ε) is the solution of (7.1) (*resp.*, (7.6)).

PROOF. This proof looks very much like the proof of Theorems 5.2 and 6.2.

(1) *A priori estimates.* From (7.6) we have

$$a(u_\varepsilon, u_\varepsilon) + j_\varepsilon(u_\varepsilon) \leq a(u_\varepsilon, v) - L(v - u_\varepsilon) + j_\varepsilon(v), \forall\, v \in V. \tag{7.10}$$

Since $j_\varepsilon(v) = \varepsilon^{-1}j(v) = 0$, $\forall\, v \in K$ [property (7.3)], we have, from (7.10),

$$a(u_\varepsilon, u_\varepsilon) + j_\varepsilon(u_\varepsilon) \leq a(u_\varepsilon, v) - L(v - u_\varepsilon), \qquad \forall\, v \in K. \tag{7.11}$$

Consider $v_0 \in K$ (since $K \neq \varnothing$, such a v_0 always exists). Taking $v = v_0$ in (7.11), from the properties of $a(\cdot, \cdot)$ and from (7.4), (7.5) we obtain

$$\alpha\|u_\varepsilon\|^2 \leq \|A\|\,\|u_\varepsilon\|\,\|v_0\| + \|L\|(\|u_\varepsilon\| + \|v_0\|), \tag{7.12}$$

$$0 \leq j(u_\varepsilon) \leq \varepsilon(\|A\|\,\|u_\varepsilon\|\,\|v_0\| + \|L\|(\|u_\varepsilon\| + \|v_0\|)) \tag{7.13}$$

(α is the ellipticity constant of $a(\cdot, \cdot)$). Then it clearly follows from (7.12) that we have

$$\|u_\varepsilon\| \leq C_1, \qquad \forall\, \varepsilon > 0, \tag{7.14}$$

which combined with (7.13) implies

$$0 \leq j(u_\varepsilon) \leq C_2 \varepsilon, \tag{7.15}$$

where in (7.14), (7.15), C_1 and C_2 denote two constants independent of ε.

(2) *Weak convergence*. It follows from (7.14) that we can extract from $\{u_\varepsilon\}_\varepsilon$ a subsequence—still denoted $\{u_\varepsilon\}_\varepsilon$—such that

$$\lim_{\varepsilon \to 0} u_\varepsilon = u^* \text{ weakly in } V, \tag{7.16}$$

where $u^* \in V$.

It then follows from (7.15), (7.16) and from the *weak lower semicontinuity* of $j(\cdot)$ that

$$j(u^*) = 0 \Rightarrow u^* \in K \quad \text{[from (7.3)]}. \tag{7.17}$$

To prove that $u^* = u$, we observe that from (7.10) we have

$$a(u_\varepsilon, u_\varepsilon) \leq a(u_\varepsilon, v) - L(v - u_\varepsilon), \qquad \forall\, v \in K, \tag{7.18}$$

which implies, at the limit as $\varepsilon \to 0$,

$$a(u^*, u^*) \leq \liminf a(u_\varepsilon, u_\varepsilon) \leq a(u^*, v) - L(v - u^*), \qquad \forall\, v \in K. \tag{7.19}$$

Combining (7.17) and (7.19), we finally obtain

$$a(u^*, v - u^*) \geq L(v - u^*), \qquad \forall\, v \in K, \quad u^* \in K; \tag{7.20}$$

we have thus proved that $u^* = u$ and that the whole $\{u_\varepsilon\}_\varepsilon$ converges weakly to u.

(3) *Strong convergence*. From (7.3), (7.4), and (7.6) we have

$$\begin{aligned} 0 &\leq \alpha\|u_\varepsilon - u\|^2 + j_\varepsilon(u_\varepsilon) \leq a(u_\varepsilon - u, u_\varepsilon - u) + j_\varepsilon(u_\varepsilon) \\ &\leq a(u_\varepsilon, u_\varepsilon) + j_\varepsilon(u_\varepsilon) - a(u, u_\varepsilon) - a(u_\varepsilon, u) + a(u, u) \\ &\leq a(u_\varepsilon, v) - L(v - u_\varepsilon) - a(u, u_\varepsilon) - a(u_\varepsilon, u) + a(u, u), \qquad \forall\, v \in K. \end{aligned} \tag{7.21}$$

The weak convergence of $\{u_\varepsilon\}_\varepsilon$ to u implies that at the limit in (7.21) we have

$$\begin{aligned} 0 &\leq \liminf_{\varepsilon \to 0} [\alpha\|u_\varepsilon - u\|^2 + j_\varepsilon(u_\varepsilon)] \leq \limsup_{\varepsilon \to 0} [\alpha\|u_\varepsilon - u\|^2 + j_\varepsilon(u_\varepsilon)] \\ &\leq a(u, v - u) - L(v - u), \qquad \forall\, v \in K. \end{aligned} \tag{7.22}$$

Taking $v = u$ in (7.22), we obtain

$$\lim_{\varepsilon \to 0} [\alpha\|u_\varepsilon - u\|^2 + j_\varepsilon(u_\varepsilon)] = 0$$

which clearly implies the convergence properties (7.8) and (7.9). □

Remark 7.2. If $a(\cdot, \cdot)$ is symmetric, then the penalized problem (7.6) is equivalent to the minimization problem:

Find $u_\varepsilon \in V$ such that

$$J_\varepsilon(u_\varepsilon) \leq J_\varepsilon(v), \qquad \forall\, v \in V, \tag{7.23}$$

where

$$J_\varepsilon(v) = \tfrac{1}{2}a(v, v) - L(v) + j_\varepsilon(v).$$

EXERCISE 7.1. Prove, if $a(\cdot, \cdot)$ is symmetric, the equivalence between problems (7.6) and (7.23).

7.4. Some examples in finite dimension

7.4.1. *Generalities*

In this section we discuss, in some detail, applications of the penalty method to the solution of some simple model problems in $\mathbb{R}^N$. In Sec. 7.4.2 we consider (resp., Sec. 7.4.3) a situation in which K is defined from *linear equality constraints* (resp., *convex inequality constraints*).

In the following $\mathbf{A}$ is a $N \times N$ real matrix, *positive definite*, possibly nonsymmetric, and $\mathbf{b} \in \mathbb{R}^N$. To $\mathbf{A}$ and $\mathbf{b}$ we associate

$$a: \mathbb{R}^N \times \mathbb{R}^N \to \mathbb{R} \quad \text{and} \quad L: \mathbb{R}^N \to \mathbb{R}$$

defined by

$$\begin{aligned} a(\mathbf{v}, \mathbf{w}) &= (\mathbf{A}\mathbf{v}, \mathbf{w}), \qquad \forall\, \mathbf{v}, \mathbf{w} \in \mathbb{R}^N, \\ L(\mathbf{v}) &= (\mathbf{b}, \mathbf{v}), \qquad \forall\, \mathbf{v} \in \mathbb{R}^N, \end{aligned} \tag{7.24}$$

where $(\cdot, \cdot)$ denotes the *usual scalar product* of $\mathbb{R}^N$, i.e.,

$$(\mathbf{v}, \mathbf{w}) = \sum_{i=1}^{N} v_i w_i, \qquad \forall\, \mathbf{v} = \{v_i\}_{i=1}^N, \quad \mathbf{w} = \{w_i\}_{i=1}^N. \tag{7.25}$$

We denote by $\|\cdot\|$ the *norm* associated to (7.25), i.e.,

$$\|\mathbf{v}\| = (\mathbf{v}, \mathbf{v})^{1/2}, \qquad \forall\, \mathbf{v} \in \mathbb{R}^N.$$

The form $L(\cdot)$ is clearly *linear* and *continuous* on $\mathbb{R}^N$; similarly $a(\cdot, \cdot)$ is *bilinear* and *continuous* on $\mathbb{R}^N \times \mathbb{R}^N$. Since $\mathbf{A}$ is positive definite, $a(\cdot, \cdot)$ is $\mathbb{R}^N$-elliptic, and we have

$$a(\mathbf{v}, \mathbf{v}) \geq \lambda_0 \|\mathbf{v}\|^2, \qquad \forall\, \mathbf{v} \in \mathbb{R}^N, \tag{7.26}$$

where λ_0 is the smallest eigenvalue of the symmetric positive-definite matrix $\mathbf{A} + \mathbf{A}^t/2$ (*with* $\mathbf{A}^t$ *the transpose matrix of* $\mathbf{A}$).

7.4.2. *A first example*

Let $\mathbf{B} \in \mathscr{L}(\mathbb{R}^N, \mathbb{R}^M)$; $\mathbf{B}$ can be identified to a $M \times N$ matrix. We define $R(\mathbf{B})$ (the *range* of $\mathbf{B}$) by

$$R(\mathbf{B}) = \{\mathbf{q} \mid \mathbf{q} \in \mathbb{R}^M, \exists\, v \in \mathbb{R}^N \text{ such that } \mathbf{q} = \mathbf{B}\mathbf{v}\}$$

and then $K \subset \mathbb{R}^N$ by

$$K = \{\mathbf{v} \in \mathbb{R}^N, \mathbf{Bv} = \mathbf{c}\} \tag{7.27}$$

where, in (7.27), we have

$$\mathbf{c} \in R(\mathbf{B})(\Rightarrow K \neq \varnothing). \tag{7.28}$$

From the above properties of $a(\cdot, \cdot)$, $L(\cdot)$, and K, the EVI problem:
Find $\mathbf{u} \in K$ *such that*

$$(\mathbf{Au}, \mathbf{v} - \mathbf{u}) \geq (\mathbf{b}, \mathbf{v} - \mathbf{u}), \qquad \forall\, \mathbf{v} \in K \tag{7.29}$$

has a unique solution since we can apply (with $V = \mathbb{R}^N$) Theorem 3.1 of Sec. 3.1.

Remark 7.3. If $\mathbf{A} = \mathbf{A}^t$, then problem (7.29) is equivalent to the minimization problem:
Find $\mathbf{u} \in K$ *such that*

$$J(\mathbf{u}) \leq J(\mathbf{v}), \qquad \forall\, \mathbf{v} \in K, \tag{7.30}$$

where

$$J(\mathbf{v}) = \tfrac{1}{2}(\mathbf{Av}, \mathbf{v}) - (\mathbf{b}, \mathbf{v}).$$

Before going on to the penalty solution of (7.29), we shall prove some properties of the solution $\mathbf{u}$ of (7.29); more precisely we have the following proposition.

Proposition 7.1. *The solution* $\mathbf{u}$ *of* (7.29) *is characterized by the existence of* $\mathbf{p} \in \mathbb{R}^M$ *such that*

$$\begin{aligned} \mathbf{Au} + \mathbf{B}^t\mathbf{p} &= \mathbf{b}, \\ \mathbf{Bu} &= \mathbf{c}. \end{aligned} \tag{7.31}$$

PROOF. (1) (7.29) *implies* (7.31). Let $\mathbf{u}$ be the solution of (7.29); we have

$$\mathbf{u} + \mathbf{w} \in K, \qquad \forall\, \mathbf{w} \in \operatorname{Ker}(\mathbf{B}), \tag{7.32}$$

where $\operatorname{Ker}(\mathbf{B}) = \{\mathbf{v} \in \mathbb{R}^N, \mathbf{Bv} = \mathbf{0}\}$.
Taking $\mathbf{v} = \mathbf{u} + \mathbf{w}$ in (7.22), we obtain

$$(\mathbf{Au} - \mathbf{b}, \mathbf{w}) \geq 0, \qquad \forall\, \mathbf{w} \in \operatorname{Ker}(\mathbf{B}), \tag{7.33}$$

and (7.33) clearly implies

$$(\mathbf{Au} - \mathbf{b}, \mathbf{w}) = 0, \qquad \forall\, \mathbf{w} \in \operatorname{Ker}(\mathbf{B})$$

i.e.,

$$\mathbf{Au} - \mathbf{b} \in \operatorname{Ker}(\mathbf{B})^{\perp}. \tag{7.34}$$

Since (it is a standard result)

$$\operatorname{Ker}(\mathbf{B})^{\perp} = R(\mathbf{B}^t),$$

we have the existence of $\mathbf{p} \in \mathbb{R}^M$ such that (7.31) holds.

(2) (7.31) *implies* (7.20). The second relation (7.31) implies that $\mathbf{u} \in K$. Letting $\mathbf{v} \in K$, we then have

$$\mathbf{v} - \mathbf{u} \in \operatorname{Ker}(\mathbf{B}), \tag{7.35}$$

and from (7.35) and from the first relation (7.31) it follows that

$$\begin{aligned}(\mathbf{b}, \mathbf{v} - \mathbf{u}) &= (\mathbf{A}\mathbf{u}, \mathbf{v} - \mathbf{u}) + (\mathbf{B}^t\mathbf{p}, \mathbf{v} - \mathbf{u}) = (\mathbf{A}\mathbf{u}, \mathbf{v} - \mathbf{u}) + (\mathbf{p}, \mathbf{B}(\mathbf{v} - \mathbf{u})) \\ &= (\mathbf{A}\mathbf{u}, \mathbf{v} - \mathbf{u}).\end{aligned} \tag{7.36}$$

We have thus proved that (7.31) implies (7.29). □

Remark 7.4. Suppose that $\mathbf{A} = \mathbf{A}^t$; then the vector $\mathbf{p}$ of Proposition 7.1 is a *Lagrange multiplier* vector for the problem (7.30), associated with the linear equality constraint $\mathbf{B}\mathbf{v} - \mathbf{c} = \mathbf{0}$ defining K.

The following proposition and its corolaries state results quite easy to prove, but of great interest in studying the behavior of the solution of the penalized problem to be defined later on.

Proposition 7.2. *Problem* (7.31) *has a unique solution in* $\mathbb{R}^N \times R(\mathbf{B})$ *if*

$$\{\mathbf{b}, \mathbf{c}\} \in \mathbb{R}^N \times R(\mathbf{B}).$$

Let us denote by $\{\mathbf{u}, \hat{\mathbf{p}}\}$ *this solution; then all solutions of* (7.31) *can be written* $\{\mathbf{u}, \hat{\mathbf{p}} + \mathbf{q}\}$, *where* $\mathbf{q}$ *is an arbitrary element of* $\operatorname{Ker}(\mathbf{B}^t)$.

Corollary 7.1. *The above vector* $\hat{\mathbf{p}}$ *has the minimal norm among all the* $\mathbf{p} \in \mathbb{R}^M$ *such that* $\{\mathbf{u}, \mathbf{p}\}$ *solves* (7.31).

Corollary 7.2. *The linear operator*

$$\mathscr{A} = \begin{pmatrix} \mathbf{A} & \mathbf{B}^t \\ -\mathbf{B} & \mathbf{0} \end{pmatrix}$$

is an isomorphism from $\mathbb{R}^N \times R(\mathbf{B})$ *onto* $\mathbb{R}^N \times R(\mathbf{B})$.

EXERCISE 7.2. Prove Proposition 7.2 and Corollaries 7.1 and 7.2.

EXERCISE 7.3. Prove that $\hat{\mathbf{p}} = \mathbf{0}$ if and only if $\mathbf{c} = \mathbf{B}\mathbf{A}^{-1}\mathbf{b}$.

In order to apply the penalty method of Sec. 7.2 to the solution of (7.29), we define $j\colon \mathbb{R}^N \to \mathbb{R}$ by

$$j(\mathbf{v}) = \tfrac{1}{2}|\mathbf{B}\mathbf{v} - \mathbf{c}|^2, \tag{7.37}$$

where $|\cdot|$ denotes the usual Euclidean norm of $\mathbb{R}^M$. We can easily see that $j(\cdot)$ obeys (7.2)–(7.4); moreover, $j(\cdot)$ is a C^∞ functional whose differential j' is given by

$$j'(\mathbf{v}) = \mathbf{B}^t(\mathbf{B}\mathbf{v} - \mathbf{c}). \tag{7.38}$$

The penalized problem associated with (7.29) and (7.37) is defined by:

Find $\mathbf{u}_\varepsilon \in \mathbb{R}^N$ *such that*

$$(\mathbf{A}\mathbf{u}_\varepsilon, \mathbf{v} - \mathbf{u}_\varepsilon) + j_\varepsilon(\mathbf{v}) - j_\varepsilon(\mathbf{u}_\varepsilon) \geq (\mathbf{b}, \mathbf{v} - \mathbf{u}_\varepsilon), \qquad \forall\, \mathbf{v} \in \mathbb{R}^N, \tag{7.39}$$

where $j_\varepsilon = (1/\varepsilon)j$ (with $\varepsilon > 0$).

From Remark 7.1 and (7.38) the penalized problem (7.39) is equivalent to the linear system

$$\left(\mathbf{A} + \frac{1}{\varepsilon}\mathbf{B}^t\mathbf{B}\right)\mathbf{u}_\varepsilon = \frac{1}{\varepsilon}\mathbf{B}^t\mathbf{c} + \mathbf{b} \tag{7.40}$$

whose matrix is positive definite (and symmetric if $\mathbf{A}$ is symmetric). It follows from Theorem 7.1 (see Sec. 7.3) that

$$\lim_{\varepsilon \to 0} \|\mathbf{u}_\varepsilon - \mathbf{u}\| = 0, \tag{7.41}$$

where $\mathbf{u}$ is the solution of (7.29). We have, in fact, $\|\mathbf{u}_\varepsilon - \mathbf{u}\| = O(\varepsilon)$; several methods can be used to prove this result; we have chosen one of them based on the *implicit function theorem.*

Define $\mathbf{p}_\varepsilon \in \mathbb{R}^M$ by

$$\mathbf{p}_\varepsilon = \frac{1}{\varepsilon}(\mathbf{B}\mathbf{u}_\varepsilon - \mathbf{c}). \tag{7.42}$$

Problem (7.40) is then equivalent to the following system:

$$\begin{aligned} \mathbf{A}\mathbf{u}_\varepsilon + \mathbf{B}^t\mathbf{p}_\varepsilon &= \mathbf{b}, \\ -\mathbf{B}\mathbf{u}_\varepsilon + \varepsilon\mathbf{p}_\varepsilon &= -\mathbf{c}, \end{aligned} \tag{7.43}$$

whose matrix

$$\mathscr{A}_\varepsilon = \begin{pmatrix} \mathbf{A} & \mathbf{B}^t \\ -\mathbf{B} & \varepsilon\mathbf{I} \end{pmatrix} \tag{7.44}$$

is a $N + M$ by $N + M$ positive-definite matrix.

EXERCISE 7.4. Prove that $\mathscr{A}_\varepsilon$ is positive definite.

Since $\mathbf{c} \in \mathbb{R}(\mathbf{B})$, we have (from (7.42)) $\mathbf{p}_\varepsilon \in R(\mathbf{B})$. About the behavior of $\{\mathbf{u}_\varepsilon, \mathbf{p}_\varepsilon\}$ as $\varepsilon \to 0$, we have the following theorem.

Theorem 7.2. *Let* $\mathbf{u}_\varepsilon$ *be the solution of* (7.39) *and let* $\mathbf{p}_\varepsilon$ *be defined by* (7.42); *we then have*

$$\|\mathbf{u}_\varepsilon - \mathbf{u}\| = O(\varepsilon), \tag{7.45}$$

$$|\mathbf{p}_\varepsilon - \hat{\mathbf{p}}| = O(\varepsilon), \tag{7.46}$$

where $\mathbf{u}$ *is the solution of* (7.29) *and* $\hat{\mathbf{p}}$ *has been defined in Proposition* 7.2.

PROOF. Define $\mathbf{F}: \mathbb{R}^N \times \mathbb{R}^M \times \mathbb{R} \to \mathbb{R}^N \times \mathbb{R}^M$ by

$$\mathbf{F}(\mathbf{v}, \mathbf{q}, \varepsilon) = \begin{pmatrix} \mathbf{A}\mathbf{v} + \mathbf{B}^t\mathbf{q} - \mathbf{b} \\ -\mathbf{B}\mathbf{v} + \varepsilon\mathbf{q} + \mathbf{c} \end{pmatrix}, \tag{7.47}$$

still with $\mathbf{c} \in R(\mathbf{B})$. We observe that $\mathbf{F}$ also maps $\mathbb{R}^N \times R(\mathbf{B}) \times \mathbb{R}$ into $\mathbb{R}^N \times R(\mathbf{B})$. We have

$$\mathbf{F}(\mathbf{u}, \hat{\mathbf{p}}, 0) = \mathbf{0}; \tag{7.48}$$

moreover, we have

$$\frac{\partial \mathbf{F}}{\partial \mathbf{v}}(\mathbf{u}, \hat{\mathbf{p}}, 0) \cdot \mathbf{v} + \frac{\partial \mathbf{F}}{\partial \mathbf{q}}(\mathbf{u}, \hat{\mathbf{p}}, 0) \cdot \mathbf{q} = \begin{pmatrix} \mathbf{A}\mathbf{v} + \mathbf{B}^t\mathbf{q} \\ -\mathbf{B}\mathbf{v} \end{pmatrix} = \mathscr{A}\begin{pmatrix} \mathbf{v} \\ \mathbf{q} \end{pmatrix}. \tag{7.49}$$

Since $\mathscr{A}$ is an isomorphism from $\mathbb{R}^N \times R(\mathbf{B})$ onto $\mathbb{R}^N \times R(\mathbf{B})$, we can apply the *implicit function theorem* in the space $\mathbb{R}^N \times R(\mathbf{B}) \times \mathbb{R}$ to define, from $\mathbf{F}(\mathbf{v}, \mathbf{q}, \varepsilon) = 0$ (i.e., from (7.43)), $\mathbf{u}_\varepsilon$ $(= \mathbf{u}(\varepsilon))$ and $\mathbf{p}_\varepsilon$ $(= \mathbf{p}(\varepsilon))$ as C^∞ functions of ε (in fact, they are analytic functions of ε) in the neighborhood of $\varepsilon = 0$.

We have $\{\mathbf{u}(0), \mathbf{p}(0)\} = \{\mathbf{u}, \hat{\mathbf{p}}\}$ and also

$$\begin{aligned} \mathbf{u}_\varepsilon &= \mathbf{u}(\varepsilon) = \mathbf{u} + \varepsilon\dot{\mathbf{u}}(0) + \cdots, \\ \mathbf{p}_\varepsilon &= \mathbf{p}(\varepsilon) = \hat{\mathbf{p}} + \varepsilon\dot{\mathbf{p}}(0) + \cdots \end{aligned} \tag{7.50}$$

(with $\{\dot{\mathbf{u}}, \dot{\mathbf{p}}\} = \{d\mathbf{u}/d\varepsilon, d\mathbf{p}/d\varepsilon\}$); (7.50) implies (7.45), (7.46). The pair $\{\dot{\mathbf{u}}(0). \dot{\mathbf{p}}(0)\}$ is clearly the unique solution in $\mathbb{R}^N \times R(\mathbf{B})$ of

$$\begin{aligned} \mathbf{A}\dot{\mathbf{u}}(0) + B^t\dot{\mathbf{p}}(0) &= \mathbf{0}, \\ \mathbf{B}\dot{\mathbf{u}}(0) &= -\hat{\mathbf{p}}. \end{aligned} \tag{7.51}$$

We have $\{\dot{\mathbf{u}}(0), \dot{\mathbf{p}}(0)\} \neq \{\mathbf{0}, \mathbf{0}\}$, unless $\hat{\mathbf{p}} = \mathbf{0}$, which corresponds (see Exercise 7.3) to the trivial situation $\mathbf{u} = \mathbf{A}^{-1}\mathbf{b}$. This proves that the estimates (7.45) and (7.46) are of *optimal order* in general. □

EXERCISE 7.5. Prove that $|p_\varepsilon| \leq |\hat{p}|$, $\forall\, \varepsilon > 0$.

Remark 7.5. It follows from Theorem 7.2 that $\mathbf{u}_\varepsilon$ and $\mathbf{p}_\varepsilon$ will be good approximations of $\mathbf{u}$ and $\mathbf{p}$, respectively, provided that we use a *sufficiently small* ε. But in this case the *condition number* $\nu(\mathbf{A}_\varepsilon)$ of the matrix

$$\mathbf{A}_\varepsilon = \mathbf{A} + \frac{1}{\varepsilon}\mathbf{B}^t\mathbf{B} \tag{7.52}$$

occurring in (7.40) will be *large*; we indeed have (we suppose $\mathbf{A} = \mathbf{A}^t$ for simplicity[2])

$$\nu(\mathbf{A}_\varepsilon) = \|\mathbf{A}_\varepsilon^{-1}\|\,\|\mathbf{A}_\varepsilon\| = \frac{1}{\varepsilon}\frac{\rho(\mathbf{B}^t\mathbf{B})}{\sigma}(1 + \beta(\varepsilon)), \tag{7.53}$$

[2] We also suppose that $\operatorname{Ker}(\mathbf{B}) \neq \varnothing$ (which is the usual case).

where, in (7.53), $\lim_{\varepsilon\to 0}\beta(\varepsilon) = 0$, $\rho(\mathbf{B}^t\mathbf{B})$ is the *spectral radius* of $\mathbf{B}^t\mathbf{B}$ (i.e., the *largest eigenvalue* of $\mathbf{B}^t\mathbf{B}$) and where σ is defined by

$$\sigma = \inf_{\mathbf{v}\in \mathrm{Ker}(\mathbf{B})-\{\mathbf{0}\}} \frac{(\mathbf{A}\mathbf{v}, \mathbf{v})}{\|\mathbf{v}\|^2}.$$

For small ε it clearly follows from (7.53) that $\mathbf{A}_\varepsilon$ is *ill conditioned*. Actually that ill-conditioning property that we pointed out for the model problem (7.29) is the main drawback of penalty methods. An elegant way to overcome this difficulty has been introduced by Hestenes [1] and Powell [1]: the so-called *augmented Lagrangian methods* in which the combined use of penalty and Lagrange multiplier methods allow larger ε and moreover produces the exact solution $\mathbf{u}$ instead of an approximated one.[3] In Chapter VI, we will discuss the solution of a particular class of variational problems by these augmented Lagrangian methods; for more details and a substantial bibliography, see Fortin and Glowinski [1] and Gabay [1].

EXERCISE 7.6. Prove (7.53).

7.4.3. *A second example*

Let $\mathbf{G}: \mathbb{R}^N \to (\overline{\mathbb{R}})^M$; we then have $\mathbf{G} = \{g_i\}_{i=1}^M$, where g_i are functionals from $\mathbb{R}^N$ to $\overline{\mathbb{R}}$. We suppose that the following properties hold:

$$\forall\, i = 1, \ldots, M,\ g_i \text{ is a convex, l.s.c., and proper functional;} \tag{7.54}$$

$$\text{the convex set } K = \{\mathbf{v} \in \mathbb{R}^N,\ g_i(\mathbf{v}) \le 0,\ \forall\, i = 1, \ldots, M\} \text{ is nonempty.} \tag{7.55}$$

Suppose that the properties of $a(\cdot, \cdot)$ and $L(\cdot)$ are those of Sec. 7.4.1. From these properties and from (7.55), the EVI problem:

Find $\mathbf{u} \in K$ *such that*

$$(\mathbf{A}\mathbf{u}, \mathbf{v} - \mathbf{u}) \ge (\mathbf{b}, \mathbf{v} - \mathbf{u}), \qquad \forall\, \mathbf{v} \in K \tag{7.56}$$

has a *unique* solution [it suffices to apply (with $V = \mathbb{R}^N$) Theorem 3.1 of Sec. 3.1].

Remark 7.6. If $\mathbf{A} = \mathbf{A}^t$, problem (7.56) is equivalent to the minimization problem:

Find $\mathbf{u} \in K$ *such that*

$$J(\mathbf{u}) \le J(\mathbf{v}), \qquad \forall\, \mathbf{v} \in K \tag{7.57}$$

with $J(\mathbf{v}) = \frac{1}{2}(\mathbf{A}\mathbf{v}, \mathbf{v}) - (\mathbf{b}, \mathbf{v})$.

---

[3] See Chapter VII, Sec. 5.8.7.3.3 for an application of augmented Lagrangian methods to the solution of (7.31) (and (7.29)).

Remark 7.7. If **G** obeys some convenient conditions (usually called *qualification conditions*), we can generalize Proposition 7.1 and associate to (7.56) the so-called *F. John–Kuhn–Tucker multipliers*; we shall not discuss this matter here[4] (the interested reader may consult Rockafellar [1], Cea [1], [2], Ekeland and Temam [1], and Aubin [1]).

In order to apply the penalty method of Sec. 7.2 to the solution of (7.56), we define $j: \mathbb{R}^N \to \mathbb{R}$ by

$$j(\mathbf{v}) = \frac{1}{2} \sum_{i=1}^{M} \alpha_i |g_i^+(\mathbf{v})|^2, \tag{7.58}$$

where, in (7.58), α_i are strictly positive and $g_i^+ = \sup(0, g_i)$. Since $j(\cdot)$ satisfies (7.2)–(7.4), the associate penalized problem (with $j_\varepsilon = (1/\varepsilon)j$, $\varepsilon > 0$) is defined by:

Find $\mathbf{u}_\varepsilon \in \mathbb{R}^N$ *such that*

$$(\mathbf{A}\mathbf{u}_\varepsilon, \mathbf{v} - \mathbf{u}_\varepsilon) + j_\varepsilon(\mathbf{v}) - j_\varepsilon(\mathbf{u}_\varepsilon) \geq (\mathbf{b}, \mathbf{v} - \mathbf{u}_\varepsilon), \qquad \forall\, \mathbf{v} \in \mathbb{R}^N. \tag{7.59}$$

Remark 7.8. Suppose that $g_i \in C^1$, $\forall\, i = 1, \ldots, M$. We then have $|g_i^+|^2 \in C^1$, $\forall\, i = 1, \ldots, M$, implying that $j \in C^1$. We have

$$j'(\mathbf{v}) = \sum_{i=1}^{M} \alpha_i g_i^+(\mathbf{v}) g_i'(\mathbf{v}), \tag{7.60}$$

and from Remark 7.1, (7.59) is equivalent to the nonlinear system in $\mathbb{R}^N$:

$$\mathbf{A}\mathbf{u}_\varepsilon + \frac{1}{\varepsilon} j'(\mathbf{u}_\varepsilon) = \mathbf{b}. \tag{7.61}$$

It follows from Theorem 7.1 (see Sec. 7.3) that

$$\lim_{\varepsilon \to 0} \|\mathbf{u}_\varepsilon - \mathbf{u}\| = 0 \tag{7.62}$$

where $\mathbf{u}$ is the solution of (7.56).

To illustrate the above penalty method, we consider its application to the solution of a *discrete obstacle problem* (see Chapter II, Sec. 2 for a mathematical and mechanical motivation). For example (with $M = N$), we have

$$g_i(\mathbf{v}) = c_i - v_i, \qquad \forall\, i = 1, \ldots, N, \tag{7.63}$$

where $\mathbf{c} = \{c_i\}_{i=1}^N$, $\mathbf{v} = \{v_i\}_{i=1}^N$; we take $\alpha_i = 1$, $\forall\, i = 1, \ldots, N$, in (7.58).

For this simple problem the equivalence property of Remark 7.8 holds and (7.61) reduces to the nonlinear system

$$\mathbf{A}\mathbf{u}_\varepsilon - \frac{1}{\varepsilon}(\mathbf{c} - \mathbf{u}_\varepsilon)^+ = \mathbf{b}, \tag{7.64}$$

[4] See, however, Exercise 7.7

where

$$(\mathbf{c} - \mathbf{v})^+ = \{(c_i - v_i)^+\}_{i=1}^N;$$

problem (7.64) can be solved by the methods described in Chapter IV, Sec. 2.6 and also in Chapter VI, Sec. 6.4.

EXERCISE 7.7. Prove that the solution $\mathbf{u}$ of (7.56), with $\mathbf{G}$ defined by (7.63), is characterized by the existence of $\mathbf{p} = \{p_i\}_{i=1}^N \in \mathbb{R}^N$ such that

$$\begin{aligned} &\mathbf{Au} - \mathbf{p} = \mathbf{b}, \\ &p_i \geq 0, \qquad \forall\, i = 1, \ldots, N, \\ &p_i(c_i - u_i) = 0, \qquad \forall\, i = 1, \ldots, N. \end{aligned} \tag{7.65}$$

Also prove that $\lim_{\varepsilon \to 0} (1/\varepsilon)(\mathbf{c} - \mathbf{u}_\varepsilon)^+ = \mathbf{p}$.

Hint: Observe that:

(i) if $\mathbf{v} \in \mathbb{R}_+^N = \{\mathbf{v} \in \mathbb{R}^N, v_i \geq 0, \forall\, i = 1, \ldots, N\}$, then $\mathbf{u} + \mathbf{v} \in K$;

(ii) $\mathbf{c} \in K$.

The vector $\mathbf{p}$ in (7.65) is precisely a F. John–Kuhn–Tucker multiplier.

7.5. Further comments

For more details on penalty methods applied to the solution of variational problems, we refer the reader to Lions [1], [4], Cea [1], [2], and also to G.L.T. [1], [3], Oden and Kikuchi [1], and Ohtake, Oden, and Kikuchi [1], [2].

Actually similar ideas can be applied to the solution of EVI of the second kind; for example, we can replace the solution of :

Find $u \in V$ such that

$$a(u, v - u) + j(v) - j(u) \geq L(v - u), \qquad \forall\, v \in V \tag{7.66}$$

(where V, $a(\cdot, \cdot)$, $L(\cdot)$, $j(\cdot)$ obey the hypotheses of Sec. 2.1) by the solution of:

Find $u_\varepsilon \in V$ such that

$$a(u_\varepsilon, v - u_\varepsilon) + j_\varepsilon(v) - j_\varepsilon(u_\varepsilon) \geq L(v - u_\varepsilon), \qquad \forall\, v \in V, \tag{7.67}$$

where j_ε is an "approximation" of j which is more regular. For example, if we suppose that j is nondifferentiable, it may be interesting from a computational point of view to replace it by j_ε differentiable. Such a process is called—for obvious reasons—a *regularization* method.

If j_ε is differentiable, (7.67) is clearly equivalent to the variational equation:

Find $u_\varepsilon \in V$ such that

$$a(u_\varepsilon, v) + \langle j'_\varepsilon(u_\varepsilon), v\rangle = L(v), \qquad \forall\, v \in V, \tag{7.68}$$

where $j'_\varepsilon(v)$ denotes the differential of j_ε at v. An application of these regularization methods is given in Chapter II, Sec. 6.6; we refer to G.L.T. [1], [2], [3] for further details and other applications of these regularization methods.

8. References

For generalities on variational inequalities from a theoretical point of view, see Lions and Stampacchia [1], Lions [1], Ekeland and Temam [1], Baiocchi and Capelo [1], [2], and Kinderlherer and Stampacchia [1].

For generalities on the approximation of variational inequalities from the numerical point of view, see Falk [1], G.L.T. [1], [2], [3], Strang [1], Brezzi, Hager, and Raviart [1], [2], Oden and Kikuchi [1], and Lions [5].

For generalities and applications of the penalty and regularization methods discussed in Sec. 7, see Lions [1], [4], Cea [1], [2], G.L.T. [1], [2], [3], and Oden and Kikuchi [1] (see also Chapter II, Sec. 6.6 and Chapter VII, Sec. 4 of this book). Some additional references will be given in the following chapters.

CHAPTER II

Application of the Finite Element Method to the Approximation of Some Second-Order EVI

1. Introduction

In this chapter we consider some examples of EVI of the first and second kinds. These EVI are related to second-order partial differential operators (for fourth-order problems, see Glowinski [2] and G.L.T. [2], [3]). The physical interpretation and some properties of the solution are given. Finite element approximations of these EVI are considered and convergence results are proved. In some particular cases we also give error estimates.

Some of the results in this chapter may be found in G.L.T. [1], [2], [3]. For the approximation of the EVI of the first kind by finite element methods, we also refer the reader to Falk [1], Strang [1], Mosco and Strang [1], Ciarlet [1], [2], [3], and Brezzi, Hager and Raviart [1], [2].

We also describe iterative methods for solving the corresponding approximate problems (cf. Cea [1], [2] and G.L.T. [1], [2], [3]).

2. An Example of EVI of the First Kind: The Obstacle Problem

Notations

All the properties of Sobolev spaces used in this chapter are proved in Lions [2], Necas [1], and Adams [1]. Usually we shall have

- Ω: a bounded domain in $\mathbb{R}^2$,
- $\Gamma = \partial\Omega$,
- $x = \{x_1, x_2\}$, a generic point of Ω,
- $\nabla = \{\partial/\partial x_1, \partial/\partial x_2\}$,
- $C^m(\bar{\Omega})$: space of m-times continuously differentiable real valued functions for which all the derivatives up to order m are continuous in $\bar{\Omega}$,
- $C_0^m(\Omega) = \{v \in C^m(\bar{\Omega}) \,|\, \mathrm{Supp}(v) \text{ is a compact subset of } \Omega\}$,
- $\|v\|_{m,p,\Omega} = \sum_{|\alpha| \le m} \|D^\alpha v\|_{L^p(\Omega)}$ for $v \in C^m(\bar{\Omega})$, where $\alpha = \{\alpha_1, \alpha_2\}$; α_1, α_2 are non-negative integers, $|\alpha| = \alpha_1 + \alpha_2$ and $D^\alpha = \partial^{|\alpha|}/\partial x_1^{\alpha_1}\, \partial x_2^{\alpha_2}$,

- $W^{m,p}(\Omega)$: completion of $C^m(\overline{\Omega})$ in the above norm,
- $W_0^{m,p}(\Omega)$: completion of $C_0^m(\Omega)$ in the above norm,
- $H^m(\Omega) = W^{m,2}(\Omega)$,
- $H_0^m(\Omega) = W_0^{m,2}(\Omega)$,
- $\mathscr{D}(\Omega) = C_0^\infty(\Omega)$.

2.1. The continuous problem

Let

$$V = H_0^1(\Omega) = \{v \in H^1(\Omega) | v|_\Gamma = \text{trace of } v \text{ on } \Gamma = 0\}$$

(cf. Lions [2] and Necas [1] for a precise definition of the trace),

$$a(u, v) = \int_\Omega \nabla u \cdot \nabla v \, dx,$$

where

$$\nabla u \cdot \nabla v = \frac{\partial u}{\partial x_1}\frac{\partial v}{\partial x_1} + \frac{\partial u}{\partial x_2}\frac{\partial v}{\partial x_2},$$

$L(v) = \langle f, v \rangle$ for $f \in V^* = H^{-1}(\Omega)$ and $v \in V$. Let $\Psi \in H^1(\Omega) \cap C^0(\overline{\Omega})$ and $\Psi|_\Gamma \leq 0$. Define $K = \{v \in H_0^1(\Omega) | v \geq \psi \text{ a.e. on } \Omega\}$.
Then the obstacle problem is a particular (P_1) problem defined by:
Find u such that

$$a(u, v - u) \geq L(v - u), \qquad \forall\, v \in K, \quad u \in K. \tag{2.1}$$

The physical interpretation of this problem is as follows: Let an elastic membrane occupy a region Ω in the x_1, x_2 plane; this membrane is fixed along the boundary Γ on Ω. If there is no obstacle, from the theory of elasticity, the vertical displacement u, obtained by applying a vertical force F, is given by the solution of the following Dirichlet problem:

$$\begin{aligned} -\Delta u &= f \ \text{ in } \Omega, \\ u|_\Gamma &= 0, \end{aligned} \tag{2.2}$$

where $f = F/t$, t being the tension of the membrane.

If there is an obstacle, we have a free boundary problem, and the displacement u satisfies the variational inequality (2.1) with ψ being the height of the obstacle. Similar EVI also occur, sometimes with nonsymmetric bilinear forms, in mathematical models for the following problems:

- Lubrication phenomena (cf. Cryer [1]).
- Filtration of liquids in porous media (cf. Baiocchi [1] and Comincioli [1]).
- Two-dimensional irrotational flows of perfect fluids (cf. Brezis and Stampacchia [1], Brezis [1], and Ciavaldini and Tournemine [1]).
- Wake problems (cf. Bourgat and Duvaut [1]).

2.2. Existence and uniqueness results

For proving the existence and uniqueness of the problem (2.1), we need the following lemmas stated below without proof (for the proofs of the lemmas, see, for instance, Lions [2], Necas [1], and Stampacchia [1]).

Lemma 2.1. *Let Ω be a bounded domain in $\mathbb{R}^N$. Then the seminorm on $H^1(\Omega)$*

$$v \to \left(\int_\Omega |\nabla v|^2 \, dx\right)^{1/2}$$

is a norm on $H_0^1(\Omega)$ and it is equivalent to the norm on $H_0^1(\Omega)$ induced from $H^1(\Omega)$.

The above Lemma 2.1 is known as the Poincaré–Friedrichs lemma.

Lemma 2.2. (Stampacchia [1]). *Let $f: \mathbb{R} \to \mathbb{R}$ be uniformly Lipschitz continuous (i.e., $\exists\, k > 0$ such that $|f(t) - f(t')| \leq k|t - t'|$, $\forall\, t, t' \in \mathbb{R}$) and such that f' has a finite number of points of discontinuity. Then the induced map f^* on $H^1(\Omega)$ defined by $v \to f(v)$ is a continuous map into $H^1(\Omega)$. Similar results hold for $H_0^1(\Omega)$ whenever $f(0) = 0$.*

Corollary 2.1. *If v^+ and v^- denote the positive and the negative parts of v for $v \in H^1(\Omega)$ (respectively, $H_0^1(\Omega)$), then the map $v \to \{v^+, v^-\}$ is continuous from $H^1(\Omega) \to H^1(\Omega) \times H^1(\Omega)$ (respectively, $H_0^1(\Omega) \to H_0^1(\Omega) \times H_0^1(\Omega)$). Also $v \to |v|$ is continuous.*

Theorem 2.1. *Problem* (2.1) *has a unique solution.*

PROOF. In order to apply Theorem 3.1 of Chapter I, we have to prove that $a(\cdot, \cdot)$ is V-elliptic and that K is a closed convex nonempty set.

The V-ellipticity of $a(\cdot, \cdot)$ follows from Lemma 2.1 and the convexity of K is trivial.

(1) *K is nonempty.* We have

$$\Psi \in H^1(\Omega) \cap C^0(\bar{\Omega}) \quad \text{with } \Psi \leq 0 \text{ on } \Gamma.$$

Hence, by Corollary 2.1, $\Psi^+ \in H^1(\Omega)$. Since $\Psi|_\Gamma \leq 0$, we have $\Psi^+|_\Gamma = 0$. This implies that $\Psi^+ \in H_0^1(\Omega)$; since

$$\Psi^+ = \max\{\Psi, 0\} \geq \Psi,$$

we have $\Psi^+ \in K$. Hence K is nonempty.

(2) *K is closed.* Let $v_n \to v$ strongly in $H_0^1(\Omega)$, where $v_n \in K$ and $v \in H_0^1(\Omega)$. Hence $v_n \to v$ strongly in $L^2(\Omega)$. Therefore we can extract a subsequence $\{v_{n_i}\}$ such that $v_{n_i} \to v$ a.e. on Ω. Then $v_{n_i} \geq \Psi$ a.e. on Ω implies that

$$v \geq \Psi \text{ a.e. on } \Omega;$$

therefore $v \in K$.

Hence, by Theorem 3.1 of Chapter I, we have a unique solution for (2.1). □

2.3. Interpretation of (2.1) as a free boundary problem

From the solution u of (2.1), we define

$$\Omega^+ = \{x \mid x \in \Omega, u(x) > \Psi(x)\},$$
$$\Omega^0 = \{x \mid x \in \Omega, u(x) = \Psi(x)\},$$
$$\gamma = \partial\Omega^+ \cap \partial\Omega^0; u^+ = u|_{\Omega^+}; u^0 = u|_{\Omega^0}.$$

Classically, problem (2.1) has been formulated as the problem of finding γ (*the free boundary*) and u such that

$$-\Delta u = f \text{ in } \Omega^+, \tag{2.3}$$
$$u = \Psi \text{ on } \Omega^0, \tag{2.4}$$
$$u = 0 \text{ on } \Gamma, \tag{2.5}$$
$$u^+|_\gamma = u^0|_\gamma. \tag{2.6}$$

The physical interpretation of these relations is the following: (2.3) means that on Ω^+ the membrane is strictly over the obstacle; (2.4) means that on Ω^0 the membrane is in contact with the obstacle; (2.6) is a transmission relation at the free boundary.

Actually (2.3)–(2.6) are not sufficient to characterise u since there are an infinite number of solutions for (2.3)–(2.6). Therefore it is necessary to add other transmission properties: for instance, if Ψ is smooth enough (say $\Psi \in H^2(\Omega)$), we require the "continuity" of ∇u at γ (we may require $\nabla u \in H^1(\Omega) \times H^1(\Omega)$).

Remark 2.1. This kind of free boundary interpretation holds for several problems modelled by EVI of the first and second kinds.

2.4. Regularity of the solutions

We state without proof the following regularity theorem for the solution of problem (2.1).

Theorem 2.2 (Brezis and Stampacchia [2]). *Let Ω be a bounded domain in $\mathbb{R}^2$ with a smooth boundary. If*

$$L(v) = \int_\Omega fv\, dx \quad \text{with } f \in L^p(\Omega), \qquad 1 < p < +\infty \tag{2.7}$$

and

$$\Psi \in W^{2,p}(\Omega), \tag{2.8}$$

then the solution of the problem (2.1) *is in* $W^{2,p}(\Omega)$.

Remark 2.2. Let $\Omega \subset \mathbb{R}^N$ have a smooth boundary. We know that

$$W^{s,p}(\Omega) \subset C^k(\overline{\Omega}) \quad \text{if } s > \frac{N}{p} + k \tag{2.9}$$

(cf. Necas [1]). It follows that the solution u of (2.1) will be in $C^1(\overline{\Omega})$ if $f \in L^p(\Omega)$, $\Psi \in W^{2,p}(\Omega)$ with $p > 2$ (take $s = 2$, $N = 2$, $k = 1$ in (2.9)).

The proof of this regularity result will be given in the following simple case:

$$L(v) = \int_\Omega fv\, dx, \qquad f \in L^2(\Omega), \tag{2.10}$$

$$\Psi = 0 \text{ on } \Omega. \tag{2.11}$$

Before proving that (2.10), (2.11) imply $u \in H^2(\Omega)$, we shall recall a classical lemma (also very useful in the analysis of fourth-order problems).

Lemma 2.3. *Let Ω be a bounded domain of $\mathbb{R}^N$ with a boundary Γ sufficiently smooth. Then $\|\Delta v\|_{L^2(\Omega)}$ defines a norm on $H^2(\Omega) \cap H_0^1(\Omega)$ which is equivalent to the norm induced by the $H^2(\Omega)$-norm.*

EXERCISE 2.1. Prove Lemma 2.3 using the following regularity result due to Agmon, Douglis, and Nirenberg [1]:

If $w \in L^2(\Omega)$ and if Γ is sufficiently smooth, then the Dirichlet problem

$$-\Delta v = w \text{ in } \Omega,$$

$$v|_\Gamma = 0,$$

has a unique solution in $H_0^1(\Omega) \cap H^2(\Omega)$ (this regularity result also holds if Ω is a convex domain with Γ Lipschitz continuous).

We shall now apply Lemma 2.3 to prove the following theorem using a method due to Brezis and Stampacchia [2].

Theorem 2.2*. *If Γ is smooth enough, if $\Psi = 0$, and if $L(v) = \int_\Omega fv\, dx$ with $f \in L^2(\Omega)$ then the solution u of the problem (2.1) satisfies*

$$u \in K \cap H^2(\Omega), \qquad \|\Delta u\|_{L^2(\Omega)} \leq \|f\|_{L^2(\Omega)}. \tag{2.12}$$

PROOF. With L and ψ as above, it follows from Theorem 2.1 that problem (2.1) has a unique solution u. Letting $\varepsilon > 0$, consider the following Dirichlet problem:

$$-\varepsilon\Delta u_\varepsilon + u_\varepsilon = u \text{ in } \Omega, \qquad u_\varepsilon|_\Gamma = 0. \tag{2.13}$$

Problem (2.13) has a unique solution in $H_0^1(\Omega)$, and the smoothness of Γ implies that u_ε belongs to $H^2(\Omega)$. Since $u \geq 0$ a.e. on Ω, by the *maximum principle* for second-order elliptic differential operators (cf. Necas [1]), we have $u_\varepsilon \geq 0$. Hence

$$u_\varepsilon \in K. \tag{2.14}$$

From (2.14) and (2.1), we obtain

$$a(u, u_\varepsilon - u) \geq L(u_\varepsilon - u) = \int_\Omega f(u_\varepsilon - u)\, dx. \tag{2.15}$$

The V-ellipticity of $a(\cdot, \cdot)$ implies

$$a(u_\varepsilon, u_\varepsilon - u) = a(u_\varepsilon - u, u_\varepsilon - u) + a(u, u_\varepsilon - u) \geq a(u, u_\varepsilon - u),$$

so that

$$a(u_\varepsilon, u_\varepsilon - u) \geq \int_\Omega f(u_\varepsilon - u)\, dx. \tag{2.16}$$

By (2.13) and (2.16), we obtain

$$\varepsilon \int_\Omega \nabla u_\varepsilon \cdot \nabla(\Delta u_\varepsilon)\, dx \geq \varepsilon \int_\Omega f \Delta u_\varepsilon\, dx,$$

so that,

$$\int_\Omega \nabla u_\varepsilon \cdot \nabla(\Delta u_\varepsilon)\, dx \geq \int_\Omega f\, \Delta u_\varepsilon\, dx. \tag{2.17}$$

By Green's formula, (2.17) implies

$$-\int_\Omega (\Delta u_\varepsilon)^2\, dx \geq \int_\Omega f\, \Delta u_\varepsilon\, dx.$$

Thus

$$\|\Delta u_\varepsilon\|_{L^2(\Omega)} \leq \|f\|_{L^2(\Omega)}, \tag{2.18}$$

using *Schwarz inequality* in $L^2(\Omega)$. By Lemma 2.3 and relations (2.13), (2.18) we obtain

$$\lim_{\varepsilon \to 0} u_\varepsilon = u \textit{ weakly} \text{ in } H^2(\Omega), \tag{2.19}$$

(which implies that $\lim u_\varepsilon = u$ strongly in $H^s(\Omega)$, for every $s < 2$ (cf. Necas [1])), so that $u \in H^2(\Omega)$ with

$$\|\Delta u\|_{L^2(\Omega)} \leq \|f\|_{L^2(\Omega)}. \tag{2.20}$$

□

2.5. Finite element approximations of (2.1)

Henceforth we shall assume that Ω is a polygonal domain of $\mathbb{R}^2$. Consider a "classical" triangulation $\mathscr{T}_h$ of Ω, i.e. $\mathscr{T}_h$ is a finite set of triangles T such that

$$T \subset \bar{\Omega} \quad \forall\, T \in \mathscr{T}_h, \qquad \bigcup_{T \in \mathscr{T}_h} T = \bar{\Omega}, \tag{2.21}$$

$$\mathring{T}_1 \cap \mathring{T}_2 = \varnothing \quad \forall\, T_1, T_2 \in \mathscr{T}_h \quad \text{and} \quad T_1 \neq T_2. \tag{2.22}$$

Figure 2.1

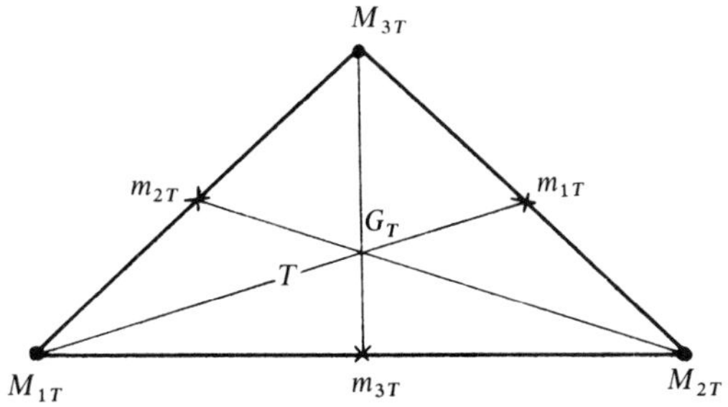

Moreover $\forall\ T_1, T_2 \in \mathscr{T}_h$ and $T_1 \neq T_2$, exactly one of the following conditions must hold

$$\begin{aligned}&(1)\quad T_1 \cap T_2 = \varnothing,\\&(2)\quad T_1 \text{ and } T_2 \text{ have only one common vertex,}\\&(3)\quad T_1 \text{ and } T_2 \text{ have only a whole common edge.}\end{aligned}\tag{2.23}$$

As usual h will be the length of the largest edge of the triangles in the triangulation.

From now on we restrict ourselves to piecewise linear and piecewise quadratic finite element approximations.

2.5.1. *Approximation of V and K*

- P_k: space of polynomials in x_1 and x_2 of degree less than or equal to k,
- $\Sigma_h = \{P \in \overline{\Omega}, P \text{ is a vertex of } T \in \mathscr{T}_h\}$,
- $\mathring{\Sigma}_h = \{P \in \Sigma_h, P \notin \Gamma\}$,
- $\Sigma'_h = \{P \in \overline{\Omega}, P \text{ is the midpoint of an edge of } T \in \mathscr{T}_h\}$,
- $\mathring{\Sigma}'_h = \{P \in \Sigma'_h, P \notin \Gamma\}$,
- $\Sigma_h^1 = \Sigma_h$ and $\Sigma_h^2 = \Sigma_h \cup \Sigma'_h$.

Figure 2.1 illustrates some further notation associated with an arbitrary triangle T. We have $m_{iT} \in \Sigma'_h$, $M_{iT} \in \Sigma_h$. The centroid of the triangle T is denoted by G_T.

The space $V = H_0^1(\Omega)$ is approximated by the family of subspaces $(V_h^k)_h$ with $k = 1$ or 2, where

$$V_h^k = \{v_h \in C^0(\overline{\Omega}), v_h|_\Gamma = 0 \text{ and } v_h|_T \in P_k, \forall\ T \in \mathscr{T}_h\}, \qquad k = 1, 2.$$

It is clear that the V_h^k are finite dimensional (cf. Ciarlet [1]). It is then quite natural to approximate K by

$$K_h^k = \{v_h \in V_h^k, v_h(P) \geq \Psi(P), \forall\ P \in \mathring{\Sigma}_h^k\}, \qquad k = 1, 2.$$

Proposition 2.1. *Then K_h^k for $k = 1, 2$ are closed convex nonempty subsets of V_h^k.*

EXERCISE 2.2. Prove Proposition 2.1.

2.5.2. *The approximate problems*

For $k = 1, 2$, the approximate problems are defined by

$$a(u_h^k, v_h - u_h^k) \geq L(v_h - u_h^k), \qquad \forall\, v_h \in K_h^k, \quad u_h^k \in K_h^k. \tag{P_{1h}^k}$$

From Theorem 3.1 of Chapter I and Proposition 2.1, it follows that:

Proposition 2.2. *(P_{1h}^k) has a unique solution for $k = 1$ and 2.*

Remark 2.3. Since the bilinear form $a(\cdot, \cdot)$ is symmetric, (P_{1h}^k) is actually equivalent to (cf. Chapter I, Remark 3.2) the *quadratic programming* problem

$$\operatorname*{Min}_{v_h \in K_h} [\tfrac{1}{2}a(v_h, v_h) - L(v_h)]. \tag{2.24}$$

2.6. Convergence results

In order to simplify the convergence proof, we shall assume in this section that

$$\Psi \in C^0(\bar{\Omega}) \cap H^1(\Omega) \quad \text{and} \quad \Psi \leq 0 \text{ in a neighborhood of } \Gamma. \tag{2.25}$$

Before proving the convergence results, we shall describe two important numerical quadrature schemes which will be used to prove the convergence theorem.

EXERCISE 2.3. With notations as in Fig. 2.1, prove the following identities for any triangle T:

$$\int_T w\, dx = \frac{\operatorname{meas}(T)}{3} \sum_{i=1}^{3} w(M_{iT}), \qquad \forall\, w \in P_1, \tag{2.26}$$

$$\int_T w\, dx = \frac{\operatorname{meas}(T)}{3} \sum_{i=1}^{3} w(m_{iT}), \qquad \forall\, w \in P_2. \tag{2.27}$$

Formula (2.26) is called the *Trapezoidal Rule* and (2.27) is known as *Simpson's Integral formula.* These formulae not only have theoretical importance, but practical utility as well.

We have the following results about the convergence of u_h^k (solution of the problem (P_{1h}^k)) as $h \to 0$.

Theorem 2.3. *Suppose that the angles of the triangles of $\mathscr{T}_h$ are uniformly bounded below by $\theta_0 > 0$ as $h \to 0$; then for $k = 1, 2$,*

$$\lim_{h \to 0} \|u_h^k - u\|_{H_0^1(\Omega)} = 0, \tag{2.28}$$

where u_h^k and u are the solutions of (P_{1h}^k) and (2.1), respectively.

PROOF. In this proof we shall use the following density result to be proved later:

$$\overline{\mathscr{D}(\Omega) \cap K} = K. \tag{2.29}$$

To prove (2.28) we shall use Theorem 5.2 of Chapter I. To do this we have to verify that the following two properties hold (for $k = 1, 2$):

(i) If $(v_h)_h$ is such that $v_h \in K_h^k$, $\forall\, h$ and converges weakly to v as $h \to 0$, then $v \in K$.
(ii) There exist χ, $\overline{\chi} = K$ and $r_h^k: \chi \to K_h^k$ such that $\lim_{h\to 0} r_h^k v = v$ strongly in V, $\forall\, v \in \chi$.

Verification of (i). Using the notation of Fig. 2.1 and considering $\phi \in \mathscr{D}(\Omega)$ with $\phi \geq 0$, we define ϕ_h by $\phi_h = \sum_{T \in \mathscr{T}_h} \phi(G_T)\chi_T$, where χ_T is the *characteristic function*[1] of T and G_T is the centroid of T. It is easy to see from the uniform continuity of ϕ that

$$\lim_{h\to 0} \phi_h = \phi \text{ strongly in } L^\infty(\Omega). \tag{2.30}$$

Then we approximate Ψ by Ψ_h such that

$$\begin{aligned} &\Psi_h \in C^0(\overline{\Omega}),\ \Psi_h|_T \in P_k, &&\forall\, T \in \mathscr{T}_h, \\ &\Psi_h(P) = \Psi(P), &&\forall\, P \in \Sigma_h^k. \end{aligned} \tag{2.31}$$

This function Ψ_h satisfies

$$\lim_{h\to 0} \Psi_h = \Psi \text{ strongly in } L^\infty(\Omega). \tag{2.32}$$

Let us consider a sequence $(v_h)_h$, $v_h \in K_h^k$, $\forall\, h$ such that

$$\lim_{h\to 0} v_h = v \text{ weakly in } V.$$

Then $\lim_{h\to 0} v_h = v$ strongly in $L^2(\Omega)$ (cf. Necas [1]) which, using (2.30) and (2.32), implies that

$$\lim_{h\to 0} \int_\Omega (v_h - \Psi_h)\phi_h\, dx = \int_\Omega (v - \Psi)\phi\, dx, \tag{2.33}$$

(actually, since $\phi_h \to \phi$ strongly in $L^\infty(\Omega)$, the weak convergence of v_h in $L^2(\Omega)$ is enough to prove (2.33)).

We have

$$\int_\Omega (v_h - \Psi_h)\phi_h\, dx = \sum_{T \in \mathscr{T}_h} \phi(G_T) \int_T (v_h - \Psi_h)\, dx. \tag{2.34}$$

From (2.26), (2.27), and from the definition of Ψ_h, we obtain for all $T \in \mathscr{T}_h$,

$$\int_T (v_h - \Psi_h)\, dx = \frac{\text{meas}\,(T)}{3} \sum_{i=1}^{3} [v_h(M_{iT}) - \psi_h(M_{iT})] \quad \text{if } k = 1, \tag{2.35}$$

$$\int_T (v_h - \Psi_h)\, dx = \frac{\text{meas}\,(T)}{3} \sum_{i=1}^{3} [v_h(m_{iT}) - \psi_h(m_{iT})] \quad \text{if } k = 2. \tag{2.36}$$

[1] $\chi_T(x) = 1, \forall\, x \in T$, $\chi_T(x) = 0$ if $x \notin T$.

Using the fact that $\phi_h \geq 0$, the definition of K_h^k and relations (2.35) and (2.36), it follows from (2.34) that

$$\int_\Omega (v_h - \Psi_h)\phi_h \, dx \geq 0, \qquad \forall \, \phi \in \mathscr{D}(\Omega), \quad \phi \geq 0,$$

so that as $h \to 0$

$$\int_\Omega (v - \Psi)\phi \, dx \geq 0, \qquad \forall \, \phi \in \mathscr{D}(\Omega), \quad \phi \geq 0$$

which in turn implies $v \geq \Psi$ a.e. in Ω. Hence (i) is verified.

Verification of (ii). From (2.29) it is natural to take $\chi = \mathscr{D}(\Omega) \cap K$. We define

$$r_h^k \colon H_0^1(\Omega) \cap C^0(\overline{\Omega}) \to V_h^k$$

as the "linear" interpolation operator if $k = 1$ and "quadratic" interpolation operator if $k = 2$, i.e.,

$$\begin{aligned} &r_h^k v \in V_h^k, \qquad \forall \, v \in H_0^1(\Omega) \cap C^0(\overline{\Omega}), \\ &(r_h^k v)(P) = v(P), \qquad \forall \, P \in \mathring{\Sigma}_h^k \quad \text{for } k = 1, 2. \end{aligned} \tag{2.37}$$

On the one hand it is known (cf., for instance, Ciarlet [1], [2] and Strang and Fix [1]) that under the assumption made on $\mathscr{T}_h$ in the statement of Theorem 2.3, we have

$$\|r_h^k v - v\|_V \leq Ch^k \|v\|_{H^{k+1}(\Omega)}, \qquad \forall \, v \in \mathscr{D}(\Omega), \quad k = 1, 2,$$

with C independent of h and v. This implies that

$$\lim_{h \to 0} \|r_h^k v - v\|_V = 0, \qquad \forall \, v \in \chi, \quad k = 1, 2.$$

On the other hand, it is obvious that

$$r_h^k v \in K_h^k, \qquad \forall \, v \in K \cap C^0(\overline{\Omega}),$$

so that

$$r_h^k v \in K_h^k, \qquad \forall \, v \in \chi \quad \text{for } k = 1, 2.$$

In conclusion, with the above χ and r_h^k, (ii) is satisfied. Hence we have proved the Theorem 2.3 modulo, the proof of the density result (2.29). □

Lemma 2.4. *Under the assumptions* (2.25), *we have* $\overline{\mathscr{D}(\Omega) \cap K} = K$.

PROOF. Let us prove the Lemma in two steps.

Step 1. Let us show that

$$\mathscr{K} = \{v \in K \cap C^0(\overline{\Omega}), v \text{ has a compact support in } \Omega\} \tag{2.38}$$

is dense in K.

Let $v \in K$; $K \subset H_0^1(\Omega)$ implies that there exists a sequence $\{\phi_n\}_n$ in $\mathscr{D}(\Omega)$ such that

$$\lim_{n \to \infty} \phi_n = v \text{ strongly in } V.$$

Define v_n by

$$v_n = \max(\Psi, \phi_n) \tag{2.39}$$

so that

$$v_n = \tfrac{1}{2}[(\Psi + \phi_n) + |\Psi - \phi_n|].$$

Since $v \in K$, from Corollary 2.1 and relation (2.39), it follows that

$$\lim_{n\to\infty} v_n = \tfrac{1}{2}[(\Psi + v) + |\Psi - v|] = \max(\Psi, v) = v \text{ strongly in } V. \tag{2.40}$$

From (2.25) and (2.39), it follows that

$$\text{each } v_n \text{ has a compact support in } \Omega, \tag{2.41}$$

$$v_n \in K \cap C^0(\overline{\Omega}). \tag{2.42}$$

From (2.40)–(2.42) we obtain (2.38).

Step 2. Let us show that:

For every $v \in \mathscr{K}$, there exists a sequence $\{v_m\}_m$ such that

$$v_m \in \mathscr{D}(\Omega) \cap K, \qquad \forall\, m \qquad \text{and} \qquad \lim_{m\to+\infty} \|v_m - v\|_{H_0^1(\Omega)} = 0.$$

From Step 1 this proves that $\mathscr{D}(\Omega) \cap K$ is dense in K. Let ρ_n be a sequence of mollifiers, i.e.,

$$\rho_n \in \mathscr{D}(\mathbb{R}^2), \qquad \rho_n \geq 0, \tag{2.43}$$

$$\int_{\mathbb{R}^2} \rho_n(y)\, dy = 1, \tag{2.44}$$

$$\bigcap_{n=1}^{\infty} \operatorname{Supp}(\rho_n) = \{0\}, \qquad \{\operatorname{Supp}(\rho_n)\} \text{ is a decreasing sequence.}$$

Let $v \in \mathscr{K}$. Let $\tilde{v}$ be an extension of v defined by

$$\tilde{v}(x) = \begin{cases} v(x) & \text{if } x \in \Omega, \\ 0 & \text{if } x \notin \Omega, \end{cases}$$

then $\tilde{v} \in H^1(\mathbb{R}^2)$. Let $\tilde{v}_n = \tilde{v} * \rho_n$, i.e.,

$$\tilde{v}_n(x) = \int_{\mathbb{R}^2} \rho_n(x - y)\tilde{v}(y)\, dy, \tag{2.45}$$

then

$$\begin{gathered} \tilde{v}_n \in \mathscr{D}(\mathbb{R}^2), \\ \operatorname{Supp}(\tilde{v}_n) \subset \operatorname{Supp}(v) + \operatorname{Supp}(\rho_n), \\ \lim_{n\to\infty} \tilde{v}_n = \tilde{v} \text{ strongly in } H^1(\mathbb{R}^2). \end{gathered} \tag{2.46}$$

Hence from (2.41) and (2.46), we have

$$\operatorname{Supp}(|\tilde{v}_n|) \subset \Omega \quad \text{for } n \text{ large enough.} \tag{2.47}$$

We also have (since supp($\tilde{v}$) is bounded)

$$\lim \tilde{v}_n = \tilde{v} \text{ strongly in } L^\infty(\mathbb{R}^2). \tag{2.48}$$

Define $v_n = \tilde{v}_n|_\Omega$; then (2.46)–(2.48) imply

$$\begin{aligned} &v_n \in \mathscr{D}(\Omega), \\ &\lim_{n\to\infty} v_n = v \text{ strongly in } H_0^1(\Omega) \cap C^0(\overline{\Omega}); \end{aligned} \tag{2.49}$$

$v \in \mathscr{K}$ and $\Psi \leq 0$ in a neighborhood of Γ imply that there exists a $\delta > 0$ such that

$$v = 0, \qquad \Psi \leq 0 \text{ on } \Omega_\delta, \tag{2.50}$$

where $\Omega_\delta = \{x \in \Omega \,|\, d(x, \Gamma) < \delta\}$ ($d(x, \Gamma)$ = distance from x to Γ).

From (2.48) and (2.50) it follows that $\forall\, \varepsilon > 0$, there exists an $n_0 = n_0(\varepsilon)$ such that $\forall\, n \geq n_0(\varepsilon)$

$$\begin{aligned} &v(x) - \varepsilon \leq v_n(x) \leq v(x) + \varepsilon, \qquad \forall\, x \in \Omega - \Omega_{\delta/2}, \\ &v_n(x) = v(x) = 0 \quad \text{for } x \in \Omega_{\delta/2} \end{aligned} \tag{2.51}$$

Since $\overline{\Omega} - \Omega_{\delta/2}$ is a compact subset of $\overline{\Omega}$, there exists a function θ (cf., for instance, H. Cartan [1]) such that

$$\begin{aligned} &\theta \in \mathscr{D}(\Omega), \qquad \theta \geq 0 \text{ in } \Omega \\ &\theta(x) = 1, \qquad \forall\, x \in \overline{\Omega} - \Omega_{\delta/2}. \end{aligned} \tag{2.52}$$

Finally, define $w_n^\varepsilon = v_n + \varepsilon\theta$.

Then from (2.49), (2.51), and (2.52), we have

$$w_n^\varepsilon \in \mathscr{D}(\Omega), \qquad \lim_{\substack{\varepsilon\to 0\\ n\to\infty\\ n\geq n_0(\varepsilon)}} w_n^\varepsilon = v \text{ strongly in } H_0^1(\Omega),$$

with $w_n^\varepsilon(x) \geq v(x) \geq \Psi(x)$, $\forall\, x \in \Omega$, so that Step 2 is proved. □

Remark 2.4. Analyzing verification (i) in the proof of Theorem 2.3, we observe that if for $k = 2$ we use, instead of K_h^2, the convex set

$$\{v_h \in V_h^2,\, v_h(P) \geq \Psi(P),\, \forall\, P \in \mathring{\Sigma}_h'\},$$

then the convergence of u_h^2 to u still holds provided $\mathscr{T}_h$ obeys the same assumptions as in the statement of Theorem 2.3.

EXERCISE 2.4. Extend the previous analysis if Ω is not a polygonal domain.

EXERCISE 2.5. Let Ω be a bounded domain of $\mathbb{R}^2$ and let Γ_0 be a "nice" subset of Γ (see Fig. 2.2). Define V by $V = \{v \in H^1(\Omega), v|_{\Gamma_0} = 0\}$. Taking the bilinear form $a(\cdot, \cdot)$ as in (2.1), and $L \in V^*$, study the following EVI:

$$a(u, v - u) \geq L(v - u), \qquad \forall\, v \in K, \quad u \in K,$$

Figure 2.2

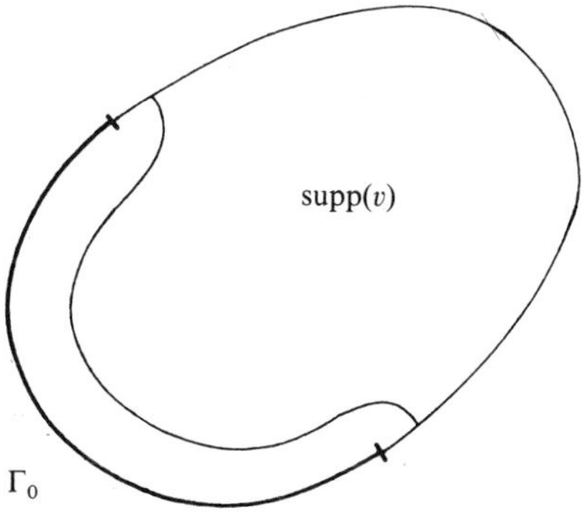

where $K = \{v \in V,\ v \geq \Psi \text{ a.e. in } \Omega\}$ and $\Psi \in C^0(\bar{\Omega}) \cap H^1(\Omega)$, $\Psi \leq 0$ in a neighborhood of Γ_0. Also study the finite element approximation of the above EVI.

Hint: Use the fact that if Γ and Γ_0 are smooth enough, then $\overline{\mathscr{V}} = V$, where (see Fig. 2.2), $\mathscr{V} = \{v \in C^\infty(\bar{\Omega}),\ v = 0 \text{ in a neighborhood of } \Gamma_0\}$.

2.7. Comments on the error estimates

We do not emphasize this subject too much since this is done in detail in Ciarlet [1, Chapter 9], [2, Chapter 5] and G.L.T. [3, Appendix 1], at least for piecewise linear approximations.

2.7.1. *Piecewise linear approximation*

Using piecewise linear finite elements and assuming that $f \in L^2(\Omega)$ and ψ, $u \in H^2(\Omega)$, $O(h)$ estimates for $\|u - u_h\|_{H^1(\Omega)}$ have been obtained by Falk [1], [2], [3], Strang [1], Mosco and Strang [1], and Brezzi, Hager, and Raviart [1]. We also refer to Ciarlet [1, Chapter 9], [2, Chapter 5] (resp., G.L.T. [3, Appendix 1]) in which the Falk (resp., Brezzi, Hager, and Raviart) analysis is given.

2.7.2. *Piecewise quadratic approximation*

Assuming more regularity for f, Ψ, and u than in the previous case (also assuming some smoothness hypotheses for the free boundary, an $O(h^{3/2-\varepsilon})$ estimate for $\|u_h - u\|_{H^1(\Omega)}$ has been obtained by Brezzi, Hager, and Raviart [1] and Brezzi and Sacchi [1] for an approximation by piecewise quadratic finite elements, similar to the one described in Sec. 2.6.

2.8. Iterative solution of the approximate problem

Once the continuous problem has been approximated and the convergence proved, it remains to effectively compute the approximate solution. In the case of the discrete obstacle problem, this can be done easily by using an *over-relaxation method with projection* as described in Cea and Glowinski [1], Cea [2], and also in Chapter V, Sec. 5 of this book.

Let us justify the use of this method. It follows from Remark 2.3 that the discrete problem is of the following type:

$$\operatorname*{Min}_{v \in C} \left[\tfrac{1}{2}(Av, v) - (b, v)\right] \tag{2.53}$$

where $(\cdot, \cdot)$ denotes the usual inner product in $\mathbb{R}^N$ and $v = \{v_1, \ldots, v_N\}$ and

$$A = (a_{ij}), \qquad 1 \leq i \leq N, \quad 1 \leq j \leq N \tag{2.54}$$

is a symmetric positive-definite $N \times N$ matrix and C is the set given by

$$C = \{v \in \mathbb{R}^N, v_i \geq \Psi_i, 1 \leq i \leq N\}. \tag{2.55}$$

Since C is the product of closed intervals of $\mathbb{R}$, the over-relaxation method with projection on C can be used. Let us describe it in detail:

$$u^0 \in C, u^0 \text{ arbitrarily chosen in } C \ (u^0 = \{\Psi_1, \ldots, \Psi_N\} \text{ may be a good guess}), \tag{2.56}$$

then u^n being known, we compute u^{n+1}, component by component using for $i = 1, 2, \ldots, N$

$$\bar{u}_i^{n+1} = \frac{1}{a_{ii}}\left(b_i - \sum_{j=1}^{i-1} a_{ij}u_j^{n+1} - \sum_{j=i+1}^{N} a_{ij}u_j^n\right), \tag{2.57$_i$}$$

$$u_i^{n+1} = P_i(u_i^n + \omega(\bar{u}_i^{n+1} - u_i^n)), \tag{2.58$_i$}$$

where

$$P_i(x) = \operatorname{Max}(x, \Psi_i), \qquad \forall\, x \in \mathbb{R}. \tag{2.59$_i$}$$

From Chapter V, Sec. 5 (see also Cea and Glowinski [1], Cea [2], and G.L.T. [1], [3]) it follows that:

Proposition 2.3. *Let $\{u^n\}_n$ be defined by* (2.56)–(2.59). *Then for every $u^0 \in C$ and $\forall\, 0 < \omega < 2$, we have $\lim_{n \to \infty} u^n = u$, where u is the unique solution of* (2.53).

Remark 2.5. In the case of the discrete obstacle problem, the components of u will be the values taken by the approximate solution at the nodes of $\mathring{\Sigma}_h$ if $k = 1$ and $\mathring{\Sigma}_h \cap \mathring{\Sigma}'_h$ if $k = 2$. Similarly, Ψ_i will be the values taken by Ψ at the nodes stated above, assuming these nodes have been ordered from 1 to N.

Remark 2.6. The optimal choice for ω is a critical and nontrivial point. However, from numerical experiments it has been observed that the so-called *Young method* for obtaining the optimal value of ω during the iterative process itself leads to a value of ω with good convergence properties. The convergence of this method has been proved for linear equations and requires special properties for the matrix of the system (see Young [1] and Varga [1]). However, an empirical justification of its success for the obstacle problem can be made, but will not be given here.

Remark 2.7. From numerical experiments it has been found that the optimal value of ω is always strictly greater than unity.

3. A Second Example of EVI of the First Kind: The Elasto-Plastic Torsion Problem

3.1. Formulation. Preliminary results

Let Ω be a bounded domain of $\mathbb{R}^2$ with a smooth boundary Γ. With the same definition for V, $a(\cdot, \cdot)$, and $L(\cdot)$ as in Sec. 2.1 of this chapter, we consider the following EVI of the first kind:

$$a(u, v - u) \geq L(v - u), \qquad \forall\, v \in K, \quad u \in K, \tag{3.1}$$

where

$$K = \{v \in H_0^1(\Omega), |\nabla v| \leq 1 \text{ a.e. in } \Omega\}. \tag{3.2}$$

Theorem 3.1. *Problem* (3.1) *has a unique solution.*

PROOF. In order to apply Theorem 3.1 of Chapter I, we only have to verify that K is a nonempty closed convex subset of V. K is nonempty because $0 \in K$, and the convexity of K is obvious. To prove that K is closed, consider a sequence $\{v_n\}$ in K such that $v_n \to v$ strongly in V. Then there exists a subsequence $\{v_{n_i}\}$ such that

$$\lim_{i \to +\infty} \nabla v_{n_i} = \nabla v \text{ a.e.}$$

Since $|\nabla v_n| \leq 1$ a.e., we get $|\nabla v| \leq 1$ a.e. Therefore $v \in K$. Hence K is closed. □

The following proposition gives a very useful property of K.

Proposition 3.1. *K is compact in $C^0(\overline{\Omega})$ and*

$$|v(x)| \leq d(x, \Gamma), \qquad \forall\, x \in \Omega \quad \text{and}\ \forall\, v \in K, \tag{3.3}$$

where $d(x, \Gamma)$ is the distance from x to Γ.

EXERCISE 3.1. Prove Proposition 3.1.

Remark 3.1. Let us define u_∞ and $u_{-\infty}$ by

$$u_\infty(x) = d(x, \Gamma),$$
$$u_{-\infty}(x) = -d(x, \Gamma).$$

Then u_∞ and $u_{-\infty}$ belong to K. We observe that u_∞ is the maximal element of K and $u_{-\infty}$ is the minimal element of K.

Remark 3.2. Since $a(\cdot, \cdot)$ is symmetric, the solution u of (3.1) is characterized (see Sec. 3.2 of Chapter I) as the unique solution of the minimization problem

$$J(u) \le J(v), \qquad \forall\, v \in K, \quad u \in K, \tag{3.4}$$

with $J(v) = \frac{1}{2}a(v, v) - L(v)$.

3.2. Physical motivation

Let us consider an infinitely long cylindrical bar of cross section Ω, where Ω is simply connected. Assume that this bar is made up of an isotropic elastic perfectly plastic material whose plasticity yields is given by the Von Mises Criterion. (For a general discussion of plasticity problems, see Koiter [1] and Duvaut and Lions [1, Chapter 5]). Starting from a zero-stress initial state, an increasing torsion moment is applied to the bar. The torsion is characterised by C, which is defined as the torsion angle per unit length. Then for all C, it follows from the Haar–Karman Principle that the determination of the stress field is equivalent (in a convenient system of physical units) to the solution of the following variational problem:

$$\operatorname*{Min}_{v \in K}\left\{\frac{1}{2}\int_\Omega |\nabla v|^2\, dx - C\int_\Omega v\, dx\right\}. \tag{3.5}$$

This is a particular case of (3.1) or (3.4) with

$$L(v) = C\int_\Omega v\, dx. \tag{3.6}$$

The stress vector σ in a cross section is related to u by $\sigma = \nabla u$, so that u is a stress potential, and we can obtain σ once the solution of (3.5) is known.

Proposition 3.2. *Let us denote by* u_C *the solution of* (3.5) *and let, as before,* $u_\infty = d(x, \Gamma)$; *then* $\lim_{C \to +\infty} u_C = u_\infty$ *strongly in* $H_0^1(\Omega) \cap C^0(\overline{\Omega})$.

PROOF. Since u_C is the solution of (3.5), it is characterized by

$$\int_\Omega \nabla u_C \cdot \nabla(v - u_C)\, dx \ge C\int_\Omega (v - u_C)\, dx, \qquad \forall\, v \in K, \quad u_C \in K. \tag{3.7}$$

Since $u_\infty \in K$, from (3.7) we have

$$\int_\Omega \nabla u_C \cdot \nabla(u_\infty - u_C)\,dx \geq C \int_\Omega (u_\infty - u_C)\,dx, \tag{3.8}$$

i.e.

$$C \int_\Omega (u_\infty - u_C)\,dx + \int_\Omega |\nabla u_C|^2\,dx \leq \int_\Omega \nabla u_\infty \cdot \nabla u_C\,dx$$

$$\leq \int_\Omega |\nabla u_\infty| \cdot |\nabla u_C|\,dx \leq \text{meas}(\Omega). \tag{3.9}$$

From (3.3) we have $u_\infty - u_C \geq 0$, so that (3.9) implies

$$\|u_\infty - u_C\|_{L^1(\Omega)} \leq C^{-1}\,\text{meas}(\Omega)$$

which in turn implies

$$\lim_{C \to +\infty} \|u_\infty - u_C\|_{L^1(\Omega)} = 0. \tag{3.10}$$

From the definition of K and from the Proposition 3.1, we find that K is bounded and weakly closed in V and hence weakly compact in V. Furthermore, K is compact in $C^0(\overline{\Omega})$.

Relation (3.10) implies

$$\begin{aligned} &\lim_{C \to +\infty} u_C = u_\infty \text{ strongly in } C^0(\overline{\Omega}), \\ &\lim_{C \to +\infty} u_C = u_\infty \text{ weakly in } V. \end{aligned} \tag{3.11}$$

It follows from (3.8) that

$$\int_\Omega \nabla u_\infty \cdot \nabla(u_\infty - u_C) \geq \int_\Omega |\nabla(u_\infty - u_C)|^2\,dx + C \int_\Omega (u_\infty - u_C)\,dx$$

$$= \|u_C - u_\infty\|_V^2 + C\|u_\infty - u_C\|_{L^1(\Omega)}. \tag{3.12}$$

It follows easily from (3.11) and (3.12) that

$$\lim_{C \to +\infty} C\|u_\infty - u_C\|_{L^1(\Omega)} = 0,$$

$$\lim_{C \to +\infty} \|u_\infty - u_C\|_V = 0. \qquad \square$$

Remark 3.3. In the case of a multiply connected cross section, the variational formulation of the torsion problem has to be redefined (see Lanchon [1], Glowinski and Lanchon [1], and Glowinski [1, Chapter 4]).

3.3. Regularity properties and exact solutions

3.3.1. *Regularity results*

Theorem 3.2 (Brezis and Stampacchia [2]). *Let u be a solution of* (3.1) *or* (3.4) *and $L(v) = \int_\Omega fv\,dx$.*

1. *If Ω is a bounded convex domain of $\mathbb{R}^2$ with Γ Lipschitz continuous and if $f \in L^p(\Omega)$ with $1 < p < +\infty$, then we have*

$$u \in W^{2,p}(\Omega). \tag{3.13}$$

2. *If Ω is a bounded domain of $\mathbb{R}^2$ with a smooth boundary Γ and if $f \in L^p(\Omega)$ with $1 < p < +\infty$, then $u \in W^{2,p}(\Omega)$.*

Remark 3.4. It will be seen in the next section that, in general, there is a limit for the regularity of the solution of (3.1) even if Γ and f are very smooth.

Remark 3.5. It has been proved by H. Brezis that under quite restrictive smoothness assumptions on Γ and f, we may have

$$u \in W^{2,\infty}(\Omega).$$

3.3.2. *Exact solutions*

In this section we are going to give some examples of problems (3.1) for which exact solutions are known.

EXAMPLE 1. We take $\Omega = \{x \mid 0 < x < 1\}$ and $L(v) = c\int_0^1 v\,dx$ with $c > 0$. Then the explicit form of (3.1) is

$$\int_0^1 u'(v' - u')\,dx \geq c\int_0^1 (v - u)\,dx, \qquad \forall\, v \in K, \quad u \in K, \tag{3.14}$$

where $K = \{v \in H_0^1(\Omega), |v'| \leq 1 \text{ a.e. on } \Omega\}$ and $v' = dv/dx$.

The exact solution of (3.14) is given by

$$u(x) = \frac{c}{2}x(1 - x), \qquad \forall\, x, \quad \text{if } c \leq 2; \tag{3.15}$$

if $c > 2$, we have

$$u(x) = \begin{cases} x & \text{if } 0 \leq x \leq \frac{1}{2} - \frac{1}{c}, \\ \frac{c}{2}\left[x(1 - x) - \left(\frac{1}{2} - \frac{1}{c}\right)^2\right] & \text{if } \frac{1}{2} - \frac{1}{c} \leq x \leq \frac{1}{2} + \frac{1}{c}, \\ 1 - x & \text{if } \frac{1}{2} + \frac{1}{c} \leq x \leq 1. \end{cases} \tag{3.16}$$

EXAMPLE 2. In this example we consider a two-dimensional problem. We take

$$\Omega = \{x \mid x_1^2 + x_2^2 < R^2\},$$

$$L(v) = c\int_\Omega v\,dx \text{ with } c > 0.$$

Then setting $r = (x_1^2 + x_2^2)^{1/2}$, the solution u of (3.1) is given by

$$u(x) = \frac{c}{4}(R^2 - r^2) \quad \text{if } c \leq \frac{2}{R}; \tag{3.17}$$

if $c > 2/R$, then

$$\begin{aligned} u(x) &= R - r \quad \text{if } \frac{2}{c} \leq r \leq R, \\ u(x) &= \frac{c}{4}\left[(R^2 - r^2) - \left(R - \frac{2}{c}\right)^2\right] \quad \text{if } 0 \leq r \leq \frac{2}{c}. \end{aligned} \tag{3.18}$$

These examples illustrate Remark 3.4. We see that for c sufficiently large, we have

$$u \in W^{2,\infty}(\Omega) \cap H_0^1(\Omega), \qquad u \notin H^3(\Omega). \tag{3.19}$$

In fact we have

$$u \in H^s(\Omega), \qquad \forall\, s < \tfrac{5}{2}.$$

EXERCISE 3.2. Verify that the u given in Examples 1 and 2 are exact solutions of the corresponding problems.

3.4. An equivalent variational formulation

In H. Brezis and M. Sibony [1] it is proved that if Ω is a bounded domain of $\mathbb{R}^2$ with a smooth boundary Γ and if

$$L(v) = c\int_\Omega v(x)\,dx \qquad (c > 0 \text{ for instance}),$$

then the solution of (3.1) is also a solution of

$$\begin{aligned} &a(u, v - u) \leq c\int_\Omega (v - u)\,dx, \qquad \forall\, v \in \hat{K}, \\ &u \in \hat{K} = \{v \in H_0^1(\Omega), |v(x)| \leq d(x, \Gamma) \text{ a.e.}\}. \end{aligned} \tag{3.20}$$

Problem (3.20) is very similar to the obstacle problem considered in Sec. 2 of this chapter. Since $a(\cdot, \cdot)$ is symmetrical, (3.20) is also equivalent to

$$J(u) \leq J(v), \qquad \forall\, v \in \hat{K}, \quad u \in \hat{K}, \tag{3.21}$$

with

$$J(v) = \tfrac{1}{2}a(v, v) - c\int_\Omega v(x)\,dx.$$

The numerical solution of (3.20) and (3.21) is considered in G.L.T. [1, Chapter 3] (see also Cea [2, Chapter 4] and Chapter V, Sec. 5.5 of this book).

EXERCISE 3.3. Study the numerical analysis of (3.21).

EXERCISE 3.4. Assume $c > 0$ in (3.20). Then prove that the solution u of (3.20) is also the solution of the EVI obtained by replacing $\hat{K}$ by $\{v \in H_0^1(\Omega), v(x) \leq d(x, \Gamma) \text{ a.e.}\}$ in (3.20).

3.5. Finite element approximation of (3.1)

In this section we consider an approximation of (3.1) by first-order finite elements. From the viewpoint of applications in mechanics (in which $f = c$), it seems that, given the equivalence of (3.1) and (3.20), it is sufficient to approximate (3.20) (using essentially the same method as in Sec. 2). However, in view of other possible applications, it seems to us that it would be interesting to consider the numerical solution of (3.1), working directly with K instead of $\hat{K}$. For the numerical analysis of (3.20) by finite differences, see G.L.T. [3, Chapter 3] and Cea, Glowinski, and Nedelec [1].

3.5.1. *Approximation of V and K*

We use the notation of Sec. 2.5 of this chapter. We assume that Ω is a polygonal domain of $\mathbb{R}^2$ (see Remark 3.8 for the nonpolygonal case), and we consider a triangulation $\mathscr{T}_h$ of Ω satisfying (2.21)–(2.23). Then V and K, respectively, are approximated by

$$V_h = \{v_h \in C^0(\overline{\Omega}), v_h = 0 \text{ on } \Gamma, v_h|_T \in P_1, \forall\, T \in \mathscr{T}_h\},$$
$$K_h = K \cap V_h.$$

Then one can easily prove:

Proposition 3.3. *K_h is a closed convex nonempty subset of V_h.*

Remark 3.6. If $v_h \in V_h$, then ∇v_h is a constant vector on every $T \in \mathscr{T}_h$.

3.5.2. *The approximate problem*

The approximate problem is defined by:

Find $u_h \in K_h$ such that

$$a(u_h, v_h - u_h) \geq L(v_h - u_h), \qquad \forall\, v_h \in K. \tag{3.22}$$

One can easily prove:

Proposition 3.4. *The approximate problem* (3.22) *has a unique solution.*

In Sec. 7 of this chapter one may find practical formulae related to finite element approximation. Using these formulae, (3.22) and the equivalent problem (3.23) can be expressed in a form more suitable for computation.

Remark 3.7. Since $a(\cdot, \cdot)$ is symmetrical, (3.22) is equivalent to the non linear programming problem

$$\operatorname*{Min}_{v_h \in K_h} [\tfrac{1}{2}a(v_h, v_h) - L(v_h)]. \tag{3.23}$$

The natural variables in (3.23) are the values taken by v_h over the set $\mathring{\Sigma}_h$ of the interior nodes of $\mathcal{T}_h$. Then the number of variables in (3.23) is $\operatorname{Card}(\Sigma_h)$. The number of constraints is the number of triangles, i.e., Card $(\mathcal{T}_h)$, and each constraint is quadratic with respect to these variables since

$$|\nabla v_h| \leq 1 \quad \text{if and only if } |\nabla v_h|^2 \leq 1 \text{ over } T. \tag{3.24}$$

Remark 3.8. If Ω is not polygonal, it is always possible to approximate Ω by a polygonal domain Ω_h in such a way that all vertices of $\Gamma_h = \partial\Omega_h$ belong to Γ. Then, instead of defining (3.22) over Ω, we define it over Ω_h.

3.5.3. *Remarks on the use of higher-order finite elements*

In this book only an approximation of (3.1) by first-order finite elements has been considered. This fact is justified by the existence of a regularity limitation for the solution of (3.2), which implies that even with very smooth data one may have $u \notin V \cap H^3(\Omega)$ (see the examples of Sec. 3.3.2).

We refer to G.L.T. [3, Chapter 3] and Glowinski [1, Chapter 4, Sec. 3.5.3] for further discussions on the use of finite elements of order ≤ 2.

3.6. Convergence results. General case

In this section we take $L(v) = \langle f, v \rangle$, with $f \in H^{-1}(\Omega) = V^*$.

3.6.1. *A density lemma*

In order to apply the general results of Chapter I, the following density lemma will be very useful.

Lemma 3.1. *We have*

$$\overline{\mathscr{D}(\Omega) \cap K} = K. \tag{3.25}$$

PROOF. We use the notation of Lemma 2.4. Let $v \in K$ and $\varepsilon > 0$; define v_ε by

$$v_\varepsilon = (v - \varepsilon)^+ - (v + \varepsilon)^-. \tag{3.26}$$

Then we have $v_\varepsilon \in H^1(\Omega)$ with $|\nabla v_\varepsilon| \leq 1$ a.e. in Ω. From the inclusion $K \subset \hat{K} = \{v \in V, |v(x)| \leq d(x, \Gamma) \text{ a.e. in } \Omega\}$, it follows that

$$\begin{aligned} v_\varepsilon(x) &= 0 \quad \text{if } d(x, \Gamma) \leq \varepsilon, \\ |v_\varepsilon(x)| &\leq d(x, \Gamma) - \varepsilon \quad \text{elsewhere} \end{aligned} \tag{3.27}$$

so that from (3.27) it follows that

$$v_\varepsilon \in K \text{ and has a compact support in } \Omega. \tag{3.28}$$

From Corollary 2.1 we have

$$\lim_{\varepsilon \to 0} v_\varepsilon = v \text{ strongly in } V. \tag{3.29}$$

From (3.28) and (3.29) it follows that if $\mathscr{K} = \{v \in K, v \text{ has a compact support in } \Omega\}$, then $\overline{\mathscr{K}} = K$.

Thus, to prove the lemma, it suffices to prove that any $v \in \mathscr{K}$ can be approximated by a sequence $(v_n)_n$ of functions in $\mathscr{D}(\Omega) \cap K$. Let $(\rho_n)_n$ be a mollifying sequence as defined in Lemma 2.4 of this chapter. Let $v \in \mathscr{K}$. Denote by $\tilde{v}$ the extension of v to $\mathbb{R}^2$ by taking zero outside Ω. Then $\tilde{v} \in H^1(\mathbb{R}^2)$.

Let $\tilde{v}_n = \tilde{v} * \rho_n$ so that

$$\tilde{v}_n(x) = \int_{\mathbb{R}^2} \rho_n(x - y)\tilde{v}(y)\, dy, \tag{3.30}$$

$$\nabla\tilde{v}_n(x) = \int_{\mathbb{R}^2} \rho_n(x - y)\nabla\tilde{v}(y)\, dy. \tag{3.31}$$

Then

$$\tilde{v}_n \in \mathscr{D}(\mathbb{R}^2) \quad \text{and} \quad \lim_{n \to \infty} \tilde{v}_n = \tilde{v} \text{ strongly in } H^1(\mathbb{R}^2). \tag{3.32}$$

Since Supp $\tilde{v} \subset \Omega$, from (3.30) we have

$$\text{Supp } \tilde{v}_n \subset \Omega \text{ for } n \text{ sufficiently large}. \tag{3.33}$$

Define $v_n = \tilde{v}_n|_\Omega$ for n sufficiently large. Then (3.32) and (3.33) imply

$$v_n \in \mathscr{D}(\Omega), \qquad \lim_{n \to \infty} v_n = v \text{ strongly in } V. \tag{3.34}$$

From (3.31), $\rho_n \geq 0$, $\int_{\mathbb{R}^2} \rho_n\, dy = 1$ and $|\nabla\tilde{v}(y)| \leq 1$ a.e. on $\mathbb{R}^2$, we obtain

$$|\nabla v_n(x)| = |\nabla\tilde{v}_n(x)| \leq \int_{\mathbb{R}^2} |\nabla\tilde{v}(y)|\, \rho_n(x - y)\, dy \leq 1, \qquad \forall\, x \in \Omega, \tag{3.35}$$

which completes the proof of the lemma. □

3.6.2. *A convergence theorem*

Theorem 3.3. *Suppose that the angles of the triangles of $\mathscr{T}_h$ are uniformly bounded by $\theta_0 > 0$ as $h > 0$. Then*

$$\lim_{h\to 0} u_h = u \text{ strongly in } V \cap C^0(\overline{\Omega}), \tag{3.36}$$

where u and u_n are, respectively, the solutions of (3.1) *and* (3.22).

PROOF. To prove the strong convergence in V, we use Theorem 5.2 of Chapter I, Sec. 5. To do this, one has to verify the following properties.

(i) If $(v_h)_h$, $v_h \in K_h$, $\forall\, h$, converges weakly to v, then $v \in K$,
(ii)′ There exists χ and r_h with the following properties:

1. $\bar{\chi} = K$,
2. $r_h: \chi \to K_h$, $\forall\, h$;
3. for each $v \in \chi$, we can find $h_0 = h_0(v)$ such that for all $h \leq h_0(v)$, $r_h v \in K_h$ and $\lim_{h\to 0} r_h v = v$ strongly in V.

Verification of (i). Since $K_h \subset K$ and K is weakly closed, (i) is obvious.

Verification of (ii)′. Let us define χ by

$$\chi = \{v \in \mathscr{D}(\Omega), |\nabla v(x)| < 1, \qquad \forall\, x \in \Omega\}.$$

Then by Lemma 3.1 and from $\lim_{\lambda\to 1} \lambda v = v$ strongly in V, $\forall\, v \in V$, it follows that $\bar{\chi} = K$. Define $r_h: V \cap C^0(\overline{\Omega}) \to V_h$ by

$$r_h v \in V_h, \qquad \forall\, v \in V \cap C^0(\overline{\Omega}),$$
$$(r_h v)(P) = v(P), \qquad \forall\, P \in \mathring{\Sigma}_h. \tag{3.37}$$

Then $r_h v$ is the "linear" interpolate of v on $\mathscr{T}_h$. From the assumptions on $\mathscr{T}_h$ we have (cf. Strang and Fix [1] and Ciarlet [1], [2])

$$|\nabla(r_h v - v)| \leq Ch\|v\|_{W^{2,\infty}(\Omega)} \text{ a.e.,} \qquad \forall\, v \in \mathscr{D}(\Omega), \tag{3.38}$$

with C independent of h and v.

This implies

$$\lim_{h\to 0} \|r_h v - v\|_V = 0, \qquad \forall\, v \in \chi, \tag{3.39}$$

$$|\nabla r_h v(x)| \leq |\nabla v(x)| + Ch\|v\|_{W^{2,\infty}(\Omega)} \text{ a.e.}. \tag{3.40}$$

Since $v \in \chi$, it follows from (3.40) that we have $|\nabla r_h v(x)| < 1$ a.e. for $h < h_0(v)$.

This implies $r_h v \in K_h$.

This completes the verification of (ii)′ and hence, by Theorem 5.2 of Chapter I, we have the strong convergence of u_h to u in V.

The strong convergence of u_h to u in the L^∞-norm follows from the convergence in V and from the compactness of K in $C^0(\overline{\Omega})$ (see Proposition 3.1). □

3.7. Error estimates

From now on we assume that $f \in L^p$ for some $p \geq 2$.

In Sec. 3.7.1 we consider a one-dimensional problem (3.1). In this case if $f \in L^2(\Omega)$ we derive an $O(h)$ error estimate in the V-norm. In Sec. 3.7.2 we consider a two-dimensional case with $f \in L^p$, $p > 2$, and Ω convex; then we derive an $O(h^{1/2-1/p})$ error estimate in the V-norm.

3.7.1. *One-dimensional case*

We assume here $\Omega = \{x \in \mathbb{R} \mid 0 < x < 1\}$ and that $f \in L^2(\Omega)$. Then problem (3.1) can be written as

$$\int_0^1 \frac{du}{dx}\left(\frac{dv}{dx} - \frac{du}{dx}\right) dx \geq \int_0^1 f(v - u)\, dx, \qquad \forall\, v \in K,$$

$$u \in K = \left\{v \in V, \left|\frac{dv}{dx}\right| \leq 1 \text{ a.e. in } \Omega\right\}. \tag{3.41}$$

Let N be a positive integer and $h = 1/N$. Let $x_i = ih$ for $i = 0, 1, \ldots, N$ and

$$e_i = [x_{i-1}, x_i], \qquad i = 1, 2, \ldots, N.$$

Let $V_h = \{v_h \in C^0(\overline{\Omega}),\ v_h(0) = v_h(1) = 0,\ v_h|_{e_i} \in P_1,\ i = 1, 2, \ldots, N\}$,

$$K_h = K \cap V_h = \{v_h \in V_h, |v_h(x_i) - v_h(x_{i-1})| \leq h \text{ for } i = 1, 2, \ldots, N\}.$$

The approximate problem is defined by

$$\int_0^1 \frac{du_h}{dx}\left(\frac{dv_h}{dx} - \frac{du_h}{dx}\right) dx \geq \int_0^1 f(v_h - u_h)\, dx, \qquad \forall\, v_h \in K_h, \quad u_h \in K_h. \tag{3.42}$$

Obviously this problem has a unique solution. Now we are going to prove:

Theorem 3.4. *Let u and u_h be the solutions of* (3.41) *and* (3.42), *respectively. If $f \in L^2(\Omega)$, then we have*

$$\|u_h - u\|_V = O(h).$$

PROOF. Since $u_h \in K_h \subset K$, from (3.41) we have

$$a(u, u_h - u) \geq \int_0^1 f(u_h - u)\, dx. \tag{3.43}$$

Adding (3.42) and (3.43), we obtain

$$a(u_h - u, u_h - u) \leq a(v_h - u, u_h - u) + a(u, v_h - u) - \int_0^1 f(v_h - u)\, dx, \qquad \forall\, v_h \in K_h$$

which in turn implies

$$\tfrac{1}{2}\|u_h - u\|_V^2 \leq \tfrac{1}{2}\|v_h - u\|_V^2 + \int_0^1 \frac{du}{dx}\left(\frac{dv_h}{dx} - \frac{du}{dx}\right) dx - \int_0^1 f(v_h - u)\, dx, \qquad \forall\, v_h \in K_h. \tag{3.44}$$

Since $u \in K \cap H^2(0, 1)$, we obtain

$$\int_0^1 \frac{du}{dx}\frac{d}{dx}(v_h - u)\,dx = \int_0^1 \left(-\frac{d^2u}{dx^2}\right)(v_h - u)\,dx$$
$$\leq \left\| \frac{d^2u}{dx^2} \right\|_{L^2} \|v_h - u\|_{L^2}.$$

But we have

$$\left\| \frac{d^2u}{dx^2} \right\|_{L^2} \leq \|f\|_{L^2}. \tag{3.45}$$

Therefore (3.44) becomes

$$\tfrac{1}{2}\|u_h - u\|_V^2 \leq \tfrac{1}{2}\|v_h - u\|_V^2 + 2\|f\|_{L^2}\|v_h - u\|_{L^2}, \qquad \forall\, v_h \in K_h. \tag{3.46}$$

Let $v \in K$. Then the usual linear interpolate $r_h v$ is defined by

$$r_h v \in V_h, \qquad (r_h v)(x_i) = v(x_i), \qquad i = 0, 1, \ldots, N. \tag{3.47}$$

We have

$$\frac{d}{dx}(r_h v)|_{e_i} = \frac{v(x_i) - v(x_{i-1})}{h} = \frac{1}{h}\int_{x_{i-1}}^{x_i} \frac{dv}{dx}\,dx.$$

Hence we obtain

$$\left|\frac{d}{dx}(r_h v)\right|_{e_i} \leq 1 \quad \text{since} \left|\frac{dv}{dx}\right| \leq 1 \text{ a.e. in } \Omega. \tag{3.48}$$

Thus $r_h v \in K_h$.

Let us replace v_h by $r_h u$ in (3.46). Then

$$\tfrac{1}{2}\|u_h - u\|_V^2 \leq \tfrac{1}{2}\|r_h u - u\|_V^2 + 2\|f\|_{L^2(\Omega)}\|r_h u - u\|_{L^2(\Omega)}. \tag{3.49}$$

From (3.45) and standard approximation results, we have

$$\|r_h u - u\|_V \leq Ch\|u\|_{H^2(\Omega)} \leq Ch\|f\|_{L^2(\Omega)}, \tag{3.50}$$

$$\|r_h u - u\|_{L^2(\Omega)} \leq Ch^2\|u\|_{H^2(\Omega)} \leq Ch^2\|f\|_{L^2(\Omega)}, \tag{3.51}$$

where C denotes constants independent of u and h. Combining (3.49)–(3.51), we get

$$\|u_h - u\|_V = O(h).$$

This proves the result. □

EXERCISE 3.5. Prove (3.45).

3.7.2. *Two-dimensional case*

In this subsection we shall assume that Ω is a convex bounded polygonal domain in $\mathbb{R}^2$ and that $f \in L^p(\Omega)$ with $p > 2$. The latter assumption is quite reasonable since in practical applications in Mechanics we have $f = $ constant.

Theorem 3.5. *Suppose that the angles of $\mathcal{T}_h$ are uniformly bounded by $\theta_0 > 0$ as $h \to 0$; then with the above assumptions on Ω and f, we have*

$$\|u_h - u\|_V = O(h^{1/2-1/p}),$$

where u and u_h are the solutions of (3.1) *and* (3.22), *respectively.*

PROOF. Since $f \in L^p(\Omega)$ with $p > 2$ and Ω is bounded, from Theorem 3.2 of this chapter we have

$$u \in W^{2,p}(\Omega).$$

Then, as in proof of Theorem 3.4 and using $K_h \subset K$, we obtain

$$\tfrac{1}{2}\|u_h - u\|_V^2 \le \tfrac{1}{2}\|v_h - u\|_V^2 + a(u, v_h - u) - \int_\Omega f(v_h - u)\,dx$$

$$\le \tfrac{1}{2}\|v_h - u\|_V^2 - \int_\Omega (-\Delta u - f)(v_h - u)\,dx, \qquad \forall\, v_h \in K_h. \tag{3.52}$$

Then, using Holder's inequality, it follows from (3.52) that

$$\tfrac{1}{2}\|u_h - u\|_V^2 \le \tfrac{1}{2}\|v_h - u\|_V^2 + \{\|\Delta u\|_{L^p(\Omega)} + \|f\|_{L^p(\Omega)}\}\|v_h - u\|_{L^{p'}(\Omega)}, \qquad \forall\, v_h \in K_h$$

$$\text{with } \frac{1}{p} + \frac{1}{p'} = 1. \tag{3.53}$$

Let $1 \le q \le \infty$. Assume that $\mathcal{T}_h$ satisfies the hypothesis of Theorem 3.5 and that $p > 2$. If $W^{2,p}(T) \subset W^{1,q}(T)$, it follows from Ciarlet [2] and the *Sobolev imbedding theorem* ($W^{2,p}(T) \subset W^{1,\infty}(T) \subset C^0(T)$) that $\forall\, T \in \mathcal{T}_h$ and $\forall\, v \in W^{2,p}(T)$, we have

$$\|\nabla(v - \pi_T v)\|_{L^q(T)\times L^q(T)} \le C h_T^{1+2(1/q-1/p)}\|v\|_{W^{2,p}(T)}. \tag{3.54}$$

In (3.54) $\pi_T v$ is the linear interpolate of v at the three vertices of T, h_T is the diameter of T, and C is a constant independent of T and v.

Let $v \in W^{2,p}(\Omega)$ and let $\pi_h\colon V \cap C^0(\bar{\Omega}) \to V_h$ be defined by

$$\pi_h v \in V_h, \qquad \forall\, v \in H_0^1(\Omega) \cap C^0(\bar{\Omega}),$$

$$(\pi_h v)(P) = v(P), \qquad \forall\, P \in \mathring{\Sigma}_h.$$

Since $p > 2$ implies $W^{2,p}(\Omega) \subset C^0(\bar{\Omega})$, one may define $\pi_h v$, but unlike the one-dimensional case, usually

$$\pi_h v \notin K_h \quad \text{for } v \in W^{2,p}(\Omega) \cap K.$$

Since $W^{2,p}(\Omega) \subset W^{1,\infty}(\Omega)$ for $p > 2$, it follows from (3.54) that a.e.

$$|\nabla(\pi_h v - v)(x)| \le r h^{1-2/p}\|v\|_{W^{2,p}(\Omega)}, \qquad \forall\, v \in W^{2,p}(\Omega)$$

which in turn implies that a.e.

$$|\nabla(\pi_h v)(x)| \le 1 + r h^{1-2/p}\|v\|_{W^{2,p}(\Omega)}, \qquad \forall\, v \in K \cap W^{2,p}(\Omega). \tag{3.55}$$

The constant r occurring in (3.55) is independent of v and h. Let us define

$$r_h\colon V \cap W^{2,p}(\Omega) \to V_h$$

by

$$r_h v = \frac{\pi_h v}{1 + rh^{1-2/p}\|v\|_{W^{2,p}(\Omega)}}. \tag{3.56}$$

It follows from (3.55) and (3.56) that

$$r_h v \in K_h, \qquad \forall\, v \in W^{2,p}(\Omega) \cap K. \tag{3.57}$$

Since $u \in W^{2,p}(\Omega) \cap K$, it follows from (3.57) that we can take $v_h = r_h u$ in (3.53) so that

$$\tfrac{1}{2}\|u_h - u\|_V^2 \leq \tfrac{1}{2}\|r_h u - u\|_V^2 + \{\|\Delta u\|_{L^p} + \|f\|_{L^p}\}\|r_h u - u\|_{L^{p'}(\Omega)}. \tag{3.58}$$

We have

$$r_h u - u = \frac{\pi_h u - u - rh^{1-2/p}\|u\|_{W^{2,p}(\Omega)} u}{1 + rh^{1-2/p}\|u\|_{W^{2,p}(\Omega)}}$$

which implies

$$\|r_h u - u\|_V \leq \|\pi_h u - u\|_V + rh^{1-2/p}\|u\|_{W^{2,p}}\|u\|_V, \tag{3.59}$$

$$\|r_h u - u\|_{L^{p'}(\Omega)} \leq \|\pi_h u - u\|_{L^{p'}(\Omega)} + rh^{1-2/p}\|u\|_{W^{2,p}(\Omega)}\|u\|_{L^{p'}}. \tag{3.60}$$

Since $p > 2$ we have $L^p(\Omega) \subset L^{p'}(\Omega)$, and from standard approximation results (see Strang and Fix [1] and Ciarlet [1], [2]) it follows that under the above assumption on $\mathcal{T}_h$ we have

$$\|\pi_h u - u\|_V \leq Ch\|u\|_{W^{2,p}(\Omega)}, \tag{3.61}$$

$$\|\pi_h u - u\|_{L^{p'}(\Omega)} \leq Ch^2\|u\|_{W^{2,p}(\Omega)}, \tag{3.62}$$

with C independent of h and u. Then the $O(h^{1/2-1/p})$ error estimate of the statement of Theorem 3.5 follows directly from (3.58)–(3.62). □

Remark 3.9. It follows from Theorem 3.5 that if $f =$ constant (which corresponds to application in mechanics) and if Ω is a *convex polygonal domain*, we have "practically" an $O(\sqrt{h})$ error estimate.

Remark 3.10. In Falk [1] one may find an analysis of the error estimate for piecewise linear approximations of (3.1) when Ω is not polygonal.

Remark 3.11. In Falk and Mercier [1] (see also G.L.T. [3, Appendix 3]) we may find a different piecewise linear approximation of (3.1). Under appropriate assumptions this approximation leads to an $O(h)$ error estimate for $\|u_h - u\|_V$. In G.L.T., *loc. cit.*, a *conjugate gradient* algorithm for solving these new types of approximate problems is also described.

3.8. A dual iterative method for solving (3.1) and (3.22)

There are several iterative methods for solving (3.1), and (3.22), and the reader who is interested in this direction of the problem may consult G.L.T. [3, Chapter 3] (see also Cea, Glowinski, and Nedelec [1]). In this section we shall use the material of Cea [2, Chapter 5, Sec. 5] to describe an algorithm of Uzawa type which has been successfully used to solve the elasto-plastic torsion problem. Another method will be described in Chapter VI, Sec. 6.2.

3.8.1. *The continuous case*

Following Cea [2] and G.L.T. [1, Chapter 3], [3, Chapter 3], we observe that K can also be written as

$$K = \{v \in V, |\nabla v|^2 - 1 \leq 0 \text{ a.e.}\}.$$

Hence it is quite natural to associate with (3.1) the following Lagrangian functional $\mathscr{L}$ defined on $H_0^1(\Omega) \times L^\infty(\Omega)$ by

$$\mathscr{L}(v, \mu) = \frac{1}{2}\int_\Omega |\nabla v|^2 \, dx - \langle f, v \rangle + \frac{1}{2}\int_\Omega \mu(|\nabla v|^2 - 1) \, dx.$$

It follows from Cea [2] and G.L.T. [1], [3] that if $\mathscr{L}$ has a saddle point $\{u, \lambda\} \in H_0^1(\Omega) \times L_+^\infty(\Omega) \, (L_+^\infty(\Omega) = \{q \in L^\infty(\Omega), q \geq 0 \text{ a.e.}\})$, then u is a solution of (3.1). Thus λ appears as an infinite-dimensional multiplier (of F. John–Kuhn–Tucker type) for (3.1). The existence of such a multiplier in L_+^∞ has been proved by H. Brezis [2] in the physical case (i.e., f = constant), but in more general cases the existence of such a multiplier in $L_+^\infty(\Omega)$ is still an open problem.

Following Cea and G.L.T., *loc. cit.*, it is then natural to use a *saddle point solver* like the following algorithm of Uzawa type for solving (3.1):

$$\lambda^0 \in L_+^\infty(\Omega) \textit{ arbitrarily given (for example, } \lambda^0 = 0); \tag{3.63}$$

then, by induction, assuming λ^n *known, we obtain* u^n *and* λ^{n+1} *by*

$$\mathscr{L}(u^n, \lambda^n) \leq \mathscr{L}(v, \lambda^n), \qquad \forall\, v \in H_0^1(\Omega), \qquad u^n \in H_0^1(\Omega), \tag{3.64}$$

$$\lambda^{n+1} = [\lambda^n + \rho(|\nabla u^n|^2 - 1)]^+ \quad \textit{with } \rho > 0. \tag{3.65}$$

Let us analyze (3.64) in detail; actually (3.64) is a linear Dirichlet problem, whose explicit form is given (in the divergence form) by

$$\begin{aligned} -\nabla \cdot ((1 + \lambda^n)\nabla u^n) &= f \text{ in } \Omega, \\ u^n|_\Gamma &= 0. \end{aligned} \tag{3.66}$$

Problem (3.66) has a unique solution in $H_0^1(\Omega)$ whenever $\lambda^n \in L_+^\infty(\Omega)$. Since we are not generally certain of the existence of a multiplier in $L_+^\infty(\Omega)$, the above algorithm is purely formal in general.

3.8.2. *The discrete case*

In this section we shall follow G.L.T. [3, Chapter 3, Sec. 9.2]. Define V_h and K_h as in Section 3.5.1 of this chapter. Define L_h (approximation of $L^\infty(\Omega)$) and

Λ_h (approximation of L_+^∞) by

$$L_h = \left\{\mu \in L^\infty(\Omega), \mu = \sum_{T \in \mathscr{T}_h} \mu_T \chi_T, \mu_T \in \mathbb{R}\right\},$$

where χ_T is the characteristic function of T, and

$$\Lambda_h = \{\mu \in L_h, \mu \geq 0 \text{ a.e. in } \Omega\}.$$

It clearly follows that for $v_h \in V_h, \nabla v_h \in L_h \times L_h$, and for $v_h \in K_h, 1 - |\nabla v_h|^2 \in \Lambda_h$.

Define the Lagrangian $\mathscr{L}$ on $V_h \times L_n$ as in Sec. 3.8.1; then we have:

Proposition 3.5. *The Lagrangian $\mathscr{L}$ has a saddle point $\{u_h, \lambda_h\}$ in $V_h \times \Lambda_h$ where*

$$u_h \text{ is the solution of } (3.22), \tag{3.67}$$

$$\lambda_h(|\nabla u_h|^2 - 1) = 0. \tag{3.68}$$

PROOF. Since V_h and L_h are finite dimensional, (3.67) and (3.68) will follow from Cea [2, Chapter 5] (cf. also Rockafellar [1, Chapter 28]) if we can prove that there exists an element of V_h in the neighborhood of which the constraints are strictly satisfied. Let us show that there exists a neighborhood $\mathscr{N}_h$ of zero in V_h such that $\forall\, v_h \in \mathscr{N}_h, |\nabla v_h|^2 - 1 < 0$. In order to show this, observe that the functional given by $v_h \to |\nabla v_h|^2 - 1$ is C^∞, and at *zero* it is equal to -1. Hence the assertion follows. □

To conclude Section 3, let us describe an algorithm of Uzawa type which is the discrete version of (3.63)–(3.65):

$$\lambda_h^0 \in \Lambda_h \text{ arbitrarily chosen (for instance, } \lambda_h^0 = 0), \tag{3.69}$$

then, by induction, once λ_h^n is known, we obtain u_h^n and λ_h^{n+1} by

$$\mathscr{L}(u_h^n, \lambda_h^n) \leq \mathscr{L}(v_h, \lambda_h^n), \qquad \forall\, v_h \in V_h, \quad u_h^n \in V_h, \tag{3.70}$$

$$\lambda_h^{n+1} = [\lambda_h^n + \rho(|\nabla u_h^n|^2 - 1)]^+ \quad \text{with } \rho > 0. \tag{3.71}$$

We observe that if λ_h^n is known, then u_h^n is the unique solution of the following approximate Dirichlet problem (given in variational form)

$$\int_\Omega (1 + \lambda_h^n)\nabla u_h^n \cdot \nabla v_h\, dx = \langle f, v_h \rangle, \qquad \forall\, v_h \in V_h, \quad u_h^n \in V_h. \tag{3.72}$$

From Cea [2, Chapter 5] and G.L.T. [3, Chapter 2], it follows that for $\rho > 0$ and sufficiently small, we have $\lim_{n \to \infty} u_h^n = u_h$, where u_h is the solution of (3.22).

Remark 3.12. The computations we have performed seem to prove that the optimal choice for ρ is almost independent of h for a given problem. Similarly, the number of iterations of Uzawa's algorithm for a given problem is almost independent of h.

4. A Third Example of EVI of the First Kind: A Simplified Signorini Problem

Most of the material in this section can be found in G.L.T. [1, Chapter 4], [3, Chapter 4].

4.1. The continuous problem: Existence and uniqueness results

As usual, let Ω be a bounded domain of $\mathbb{R}^2$ with a smooth boundary Γ. We define

$$V = H^1(\Omega), \tag{4.1}$$

$$a(u, v) = \int_\Omega \nabla u \cdot \nabla v \, dx + \int_\Omega uv \, dx, \tag{4.2}$$

$$L(v) = \langle f, v \rangle, \qquad f \in V^* \tag{4.3}$$

$$K = \{v \in H^1(\Omega), \gamma v \geq 0 \text{ a.e. on } \Gamma\}, \tag{4.4}$$

where γv denotes the *trace* of v on Γ. We then have the following:

Theorem 4.1. *The variational inequality*

$$a(u, v - u) \geq L(v - u), \qquad \forall\, v \in K, \quad u \in K \tag{4.5}$$

has a unique solution.

PROOF. Since the bilinear form $a(\cdot, \cdot)$ is the usual scalar product in $H^1(\Omega)$ and L is continuous, it follows from Theorem 3.1 of Chapter 1 that (4.5) has a unique solution provided we show that K is a closed convex nonempty subset of V.

Since $0 \in K$ (actually, $H_0^1(\Omega) \subset K$), K is nonempty. The convexity of K is obvious. If $(v_n)_n \subset K$ and $v_n \to v$ in $H^1(\Omega)$, then $\gamma v_n \to \gamma v$, since $\gamma: H^1(\Omega) \to L^2(\Gamma)$ is continuous. Since $v_n \in K, \gamma v_n \geq 0$ a.e. on Γ. Therefore $\gamma v \geq 0$ a.e. on Γ. Hence $v \in K$ which shows that K is closed. □

Remark 4.1. Since $a(\cdot, \cdot)$ is symmetric, the solution u of (4.5) is characterized (see Chapter 1, Section 3.2) as the unique solution of the minimization problem

$$J(u) \leq J(v), \qquad \forall\, v \in K, \quad u \in K, \tag{4.6}$$

where $J(v) = \frac{1}{2}a(v, v) - L(v)$. (4.7)

Remark 4.2. Actually (4.5) or (4.6) is a simplified version of a problem occurring in elasticity, called the *Signorini problem* for which we refer to Duvaut and Lions [1, Chapter 3] and to the references therein. We also refer to Duvaut and Lions, *loc. cit.*, Chapters 1 and 2 for other physical and mechanical interpretations of (4.5) and (4.6).

Remark 4.3. Assuming that Ω is bounded (at least in one direction of $\mathbb{R}^2$), we consider

$$\hat{V} = \{v \in H^1(\Omega), v = 0 \text{ a.e. on } \Gamma_0\}, \tag{4.8}$$

$$\hat{a}(u, v) = \int_\Omega \nabla u \cdot \nabla v \, dx, \tag{4.9}$$

$$\hat{L}(v) = \langle f, v \rangle \quad \text{with } f \in \hat{V}^*, \tag{4.10}$$

$$\hat{K} = \{v \in V, \gamma v \geq g \text{ a.e. on } \Gamma_1\}, \tag{4.11}$$

where Γ_0 and Γ_1 are "good" subsets of Γ such that $\Gamma_1 \cap \Gamma_0 = \emptyset, \Gamma = \Gamma_1 \cup \Gamma_0$ (see Fig. 4.1).

Assuming that the measure of Γ_0 is positive and that g is sufficiently smooth, it can be proved that the following variant of (4.5),

$$\hat{a}(u, v - u) \geq \hat{L}(v - u), \qquad \forall\, v \in \hat{K}, \quad u \in \hat{K}, \tag{4.12}$$

has a unique solution.

In the proof of this result, one uses the fact that $\hat{a}(v, v)$ defines a norm on $\hat{V}$ which is equivalent to the norm induced by $H^1(\Omega)$.

EXERCISE 4.1. Prove that $\hat{a}(v, v)$ defines a norm equivalent to the norm induced by $H^1(\Omega)$.

4.2. Regularity of the solution

Theorem 4.2. (H. Brezis [3]). *Let Ω be a bounded domain of $\mathbb{R}^2$ with a smooth boundary Γ (or Ω is a convex polygonal domain). If $L(v) = \int_\Omega fv \, dx$ with $f \in L^2(\Omega)$, then the solution u of (4.5) is in $H^2(\Omega)$.*

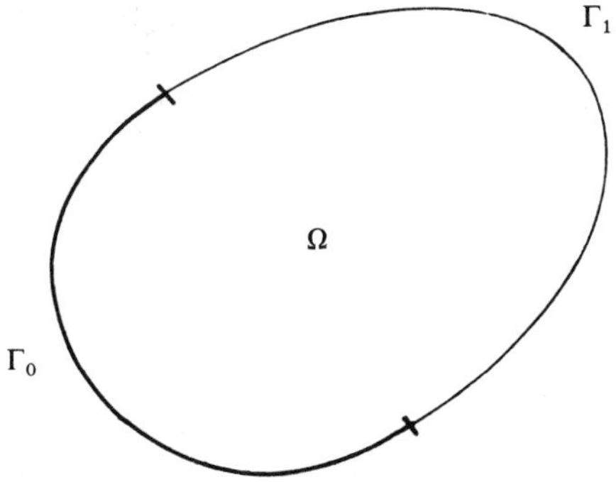

Figure 4.1

4.3. Interpretation of (4.5) as a free boundary problem

Let us recall some definitions and results related to cones.

Definition 4.1. Let X be a vector space, $C \subset X$ and $x \in C$; then C is called a cone with vertex at x if for all $y \in C$, $t \geq 0$ implies $x + t(y - x) \in C$.

Lemma 4.1. *Let H be a real Hilbert space, let $b(\cdot, \cdot)$ be a bilinear form on $H \times H$, let λ be a linear form on H, and let C be a convex cone contained in H with vertex at 0. Then every solution of*

$$b(u, v - u) \geq \lambda(v - u), \qquad \forall\, v \in C, \quad u \in C \tag{4.13}$$

is a solution of

$$\begin{aligned} b(u, v) &\geq \lambda(v), \qquad \forall\, v \in C, \\ b(u, u) &= \lambda(u), \\ u &\in C, \end{aligned} \tag{4.14}$$

and conversely.

EXERCISE 4.2. Prove Lemma 4.1.

Proposition 4.1. *Assume that*

$$L(v) = \int_\Omega fv\, dx + \int_\Gamma g\gamma v\, d\Gamma, \tag{4.15}$$

with f and g sufficiently smooth. Then the solution u of (4.5) is characterised by

$$\begin{aligned} -\Delta u + u &= f \text{ a.e. in } \Omega, \\ \gamma u \geq 0, \quad \frac{\partial u}{\partial n} &\geq g \text{ a.e. on } \Gamma, \\ \gamma u\left(\frac{\partial u}{\partial n} - g\right) &= 0 \text{ a.e. on } \Gamma. \end{aligned} \tag{4.16}$$

PROOF. First we will prove that (4.5) implies (4.16).

Since K is a convex cone with vertex at 0, it follows from Lemma 4.1 that

$$a(u, v) \geq L(v), \qquad \forall\, v \in K, \tag{4.17}$$

$$a(u, u) = L(u). \tag{4.18}$$

Since $\mathscr{D}(\Omega) \subset K$, from (4.17) we have

$$\int_\Omega \nabla u \cdot \nabla \phi\, dx + \int_\Omega u\phi\, dx = \int_\Omega f\phi\, dx, \qquad \forall\, \phi \in \mathscr{D}(\Omega). \tag{4.19}$$

From (4.19) it follows that

$$-\Delta u + u = f \text{ a.e. in } \Omega. \tag{4.20}$$

Let $v \in K$. Multiplying (4.20) by v and using *Green's formula*, it follows that

$$a(u, v) = \int_\Omega fv\,dx + \int_\Gamma \gamma v \frac{\partial u}{\partial n}\,d\Gamma, \qquad \forall\, v \in K. \tag{4.21}$$

From (4.17) and (4.21) we obtain

$$\int_\Gamma \left(\frac{\partial u}{\partial n} - g\right)\gamma v\,d\Gamma \geq 0, \qquad \forall\, v \in K. \tag{4.22}$$

Since the cone γK is dense in $L^2_+(\Gamma) = \{v \in L^2(\Gamma), v \geq 0 \text{ a.e. on } \Gamma\}$, from (4.22) it follows that

$$\frac{\partial u}{\partial n} - g \geq 0 \text{ a.e. on } \Gamma. \tag{4.23}$$

Taking $v = u$ in (4.21) and using (4.18), we obtain

$$\int_\Gamma \gamma u\left(\frac{\partial u}{\partial n} - g\right) d\Gamma = 0. \tag{4.24}$$

Since $\gamma u \geq 0$, and using (4.23), we obtain $\gamma u(\partial u/\partial n - g) = 0$ a.e. on Γ. This shows that (4.5) implies (4.16).

(2) Let us show that (4.16) implies (4.5). Starting from (4.20) and using Green's formula, one can easily prove (4.17) and (4.18). These two relations, in turn, imply, from Lemma 4.1, that u is the solution of (4.5). □

Remark 4.4. Similar results may be proved for the variant (4.12) of (4.5) (see Remark 4.3).

Remark 4.5. From the equivalent formulation (4.16) of (4.5), it appears that the solution u of (4.5) is the solution of a *free boundary problem*, namely:

Find a sufficiently smooth function u and two subsets Γ_0 and Γ_+ such that

$$\Gamma_0 \cup \Gamma_+ = \Gamma, \qquad \Gamma_0 \cap \Gamma_+ = \varnothing, \tag{4.25}$$

$$-\Delta u + u = f \textit{ in } \Omega$$

$$\gamma u = 0 \textit{ on } \Gamma_0, \frac{\partial u}{\partial n} \geq g \textit{ on } \Gamma_0, \tag{4.26}$$

$$\gamma u > 0 \textit{ on } \Gamma_+, \frac{\partial u}{\partial n} = g \textit{ on } \Gamma_+.$$

4.4. Finite-element approximation of (4.5)

In this section we consider the approximation of (4.5) by piecewise linear and piecewise quadratic finite elements. We assume that Ω is a bounded polygonal domain of $\mathbb{R}^2$, and we consider a triangulation $\mathcal{T}_h$ of Ω obeying (2.21)–(2.23) (see Sec. 2.5, Chapter II); we use the notation of Secs. 2.5.1 and 3.6 of this chapter.

4.4.1. *Approximation of V and K*

The space $V = H^1(\Omega)$ may be approximated by the spaces V_h^k, where

$$V_h^k = \{v_h \in C^0(\overline{\Omega}),\ v_h|_T \in P_k,\ \forall\ T \in \mathscr{T}_h\}, \qquad k = 1, 2.$$

Define $\gamma_h = \{P \in \Sigma_h \cap \Gamma\} = \Sigma_h - \mathring{\Sigma}_h$,

$$\gamma'_h = \{P \in \Sigma'_h \cap \Gamma\} = \Sigma'_h - \mathring{\Sigma}'_h,$$

$$\gamma_h^k = \begin{cases} \gamma_h & \text{if } k = 1 \\ \gamma_h \cup \gamma'_h & \text{if } k = 2. \end{cases}$$

Then we approximate K by

$$K_h^k = \{v_h \in V_h^k,\ v_h(P) \geq 0,\ \forall\ P \in \gamma_h^k\}.$$

We then have the obvious:

Proposition 4.2. *For $k = 1, 2$, the K_h^k are closed convex nonempty subsets of V_h^k and $K_h^1 \subset K, \forall\ h$.*

4.4.2. *The approximate problems*

For $k = 1, 2$, the approximate problems are defined by

$$a(u_h^k, v_h - u_h^k) \geq L(v_h - u_h^k), \qquad \forall\ v_h \in K_h^k, \quad u_h^k \in K_h^k. \tag{P_{1h}^k}$$

Then one can easily prove:

Proposition 4.3. *The problem* (P_{1h}^k) $(k = 1, 2)$ *has a unique solution.*

Remark 4.6. Since $a(\cdot, \cdot)$ is symmetric, (P_{1h}^k) is equivalent (see Chapter I, Sec. 3.2) to the *quadratic programming* problem

$$\min_{v_h \in K_h^k} [\tfrac{1}{2}a(v_h, v_h) - L(v_h)].$$

Remark 4.7. Using the formulae of Sec. 7, one may express (4.5) and the equivalent quadratic problem in a form more suitable for computation.

4.5. Convergence results

4.5.1. *A density lemma*

To prove the convergence results of Sec. 4.5.2, we shall use the following:

Lemma 4.2. *Under the above assumptions on Ω, we have*

$$\overline{K \cap C^\infty(\overline{\Omega})} = K.$$

PROOF. Since Γ is Lipschitz continuous, we have (see Necas [1])

$$H^1(\Omega) = \overline{C^\infty(\bar{\Omega})};$$

using the standard decomposition $v = v^+ - v^-$, from Corollary 2.1 it follows that

$$v \in K \quad \text{if and only if } v^- \in H_0^1(\Omega). \tag{4.27}$$

Since $\overline{\mathscr{D}(\Omega)} = H_0^1(\Omega)$ in the $H^1(\Omega)$-topology, from (4.27) it follows that we have only to prove

$$\overline{\hat{K} \cap C^\infty(\bar{\Omega})} = \hat{K}, \tag{4.28}$$

where $\hat{K} = \{v \in H^1(\Omega), v \geq 0 \text{ a.e. in } \Omega\}$.

Since Γ is Liptchitz continuous, Ω has (see Lions [2] and Necas [1]) the so-called 1-extension property which implies

$$\forall\, v \in H^1(\Omega), \exists\, \tilde{v} \in H^1(\mathbb{R}^2) \quad \text{such that } \tilde{v}|_\Omega = v. \tag{4.29}$$

Let $v \in K$ and let $\tilde{v} \in H^1(\mathbb{R}^2)$ be an extension of v obeying (4.29). From $v \geq 0$ a.e. in Ω and Corollary 2.1, it follows that $|\tilde{v}|$ is also an extension of v obeying (4.29). Therefore if $v \in \hat{K}$, it always has an extension $\tilde{v} \geq 0$ a.e. obeying (4.29). Consider such a non-negative extension $\tilde{v}$ and a mollifying sequence ρ_n (as in Lemma 2.4 of this Chapter) Define $\tilde{v}_n$ by

$$\tilde{v}_n = \tilde{v} * \rho_n. \tag{4.30}$$

We have

$$\tilde{v}_n \in \mathscr{D}(\mathbb{R}^2), \qquad \lim \tilde{v}_n = \tilde{v} \text{ strongly in } H^1(\mathbb{R}^2). \tag{4.31}$$

From $\rho_n \geq 0$ and $\tilde{v} \geq 0$ a.e., we obtain, from (4.30),

$$\tilde{v}_n(x) \geq 0, \qquad \forall\, x \in \mathbb{R}^2.$$

Define v_n by

$$v_n = \tilde{v}_n|_\Omega; \tag{4.32}$$

from (4.31) and (4.32), it follows that

$$v_n \in C^\infty(\bar{\Omega}), \lim_{n \to \infty} v_n = v \text{ strongly in } H^1(\Omega), v_n \geq 0 \text{ a.e. in } \Omega.$$

This proves the lemma. □

4.5.2. *Convergence theorem*

Theorem 4.3. *Suppose that the angles of $\mathscr{T}_h$ are uniformly bounded below by $\theta_0 > 0$ as $h \to 0$; then*

$$\lim_{h \to 0} u_h^k = u \text{ strongly in } H^1(\Omega), \tag{4.33}$$

where u, u_h^k are the solutions of (4.5) and (P_{1h}^k), respectively, for $k = 1, 2$.

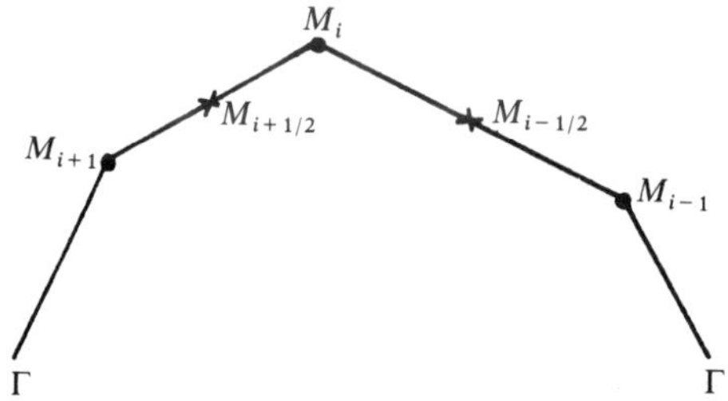

Figure 4.2

PROOF. To prove (4.33), we use Theorem 5.2 of Chapter I. To do this we only have to verify that the following two properties hold:

(i) If $(v_h)_h$, $v_h \in K_h^k$, converges weakly to v, then $v \in K$.
(ii) There exist $\chi \subset K$ and $r_h^k\colon \chi \to K_h^k$ such that $\bar{\chi} = K$ and $\lim_{h\to 0} r_h^k v = v$ strongly in V, $\forall\, v \in \chi$.

Verification of (i). If $k = 1$, then (i) is trivially satisfied, since $K_h^1 \subset K$.

If $k = 2$, using the notation of Fig. 4.2, we consider $\phi \in C^0(\Gamma)$, $\phi \geq 0$, and we define ϕ_h by

$$\phi_h = \sum_i \phi(M_{i+1/2})\chi_{i+1/2}, \tag{4.34}$$

where $\chi_{i+1/2}$ denotes the characteristic function of the open segment $]M_i, M_{i+1}[$. Then

$$\phi_h \geq 0 \text{ a.e. on } \Gamma, \qquad \lim_{h\to 0} \|\phi_h - \phi\|_{L^\infty(\Gamma)} = 0. \tag{4.35}$$

Let us consider a sequence $(v_h)_h$, $v_h \in K_h^2$, $\forall\, h$, such that

$$\lim v_h = v \text{ weakly in } V. \tag{4.36}$$

From (4.36) (see Necas [1]) it follows that $\lim_{h\to 0} \gamma v_h = \gamma v$ strongly in $L^2(\Gamma)$. This in turn implies that

$$\lim_{h\to 0} \int_\Gamma \gamma v_h \phi_h \, d\Gamma = \int_\Gamma \gamma v \phi \, d\Gamma. \tag{4.37}$$

From Simpson's rule it follows that

$$\int_\Gamma \gamma v_h \phi_h \, d\Gamma = \frac{1}{6} \sum_i |\overrightarrow{M_i M_{i+1}}| \phi(M_{i+1/2})[v_h(M_i) + 4v_h(M_{i+1/2}) + v_h(M_{i+1})] \geq 0,$$

$$\forall\, v_h \in K_h^2, \quad \forall\, \phi \in C^0(\Gamma), \quad \phi \geq 0. \tag{4.38}$$

From (4.37) and (4.38) we obtain

$$\int_\Gamma \gamma v \phi \, d\Gamma \geq 0, \qquad \forall\, \phi \in C^0(\Gamma), \quad \phi \geq 0,$$

which implies $\gamma v \geq 0$ a.e. on Γ. This proves (i).

Verification of (ii). From Lemma 4.2, it is natural to take $\chi = K \cap C^\infty(\bar{\Omega})$. Define $r_h^k: H^1(\Omega) \cap C^0(\bar{\Omega}) \to V_h^k$ by

$$r_h^k v \in V_h^k, \qquad \forall\, v \in H^1(\Omega) \cap C^0(\bar{\Omega}),$$

$$r_h^k v(P) = v(P), \qquad \forall\, P \in \Sigma_h^k, \quad k = 1, 2. \tag{4.39}$$

On one hand, under the assumptions made on $\mathscr{T}_h$, we have (see Strang and Fix [1])

$$\|r_h^k v - v\|_V \le Ch^k \|v\|_{H^{k+1}(\Omega)}, \qquad \forall\, v \in C^\infty(\bar{\Omega}), \quad k = 1, 2, \tag{4.40}$$

with C independent of h and v.

This implies

$$\lim_{h \to 0} \|r_h^k v - v\|_V = 0, \qquad \forall\, v \in \chi, \quad k = 1, 2. \tag{4.41}$$

On the other hand, it is obvious that $r_h^k v \in K_h^k$, $\forall\, v \in K \cap C^0(\bar{\Omega})$, so that $r_h^k v \in K_h^k$, $\forall\, v \in \chi$, $k = 1, 2$.

In conclusion, with the above χ and r_h^k, (ii) is satisfied. □

Remark 4.8. For error estimates in the approximation of (4.5) by piecewise linear finite elements, it has been shown by Brezzi, Hager, and Raviart [1] that we have

$$\|u_h - u\|_{H^1(\Omega)} = O(h),$$

assuming reasonable smoothness for u on Ω.

4.6. Iterative methods for solving the discrete problem

We shall briefly describe two types of methods which seem to be appropriate for solving the approximate problems of Sec. 4.4.

4.6.1. *Solution by an over-relaxation method*

The approximate problems (P_{1h}^k) are, for $k = 1, 2$, equivalent to the quadratic programming problems described in Remark 4.6. By virtue of the properties of K_h^k (see Sec. 4.4.1), for the solution of (P_{1h}^k), we can use the over-relaxation method with projection, which has already been used in Sec. 2.8 to solve the approximate obstacle problem and which is described in Chapter V, Sec. 5. From the properties of our problem, the method will converge, provided $0 < \omega < 2$.

4.6.2. *Solution by a duality method*

We first consider the *continuous case*. Let us define a Lagrangian $\mathscr{L}$ by

$$\mathscr{L}(v, q) = \tfrac{1}{2}a(v, v) - L(v) - \int_\Gamma q\gamma v \, d\Gamma, \tag{4.42}$$

and let Λ be the *positive cone* of $L^2(\Gamma)$, i.e.,

$$\Lambda = \{q \in L^2(\Gamma), q \geq 0 \text{ a.e. on } \Gamma\}.$$

Then we have:

Theorem 4.4. *Let* $L(v) = \int_\Omega fv\,dx + \int_\Gamma g\gamma v\,d\Gamma$ *with* f *and* g *sufficiently smooth. Suppose that the solution* u *of* (4.5) *and* (4.6) *belongs to* $H^2(\Omega)$; *then* $\{u, \partial u/\partial n - g\}$ *is the unique saddle point of* $\mathscr{L}$ *over* $H^1(\Omega) \times \Lambda$.

PROOF. We divide the proof into two parts. In the first part we will show that $\{u, \partial u/\partial n - g\}$ is a saddle point of $\mathscr{L}$ over $H^1(\Omega) \times \Lambda$, and in the second part we will prove uniqueness.

(1) Let $p = \partial u/\partial n - g$. From the definition of a saddle point, we have to prove that

$$\mathscr{L}(u, q) \leq \mathscr{L}(u, p) \leq \mathscr{L}(v, p), \qquad \forall \{v, q\} \in V \times \Lambda, \quad \{u, p\} \in V \times \Lambda. \tag{4.43}$$

Since $u \in H^2(\Omega)$, we have $\partial u/\partial n \in H^{1/2}(\Gamma) \subset L^2(\Gamma)$ (see Lions and Magenes [1]). Then if g is smooth enough, we have $p = \partial u/\partial n - g \in L^2(\Gamma)$. From Proposition 4.1 we have

$$p = \frac{\partial u}{\partial n} - g \geq 0 \text{ on } \Gamma,$$

$$p\gamma u = 0 \text{ a.e. on } \Gamma. \tag{4.44}$$

This implies that we have $\{u,p\} \in H^1(\Omega) \times \Lambda$ and $\int_\Gamma p\gamma u\,d\Gamma = 0$. Since $\gamma u \geq 0$ on Γ, we have

$$\int_\Gamma q\gamma u\,d\Gamma \geq 0, \qquad \forall\, q \in \Lambda. \tag{4.45}$$

From (4.44) and (4.45) it follows that

$$\mathscr{L}(u, q) = \tfrac{1}{2}a(u, u) - L(u) - \int_\Gamma q\gamma u\,d\Gamma \leq \tfrac{1}{2}a(u, u) - L(u)$$

$$= \tfrac{1}{2}a(u, u) - L(u) - \int_\Gamma p\gamma u\,d\Gamma = \mathscr{L}(u, p), \qquad \forall\, q \in \Lambda$$

which proves the first inequality of (4.43).

To prove the second inequality of (4.43), we observe that the solution u^* of the minimization problem

$$\mathscr{L}(u^*, p) \leq \mathscr{L}(v, p), \qquad \forall\, v \in H^1(\Omega), \quad u^* \in H^1(\Omega) \tag{4.46}$$

is unique and is actually the solution of the linear variational equation

$$a(u^*, v) = L(v) + \int_\Gamma p\gamma v\,d\Gamma, \qquad \forall\, v \in H^1(\Omega), \quad u^* \in H^1(\Omega). \tag{4.47}$$

Since $L(v) = \int_\Omega fv\,dx + \int_\Gamma g\gamma v\,d\Gamma$, u^* is actually the solution of the Neumann problem

$$-\Delta u^* + u^* = f \text{ in } \Omega,$$

$$\frac{\partial u^*}{\partial n} = p + g = \frac{\partial u}{\partial n} \text{ on } \Gamma. \tag{4.48}$$

Since from Proposition 4.1 we obviously have

$$-\Delta u + u = f \text{ in } \Omega,$$

$$\frac{\partial u}{\partial n} = \frac{\partial u^*}{\partial n} \text{ on } \Gamma,$$

it follows from the uniqueness property of the Neumann problem (4.48) that $u = u^*$. Using (4.46) and $u = u^*$, we obtain the second inequality in (4.43). This proves that $\{u, p\}$ is a saddle point of $\mathscr{L}$ over $H^1(\Omega) \times \Lambda$.

(2) *Uniqueness.* Let $\{u^*, p^*\}$ be a saddle point of $\mathscr{L}$ over $H^1(\Omega) \times \Lambda$. We will show that $u^* = u$, $p^* = p$. From (4.42) and (4.43) it follows that

$$\int_\Gamma (p - q)\gamma u \, d\Gamma \le 0, \qquad \forall\, q \in \Lambda. \tag{4.49}$$

Similarly, we have

$$\int_\Gamma (p^* - q)\gamma u^* \, d\Gamma \le 0, \quad \forall\, q \in \Lambda. \tag{4.50}$$

Taking $q = p^*$ (respectively, $q = p$) in (4.49) (respectively, (4.50)), we obtain

$$\int_\Gamma (p^* - p)\gamma(u^* - u) \, d\Gamma \le 0. \tag{4.51}$$

From the second inequality of (4.43) it follows that u is the solution of

$$a(u, v) = L(v) + \int_\Gamma p\gamma v \, d\Gamma, \qquad \forall\, v \in H^1(\Omega), \quad u \in H^1(\Omega), \tag{4.52}$$

and similarly,

$$a(u^*, v) = L(v) + \int_\Gamma p^*\gamma v \, d\Gamma, \qquad \forall\, v \in H^1(\Omega), \quad u^* \in H^1(\Omega). \tag{4.53}$$

Taking $v = u^* - u$ (respectively, $v = u - u^*$) in (4.52) (respectively, (4.53)), we obtain

$$a(u^* - u, u^* - u) = \int_\Gamma (p^* - p)\gamma(u^* - u) \, d\Gamma. \tag{4.54}$$

Using the V-ellipticity of $a(\cdot, \cdot)$, from (4.51)–(4.54) it then follows that $u^* = u$ and

$$\int_\Gamma (p^* - p)\gamma v \, d\Gamma = 0, \qquad \forall\, v \in H^1(\Omega),$$

which implies that $p^* = p$.

Hence $\{u, p\}$ is the unique saddle point of $\mathscr{L}$ over $H^1(\Omega) \times \Lambda$. □

From Theorem 4.4 it follows that we can apply Uzawa's algorithm to solve (4.5) (see Cea [2, Chapter 5] and G.L.T. [1, Chapter 2], [2, Chapter 4, Sec. 3.6], [3]). In the present case this algorithm is written as follows:

$$p^0 \in \Lambda \textit{ is arbitrarily chosen (for instance, } p^0 = 0); \tag{4.55}$$

then p^n *being known, we compute, for* $n \geq 0$, $\{u^n, p^{n+1}\}$ *by*

$$\mathscr{L}(u^n, p^n) \leq \mathscr{L}(v, p^n), \qquad \forall\, v \in H^1(\Omega), \quad u^n \in H^1(\Omega), \tag{4.56}$$

$$p^{n+1} = P_\Lambda(p^n - \rho\gamma u^n), \tag{4.57}$$

where P_Λ is the *projection operator* from $L^2(\Gamma)$ to Λ in the $L^2(\Gamma)$ norm, and $\rho > 0$. From (4.56) it follows that u^n is in fact the solution of the Neumann problem

$$\begin{aligned} -\Delta u^n + u^n &= f \text{ in } \Omega, \\ \left.\frac{\partial u^n}{\partial n}\right|_\Gamma &= p^n + g. \end{aligned} \tag{4.58}$$

The projection P_Λ is given by

$$P_\Lambda(q) = q^+, \qquad \forall\, q \in L^2(\Gamma). \tag{4.59}$$

Since $\gamma\colon H^1(\Omega) \to L^2(\Gamma)$ is a continuous linear mapping, we have

$$\|\gamma v\|_{L^2(\Gamma)} \leq \|\gamma\| \cdot \|v\|_{H^1(\Omega)}, \qquad \forall\, v \in H^1(\Omega). \tag{4.60}$$

From Cea and G.L.T., *loc. cit.*, it then follows that

$$\lim_{n\to\infty} u^n = u \text{ strongly in } H^1(\Omega), \tag{4.61}$$

where u is the solution of the problem (4.5) provided that $0 < \rho < 2/\|\gamma\|^2$.

Let us give a direct proof for this convergence result. This proof will use the characterization (4.5) given in Proposition 4.1 (even if $a(\cdot, \cdot)$ is not symmetrical, the same result follows). It will be convenient to take (4.56), (4.58) in the following equivalent form:

$$a(u^n, v) = L(v) + \int_\Gamma p^n \gamma v \, d\Gamma, \qquad \forall\, v \in H^1(\Omega), \quad u^n \in H^1(\Omega). \tag{4.62}$$

Let u be the solution of (4.5) and $p = \partial u/\partial n - g$. From Proposition 4.1 it follows that

$$a(u, v) = L(v) + \int_\Gamma p\gamma v \, d\Gamma, \qquad \forall\, v \in H^1(\Omega), \quad u \in H^1(\Omega), \tag{4.63}$$

$$\int_\Gamma (q - p)\gamma u \, d\Gamma \geq 0, \qquad \forall\, q \in \Lambda, \quad p \in \Lambda. \tag{4.64}$$

Relation (4.64) can also be written as

$$\int_\Gamma (q - p)(p - \rho\gamma u - p)\, d\Gamma \leq 0, \qquad \forall\, q \in \Lambda, \quad \rho > 0 \quad \text{and} \quad p \in \Lambda,$$

which is classically equivalent to

$$p = P_\Lambda(p - \rho\gamma u). \tag{4.65}$$

Let us consider

$$\bar{u}^n = u^n - u, \qquad \bar{p}^n = p^n - p.$$

Since P_Λ is a contraction, from (4.57) and (4.65) we have

$$\|\bar{p}^{n+1}\|_{L^2(\Gamma)} \leq \|\bar{p}^n - \rho\gamma\bar{u}^n\|_{L^2(\Gamma)}. \tag{4.66}$$

From (4.66) it follows that

$$\|\bar{p}^n\|^2_{L^2(\Gamma)} - \|\bar{p}^{n+1}\|_{L^2(\Gamma)} \geq 2\rho \int_\Gamma \gamma\bar{u}^n\bar{p}^n\, d\Gamma - \rho^2\|\gamma\bar{u}^n\|^2_{L^2(\Gamma)}. \tag{4.67}$$

Taking $v = \bar{u}^n$ in (4.62) and (4.63), we obtain

$$a(\bar{u}^n, \bar{u}^n) = \int_\Gamma \bar{p}^n\gamma\bar{u}^n\, d\Gamma. \tag{4.68}$$

From (4.67) and (4.68) it then follows that

$$\|\bar{p}^n\|^2_{L^2(\Gamma)} - \|\bar{p}^{n+1}\|^2_{L^2(\Gamma)} \geq \rho(2 - \rho\|\gamma\|^2)\|\bar{u}^n\|^2_{H^1(\Omega)}. \tag{4.69}$$

If $0 < \rho < 2/\|\gamma\|^2$, we observe that the sequence $\{\|\bar{p}^n\|^2_{L^2(\Gamma)}\}_n$ is decreasing and hence converges. Therefore we have

$$\lim_{n\to\infty} (\|\bar{p}^n\|^2_{L^2(\Gamma)} - \|\bar{p}^{n+1}\|^2_{L^2(\Gamma)}) = 0$$

so that

$$\lim_{n\to\infty} \|\bar{u}^n\|_{H^1(\Omega)} = 0.$$

Since $\bar{u}^n = u^n - u$, we have proved the convergence.

Similarly we can solve the approximate problems (P^k_{1h}), $k = 1, 2$, using the discrete version of algorithm (4.55)–(4.57). We shall limit ourselves to $k = 1$, since the extension to $k = 2$ is trivial.

Here we use the notations of Sec. 4.1. Assume that $\gamma_h = \Sigma_h - \mathring{\Sigma}_h$ has been ordered.

Let $\gamma_h = \{M_i\}_i$.

We approximate Λ and $\mathscr{L}$ by

$$\Lambda_h^1 = \{q_h | q_h = \{q_i\}_i, q_i \geq 0\}$$

and

$$\mathscr{L}_h^1(v_h, q_h) = \tfrac{1}{2}a(v_h, v_h) - L(v_h) - \tfrac{1}{2}\sum_i |M_i M_{i+1}|[q_i v_h(M_i) + q_{i+1} v_h(M_{i+1})]. \qquad (4.70)$$

We can prove that $\mathscr{L}_h^1$ has a unique saddle point $\{u_h, p_h\}$, where p_h is a F. John–Kuhn–Tucker vector for (P_{1h}^1) over $V_h^1 \times \Lambda_h^1$ and u_h is precisely the solution of (P_{1h}^1). The discrete analogue of (4.55)–(4.57) is then

$$p_h^0 \in \Lambda_h^1, \qquad (4.71)$$

$$\mathscr{L}_h^1(u_h^n, p_h^n) \leq \mathscr{L}_h^1(v_h, p_h^n), \qquad \forall\, v_h \in V_h^1, \quad u_h^n \in V_h^1, \qquad (4.72)$$

$$p_i^{n+1} = [p_i^n - \rho u_h^n(M_i)]^+, \qquad \forall\, i, \quad \rho > 0. \qquad (4.73)$$

One can prove that if $0 < \rho < \beta$, with β small enough, then $\lim_{n\to+\infty} u_h^n = u_h$, where u_h is the solution of (P_{1h}^1). In G.L.T. [3, Chapter 4] one may find numerical applications of the above iterative methods for piecewise linear and piecewise quadratic approximations of (4.5).

EXERCISE 4.3. Extend the above considerations to (P_{1h}^2).

5. An Example of EVI of the Second Kind: A Simplified Friction Problem

5.1. The continuous problem. Existence and uniqueness results

Let Ω be a bounded domain of $\mathbb{R}^2$ with a smooth boundary $\Gamma = \partial\Omega$. Using the same notations as in Sec. 4, we define

$$V = H^1(\Omega), \qquad (5.1)$$

$$a(u, v) = \int_\Omega \nabla u \cdot \nabla v\, dx + \int_\Omega uv\, dx, \qquad (5.2)$$

$$L(v) = \langle f, v\rangle, \qquad f \in V^*, \qquad (5.3)$$

$$j(v) = g\int_\Gamma |\gamma v|\, d\Gamma, \quad \text{where } g > 0. \qquad (5.4)$$

We have then the following:

Theorem 5.1. *The variational inequality*

$$a(u, v - u) + j(v) - j(u) \geq L(v - u), \qquad \forall\, v \in V, \quad u \in V, \tag{5.5}$$

has a unique solution.

PROOF. In order to apply Theorem 4.1 of Chapter I, it is enough to verify that $j(\cdot)$ is convex, proper, and l.s.c.. Actually $j(\cdot)$ is a seminorm on V. Therefore, using the Schwarz inequality in $L^2(\Gamma)$ and the fact the $\gamma \in \mathscr{L}(H^1(\Omega), L^2(\Gamma))$, we have

$$|j(u) - j(v)| \leq |j(u - v)| \leq g(\text{meas}(\Gamma))^{1/2} \|\gamma(v - u)\|_{L^2(\Gamma)} \leq C\|u - v\|_V \tag{5.6}$$

for some constant C.

Hence $j(\cdot)$ is Lipschitz continuous on V, so that $j(\cdot)$ is l.s.c.; $j(\cdot)$ is obviously convex and proper. Hence the theorem is proved. □

Remark 5.1. If $g = 0$, it is easy to prove that (5.5) reduces to the variational equation

$$a(u, v) = L(v), \qquad \forall\, v \in V, \quad u \in V.$$

This is related to the variational formulation of the Neumann problem.

Remark 5.2. Since $a(\cdot, \cdot)$ is symmetrical, the solution u of (5.5) is characterized, using Lemma 4.1 of Chapter 1, as the unique solution of the minimization problem

$$J(u) \leq J(v), \qquad \forall\, v \in V, \quad u \in V, \tag{5.7}$$

where $J(v) = \frac{1}{2}a(v, v) + j(v) - L(v)$.

Remark 5.3. Problem (5.5) (and (5.7)) is the simplified version of a friction problem occurring in elasticity. For this type of problem we refer to Duvaut and Lions [1] and the bibliography therein.

EXERCISE 5.1. Let us denote the solution of (5.5) by u_g. Then prove that

$$\lim_{g \to +\infty} u_g = \hat{u} \text{ strongly in } H^1(\Omega),$$

where $\hat{u}$ is the unique solution of

$$a(\hat{u}, v) = L(v), \qquad \forall\, v \in H_0^1(\Omega), \quad \hat{u} \in H_0^1(\Omega).$$

5.2. Regularity of the solution

Theorem 5.2 (H. Brezis [3]). *If Ω is a bounded domain with a smooth boundary and if $L(v) = \int_\Omega fv\, dx$ with $f \in L^2(\Omega)$, then the solution u of* (5.5) *is in $H^2(\Omega)$.*

5.3. Existence of a multiplier

Let us define Λ by

$$\Lambda = \{\mu \in L^2(\Gamma), |\mu(x)| \leq 1 \text{ a.e. on } \Gamma\}.$$

Then we have:

Theorem 5.3. *The solution u of (5.5) is characterized by the existence of λ such that*

$$a(u, v) + g \int_\Gamma \lambda \gamma v \, d\Gamma = L(v), \qquad \forall\, v \in V, \quad u \in V, \tag{5.8}$$

$$\lambda \in \Lambda, \qquad \lambda \gamma u = |\gamma u| \text{ a.e. on } \Gamma. \tag{5.9}$$

PROOF. We will prove first that (5.5) implies (5.8) and (5.9).

Taking $v = 0$ and $v = 2u$ in (5.5), we have

$$a(u, u) + j(u) = L(u). \tag{5.10}$$

It follows then from (5.5), (5.10) that

$$L(v) - a(u, v) \leq j(v), \qquad \forall\, v \in V,$$

which implies

$$|L(v) - a(u, v)| \leq j(v) = g \int_\Gamma |\gamma v| \, d\Gamma, \qquad \forall\, v \in V. \tag{5.11}$$

We have $H^1(\Omega) = H_0^1(\Omega) \oplus [H_0^1(\Omega)]^\perp$, where $[H_0^1(\Omega)]^\perp$ is the orthogonal complement of $H_0^1(\Omega)$ in $H^1(\Omega)$.

Since $\gamma: [H_0^1(\Omega)]^\perp \to H^{1/2}(\Gamma)$ is an isomorphism, it follows from (5.11) that

$$L(v) - a(u, v) = l(\gamma v), \qquad \forall\, v \in V, \tag{5.12}$$

where $l(\cdot)$ is a continuous linear functional on $H^{1/2}(\Gamma)$. It then follows from (5.11), (5.12) that

$$|l(\mu)| \leq g \|\mu\|_{L^1(\Gamma)}, \qquad \forall\, \mu \in H^{1/2}(\Gamma). \tag{5.13}$$

Since $H^{1/2}(\Gamma) \subset L^1(\Gamma)$, it follows from (5.13) that we can apply to $l(\cdot)$ the *Hahn–Banach Theorem* (see, for instance, Yosida [1]) to obtain the existence of $\lambda \in L^\infty(\Gamma)$, $|\lambda(x)| \leq 1$ a.e. on Γ such that

$$l(\mu) = g \int_\Gamma \lambda \mu \, d\Gamma, \qquad \forall\, \mu \in H^{1/2}(\Gamma). \tag{5.14}$$

Therefore it follows from (5.12) and (5.14) that

$$a(u, v) + g \int_\Gamma \lambda \gamma v \, d\Gamma = L(v), \qquad \forall\, v \in V,$$

which proves (5.8).

Taking $v = u$ in (5.8), we obtain

$$a(u, u) + g \int_\Gamma \lambda \gamma u \, d\Gamma = L(u).$$

Using (5.10) and the above equations, we obtain

$$\int_\Gamma (|\gamma u| - \lambda \gamma u) \, d\Gamma = 0. \tag{5.15}$$

Since $|\lambda| \leq 1$ a.e., we have

$$|\gamma u| - \lambda \gamma u \geq 0 \text{ a.e..} \tag{5.16}$$

From (5.15) and (5.16) it follows that

$$|\gamma u| = \lambda \gamma u \text{ a.e..}$$

This completes the proof of (5.8) and (5.9). Assuming (5.8) and (5.9), we will show that (5.5) holds.

Let $\{u, \lambda\}$ be a solution of (5.8), (5.9). From (5.8) it follows that

$$a(u, v - u) + g \int_\Gamma \lambda \gamma (v - u) \, d\Gamma = L(v - u), \qquad \forall \, v \in V,$$

which can also be written as

$$a(u, v - u) + g \int_\Gamma \lambda \gamma v \, d\Gamma - g \int_\Gamma \lambda \gamma u \, d\Gamma = L(v - u), \qquad \forall \, v \in V. \tag{5.17}$$

From (5.9) and (5.17) we obtain

$$a(u, v - u) + g \int_\Gamma \lambda \gamma v \, d\Gamma - g \int_\Gamma |\gamma u| \, d\Gamma = L(v - u), \qquad \forall \, v \in V. \tag{5.18}$$

But since $\lambda \gamma v \leq |\gamma v|$ a.e. in Γ, it follows from (5.18) that

$$a(u, v - u) + j(v) - j(u) \geq L(v - u), \qquad \forall \, v \in V.$$

This proves the characterization. □

Remark 5.4. Assuming that

$$L(v) = \int_\Omega f_0 v \, dx + \int_\Gamma f_1 \gamma v \, d\Gamma,$$

with f_0, f_1 sufficiently smooth, we can express (5.8) by

$$-\Delta u + u = f_0 \text{ in } \Omega$$

$$\frac{\partial u}{\partial n} + g\lambda = f_1 \text{ a.e. on } \Gamma. \tag{5.19}$$

From (5.19) it follows that λ is unique.

EXERCISE 5.2. Prove that λ is unique.

5.4. Finite element approximation of (5.5)

Let Ω be a bounded polygonal domain of $\mathbb{R}^2$. The notation used here is largely the same as in Sec. 4.4 of this chapter.

5.4.1. *Approximation of V*

We use the piecewise linear and piecewise quadratic approximations of $V = H^1(\Omega)$ described in Section 4.4.1 of this chapter.

5.4.2. *Approximation of $j(\cdot)$*

We use the notation of Figure 4.2. Then we approximate $j(\cdot)$ by

$$j_h^1(v_h) = \frac{g}{2}\sum_i |\overrightarrow{M_i M}_{i+1}|(|\gamma v_h(M_i)| + |\gamma v_h(M_{i+1})|), \qquad \forall\, v_h \in V_h^1, \tag{5.20}$$

$$\begin{aligned} j_h^2(v_h) = \frac{g}{6}\sum_i |M_i M_{i+1}|(|\gamma v_h(M_i)| + 4|\gamma v_h(M_{i+1/2})| \\ + |\gamma v_h(M_{i+1})|), \qquad \forall\, v_h \in V_h^2. \end{aligned} \tag{5.21}$$

In (5.20) and (5.21) we have $M_i \in \gamma_h$ and $M_{i+1/2} \in \gamma_h'$.

Remark 5.5. Clearly (5.20) and (5.21) are obtained from $j(\cdot)$ by using trapezoidal and Simpson's numerical integration formulae, respectively.

5.4.3. *The approximate problem*

For $k = 1, 2$, problem (5.5) is approximated by

$$a(u_h^k, v_h - u_h^k) + j_h^k(v_h) - j_h^k(u_h^k) \geq L(v_h - u_h^k), \qquad \forall\, v_h \in V_h^k, \quad u_h^k \in V_h^k. \tag{P_{2h}^k}$$

Then:

Proposition 5.1. *Problem* (P_{2h}^k) *has a unique solution.*

Remark 5.6. Since $a(\cdot, \cdot)$ is symmetrical, (P_{2h}^k) is equivalent to the nonlinear programming problem

$$\operatorname*{Min}_{v_h \in V_h^k}[\tfrac{1}{2}a(v_h, v_h) + j_h^k(v_h) - L(v_h)]. \tag{5.22}$$

Remark 5.7. Using (5.20), (5.21), and (7.1)–(7.4) of Sec. 7 of this chapter, we may express (P_{2k}^k) and (5.22) in a form more suitable for computations.

5.5. Convergence results

Theorem 5.4. *Suppose that the angles of $\mathcal{T}_h$ are uniformly bounded below by $\theta_0 > 0$ as $h \to 0$; then*

$$\lim_{h\to 0} u_h^k = u \text{ strongly in } H^1(\Omega), \tag{5.23}$$

where u and u_h^k are, respectively, the solutions of (5.5) and (P_{2h}^k) for $k = 1, 2$.

PROOF. To prove (5.23) it suffices to verify the following properties (see Theorem 6.3 of Chapter I).

(i) There exists $U \subset V$, $\overline{U} = V$ and $r_h^k \colon U \to V_h^k$, such that

$$\lim_{h\to 0} r_h^k v = v \text{ strongly in } V, \qquad \forall\, v \in U.$$

(ii) If $v_h \to v$ weakly in V, then

$$\liminf_{h\to 0} j_h^k(v_h) \geq j(v).$$

(iii)

$$\lim_{h\to 0} j_h^k(r_h^k v) = j(v), \qquad \forall\, v \in U.$$

Verification of (i). Since Γ is Lipschitz continuous, we have (see Necas [1])

$$\overline{C^\infty(\overline{\Omega})} = H^1(\Omega). \tag{5.24}$$

Therefore it is natural to take $U = C^\infty(\overline{\Omega})$. Define r_h^k by (4.39) of Theorem 4.3, Chapter II; under the above assumptions on $\mathcal{T}_h$, it follows from Strang and Fix [1] that

$$\|r_h^k v - v\|_V \leq C h^k \|v\|_{H^{k+1}(\Omega)}, \qquad \forall\, v \in V, \tag{5.25}$$

where C is a constant independent of h and v. This implies (i).

Verification of (ii)

(1) *Case* $k = 1$. We again use the notation of Fig. 4.2. Since the trace of v_h restricted to $[M_i, M_{i+1}]$ is affine, it follows that

$$\gamma v_h(M) = \frac{1}{|\overrightarrow{M_iM}_{i+1}|}\left(|\overrightarrow{MM}_{i+1}|\gamma v_h(M_i) + |\overrightarrow{MM}_i|\gamma v_h(M_{i+1})\right),$$
$$\forall\, v_h \in V_h^1, \quad \forall\, M \in [M_i, M_{i+1}]. \tag{5.26}$$

Since

$$\frac{|\overrightarrow{MM}_{i+1}|}{|\overrightarrow{M_iM}_{i+1}|} + \frac{|\overrightarrow{MM}_i|}{|\overrightarrow{M_iM}_{i+1}|} = 1,$$

the convexity of $\xi \to |\xi|$ implies

$$|\gamma v_h(M)| \leq \frac{1}{|\overrightarrow{M_iM}_{i+1}|}\left(|\overrightarrow{MM}_{i+1}||\gamma v_h(M_i)|\right.$$
$$\left. + |\overrightarrow{M_iM}|\, |\gamma v_h(M_{i+1})|\right), \qquad \forall\, v_h \in V_h^1, \quad \forall\, M \in [M_i, M_{i+1}]. \tag{5.27}$$

Integrating (5.27) on $\widehat{M_i M_{i+1}}$, we obtain

$$\int_{\widehat{M_i M_{i+1}}} |\gamma v_h| \, d\Gamma \leq \frac{|\overrightarrow{M_i M_{i+1}}|}{2} (|\gamma v_h(M_i)| + |\gamma v_h(M_{i+1})|),$$

which implies that, $\forall\, v_h \in V_h^1$, we have

$$j(v_h) = g \int_\Gamma |\gamma v_h| \, d\Gamma = g \sum_i \int_{\widehat{M_i M_{i+1}}} |\gamma v_h| \, d\Gamma$$

$$\leq \frac{g}{2} \sum_i |\overrightarrow{M_i M_{i+1}}| (|\gamma v_h(M_i)| + |\gamma v_h(M_{i+1})|) = j_h^1(v_h).$$

Thus we have proved

$$j(v_h) \leq j_h^1(v_h), \qquad \forall\, v_h \in V_h^1. \tag{5.28}$$

Let $v_h \to v$ weakly in V. Then $\lim_{h \to 0} \gamma v_h = \gamma v$ strongly in $L^2(\Gamma)$, which implies

$$\lim_{h \to 0} j(v_h) = j(v). \tag{5.29}$$

It then follows from (5.28) and (5.29) that $\liminf_{h \to 0} j_h^1(v_h) \geq j(v)$, which proves (ii) if $k = 1$.

(2) *Case* $k = 2$. Let us define $M_{i+1/6}, M_{i+5/6}$ by (see Fig. 5.1)

$$\overrightarrow{M_i M_{i+1/6}} = \tfrac{1}{6} \overrightarrow{M_i M_{i+1}}, \quad \overrightarrow{M_i M_{i+5/6}} = \tfrac{5}{6} \overrightarrow{M_i M_{i+1}}.$$

Then we define $q_h: C^0(\Gamma) \to L^\infty(\Gamma)$ by

$$q_h(\mu) = \sum_{M_i \in \gamma_h} \mu(M_i) X_i + \sum_{M_{i+1/2} \in \gamma_h'} \mu(M_{i+1/2}) X_{i+1/2}, \qquad \forall\, \mu \in C^0(\Gamma), \tag{5.30}$$

where X_i (*respectively*, $X_{i+1/2}$) is the characteristic function of $\widehat{M_{i-1/6} M_i M_{i+1/6}}$ (respectively, $\widehat{M_{i+1/6} M_{i+5/6}}$). We then have the following obvious properties:

$$\lim_{h \to 0} q_h(\mu) = \mu \text{ strongly in } L^\infty(\Gamma), \qquad \forall\, \mu \in C^0(\Gamma), \tag{5.31}$$

$$j_h^2(v_h) = g \int_\Gamma |q_h \gamma v_h| \, d\Gamma = g \|q_h \gamma v_h\|_{L^1(\Gamma)}, \qquad \forall\, v_h \in V_h^2, \tag{5.32}$$

$$C_1 \|\gamma v_h\|_{L^2(\Gamma)} \leq \|q_h \gamma v_h\|_{L^2(\Gamma)} \leq C_2 \|\gamma v_h\|_{L^2(\Gamma)}, \qquad \forall\, v_h \in V_h^2, \tag{5.33}$$

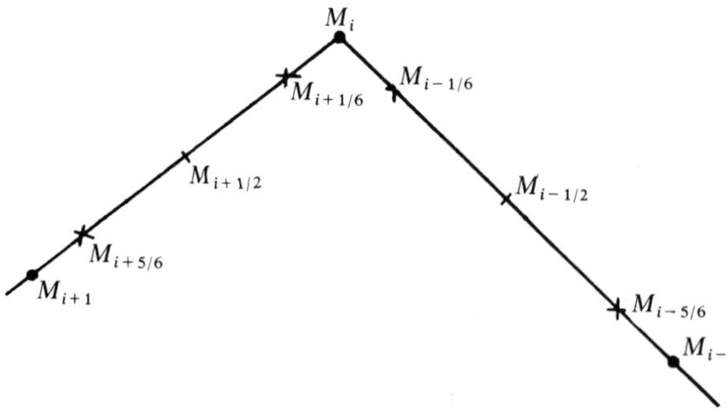

Figure 5.1

where in (5.33), C_1 and C_2 are positive constants independent of v_h, h, and Γ (values for C_1 and C_2 may be found in G.L.T. [3, Chapter 4]).

Now, taking $\mu \in C^0(\Gamma)$, we define $s_h(\mu)$ by

$$s_h(\mu) \in L^\infty(\Gamma), \qquad s_h(\mu)|_{]M_i, M_{i+1}[} = \mu(M_{i+1/2}). \tag{5.34}$$

Then

$$\lim_{h\to 0} s_h(\mu) = \mu \text{ strongly in } L^\infty(\Gamma), \qquad \forall\, \mu \in C^0(\Gamma), \tag{5.35}$$

and from Simpson's integration formula, we have

$$\int_\Gamma s_h(\mu) q_h \gamma v_h \, d\Gamma = \int_\Gamma s_h(\mu)\gamma v_h \, d\Gamma, \qquad \forall\, \mu \in C^0(\Gamma), \quad \forall\, v_h \in V_h^2. \tag{5.36}$$

Let $v_h \to v$ weakly in V, $v_h \in V_h^2$, $\forall\, h$; then

$$\lim_{h\to 0} \gamma v_h = \gamma v \text{ strongly in } L^2(\Gamma). \tag{5.37}$$

On the one hand, it follows from (5.33) that

$$\|q_h \gamma v_h\|_{L^2(\Gamma)} \le C, \tag{5.38}$$

where C is independent of h.

On the other hand, (5.31), (5.35)–(5.38) imply that

$$\lim_{h\to 0} \int_\Gamma s_h(\mu) q_h \gamma v_h \, d\Gamma = \int_\Gamma \mu\gamma v \, d\Gamma, \qquad \forall\, \mu \in C^0(\Gamma). \tag{5.39}$$

In turn, (5.38) and (5.39) imply that

$$\lim_{h\to 0} q_h \gamma v_h = v \text{ weakly in } L^2(\Gamma). \tag{5.40}$$

Since the functional $\mu \to \|\mu\|_{L^1(\Omega)}$ is convex and continuous on $L^1(\Gamma)$, it follows from (5.40) that

$$\liminf_{h\to 0} \|q_h \gamma v_h\|_{L^1(\Gamma)} \ge \|\gamma v\|_{L^1(\Gamma)}. \tag{5.41}$$

Combining (5.41) with (5.32), we obtain $\liminf_{h\to 0} j_h^2(v_h) \ge j(v)$ which proves (ii) when $k = 2$.

Verification of (iii). Let $v \in U = C^\infty(\bar{\Omega})$. From (5.25) and from the uniform continuity of γv on Γ, it follows almost immediately that

$$\lim_{h\to 0} j_h^k(r_h^k v_h) = j(v), \qquad k = 1, 2.$$

Since conditions (i), (ii), and (iii) are satisfied, the strong convergence of u_h to u follows from the Theorem 6.2 of Chapter I. □

Remark 5.8. It is proved in G.L.T. [3, Chapter 4] that $v_h \to v$ weakly in V, $v_h \in V_h^k$, implies $\lim_{h\to 0} j_h^k(v_h) = j(v)$, $k = 1, 2$.

Since the proof of this result is rather technical we have used a simpler approach in this book, from which it follows that

$$\liminf_{h \to 0} j_h^k(v_h) \geq j(v), \qquad k = 1, 2.$$

As we have seen before, this result was sufficient for proving Theorem 5.4.

5.6. Iterative methods for solving (P_{2h}^k)

In this section we briefly describe some iterative methods which may be useful for solving the approximate problems (P_{2h}^k).

5.6.1. *Solution of* (P_{2h}^k) *by relaxation methods*

It follows from (5.20)–(5.22) (see Remark 5.6) that (P_{2h}^k), $k = 1, 2$, are particular cases of

$$\operatorname*{Min}_{v \in \mathbb{R}^N} f(v), \tag{5.42}$$

where, with $v = \{v_1, \ldots, v_N\}$,

$$f(v) = \tfrac{1}{2}(Av, v) - (b, v) + \sum_{i=1}^{N} \alpha_i |v_i|. \tag{5.43}$$

In (5.43), $(\cdot, \cdot)$ denotes the usual inner product of $\mathbb{R}^N$, A is a $N \times N$ symmetrical positive-definite matrix, and $\alpha_i \geq 0$, $\forall\, i = 1, \ldots, N$.

It then follows from Chapter V, Sec. 5, Cea [2, Chapter 4], Cea and Glowinski [1], and G.L.T. [3, Chapter 2] that we can use a relaxation method for solving (5.42). Actually, from the computations we have done, it appears that the introduction of an over-relaxation parameter ω, $\omega > 1$, will increase the speed of convergence.

Finally the algorithm we used is the following:

$$u^0 \textit{ arbitrarily given in } \mathbb{R}^N; \tag{5.44}$$

then for $i = 1, 2, \ldots, N$,

$$f(u_1^{n+1}, \ldots, u_{i-1}^{n+1}, \bar{u}_i^{n+1}, u_{i+1}^n, \ldots) \leq f(u_1^{n+1}, \ldots, u_{i-1}^{n+1}, v_i, u_{i+1}^n, \ldots), \quad \forall\, v_i \in \mathbb{R}, \tag*{(5.45)$_i$}$$

$$u_i^{n+1} = u_i^n + \omega(\bar{u}_i^{n+1} - u_i^n). \tag*{(5.46)$_i$}$$

If $\omega = 1$, (5.44)–(5.46) reduces to a relaxation method. Numerical solutions of (5.5) using (5.44)–(5.46) are given in G.L.T. [2, Chapter 4], [3, Chapter 4].

Remark 5.9. If $\alpha_i > 0$, $\bar{u}_i^{n+1}$ is the solution of a *single variable nondifferentiable minimization problem* which can be exactly solved by simple calculations.

EXERCISE 5.3. Express $\bar{u}_i^{n+1}$ as a function of A, b, u^n, u^{n+1}.

5.6.2. *Solution of* (P_{2h}^k) *by a duality method*

We first examine the continuous case. Define a Lagrangian $\mathscr{L}: H^1(\Omega) \times L^2(\Gamma) \to \mathbb{R}$ by

$$\mathscr{L}(v, \mu) = \tfrac{1}{2}a(v, v) - L(v) + g \int_\Gamma \mu \gamma v \, d\Gamma. \tag{5.47}$$

Then, using the notation of Sec. 5.3, the following theorem follows from Theorem 5.3.

Theorem 5.5. *Let* $\{u, \lambda\}$ *be a solution of* (5.8), (5.9); *then* $\{u, \lambda\}$ *is the unique saddle point of* $\mathscr{L}$ *over* $H^1(\Omega) \times \Lambda$.

EXERCISE 5.4. Prove Theorem 5.5.

From Theorem 5.5 it follows that to solve (5.5) we can use the following Uzawa algorithm:

$$\lambda^0 \in \Lambda \textit{ arbitrarily chosen (for instance, } \lambda^0 = 0); \tag{5.48}$$

then by induction, knowing λ^n, *we compute* u^n *and* λ^{n+1} *by*

$$\mathscr{L}(u^n, \lambda^n) \le \mathscr{L}(v, \lambda^n), \qquad \forall\, v \in H^1(\Omega), \quad u^n \in V, \tag{5.49}$$

$$\lambda^{n+1} = P_\Lambda(\lambda^n + \rho g \gamma u^n), \qquad \rho > 0. \tag{5.50}$$

The minimization problem (5.49) is actually equivalent to the Neumann variational problem

$$a(u^n, v) = L(v) - g \int_\Gamma \lambda^n \gamma v \, d\Gamma, \quad \forall\, v \in H^1(\Omega), \quad u^n \in H^1(\Omega). \tag{5.51}$$

In (5.50), P_Λ is the projection operator from $L^2(\Gamma)$ to Λ in the L^2-norm; then

$$P_\Lambda(\mu) = \operatorname{Sup}(-1, \operatorname{Inf}(1, \mu)), \quad \forall\, \mu \in L^2(\Gamma). \tag{5.52}$$

Using Cea [2] and G.L.T. [1, Chapter 2], [3, Chapter 2], it follows that for $0 < \rho < 2/g^2 \|\gamma\|^2$ we have

$$\lim_{n \to \infty} u^n = u \text{ strongly in } H^1(\Omega); \qquad u \text{ is solution of (5.5), (5.7).}$$

As in Sec. 4.6.2, a direct proof of the convergence of (5.48)–(5.50) can be given; however, it will use the results of Theorem 5.3.

EXERCISE 5.5. Using Theorem 5.3, give a direct proof of the convergence of (5.48)–(5.50).

The adaptation of (5.48)–(5.50) to the discrete problems (P_{2h}^k), $k = 1, 2$, is straightforward (see G.L.T. [3, Chapter 4]), since it is a simple variant of the discrete algorithm described in Sec. 4.6.2.

EXERCISE 5.6. Study the discrete analogues of (5.48)–(5.50) related to (P_{2h}^k), $k = 1, 2$.

6. A Second Example of EVI of the Second Kind: The Flow of a Viscous Plastic Fluid in a Pipe

Most of the material in this section can be found in G.L.T. [2, Chapter 5], [3, Chapter 5] and in Glowinski [1], [3].

6.1. The continuous problem. Existence and uniqueness results

Let Ω be a bounded domain of $\mathbb{R}^2$ with a smooth boundary Γ. We define

$$V = H_0^1(\Omega),$$

$$a(u, v) = \int_\Omega \nabla u \cdot \nabla v \, dx,$$

$$L(v) = \langle f, v \rangle, \qquad f \in V^*,$$

$$j(v) = \int_\Omega |\nabla v| \, dx.$$

Let μ, g be two positive parameters; then:

Theorem 6.1. *The variational inequality*

$$\mu a(u, v - u) + gj(v) - gj(u) \geq L(v - u), \qquad \forall\, v \in V, \quad u \in V \tag{6.1}$$

has a unique solution.

PROOF. In order to apply Theorem 4.1 of Chapter I, we only have to verify that $j(\cdot)$ is convex, proper, and l.s.c..

It is obvious that $j(\cdot)$ is convex and proper. Let $u, v \in V$; then

$$|j(v) - j(u)| \leq \sqrt{\operatorname{meas}(\Omega)}\, \|u - v\|_V; \tag{6.2}$$

hence $j(\cdot)$ is l.s.c.

This proves the theorem. □

EXERCISE 6.1. Prove that $j(\cdot)$ is a norm on V.

Remark 6.1. If we take $g = 0$ in (6.1), we recover the variational formulation of the Dirichlet problem

$$-\mu\Delta u = f \text{ in } \Omega,$$
$$u = 0 \text{ on } \Gamma.$$

Remark 6.2. Since $a(\cdot, \cdot)$ is symmetrical, the solution u of (6.1) is characterized (using Lemma 4.1 of Chapter I) as the unique solution of the minimization problem

$$J(u) \leq J(v), \qquad \forall\, v \in V, \quad u \in V, \tag{6.3}$$

where $J(v) = (\mu/2)a(v, v) + gj(v) - L(v)$.

6.2. Physical motivation

If $L(v) = C \int_\Omega v\, dx$ (for instance, $C > 0$), as proved in Duvaut and Lions [1, Chapter 6], (6.1) models the *laminar stationary flow* of a *Bingham fluid* in a cylindrical pipe of cross section Ω, $u(x)$ being the velocity at $x \in \Omega$ (we refer to Prager [1], Germain [1], and Duvaut and Lions [1, Chapter 6] for the definition of Bingham fluids). The constant C is the *linear decay of pressure* and μ, g are, respectively, the *viscosity* and *plasticity yield* of the fluid. The above medium behaves like a viscous fluid (of viscosity μ) in

$$\Omega^+ = \{x \in \Omega, |\nabla u(x)| > 0\}$$

and like a rigid medium in $\Omega^0 = \{x \in \Omega, \nabla u(x) = 0\}$. We refer to Mossolov and Miasnikov [1], [2], [3] for a detailed study of the properties of Ω^+ and Ω^0. We observe that (6.1) also appears as a free boundary problem.

6.3. Regularity properties

Theorem 6.2 (H. Brezis [4]). *If $L(v) = \int_\Omega fv\, dx$, $f \in L^2(\Omega)$, then the solution u of* (6.1) *satisfies*

$$u \in V \cap H^2(\Omega),$$

and if Ω is convex, we have

$$\|u\|_{H^2(\Omega)} \leq \frac{\gamma(\Omega)}{\mu} \|f\|_{L^2(\Omega)}. \tag{6.4}$$

6.4. Further properties

Let us denote by α the following quantity:

$$\alpha = \inf_{\substack{v \in H^1(\Omega) \\ v \neq 0}} \frac{j(v)}{\|v\|_{L^1(\Omega)}}. \tag{6.5}$$

Then $\alpha > 0$.
From this we derive the following:

Proposition 6.1. *Let u be the solution of (6.1) and $f \in L^\infty(\Omega)$; then $u = 0$ if*

$$\|f\|_{L^\infty(\Omega)} \leq g\alpha. \tag{6.6}$$

Proof. By taking $v = 0$, $v = 2u$ in (6.1), we obtain

$$\mu a(u, u) + gj(u) = \int_\Omega fu\, dx. \tag{6.7}$$

It then follows from (6.5) and from $\int_\Omega fu\, dx \leq \|f\|_{L^\infty(\Omega)}\|u\|_{L^1(\Omega)}$ that

$$\mu a(u, u) + (g\alpha - \|f\|_{L^\infty(\Omega)})\|u\|_{L^1(\Omega)} \leq 0. \tag{6.8}$$

If f obeys (6.6), it then follows from (6.8) that $u = 0$. □

We also have:

Proposition 6.2. *Let u be the solution of (6.1) and $f \in L^p(\Omega)$, $p > 1$. Then if $f \geq 0$, we have $u \geq 0$.*

Exercise 6.2. Prove Proposition 6.2.

Proposition 6.3. *Let u be the solution of (6.2) and $f = c$, a constant. Then $u = 0$ if and only if $c \leq g\alpha$ and $u \neq 0$ if $c > g\alpha$.*

Exercise 6.3. Prove Proposition 6.3.

Proposition 6.4. *Let u be the solution of (6.1) and $f \in L^2(\Omega)$; then $u = 0$ if*

$$\|f\|_{L^2(\Omega)} \leq g\beta, \tag{6.9}$$

where

$$\beta = \inf_{\substack{v \in H_0^1(\Omega) \\ v \neq 0}} \frac{j(v)}{\|v\|_{L^2(\Omega)}}.$$

Proof. From a result of L. Nirenberg and M. Strauss (cf. Strauss [1]), it follows that $\beta > 0$.

By taking $v = 0$ and $v = u$ in (6.1), we obtain

$$\mu a(u, u) + gj(u) = \int_\Omega fu\, dx. \tag{6.10}$$

Using $\int_\Omega fu\,dx \leq \|f\|_{L^2(\Omega)}\|u\|_{L^2(\Omega)}$ and $\beta > 0$, we obtain

$$\mu a(u, u) + (g\beta - \|f\|_{L^2(\Omega)})\|u\|_{L^2(\Omega)} \leq 0. \tag{6.11}$$

Hence, if f satisfies (6.9), from (6.11) we have $a(u, u) = 0$. This implies $u = 0$, hence, the proposition is proved. □

6.5. Exact solutions

6.5.1. *Example* 1

We take $\Omega =]0, 1[$ and $f = c$, c a positive constant.

In this case the solution of

$$\mu \int_0^1 u(v' - u')\,dx + g \int_0^1 |v'|\,dx - g \int_0^1 |u'|\,dx \geq c \int_0^1 (v - u)\,dx,$$
$$\forall\, v \in H_0^1(\Omega), \quad u \in H_0^1(\Omega), \tag{6.12}$$

(where $v' = dv/dx$) is given by

$$u = 0 \text{ if } g \geq \frac{c}{2}. \tag{6.13}$$

If $g \leq c/2$,

$$\begin{aligned} u(x) &= \frac{c}{2\mu}x(1 - x) - \frac{gx}{\mu} && \text{if } 0 \leq x \leq \frac{1}{2} - \frac{g}{c}, \\ u(x) &= \frac{c}{2\mu}\left(\frac{1}{2} - \frac{g}{c}\right)^2 && \text{if } \frac{1}{2} - \frac{g}{c} \leq x \leq \frac{1}{2} + \frac{g}{c}, \\ u(x) &= \frac{c}{2\mu}x(1 - x) - \frac{g}{\mu}(1 - x) && \text{if } \frac{1}{2} + \frac{g}{c} \leq x \leq 1. \end{aligned} \tag{6.14}$$

We observe that if $g < c/2$, $u \in H_0^1(\Omega) \cap W^{2,\infty}(\Omega)$, but $u \notin H^3(\Omega)$.

6.5.2. *Example* 2

Let $\Omega = \{x \mid x_1^2 + x_2^2 < R^2\}$, $f = C$, C a positive constant. Then the solution of (6.1) is given by

$$u = 0 \quad \text{if } g \geq \frac{CR}{2}, \tag{6.15}$$

If $g \leq CR/2$, *then*

$$\begin{aligned} u(x) &= \left(\frac{R - r}{2\mu}\right)\left[\frac{C}{2}(R + r) - 2g\right] && \text{if } R' \leq r \leq R, \\ u(x) &= \left(\frac{R - R'}{2\mu}\right)\left[\frac{C}{2}(R + R') - 2g\right] && \text{if } 0 \leq r \leq R', \end{aligned} \tag{6.16}$$

where

$$r = \sqrt{x_1^2 + x_2^2}, \quad R' = \frac{2g}{C}.$$

We also observe that if $g < CR/2$, then $u \in H_0^1(\Omega) \cap W^{2,\infty}(\Omega)$, but $u \notin H^3(\Omega)$ (we actually have $u \in H^s(\Omega)$, $s < \frac{5}{2}$).

EXERCISE 6.4. Verify that (6.13), (6.14) and (6.15), (6.16) are solutions of (6.12) and (6.1), respectively.

6.6. Existence of multipliers

Let us define Λ by

$$\Lambda = \{q \mid q \in L^2(\Omega) \times L^2(\Omega), |q(x)| \leq 1 \text{ a.e.}\}$$

with

$$|q(x)| = \sqrt{q_1^2(x) + q_2^2(x)};$$

then we have:

Theorem 6.3. *The solution u of* (6.1) *is characterized by the existence of p such that*

$$\mu a(u, v) + g \int_\Omega p \cdot \nabla v \, dx = \langle f, v \rangle, \qquad \forall\, v \in V, \quad u \in V, \tag{6.17}$$

$$p \cdot \nabla u = |\nabla u| \text{ a.e.}, \qquad p \in \Lambda. \tag{6.18}$$

PROOF. There are classical proofs of the above theorem using Min-Max or Hahn–Banach Theorems (see, for instance, Cea [2, Chapter 5], G.L.T. [1, Chapter 1], and Ekeland and Temam [1]). In the sequel following G.L.T. [3, Chapter 5] we shall give an "almost constructive" proof, making use of a regularization technique.

(1) We first prove that (6.17), (6.18) imply (6.1). It follows from (6.17) that

$$\begin{aligned} \mu a(u, v - u) + g \int_\Omega p \cdot \nabla(v - u) \, dx &= \mu a(u, v - u) + g \int_\Omega p \cdot \nabla v \, dx - g \int_\Omega p \cdot \nabla u \, dx \\ &= \langle f, v - u \rangle, \qquad \forall\, v \in V. \end{aligned} \tag{6.19}$$

It follows from (6.18) that

$$\int_\Omega p \cdot \nabla u \, dx = \int_\Omega |\nabla u| \, dx, \tag{6.20}$$

and from the definition of Λ that

$$\int_\Omega p \cdot \nabla v \, dx \leq \int_\Omega |p||\nabla v| \, dx \leq \int_\Omega |\nabla v| \, dx, \qquad \forall\, v \in V. \tag{6.21}$$

Then, from (6.17), (6.19)–(6.21), we obtain

$$\mu a(u, v - u) + gj(v) - gj(u) \geq \langle f, v - u \rangle, \qquad \forall\, v \in V, \quad u \in V.$$

Thus (6.17) and (6.18) implies (6.1).

(2) *Necessity of* (6.17) *and* (6.18). Taking $u = 0$ and $v = 2u$ in (6.1), we obtain

$$\mu a(u, u) + gj(u) = \langle f, u \rangle. \tag{6.22}$$

Let $\varepsilon > 0$. Regularize $j(\cdot)$ by $j_\varepsilon(\cdot)$ defined by $j_\varepsilon(v) = \int_\Omega \sqrt{\varepsilon^2 + |\nabla v|^2}\, dx$. Since $j_\varepsilon(\cdot)$ is convex and continuous on V, the regularized problem

$$\mu a(u_\varepsilon, v - u_\varepsilon) + gj_\varepsilon(v) - gj_\varepsilon(u_\varepsilon) \geq L(v - u_\varepsilon), \qquad \forall\, v \in V, \quad u_\varepsilon \in V, \tag{6.23}$$

has a unique solution. Let us show that $\lim_{\varepsilon \to 0} u_\varepsilon = u$ strongly in V.

From (6.1) and (6.23) it follows that

$$\mu a(u_\varepsilon, u - u_\varepsilon) + gj_\varepsilon(u) - gj_\varepsilon(u_\varepsilon) \geq L(u - u_\varepsilon),$$

$$\mu a(u, u_\varepsilon - u) + gj(u_\varepsilon) - gj(u) \geq L(u_\varepsilon - u).$$

Adding these inequalities we obtain

$$\mu a(u_\varepsilon - u, u_\varepsilon - u) + g(j_\varepsilon(u_\varepsilon) - j(u_\varepsilon)) \leq g(j_\varepsilon(u) - j(u)). \tag{6.24}$$

From

$$0 < \sqrt{t^2 + \varepsilon^2} - |t| = \frac{\varepsilon^2}{\sqrt{\varepsilon^2 + t^2} + |t|} \leq \varepsilon, \qquad \forall\, t \in \mathbb{R},$$

it follows that $0 < j_\varepsilon(v) - j(v) \leq \varepsilon \text{ meas}(\Omega)$, $\forall\, v \in V$, which using (6.24), implies that $\mu a(u_\varepsilon - u, u_\varepsilon - u) \leq g\varepsilon \text{ meas}(\Omega)$, so that

$$\|u_\varepsilon - u\|_V \leq \sqrt{\text{meas}(\Omega)} \left(\frac{g}{\mu}\varepsilon\right)^{1/2}. \tag{6.25}$$

From (6.25) we obtain

$$\lim_{\varepsilon \to 0} u_\varepsilon = u \text{ strongly in } V. \tag{6.26}$$

Since $j_\varepsilon(\cdot)$ is differentiable on V, problem (6.23) is equivalent (see, e.g., Cea [1]) to the following nonlinear variational equation:

$$\mu a(u_\varepsilon, v) + g\langle j'_\varepsilon(u), v \rangle = L(v), \qquad \forall\, v \in V, \quad u_\varepsilon \in V, \tag{6.27}$$

with

$$\langle j'_\varepsilon(w), v \rangle = \int_\Omega \frac{\nabla w \cdot \nabla v}{\sqrt{\varepsilon^2 + |\nabla w|^2}}\, dx, \qquad \forall\, v, w \in V. \tag{6.28}$$

If we define p_ε by

$$p_\varepsilon = \frac{\nabla u_\varepsilon}{\sqrt{\varepsilon^2 + |\nabla u_\varepsilon|^2}}, \tag{6.29}$$

then

$$p_\varepsilon \in \Lambda. \tag{6.30}$$

From (6.27)–(6.30) we obtain

$$\mu a(u_\varepsilon, v) + g \int_\Omega p_\varepsilon \cdot \nabla v \, dx = L(v), \qquad \forall\, v \in V, \quad u_\varepsilon \in V. \tag{6.31}$$

Since Λ is a bounded closed convex subset of $L^2(\Omega) \times L^2(\Omega)$, it is *weakly compact*, so that from $(p_\varepsilon)_\varepsilon$ we can extract a subsequence, still denoted by $(p_\varepsilon)_\varepsilon$, such that

$$\lim_{\varepsilon \to 0} p_\varepsilon = p \text{ weakly in } L^2(\Omega) \times L^2(\Omega), \qquad p \in \Lambda. \tag{6.32}$$

Actually we have $p_\varepsilon \to p$ in $L^\infty(\Omega) \times L^\infty(\Omega)$ weakly $*$.

Taking the limit as $\varepsilon \to 0$ in (6.31), from (6.26) and (6.32) we have

$$\mu a(u, v) + g \int_\Omega p \cdot \nabla v \, dx = L(v), \qquad \forall\, v \in V, \quad u \in V, \tag{6.33}$$

so that (6.17) is proved.

To complete the proof of the theorem, we need only prove that

$$p \cdot \nabla u = |\nabla u| \text{ a.e.}. \tag{6.34}$$

Taking $v = u$ in (6.33) and comparing with (6.22), we obtain

$$\int_\Omega |\nabla u| \, dx - \int_\Omega p \cdot \nabla u \, dx = \int_\Omega (|\nabla u| - p \cdot \nabla u) \, dx = 0. \tag{6.35}$$

Since $p \in \Lambda$, it follows from Schwarz inequality in $\mathbb{R}^2$ that

$$p \cdot \nabla u \leq |\nabla u| \text{ a.e.}.$$

Combining (6.35) and this inequality we obtain

$$p \cdot \nabla u = |\nabla u| \text{ a.e.}.$$

This proves (6.18) and, hence, the theorem. □

Remark 6.3. The function p occurring in (6.17), (6.18) is *not unique* if $\Omega \in \mathbb{R}^2$; this is shown in G.L.T. [2, Chapter 5], [3, Chapter 5].

Remark 6.4. Relation (6.17) implies

$$\begin{aligned} -\mu \Delta u - g \nabla \cdot p &= f \text{ in } \Omega \\ u|_\Gamma &= 0. \end{aligned} \tag{6.36}$$

6.7. Finite element approximation of (6.1)

In this section we follow G.L.T. [3, Chapter 5]. For the sake of simplicity, we shall assume that Ω is a *polygonal* domain of $\mathbb{R}^2$.

6.7.1. *Definition of the approximate problem*

Let $\mathscr{T}_h$ be as in Sec. 2 of this chapter. We approximate V by

$$V_h = \{v_h \in C^0(\bar{\Omega}),\ v_h = 0 \text{ on } \Gamma,\ v_h|_T \in P_1, \quad \forall\, T \in \mathscr{T}_h\}$$

and (6.1) by

$$\mu a(u_h, v_h - u_h) + gj(v_h) - gj(u_h) \geq \langle f, v_h - u_h \rangle, \qquad \forall\, v_h \in V_h, \quad u_h \in V_h. \tag{6.37}$$

Then:

Theorem 6.4. *The approximate problem* (6.37) *has a unique solution.*

Remark 6.5. So far, only an approximation by piecewise linear finite elements has been considered. This fact is justified by the existence of a regularity limitation for the solutions of (6.1), which implies that even with very smooth data we may have $u \notin H^3(\Omega)$ (see Sec. 6.5). Nevertheless, in Fortin [1], Bristeau [1], G.L.T. [3, Chapter 5], and Bristeau and Glowinski [1], one may find applications of piecewise quadratic finite elements, straight or isoparametric, for solving (6.1). The numerical results which have been obtained seem to prove that for the same *number of degrees of freedom* the accuracies at the nodes are of the same order for the finite elements of order 1 and 2. From the above works it also appears that the second-order finite elements are much more costly to use (storage, computational time, etc.). We must also note that when using first-order finite elements, $\int_\Omega |\nabla v_h|\, dx$ can be expressed exactly with respect to the values of v_h at the nodes of $\mathcal{T}_h$, while with second-order finite elements we need a numerical integration procedure.

Remark 6.6. From the symmetry of $a(\cdot, \cdot)$, (6.37) is equivalent to the minimization problem

$$J(u_h) \leq J(v_h), \qquad \forall\, v_h \in V_h, \quad u_h \in V_h, \tag{6.38}$$

where

$$J(v_h) = \frac{\mu}{2} a(v_h, v_h) + g \int_\Omega |\nabla v_h|\, dx - \langle f, v_h \rangle. \tag{6.39}$$

6.7.2. *Convergence of the approximate solutions (general case)*

We use the notations of the previous sections.

Theorem 6.5. *If, as* $h \to 0$, *the angles of* $\mathcal{T}_h$ *are bounded from below uniformly in* h, *by* $\theta_0 > 0$, *then*

$$\lim_{h \to 0} \|u_h - u\|_V = 0, \tag{6.40}$$

where u *and* u_h *are, respectively, the solutions of* (6.1) *and* (6.37).

PROOF. In order to prove (6.40), we use Theorem 6.3 of Chapter I. Here we have to verify that the following three properties hold:

(i) There exist $U \subset V$, $\bar{U} = V$, and $r_h: U \to V_h$ such that $\lim_{h\to 0} r_h v = v$ strongly in V, $\forall\, v \in U$.
(ii) If $v_h \to v$ weakly in V as $h \to 0$, then $\liminf j_h(v_h) \geq j(v)$.
(iii) $\lim_{h\to 0} j_h(r_h v) = j(v)$, $\forall\, v \in U$.

Verification of (i). We take $U = \mathscr{D}(\Omega)$. Then $\bar{U} = H_0^1(\Omega) = V$. Define $r_h v$ by

$$\begin{aligned} r_h v \in V_h, &\qquad \forall\, v \in H_0^1(\Omega) \cap C^0(\bar{\Omega}),\\ (r_h v)(P) = v(P), &\qquad \forall\, P \in \mathring{\Sigma}_h. \end{aligned} \tag{6.41}$$

Then since $r_h v$ is the "linear" interpolate of v on $\mathscr{T}_h$, it follows from Ciarlet [1], [2], and Strang and Fix [1] that under the above assumptions on $\mathscr{T}_h$, we have

$$\|r_h v - v\|_{H^1(\Omega)} \leq Ch\|v\|_{W^{2,\infty}(\Omega)} \tag{6.42}$$

Then from (6.42) we find that $\lim_{h\to 0} r_h v = v$ strongly in $H_0^1(\Omega)$, $\forall\, v \in U$.

Verification of (ii). Since $j_h(v_h) = j(v_h)$, $\forall\, v_h \in V_h$, (ii) is trivially satisfied.

Verification of (iii). Since $j_h(v_h) = j(v_h)$, $\forall\, v_h \in V_h$ and from the continuity of $j(\cdot)$ on V, (iii) is trivially satisfied.

Hence from (i), (ii), and (iii), it follows that $u_h \to u$ strongly in V. □

6.7.3. *Convergence of the approximate solutions* ($f \in L^2(\Omega)$)

From the regularity theorem, Theorem 6.2 of this chapter, we have

$$\|u\|_{H^2(\Omega)} \leq \frac{\gamma_0(\Omega)}{\mu}\|f\|_{L^2(\Omega)} \tag{6.43}$$

if Ω is convex and Γ is sufficiently smooth. This property still holds if Ω is a convex polygonal set.

In this section we will always assume that Ω is a convex polygonal domain. We have the following:

Theorem 6.6. *With assumptions on $\mathscr{T}_h$ as in Theorem 6.5, we have*

$$\|u_h - u\|_V = O(h^{1/2}). \tag{6.44}$$

PROOF. From (6.1) and (6.37) it follows that $\mu a(u, u_h - u) + gj(u_h) - gj(u) \geq (f, u_h - u)$, $\mu a(u_h, v_h - u_h) + gj(v_h) - gj(u_h) \geq (f, v_h - u_h)$, $\forall\, v_h \in V_h$, which imply, by addition,

$$\begin{aligned} \mu a(u_h - u, u_h - u) \leq \mu a(u_h - u, v_h - u) + gj(v_h) - gj(u)\\ + \mu a(u, v_h - u) - (f, v_h - u), \qquad \forall\, v_h \in V_h. \end{aligned} \tag{6.45}$$

Since $j(\cdot)$ is a norm on V, using the Schwarz inequality in V and (6.45), we obtain

$$\frac{\mu}{2}\|u_h - u\|_V^2 \leq \frac{\mu}{2}\|v_h - u\|_V^2 + gj(v_h - u) + \mu a(u, v_h - u) - (f, v_h - u), \qquad \forall\, v_h \in V_h. \tag{6.46}$$

Since $u \in H_0^1(\Omega) \cap H^2(\Omega)$, we have

$$a(u, v) = -\int_\Omega \Delta uv\, dx, \qquad \forall\, v \in V.$$

Hence, from (6.46) we have

$$\frac{\mu}{2}\|u_h - u\|_V^2 \le \frac{\mu}{2}\|v_h - u\|_V^2 + gj(v_h - u) + \int_\Omega (-\mu\Delta u - f)(v_h - u)\, dx, \qquad \forall\, v_h \in V_h,$$

so that

$$\frac{\mu}{2}\|u_h - u\|_V^2 \le \frac{\mu}{2}\|v_h - u\|_V^2 + gj(v_h - u)$$

$$+ (\|\Delta u\|_{L^2(\Omega)} + \|f\|_{L^2(\Omega)})\|v_h - u\|_{L^2(\Omega)}, \qquad \forall\, v_h \in V_h. \tag{6.47}$$

Since $\|\Delta u\|_{L^2(\Omega)}$ is a norm equivalent to the $H^2(\Omega)$-norm over $H^2(\Omega) \cap H_0^1(\Omega)$, it follows from (6.2), (6.43) and (6.47) that

$$\frac{\mu}{2}\|u_h - u\|_V^2 \le \frac{\mu}{2}\|v_h - u\|_V^2 + g\sqrt{\text{meas}(\Omega)}\,\|v_h - u\|_V$$

$$+ (1 + \gamma_0(\Omega))\|f\|_{L^2(\Omega)}\|v_h - u\|_{L^2(\Omega)}, \qquad \forall\, v_h \in V_h. \tag{6.48}$$

Since Γ is Lipschitz continuous, we have (cf. Necas [1]) $H^2(\Omega) \subset C^0(\bar{\Omega})$. The solution u of (6.1) $\in H^2(\Omega) \subset C^0(\bar{\Omega})$; using the above inclusion, we can define $r_h u$ by

$$r_h u \in V_h, \qquad (r_h u)(P) = v(P), \qquad \forall\, P \in \mathring{\Sigma}_h.$$

From the above assumptions on $\mathscr{T}_h$, we have (cf. Ciarlet [1], [2] and Strang and Fix [1])

$$\|r_h u - u\|_V \le \gamma_1 h \|u\|_{H^2(\Omega)}, \tag{6.49}$$

$$\|r_h u - u\|_{L^2(\Omega)} \le \gamma_2 h^2 \|u\|_{H^2(\Omega)}, \tag{6.50}$$

where γ_1 and γ_2 are constants independent of h and u.

Taking $v_h = r_h u$ in (6.48), it follows from (6.43), (6.49), and (6.50) that

$$\frac{\mu}{2}\|u_h - u\|_V^2 \le \frac{\gamma_0\|f\|_{L^2(\Omega)}}{\mu}\left[\left(\frac{\gamma_1^2\gamma_0}{2} + (1 + \gamma_0)\gamma_2\right)\|f\|_{L^2}h^2 + g\sqrt{M(\Omega)}\gamma_1 h\right] \tag{6.51}$$

with $\gamma_0 = \gamma_0(\Omega)$ and $M(\Omega) = \text{meas}(\Omega)$. Hence, from (6.51) we have

$$\|u_h - u\| = O(h^{1/2}).$$

This proves the theorem. □

Remark 6.7. If Ω is an open interval of $\mathbb{R}$, if $f \in L^2(\Omega)$, and if piecewise linear elements are used, then an $O(h)$ estimate can be obtained for the approximation error; the proof of the above estimate is very close to the proof of Theorem 3.4 of Sec. 3.7.1 and is given in detail in G.L.T. [3, Appendix 5].

6.8. The case of a circular domain with f = constant

In this section we consider a particular case of the general problem (6.1) by taking

$$\Omega = \{x \in \mathbb{R}^2, \sqrt{x_1^2 + x_2^2} < R\}, \tag{6.52}$$

$$L(v) = C \int_\Omega v \, dx, \qquad C > 0. \tag{6.53}$$

6.8.1. *Exact solutions and regularity properties*

The solution of (6.1) corresponding to (6.52), (6.53) is given in Sec. 6.5.2 of this chapter. We recall that if $g < CR/2$, then

$$u \in V \cap W^{2,\infty}(\Omega), \qquad u \notin V \cap H^3(\Omega). \tag{6.54}$$

In the sequel we assume that $g < CR/2$.

6.8.2. *Approximation by finite element of order* 1

Let $\mathcal{T}_h$ be a finite triangulation of Ω satisfying (2.22), (2.23) of Sec. 2.5, Chapter II and

$$\forall\, T \in \mathcal{T}_h, \qquad T \subset \overline{\Omega}. \tag{6.55}$$

Define Ω_h and Γ_h by

$$\Omega_h = \bigcup_{T \in \mathcal{T}_h} T, \qquad \Gamma_h = \partial\Omega_h.$$

Then $\Omega_h \subset \Omega$, and in the sequel we assume that Γ_h satisfies:

$$\text{all the vertices of } \Gamma_h \text{ belong to } \Gamma. \tag{6.56}$$

Then we approximate V by

$$V_h = \{v_h \in C^0(\overline{\Omega}_h),\ v_h = 0 \text{ on } \Gamma_h,\ v_h|_T \in P_1,\ \forall\, T \in \mathcal{T}_h\}.$$

Now V_h can be considered to be a subspace of V, obtained by extending $v_h \in V_h$ to Ω by taking zero in $\Omega - \Omega_h$. It is then possible to approximate (6.1) by

$$\mu a(u_h, v_h - u_h) + gj(v_h) - gj(u_h) \geq C \int_\Omega (v_h - u_h)\, dx, \qquad \forall\, v_h \in V_h, \quad u_h \in V_h. \tag{6.57}$$

This is a finite-dimensional problem which has a unique solution.

6.8.3. *Error estimate*

In this section we will obtain an error estimate of order $h\sqrt{-\log h}$. The following three lemmas play an important role in obtaining the above error estimate.

Lemma 6.1. *Let* $\mathbf{p}, \mathbf{q} \in \mathbb{R}^2 - \{\mathbf{0}\}$. *Then*

$$\left|\frac{\mathbf{p}}{|\mathbf{p}|} - \frac{\mathbf{q}}{|\mathbf{q}|}\right| \leq 2\frac{|\mathbf{p} - \mathbf{q}|}{|\mathbf{p}| + |\mathbf{q}|}. \tag{6.58}$$

PROOF. We have

$$(|\mathbf{p}| + |\mathbf{q}|)\left(\frac{\mathbf{p}}{|\mathbf{p}|} - \frac{\mathbf{q}}{|\mathbf{q}|}\right) = (\mathbf{p} - \mathbf{q}) + \left(\frac{|\mathbf{q}|}{|\mathbf{p}|}\mathbf{p} - \frac{|\mathbf{p}|}{|\mathbf{q}|}\mathbf{q}\right).$$

But

$$\left|\frac{|\mathbf{q}|}{|\mathbf{p}|}\mathbf{p} - \frac{|\mathbf{p}|}{|\mathbf{q}|}\mathbf{q}\right|^2 = |\mathbf{p}|^2 + |\mathbf{q}|^2 - 2\mathbf{p}\cdot\mathbf{q} = |\mathbf{p} - \mathbf{q}|^2.$$

Consequently,

$$(|\mathbf{p}| + |\mathbf{q}|)\left|\frac{\mathbf{p}}{|\mathbf{p}|} - \frac{\mathbf{q}}{|\mathbf{q}|}\right| \leq 2|\mathbf{p} - \mathbf{q}|$$

which obviously implies (6.58). □

Remark 6.8. In (6.58), 2 is the best possible constant (take $\mathbf{p} = -\mathbf{q}$). Moreover, (6.58) is also true in $\mathbb{R}^N$, $\forall\ N$.

Lemma 6.2. *Let u and u_h be the solutions of* (6.1) *and* (6.57), *respectively. Let p satisfy* (6.17), (6.18); *then we have*

$$\mu a(u_h - u, u_h - u) \leq \mu a(u_h - u, v_h - u) + g\int_\Omega (p_h - p)\cdot\nabla(v_h - u)\,dx,$$

$$\forall\ v_h \in V_h, \quad \text{and} \quad \forall\ p_h \in \Lambda \quad \text{such that } p_h\cdot\nabla v_h = |\nabla v_h| \text{ a.e.}. \tag{6.59}$$

PROOF. We shall prove (6.59) with $f \in V^*$. From (6.45) we have

$$\mu a(u_h - u, u_h - u) \leq \mu a(u_h - u, v_h - u) + gj(v_h) - gj(u) + \mu a(u, v_h - u) - \langle f, v_h - u\rangle, \qquad \forall\ v_h \in V_h.$$

Taking into account (6.17), we obtain

$$\mu a(u_h - u, u_h - u) \leq \mu a(u_h - u, v_h - u) + gj(v_h) - gj(u) - g\int_\Omega p\cdot\nabla(v_h - u)\,dx, \qquad \forall\ v_h \in V_h. \tag{6.60}$$

Let $v_h \in V_h$ and $p_h \in \Lambda$ such that $p_h\cdot\nabla v_h = |\nabla v_h|$ a.e.; such a p_h always exists. Substituting this in (6.60) and using the relations

$$j(v_h) = \int_\Omega p_h\cdot\nabla v_h\,dx, \tag{6.61}$$

$$j(u) = \int_\Omega |\nabla u|\,dx \geq \int_\Omega p_h\cdot\nabla u\,dx, \tag{6.62}$$

we obtain (6.59).

This proves the lemma. □

Let u be the solution of (6.1) and $\delta > 0$. Define $\Omega^\delta \subset \Omega$ by

$$\Omega^\delta = \{x \in \Omega, |\nabla u(x)| > \delta\}.$$

In the case of the problem (6.1) associated with (6.52), (6.53) (assuming $g < CR/2$), we have:

Lemma 6.3. *We have the following identity*

$$\int_{\Omega^\delta} \frac{dx}{|\nabla u|} = \frac{4\pi\mu}{C}\left[-\frac{2\mu}{C}\delta + \left(R - \frac{2g}{C}\right) + \frac{2g}{C}\log\left(R - \frac{2g}{C}\right) - \frac{2g}{C}\log\frac{2\mu}{C}\delta\right]. \tag{6.63}$$

PROOF. From (6.16) we obtain

$$\left|\frac{du}{dr}\right| = \frac{1}{\mu}\left(\frac{Cr}{2} - g\right) \quad \text{if } \frac{2g}{C} \le r \le R,$$

so that

$$\int_{\Omega^\delta} \frac{dx}{|\nabla u|} = 2\pi\mu \int_{(2/C)(\mu\delta+g)}^{R} \frac{r\,dr}{Cr/2 - g},$$

which implies (6.63). □

From the above lemmas we deduce:

Theorem 6.7. *Let u be the solution of the problem (6.1) associated with (6.52), (6.53). Let u_h be the solution of the problem (6.57) with $\mathcal{T}_h$ satisfying (6.55), (6.56). Assume that as $h \to 0$, the angles of $\mathcal{T}_h$ are bounded from below uniformly in h by $\theta_0 > 0$. Then we have*

$$\|u_h - u\|_V = O(h\sqrt{-\log h}). \tag{6.64}$$

PROOF. Starting from Lemma 6.2, from (6.59) we obtain

$$\frac{\mu}{2}\|u_h - u\|_V^2 \le \frac{\mu}{2}\|r_h u - u\|_V^2 + g\int_\Omega |p_h - p||\nabla(r_h u - u)|\,dx, \qquad \forall\, p_h \in \Lambda$$

$$\text{such that } p_h \cdot \nabla r_h u = |\nabla r_h u|, \tag{6.65}$$

where $r_h u$ is defined by

$$r_h u \in V_h, \qquad (r_h u)(P) = u(P), \qquad \forall\, P \in \text{vertex of } \mathcal{T}_h.$$

We have $r_h u = 0$ on $\Omega - \Omega_h$ so that

$$\|r_h u - u\|_V^2 = \int_\Omega |\nabla(r_h u - u)|^2\,dx = \int_{\Omega-\Omega_h} |\nabla u|^2\,dx + \int_{\Omega_h} |\nabla(r_h u - u)|^2\,dx. \tag{6.66}$$

Let us define

$$X_1 = \frac{\mu}{2}\int_{\Omega-\Omega_h} |\nabla u|^2\, dx,$$

$$X_2 = \frac{\mu}{2}\int_{\Omega_h} |\nabla(r_h u - u)|^2\, dx.$$

It is easily shown that

$$\text{meas}(\Omega - \Omega_h) < \frac{\pi}{4} h^2. \tag{6.67}$$

Furthermore, (6.16) implies

$$|\nabla u(x)| \leq \frac{C}{2\mu}\left(R - \frac{2g}{C}\right), \qquad \forall\, x \in \Omega. \tag{6.68}$$

From (6.67) and (6.68) it follows that

$$X_1 \leq \frac{\pi}{32\mu} C^2\left(R - \frac{2g}{C}\right)^2 h^2. \tag{6.69}$$

Since $u \in W^{2,\infty}(\Omega)$, on each triangle $T \in \mathscr{T}_h$, we have (cf. Ciarlet and Wagshal [1])

$$|\nabla(r_h u - u)(x)| \leq \frac{2h}{\sin\theta_0} \|\rho(D_2 u)\|_{L^\infty(\Omega)}. \tag{6.70}$$

where $D_2 u(x)$ is the *Hessian matrix* of u at x, defined by

$$D_2 u(x) = \begin{pmatrix} \dfrac{\partial^2 u}{\partial x_1^2}(x) & \dfrac{\partial^2 u}{\partial x_1\, \partial x_2}(x) \\[2ex] \dfrac{\partial^2 u}{\partial x_1\, \partial x_2}(x) & \dfrac{\partial^2 u}{\partial x_2^2}(x) \end{pmatrix}$$

and $\rho(D_2 u(x))$ is the *spectral radius* of $D_2 u(x)$.

We have

$$D_2 u(x) = 0 \quad \text{if } 0 < r < \frac{2g}{C}, \tag{6.71}$$

so that $\rho(D_2 u(x)) = 0$, and it is easily verified that

$$\rho(D_2 u(x)) = \frac{C}{2\mu} \quad \text{if } \frac{2g}{C} < r < R. \tag{6.72}$$

Then (6.70)–(6.72) imply

$$X_2 \leq \frac{\pi}{\mu} R\left(R - \frac{2g}{C} + h\right) C^2 \left(\frac{h}{\sin\theta_0}\right)^2. \tag{6.73}$$

So it remains to estimate the term $g \int_\Omega |p_h - p||\nabla(r_h u - u)|\, dx$ in (6.65). We have

$$\bar{\Omega} = \bigcup_{i=3}^{6} \bar{\Omega}_i, \tag{6.74}$$

where

$$\Omega_3 = \Omega - \Omega_h,$$

$$\Omega_4 = \{x \in \Omega_h, r > 2g/C + h\},$$

$$\Omega_5 = \{x \in \Omega_h, 2g/C - h < r < 2g/C + h\},$$

$$\Omega_6 = \{x \in \Omega_h, 0 \le r < 2g/C - h\}.$$

Let us define, for $3 \le i \le 6$,

$$X_i = g \int_{\Omega_i} |p_h - p||\nabla(r_h u - u)|\,dx. \tag{6.75}$$

We have $r_h u = 0$ over $\Omega - \Omega_h$, so that in (6.65) we can take

$$p_h = 0 \quad \text{over } \Omega - \Omega_h. \tag{6.76}$$

From (6.67), (6.68), (6.70), and $p \in \Lambda$, it follows that

$$X_3 \le g \int_{\Omega - \Omega_h} |\nabla u|\,dx \le \frac{\pi}{8\mu}\, gC\left(R - \frac{2g}{C}\right)h^2. \tag{6.77}$$

From (6.70)–(6.72), and since p, $p_h \in \Lambda$, we have

$$X_5 \le 2g\,\frac{C}{\mu}\,\text{meas}(\Omega_5)\,\frac{h}{\sin\theta_0},$$

so that

$$X_5 \le \frac{16\pi}{\mu}\, g^2\,\frac{h^2}{\sin\theta_0}. \tag{6.78}$$

From the definition of h (h = maximal length of the edges of $T \in \mathscr{T}_h$), we have $r_h u = u =$ constant over Ω_6, so that

$$X_6 = 0. \tag{6.79}$$

It remains to estimate X_4. Since the equipotential lines of u in Ω_4 are circular, for h sufficiently small, from (6.16) we have

$$r_h u|_T \ne \text{constant}, \qquad \forall\, T \in \mathscr{T}_h, \quad T \subset \bar{\Omega}_4,$$

$$\text{so that } \nabla r_h u_T \ne 0, \qquad \forall\, T \in \mathscr{T}_h, \quad T \subset \bar{\Omega}_4. \tag{6.80}$$

Taking into account (6.80), it follows from (6.65) that

$$p_h|_T = \left.\frac{\nabla r_h u}{|\nabla r_h u|}\right|_T, \qquad \forall\, T \in \mathscr{T}_h, \quad T \subset \bar{\Omega}_4. \tag{6.81}$$

Furthermore, we observe that (6.16) implies

$$|\nabla u(x)| \ge \frac{Ch}{2\mu} > 0, \qquad \forall\, x \in \Omega_4. \tag{6.82}$$

This, in turn, implies

$$p = \frac{\nabla u(x)}{|\nabla u(x)|}, \qquad \forall\, x \in \Omega_4.$$

Applying Lemma 6.1 to the pair $\{\nabla r_h u, \nabla u\}$ and Lemma 6.3 with $\delta = Ch/2\mu$, it follows from (6.80)–(6.82) that

$$X_4 \le g\|\nabla(r_h u - u)\|_\infty^2 \int_{\Omega_4} \frac{dx}{|\nabla u| + |\nabla r_h u|},$$

where

$$\|\nabla v\|_\infty = \|\nabla v\|_{L^\infty(\Omega) \times L^\infty(\Omega)}.$$

From (6.70) and (6.82) it follows that

$$X_4 \le g \frac{C^2}{\mu^2}\left(\frac{Ch}{\sin\theta_0}\right)^2 \int_{\Omega^\delta} \frac{dx}{|\nabla u|},$$

which, using (6.63), implies

$$X_4 \le \frac{4\pi}{\mu} gC\left(\frac{h}{\sin\theta_0}\right)^2\left[-h + \left(R - \frac{2g}{C}\right) + \frac{2g}{C}\log\left(R - \frac{2g}{C}\right) - \frac{2g}{C}\log h\right],$$

or, more simply,

$$X_4 \le \frac{4\pi}{\mu} gC\left(\frac{h}{\sin\theta_0}\right)^2\left(R - \frac{2g}{C}\log\frac{h}{R}\right). \tag{6.83}$$

Taking (6.65) into account, the estimate (6.64) is obtained by addition of the X_i, $i = 1, \ldots, 6$. More precisely, for sufficiently small h,

$$\|u_h - u\|_V \le 4\frac{g}{\mu}\sqrt{\pi}\frac{h}{\sin\theta_0}\sqrt{-\log h}. \tag{6.84}$$

□

6.8.4. *Generalization*

From the numerical experiment we have done, it seems that in a great number of cases (important from the point of view of application) we have the following properties for u:

(1) $u \in V \cap W^{2,\infty}(\Omega)$,
(2) $\Omega_0 = \{x \mid \nabla u(x) = 0\}$ is a compact subset of Ω with a smooth boundary,
(3) Ω_0 has a finite number of connected components.

Moreover, it seems that in the above cases we can conjecture that for $\delta > 0$, we still have

$$\int_{\Omega^\delta} \frac{dx}{|\nabla u(x)|} = O(-\log\delta). \tag{6.85}$$

With these properties we can easily prove the following error estimate:

$$\|u_h - u\|_V = O(h\sqrt{-\log h}).$$

Remark 6.9. Using an equivalent formulation of (6.1) (less suitable for computations) Falk and Mercier [1] have obtained an $O(h)$ estimate for $\|u_2 - u\|_{H^1(\Omega)}$ for a piecewise linear approximation (see also G.L.T. [3, Appendix 5]).

6.9. Iterative solution of the continuous and approximate problems by Uzawa's algorithm

We begin with the continuous problem (6.1). Let us define $\mathscr{L}: V \times H \to \mathbb{R}$ by

$$\mathscr{L}(v, q) = \frac{\mu}{2} a(v, v) - L(v) + g \int_\Omega q \cdot \nabla v \, dx, \qquad \forall\, v \in V, \quad \forall\, q \in H,$$

where $H = L^2(\Omega) \times L^2(\Omega)$.

Let $\{u, p\}$ be the solution of (6.17), (6.18). Then we have:

Theorem 6.8. *The pair $\{u, p\}$ is a saddle point of $\mathscr{L}$ over $V \times \Lambda$ if and only if $\{u, p\}$ satisfies* (6.17) *and* (6.18).

EXERCISE 6.5. Prove Theorem 6.8.

From Cea [2, Chapter 5] (see also G.L.T. [3, Chapter 5]) it follows that to solve (6.1) we can use the following Uzawa algorithm:

$$p^0 \in \Lambda \textit{ arbitrarily chosen (for example, } p^0 = 0\textit{):} \tag{6.86}$$

then, by induction, knowing p^n, we compute u^n and p^{n+1} by

$$\mu a(u^n, v) = \langle f, v \rangle - g \int_\Omega p^n \cdot \nabla v \, dx, \qquad \forall\, v \in V, \quad u^n \in V, \tag{6.87}$$

$$p^{n+1} = P_\Lambda(p^n + \rho g \nabla u^n), \tag{6.88}$$

where $P_\Lambda: H \to \Lambda$ is the projection operator in the H-norm, defined by

$$P_\Lambda(q) = \frac{q}{\operatorname{Sup}(1, |q|)}.$$

Since u^n is a solution of (6.87), u^n is actually the unique solution in V of

$$\begin{aligned} -\mu \Delta u^n &= f + g \nabla \cdot p^n, \\ u^n|_\Gamma &= 0. \end{aligned} \tag{6.89}$$

We shall give a direct proof for the convergence of (6.86)–(6.88) based on Theorem 6.3 of Sec. 6.6.

Theorem 6.9. *Let u^n be the solution of* (6.87). *Then if*

$$0 < \rho < \frac{2\mu}{g^2}, \tag{6.90}$$

we have

$$\lim_{n\to\infty} \|u_n - u\|_V = 0, \tag{6.91}$$

where u is the solution of (6.1).

PROOF. Let $\{u, p\}$ satisfies (6.17) and (6.18). Then (6.18) implies

$$p = P_\Lambda(p + \rho g \nabla u). \tag{6.92}$$

We define $\bar{u}^n = u^n - u$, $\bar{p}^n = p^n - p$. Using the fact that P_Λ is a contraction mapping, and from (6.88), (6.92), we obtain

$$|\bar{p}^{n+1}|^2 \leq |\bar{p}^n|^2 + 2\rho g \int_\Omega \bar{p}^n \cdot \nabla \bar{u}^n \, dx + \rho^2 g^2 \int_\Omega |\nabla \bar{u}^n|^2 \, dx, \tag{6.93}$$

where

$$|q| = \|q\|_{L^2(\Omega) \times L^2(\Omega)}.$$

From (6.17) and (6.87) it follows that

$$\mu a(\bar{u}^n, v) + g \int_\Omega \bar{p}^n \cdot \nabla v \, dx = 0, \qquad \forall\, v \in V. \tag{6.94}$$

Replacing v by $\bar{u}^n$ in (6.94), we obtain

$$\mu a(\bar{u}^n, \bar{u}^n) + g \int_\Omega \bar{p}^n \cdot \nabla \bar{u}^n \, dx = 0. \tag{6.95}$$

From (6.93) and (6.95) we have

$$|\bar{p}^n|^2 - |\bar{p}^{n+1}|^2 \geq \rho(2\mu - \rho g^2)\|\bar{u}^n\|_V^2. \tag{6.96}$$

If $0 < \rho < 2\mu/g^2$, then using standard reasoning, we obtain

$$\lim_{n\to\infty} \|\bar{u}^n\|_V = 0,$$

which proves the theorem. □

Let us describe the adaptation of (6.86)–(6.88) to the approximate problem (6.37). We define $L_h \subset L^2(\Omega) \times L^2(\Omega)$ by

$$L_h = \left\{ q_h \,|\, q_h = \sum_{T \in \mathscr{T}_h} q_T X_T,\, q_T \in \mathbb{R}^2,\, \forall\, T \in \mathscr{T}_h \right\}$$

where X_T is the characteristic function of T.

It is then clear that $\forall\, v_h \in V_h$, $\nabla v_h \in L_h$. We also define Λ_h by $\Lambda_h = \Lambda \cap L_h$. We can easily prove that

$$P_{\Lambda_h}(q_h) = P_\Lambda(q_h), \qquad \forall\, q_h \in L_h.$$

Then (6.86)–(6.88) is approximated by:

$$p_h^0 \in \Lambda_h \text{ arbitrarily chosen,} \tag{6.97}$$

by induction, knowing p_h^n, we obtain u_h^n and p_h^{n+1} by

$$\mu a(u_h^n, v_h) = L(v_h) - g \int_\Omega p_h^n \cdot \nabla v_h \, dx, \quad \forall v_h \in V_h, \quad u_h^n \in V_h, \tag{6.98}$$

$$p_h^{n+1} = P_\Lambda(p_h^n + \rho g \nabla u_h^n). \tag{6.99}$$

Then for $0 < \rho < 2\mu/g^2$ we obtain the convergence of u_h^n to u_h.

EXERCISE 6.6. Study the convergence of (6.97)–(6.99).

Remark 6.10. The above methods have been numerically applied for solving (6.1) in Cea and Glowinski [2], Fortin [1], Bristeau [1], Bristeau and Glowinski [1], and G.L.T. [3, Chapter 5]. They appear to be very efficient and particularly well suited for taking nondifferentiable functionals like $\int_\Omega |\nabla v| \, dx$ into account.

7. On Some Useful Formulae

Let T be the triangle of Fig. 7.1. We denote by $M(T)$ the measure of T.

Let v be a smooth function defined on T. We define v_i and v_{jk} by

$$v_i = v(M_i), \qquad v_{jk} = v(M_{jk}).$$

Then we have the following formulae:

$$\int_T uv \, dx = \frac{M(T)}{12} \{(u_1 + u_2)(v_1 + v_2) + (u_2 + u_3)(v_2 + v_3) + (u_3 + u_1)(v_3 + v_1)\}, \quad \forall u, v \in P_1, \tag{7.1}$$

$$|\nabla v|^2 = \frac{1}{4M(T)^2} \{|\overrightarrow{M_2M_3}|^2 v_1^2 + |\overrightarrow{M_3M_1}|^2 v_2^2 + |\overrightarrow{M_1M_2}|^2 v_3^2 + \overrightarrow{2M_2M_3} \cdot \overrightarrow{M_3M_1} v_1 v_2 + \overrightarrow{2M_1M_2} \cdot \overrightarrow{M_2M_3} v_3 v_1 + \overrightarrow{2M_3M_1} \cdot \overrightarrow{M_1M_2} v_2 v_3\}, \quad \forall v \in P_1, \tag{7.2}$$

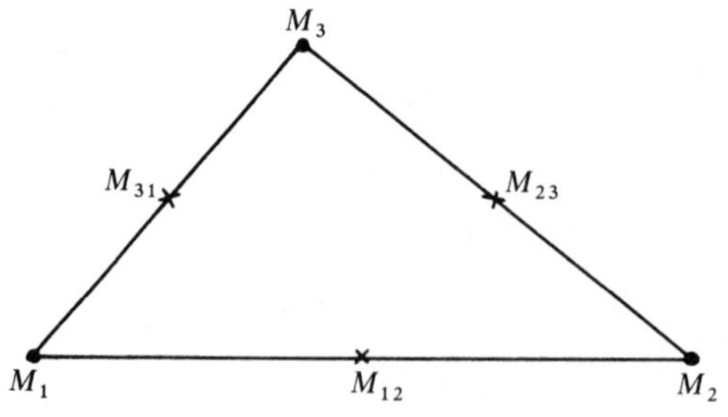

Figure 7.1

$$\int_T |v|^2\,dx = \frac{M(T)}{3}\{\tfrac{1}{10}(v_1^2 + v_2^2 + v_3^2) + \tfrac{8}{15}(v_{12}^2 + v_{23}^2 + v_{31}^2)$$

$$- \tfrac{1}{30}(v_1v_2 + v_2v_3 + v_3v_1) + \tfrac{8}{15}(v_{12}v_{23} + v_{23}v_{31} + v_{31}v_{12})$$

$$- \tfrac{2}{15}(v_1v_{23} + v_2v_{31} + v_3v_{12})\}, \qquad \forall\, v \in P_2, \tag{7.3}$$

$$\int_T |\nabla v|^2\,dx = \frac{1}{12M(T)}\{|v_1\overrightarrow{M_2M_3} - v_2\overrightarrow{M_3M_1} + v_3\overrightarrow{M_1M_2}$$

$$+ 2(v_{12} + v_{23} - v_{31})\overrightarrow{M_3M_1}|^2 + |v_1\overrightarrow{M_2M_3} + v_2\overrightarrow{M_3M_1}$$

$$- v_3\overrightarrow{M_1M_2} + 2(v_{23} + v_{31} - v_{12})\overrightarrow{M_1M_2}|^2 + |-v_1\overrightarrow{M_2M_3}$$

$$+ v_2\overrightarrow{M_3M_1} + v_3\overrightarrow{M_1M_2} + 2(v_{31} + v_{12} - v_{23})\overrightarrow{M_2M_3}|^2\},$$

$$\forall\, v \in P_2. \tag{7.4}$$

The above formula may be useful for expressing the approximations of the problems of this chapter, in a form suitable for computations.

CHAPTER III

On the Approximation of Parabolic Variational Inequalities

1. Introduction: References

In this chapter we would like to give some indications of the approximation of *parabolic variational inequalities* (PVI) (mostly without proof). For a detailed reference, see Fortin [1], Bristeau [1], Bristeau and Glowinski [1], C. Johnson [1], and A. Berger [1].[1] See also Richtmyer and Morton [1], Kreiss [1], and Lascaux [1] for the numerical analysis of *time-dependent equations*.

2. Formulation and Statement of the Main Results

Let H and V be two real Hilbert spaces such that $V \subset H$, $\overline{V} = H$. Assuming that $H = H^*$, we have $V \subset H \subset V^*$.

The scalar product in H (resp., in V) and the corresponding norms are denoted by $(\cdot, \cdot)$, $|\cdot|$ (resp., $((\cdot, \cdot))$, $\|\cdot\|$). Moreover, we also use $(\cdot, \cdot)$ for the duality between V^* and V.

We now introduce:

- A time interval $[0, T]$ with $0 < T < \infty$, a bilinear form $a: V \times V \to \mathbb{R}$, continuous and *elliptical* in the following sense: $\exists\, \alpha > 0$ and $\lambda \geq 0$ such that

$$a(v, v) + \lambda |v|^2 \geq \alpha \|v\|^2, \qquad \forall\, v \in V,$$

- $f \in L^2(0, T; V^*)$, $u^0 \in H$ (for the definition of $L^2(0, T; X)$, see Lions [1], [3]);
- K: closed convex nonempty subset of V,
- $j: V \to \overline{\mathbb{R}}$ convex, proper, l.s.c.

We then consider the following two families of PVI:
Find $u(t)$ such that

$$\left(\frac{\partial u}{\partial t}, v - u\right) + a(u, v - u) \geq (f, v - u), \qquad \forall\, v \in K, \text{ a.e. } t \in]0, T[, \quad (2.1)$$

$$u(t) \in K \text{ a.e. } t \in]0, T[,\ u(0) = u_0.$$

[1] See also G.L.T. [3, Appendix 6] and the references therein.

Find $u(t)$ such that

$$\left(\frac{\partial u}{\partial t}, v - u\right) + a(u, v - u) + j(v) - j(u) \geq (f, v - u),$$
$$\forall\, v \in V, \text{ a.e. } t \in]0, T[,$$
$$u(t) \in V \text{ a.e. } t \in]0, T[,\ u(0) = u_0. \tag{2.2}$$

Remark 2.1. If $K = V$ and $j \equiv 0$, then (2.1) and (2.2) reduce to the standard parabolic variational equation:

$$\left(\frac{\partial u}{\partial t}, v\right) + a(u, v) = (f, v), \qquad \forall\, v \in V \text{ a.e. in } t \in]0, T[,$$
$$u(t) \in V \text{ a.e. } t \in]0, T[, \qquad u(0) = u_0. \tag{2.3}$$

Under appropriate assumptions on u_0, K, and $j(\cdot)$, it is proved that (2.1), (2.2) have unique solutions in $L^2(0, T; V) \cap C^0([0, T], H)$. For the proof we refer to Brezis [4], [5], Lions [1], and Duvaut and Lions [1].

In the following sections of this chapter, we would like to give some discretization schemes for (2.1) and (2.2), and then, in Sec. 6, study the asymptotic properties over time of a specific example, for the continuous and discrete cases.

3. Numerical Schemes for Parabolic Linear Equations

Let us assume that V and H have been approximated (as $h \to 0$) by the same family $(V_h)_h$ of closed subspaces of V (in practice, V_h are finite dimensional). We also approximate $(\cdot, \cdot)$, $a(\cdot, \cdot)$ by $(\cdot, \cdot)_h$, $a_h(\cdot, \cdot)$ in such a way that ellipticity, symmetry, etc. are preserved. We also assume that u_0 is approximated by $(u_{0h})_h$ such that $u_{0h} \in V_h$ and $\lim_{h\to 0} u_{0h} = u_0$ strongly in H.

We now introduce a *time step* Δt; then, denoting u_h^n the approximation of u at time $t = n\Delta t$ $(n = 0, 1, 2, \ldots)$, we approximate (2.3) using the classical step-by-step numerical schemes (i.e., we describe how to compute u_h^{n+1} if u_h^n and u_h^{n-1} are known).

3.1. Explicit scheme (Euler's scheme)

$$\left(\frac{u_h^{n+1} - u_h^n}{\Delta t}, v_h\right)_h + a_h(u_h^n, v_h) = (f_h^n, v_h)_h, \qquad \forall\, v_h \in V_h,$$
$$n = 0, 1, \ldots;\quad u_h^0 = u_{0h}. \tag{3.1}$$

Stability: *Conditional* (See Lascaux [1] for the terminology).
Accuracy: $O(\Delta t)$ (we consider only the influence of the time discretization).

3.2. Ordinary implicit scheme (backward Euler's scheme)

$$\left(\frac{u_h^{n+1} - u_h^n}{\Delta t}, v_h\right)_h + a_h(u_h^{n+1}, v_h) = (f_h^{n+1}, v_h)_h, \qquad \forall\, v_h \in V_h,$$

$$n = 0, 1, 2, \ldots; \quad u_h^0 = u_{0h}. \tag{3.2}$$

Stability: *Unconditional.*
Time accuracy: $O(\Delta t)$.

3.3. Crank–Nicholson scheme

$$\left(\frac{u_h^{n+1} - u_h^n}{\Delta t}, v_h\right)_h + a_h\left(\frac{u_h^{n+1} + u_h^n}{2}, v_h\right)$$

$$= \left(\frac{f_h^{n+1} + f_h^n}{2}, v_h\right)_h, \qquad \forall\, v_h \in V_h,$$

$$= (f_h^{n+1/2}, v_h)_h, \qquad \forall\, v_h \in V_h, \quad n = 0, 1, 2, \ldots; \quad u_h^0 = u_{0h}. \tag{3.3}$$

Stability: *Unconditional.*
Time accuracy: $O(|\Delta t|^2)$.

3.4. Two-step implicit scheme

$$\left(\frac{\frac{3}{2}u_h^{n+1} - 2u_h^n + \frac{1}{2}u_h^{n-1}}{\Delta t}, v_h\right)_h + a_h(u_h^{n+1}, v_h)$$

$$= (f_h^{n+1}, v_h)_h, \qquad \forall\, v_h \in V_h, \quad n = 1, 2, \ldots; \quad u_h^0 = u_{0h},\ u_h^1 \text{ given.} \tag{3.4}$$

Stability: *Unconditional.*
Time accuracy: $O(|\Delta t|^2)$.

Unlike the three previous schemes, this latter scheme requires the use of a starting procedure to obtain u_h^1 from $u_h^0 = u_{0h}$; to compute u_h^1 we can use, for example, one of the schemes (3.1), (3.2), or (3.3); we recommend (3.3), since it is also an $O(|\Delta t|^2)$-scheme. Similarly, the generalizations of the scheme (3.4) discussed in Secs. 4 and 5 will require the use of a starting procedure which can be the corresponding generalization of schemes (3.1), (3.2), or (3.3).

3.5. Remarks

Remark 3.1. The function f_h^n (or $f_h^{n+1/2}$) occurring in the right-hand sides of (3.1)–(3.4) is a convenient approximation of f at $t = n\Delta t$ (or $t = (n + \frac{1}{2})\Delta t$).

In some cases it may be defined as follows (we only consider f_h^n since the technique described below is also applicable to $f_h^{n+1/2}$).

First we define $f^n \in V^*$ by

$$f^n = f(n\,\Delta t) \quad \text{if } f \in C^0[0, T; V^*].$$

In the general case, it is defined by

$$f^0 = \frac{2}{\Delta t}\int_0^{\Delta t/2} f(t)\,dt,$$

$$f^n = \frac{1}{\Delta t}\int_{(n-1/2)\,\Delta t}^{(n+1/2)\,\Delta t} f(t)\,dt \quad \text{if } n \geq 1.$$

Then, since $(\cdot, \cdot)_h$ is a scalar product on V_h, one may define f_h^n by

$$(f_h^n, v_h)_h = (f^n, v_h), \qquad \forall\, v_h \in V_h, \quad f_h^n \in V_h.$$

In some cases we have to use more sophisticated methods to define f_h^n.

Remark 3.2. At each step $(n + 1)$, we have to solve a linear system to compute u_h^{n+1}; however, if we can use a scalar product $(\cdot, \cdot)_h$ leading to a *diagonal matrix*, with regard to the variables defining v_h, then the use of the explicit scheme will only require us to solve *linear equations in one variable* at each step.

Remark 3.3. We can also use nonconstant time steps Δt_n.

Remark 3.4. If we are interested in the numerical integration of *stiff phenomenon* or in *long-range integration*, we can briefly say:

- Schemes (3.1), (3.2) are *too* dissipative; moreover the stability condition in (3.1) may be a serious drawback.
- Scheme (3.3) is, in some sense, *not sufficiently dissipative.*
- Scheme (3.4) avoids the above inconveniences and is highly recommended for "stiff" problems and long-range integration. In most cases the extra storage it requires is not a serious drawback.

Remark 3.5. There are many works related to the numerical analysis of parabolic equations via finite differences in time and finite elements in space approximations. We refer to Raviart [1], [2], Crouzeix [1], Strang and Fix [1], Oden and Reddy [1, Chapter 9], and the bibliographies therein.

4. Approximation of PVI of the First Kind

We assume that K in (2.1) has been approximated by $(K_h)_h$, $K_h \subset V_h$, $\forall\, h$, as in the elliptic case (see Chapter I). We also suppose that the bilinear form $a(\cdot, \cdot)$ is possibly dependent on the time t and has been approximated by $a_h(t; u_h, v_h)$.

4.1. Explicit scheme

$$\left(\frac{u_h^{n+1}-u_h^n}{\Delta t}, v_h - u_h^{n+1}\right)_h + a_h(n\,\Delta t; u_h^n, v_h - u_h^{n+1})$$
$$\geq (f_h^n, v_h - u_h^{n+1})_h, \qquad \forall\, v_h \in K_h, \quad u_h^{n+1} \in K_h, \quad n = 0, 1, 2, \ldots; \quad u_h^0 = u_{0h}. \tag{4.1}$$

Stability: *Conditional* (see G.L.T. [3, Chapter 6]).
This scheme is almost never used in practice, since it is conditionally stable and the computation of u_h^{n+1} will generally, require the use of an iterative method, *even* if the matrix corresponding to $(\cdot, \cdot)_h$ is *diagonal*.

4.2. Ordinary implicit scheme

$$\left(\frac{u_h^{n+1}-u_h^n}{\Delta t}, v_h - u_h^{n+1}\right)_h + a_h((n+1)\,\Delta t; u_h^{n+1}, v_h - u_h^{n+1})$$
$$\geq (f_h^{n+1}, v_h - u_h^{n+1})_h,$$
$$\forall\, v_h \in K_h, \quad u_h^{n+1} \in K_h, \quad n = 0, 1, 2, \ldots; \quad u_h^0 = u_{0h}. \tag{4.2}$$

Stability: *Unconditional*.
At each step we have to solve an EVI of the first kind related to K_h to compute u_h^{n+1}. This scheme is often used in practice.

4.3. Crank–Nicholson scheme

$$\left(\frac{u_h^{n+1}-u_h^n}{\Delta t}, v_h - u_h^{n+1/2}\right)_h + a_h((n+\tfrac{1}{2})\,\Delta t; u_h^{n+1/2}, v_h - u_h^{n+1/2})$$
$$\geq (f_h^{n+1/2}, v_h - u_h^{n+1/2})_h, \qquad \forall\, v_h \in K_h,$$
$$u_h^{n+1/2} \in K_h, \quad u_h^{n+1/2} = \frac{u_h^n + u_h^{n+1}}{2}, \quad n = 0, 1, 2, \ldots; \quad u_h^0 = u_{0h}. \tag{4.3}$$

Stability: *Unconditional*.
Since $(u_h^{n+1} - u_h^n)/\Delta t = (u_h^{n+1/2} - u_h^n)/(\Delta t/2)$, we observe that at each step we have to solve an EVI of the first kind to compute $u_h^{n+1/2}$. We also observe that possibly $u_h^n \notin K_h$. We do not recommend this scheme if the regularity over time of the continuous solution is poor.

4.4. Two-step implicit scheme

$$\left(\frac{\frac{3}{2}u_h^{n+1} - 2u_h^n + \frac{1}{2}u_h^{n-1}}{\Delta t}, v_h - u_h^{n+1}\right)_h + a_h((n+1)\,\Delta t; u_h^{n+1}, v_h - u_h^{n+1})$$
$$\geq (f_h^{n+1}, v_h - u_h^{n+1})_h, \qquad \forall\, v_h \in K_h, u_h^{n+1} \in K_h,$$
$$n = 1, 2, \ldots; \quad u_h^0 = u_{0h}, u_h^1 \text{ given}. \tag{4.4}$$

Stability: *Unconditional.*

At each step we have to solve an EVI of the first kind in K_h to compute u_h^{n+1}. Remark 3.4 also applies to this scheme.

5. Approximation of PVI of the Second Kind

5.1. Explicit scheme

$$\left(\frac{u_h^{n+1} - u_h^n}{\Delta t}, v_h - u_h^{n+1}\right) + a_h(n\,\Delta t; u_h^n, v_h - u_h^{n+1}) + j_h(v_h) - j_h(u_h^{n+1})$$
$$\geq (f_h^n, v_h - u_h^{n+1})_h, \qquad \forall\, v_h \in V_h, \quad u_h^{n+1} \in V_h,$$
$$n = 0, 1, 2, \ldots; \quad u_h^0 = u_{0h}. \tag{5.1}$$

Stability: *Conditional.*

This scheme also, is, almost never used in practice since it is conditionally stable and the computation of u_h^{n+1} will require the solution of an EVI of the second kind in V_h (in general, by an iterative method) even if the matrix corresponding to $(\cdot, \cdot)_h$ is diagonal.

5.2. Implicit scheme

$$\left(\frac{u_h^{n+1} - u_h^n}{\Delta t}, v_h - u_h^{n+1}\right)_h + a_h((n + 1)\,\Delta t; u_h^{n+1}, v_h - u_h^{n+1}) + j_h(v_h) - j_h(u_h^{n+1})$$
$$\geq (f_h^{n+1}, v_h - u_h^{n+1})_h, \qquad \forall\, v_h \in V_h, \quad u_h^{n+1} \in V_h,$$
$$n = 0, 1, 2, \ldots; \quad u_h^0 = u_{0h}. \tag{5.2}$$

Stability: *Unconditional.*

At each step we have to solve in V_h an EVI of the second kind to compute u_h^{n+1}.

5.3. Crank–Nicholson scheme

$$\left(\frac{u_h^{n+1} - u_h^n}{\Delta t}, v_h - u_h^{n+1/2}\right)_h + a_h((n + \tfrac{1}{2})\,\Delta t; u_h^{n+1/2}, v_h - u_h^{n+1/2}) + j_h(v_h)$$
$$- j_h(u_h^{n+1/2}) \geq (f_h^{n+1/2}, v_h - u_h^{n+1/2})_h, \quad \forall\, v_h \in V_h, \quad u_h^{n+1/2} \in V_h,$$
$$u_h^{n+1/2} = \frac{u_h^n + u_h^{n+1}}{2}, \quad n = 0, 1, 2, \ldots; \quad u_0^h = u_{0h}. \tag{5.3}$$

Stability: Conditional.

Since $(u_h^{n+1} - u_h^n)/\Delta t = (u_h^{n+1/2} - u_h^n)/(\Delta t/2)$, we observe that at each step we have to solve an EVI of the second kind to compute $u_h^{n+1/2}$. If the regularity over time of the solution is poor, we do not recommend this scheme.

5.4. Two-step implicit scheme

$$\left(\frac{\frac{3}{2}u_h^{n+1} - 2u_h^n + \frac{1}{2}u_h^{n-1}}{\Delta t}, v_h - u_h^{n+1}\right)_h + j_h(v_h) - j_h(u_h^{n+1})$$
$$+ a_h((n+1)\,\Delta t; u_h^{n+1}, v_h - u_h^{n+1}) \geq (f_h^{n+1}, v_h - u_h^{n+1}),$$
$$\forall\, v_h \in V_h, \quad n = 0, 1, 2, \ldots; \quad u_h^0 = u_{0h}, u_h^1 \text{ given.} \tag{5.4}$$

We use one of the above schemes (5.1)–(5.3) to compute u_h^1, starting from $u_h^0 = u_{0h}$.

Stability: Unconditional.

At each step we have to solve, in V_h, an EVI of the second kind to compute u_h^{n+1}. Remark 3.4 applies to this scheme as well.

5.5. Comments

The properties of stability and convergence of the various schemes of Secs. 4 and 5 are studied in the references given in Sec. 1. In some cases, error estimates also have been obtained.

In Fortin [1] and G.L.T. [2, Chapter 6], [3, Chapter 6 and Appendix 6], applications to more complicated PVI than (2.1), (2.2) are also given. For the numerical analysis of hyperbolic variational inequalities, see G.L.T. [2, Chapter 6], [3, Chapter 6] and Tremolieres [1].

6. Application to a Specific Example: Time-Dependent Flow of a Bingham Fluid in a Cylindrical Pipe

Following Glowinski [4], we consider the time-dependent problem associated with the EVI of Chapter II, Sec. 6 and study its asymptotic properties.

6.1. Formulation of the problem. Existence and uniqueness theorem

Let Ω be a bounded domain of $\mathbb{R}^2$ with a smooth boundary Γ. We consider:

- $V = H_0^1(\Omega)$, $H = L^2(\Omega)$, $V^* = H^{-1}(\Omega)$,
- $a(u, v) = \int_\Omega \nabla u \cdot \nabla v \, dx$,
- A time interval $[0, T]$, $0 < T < \infty$,

- $f \in L^2(0, T; V^*)$, $u_0 \in H$,
- $j(v) = \int_\Omega |\nabla v| \, dx$,
- $\mu > 0, g > 0$.

We then have the following:

Theorem 6.1. *The PVI*

$$\left(\frac{\partial u}{\partial t}, v - u\right) + \mu a(u, v - u) + gj(v) - gj(u) \geq (f, v - u),$$

$$\forall\, v \in V \text{ a.e. } t \in]0, T[, \quad u(x, 0) = u_0(x), \quad (6.1)$$

has a unique solution u such that

$$u \in L^2(0, T; V) \cap C^0([0, T]; H), \qquad \frac{\partial u}{\partial t} \in L^2(0, T, V^*)$$

and this $\forall\, u_0 \in H$, $\forall\, f \in L^2(0, T; V^*)$.

For a proof, see Lions and Duvaut [1, Chapter 6].

6.2. The asymptotic behavior of the continuous solution

Assume that f is independent of t and that $f \in L^2(\Omega)$. We consider the following stationary problem:

$$\mu a(u, v - u) + gj(v) - gj(u) \geq (f, v - u), \qquad \forall\, v \in V, \quad u \in V. \qquad (6.2)$$

In Lions and Duvaut [1, Chapter 6] (see also Chapter II, Sec. 6 of this book), it is proved that

$$u \equiv 0 \quad \text{if } g\beta \geq \|f\|_{L^2(\Omega)}, \qquad (6.3)$$

where

$$\beta = \inf_{\substack{v \in V \\ v \neq 0}} \frac{j(v)}{\|v\|_{L^2(\Omega)}}. \qquad (6.4)$$

Then we can prove the following:

Theorem 6.2. *Assume that* $f \in L^2(\Omega)$ *with* $\|f\|_{L^2(\Omega)} < \beta g$; *then if u is the solution of* (6.1), *we have*

$$u(t) = 0 \quad \text{for } t \geq \frac{1}{\lambda_0 \mu} \operatorname{Log}\left(1 + \lambda_0 \mu \frac{\|u_0\|_{L^2}}{\beta g - \|f\|_{L^2}}\right), \qquad (6.5)$$

where λ_0 *is the smallest eigenvalue of* $-\Delta$ *in* $H_0^1(\Omega)$ $(\lambda_0 > 0)$.

PROOF. We use $|\cdot|$ for the $L^2(\Omega)$-norm and $\|\cdot\|$ for the $H_0^1(\Omega)$-norm. Since

$$f \in L^\infty(\mathbb{R}_+; L^2(\Omega)),$$

it follows from Theorem 6.1 that the solution of (6.1) is defined on the whole $\mathbb{R}_+$.

We now observe that if $g\beta > |f|$, then zero is the unique solution of (6.2); it then follows from Theorem 6.1 that if $u(t_0) = 0$ for some $t_0 \geq 0$, then

$$u(t) = 0, \qquad \forall\, t \geq t_0. \tag{6.6}$$

Taking $v = 0$ and $v = 2u$ in (6.1), we obtain

$$\left(\frac{\partial u}{\partial t}, u\right) + \mu a(u, u) + gj(u) = (f, u) \text{ a.e. in } t. \tag{6.7}$$

But since $v \in L^2(0, T; V)$, $v' \in L^2(0, T; V^*)$ implies (this is a general result) that $t \to |v(t)|^2$ is *absolutely continuous* with $(d/dt)|v|^2 = 2(dv/dt, v)$, from (6.7) we obtain

$$\frac{1}{2}\frac{d}{dt}|u|^2 + \mu a(u, u) + gj(u) = (f, u) \leq |f||u| \text{ a.e. in } t. \tag{6.8}$$

Since $a(v, v) \geq \lambda_0|v|^2$, $\forall\, v \in V$, and $j(v) \geq \beta|v|$, $\forall\, v \in V$ (from (6.4)), from (6.8) we obtain

$$\frac{1}{2}\frac{d}{dt}|u|^2 + \mu\lambda_0|u|^2 + (g\beta - |f|)|u| \leq 0 \text{ a.e. in } t \in \mathbb{R}_+. \tag{6.9}$$

Assume that $u(t) \neq 0, \forall\, t \geq 0$; since $t \to |u(t)|^2$ is absolutely continuous with $|u(t)| > 0$, it follows that $t \to |u(t)|$ is also absolutely continuous. Therefore from (6.9) we obtain

$$\frac{d}{dt}|u(t)| + \mu\lambda_0|u(t)| + (g\beta - |f|) \leq 0 \text{ a.e. } t \in \mathbb{R}_+. \tag{6.10}$$

From (6.10) it follows that

$$\frac{d/dt|u(t)|}{|u(t)| + (g\beta - |f|)/\mu\lambda_0} \leq -\mu\lambda_0 \text{ a.e. } t \in \mathbb{R}_+. \tag{6.11}$$

Define γ by $\gamma = (g\beta - |f|)/\mu\lambda_0$; then $\gamma > 0$. By integrating (6.11) it then follows that

$$|u(t)| + \gamma \leq (|u_0| + \gamma)e^{-\mu\lambda_0 t}, \qquad \forall\, t \in \mathbb{R}_+; \tag{6.12}$$

(6.12) is absurd for t large enough. Actually we have $u(t) = 0$ if

$$\gamma \geq (|u_0| + \gamma)e^{-\mu\lambda_0 t},$$

i.e.,

$$t \geq \frac{1}{\lambda_0\mu}\operatorname{Log}\left(1 + \frac{\lambda_0\mu\|u_0\|_{L^2(\Omega)}}{g\beta - \|f\|_{L^2(\Omega)}}\right). \tag{6.13}$$

EXERCISE 6.1. Let $f \in L^2(\Omega)$, possibly with $|f| \geq g\beta$. Let us denote by u_∞ the solution of (6.2); then prove that

$$|u(t) - u_\infty| \leq |u_0 - u_\infty|e^{-\lambda_0\mu t},$$

where $u(t)$ is the solution of (6.1).

6.3. On the asymptotic behavior of the discrete solution

We still assume that $f \in L^2(\Omega)$. To approximate (6.1) we proceed as follows: Assuming that Ω is a polygonal domain, we use the same approximation with regard to the space variables as in Chapter II, Sec. 6 (i.e., by means of piecewise linear finite elements, see Chapter II, Sec. 6). Hence we have

$$a_h(u_h, v_h) = a(u_h, v_h), \qquad \forall\, u_h, v_h \in V_h,$$

$$j_h(v_h) = j(v_h), \qquad \forall\, v_h \in V_h,$$

and from the formula of Chapter II, Sec. 7, we can also take

$$(u_h, v_h)_h = (u_h, v_h), \qquad \forall\, u_h, v_h \in V_h.$$

Then we approximate (6.1) by the implicit scheme (5.2) and obtain

$$\left(\frac{u_h^{n+1} - u_h^n}{\Delta t}, v_h - u_h^{n+1}\right) + \mu \int_\Omega \nabla u_h^{n+1} \cdot \nabla(v_h - u_h^{n+1})\, dx + gj(v_h)$$
$$- gj(u_h^{n+1}) \geq (f_h, v_h - u_h^{n+1}),$$
$$\forall\, v_h \in V_h, \quad u_h^{n+1} \in V_h; \quad n = 0, 1, 2, \ldots; \quad u_h^0 = u_{0h}. \tag{6.14}$$

We assume that $u_{0h} \in V_h$, $\forall\, h$ and

$$\lim_{h \to 0} u_{0h} = u_0 \text{ strongly in } L^2(\Omega). \tag{6.15}$$

Similarly, we assume that f is approximated by $(f_h)_h$ in such a way that (f_h, v_h) can be computed easily and

$$\lim_{h \to 0} f_h = f \text{ strongly in } L^2(\Omega). \tag{6.16}$$

Theorem 6.3. *Let $|f| < \beta g$. If (6.15) and (6.16) hold, then if h is sufficiently small, we have $u_h^n = 0$ for n large enough.*

PROOF. As in the proof of Theorem 6.2, taking $v_h = 0$ and $v_h = 2u_h^{n+1}$ in (6.14), we obtain

$$\left(\frac{u_h^{n+1} - u_h^n}{\Delta t}, u_h^{n+1}\right) + \mu \int_\Omega |\nabla u_h^{n+1}|^2\, dx + g \int_\Omega |\nabla u_h^{n+1}|\, dx = \int_\Omega f_h u_h^{n+1}\, dx, \qquad \forall\, n \geq 0; \tag{6.17}$$

using the Schwarz inequality in $L^2(\Omega)$, it follows from (6.17) that

$$\frac{|u_h^{n+1}| - |u_h^n|}{\Delta t} |u_h^{n+1}| + \mu\lambda_0 |u_h^{n+1}|^2 + (g\beta - |f_h|)|u_h^{n+1}| \leq 0, \qquad \forall\, n \geq 0. \tag{6.18}$$

Since $f_h \to f$ strongly in $L^2(\Omega)$, we have

$$g\beta - |f_h| > 0 \quad \text{for } h \text{ sufficiently small} \tag{6.19}$$

From (6.18), (6.19) it then follows that

$$u_h^{n_0} = 0 \Rightarrow u_h^n = 0 \quad \text{for } n \geq n_0 \text{ if } h \text{ is small enough.} \tag{6.20}$$

Assume that $u_h^n \neq 0$, $\forall\, n$; then (6.18) implies

$$\frac{|u_h^{n+1}| - |u_h^n|}{\Delta t} + \mu\lambda_0 |u_h^{n+1}| + g\beta - |f_h| \leq 0, \qquad \forall\, n \geq 0. \tag{6.21}$$

We define γ_h by $\gamma_h = g\beta - |f_h|$; then

$$\gamma_h > 0 \quad \text{for } h \text{ small enough and} \quad \lim_{h\to 0} \gamma_h = \gamma = g\beta - |f|. \tag{6.22}$$

From (6.21) it follows that

$$\left(|u_h^{n+1}| + \frac{\gamma_h}{\lambda_0 \mu}\right)(1 + \lambda_0 \mu\, \Delta t) \leq |u_h^n| + \frac{\gamma_h}{\lambda_0 \mu}, \qquad \forall\, n \geq 0,$$

which implies that

$$\left(|u_h^n| + \frac{\gamma_h}{\lambda_0 \mu}\right) \leq (1 + \lambda_0 \mu\, \Delta t)^{-n}\left(|u_h^0| + \frac{\gamma_h}{\lambda_0 \mu}\right). \tag{6.23}$$

Since $\gamma_h > 0$ for h small enough, (6.23) is impossible for n large enough. More precisely, we shall have $u_h^n = 0$ if

$$\frac{\gamma_h}{\lambda_0 \mu} \geq (1 + \lambda_0 \mu\, \Delta t)^{-n}\left(|u_h^0| + \frac{\gamma_h}{\lambda_0 \mu}\right),$$

which implies:

$$\textit{If } h \textit{ is small enough, then } u_h^n = 0 \textit{ if } n \geq \frac{\mathrm{Log}[1 + \lambda_0 \mu(|u_h^0|/\gamma_h)]}{\mathrm{Log}(1 + \lambda_0 \mu\, \Delta t)}. \tag{6.24}$$

Relation (6.24) makes the statement of Theorem 6.3 more precise. Moreover, in terms of time, (6.24) implies that u_h^n is equal to zero if

$$n\, \Delta t \geq \Delta t\, \frac{\mathrm{Log}[1 + \lambda_0 \mu(|u_h^0|/\gamma_h)}{\mathrm{Log}(1 + \lambda_0 \mu\, \Delta t)}. \tag{6.25}$$

We observe that

$$\lim_{\substack{h\to 0\\ \Delta t \to 0}} \Delta t\, \frac{\mathrm{Log}[1 + \lambda_0 \mu(|u_h^0|/\gamma_h)]}{\mathrm{Log}(1 + \lambda_0 \mu\, \Delta t)} = \frac{1}{\lambda_0 \mu}\, \mathrm{Log}\left(1 + \lambda_0 \mu\, \frac{|u^0|}{\gamma}\right)$$

Hence, taking the limit in (6.25), we obtain another proof (assuming that u_h^n converges to u in some topology) of the estimate (6.5) given in the statement of Theorem 6.2.

EXERCISE 6.2. Let u_h^∞ be the solution of the time-independent problem associated with f_h, possibly with $|f_h| \geq \beta g$; then prove that

$$|u_h^n - u_h^\infty| \leq (1 + 2\mu\lambda_0\, \Delta t)^{-n/2} |u_h^0 - u_h^\infty|, \qquad \forall\, n \geq 0.$$

6.4. Remarks

Remark 6.1. We can generalize Theorem 6.2 to the case of a Bingham flow in a two-dimensional bounded cavity.

Remark 6.2. In Glowinski [4], Bristeau [1], and Begis [1], numerical verifications of the above asymptotic properties have been performed and found to be consistent with the theoretical predictions.

Remark 6.3. In H. Brezis [5], one may find many results on the asymptotic behavior of various PVI as $t \to +\infty$.

CHAPTER IV

Applications of Elliptic Variational Inequality Methods to the Solution of Some Nonlinear Elliptic Equations

1. Introduction

For solving some nonlinear elliptic equations, it may be convenient, from the theoretical and numerical points of view, to view them as EVI's.

In this chapter we shall consider two examples of such situations:

(1) a family of mildly nonlinear elliptic equations;

(2) a nonlinear elliptic equation modeling the subsonic flow of an *inviscid compressible* fluid.

2. Theoretical and Numerical Analysis of Some Mildly Nonlinear Elliptic Equations

2.1. Formulation of the continuous problem

Let Ω be a bounded domain of $\mathbb{R}^N$ $(N \geq 2)$ with a smooth boundary Γ. We consider

- $V = H_0^1(\Omega)$;
- $L(v) = \langle f, v \rangle, f \in V^* = H^{-1}(\Omega)$;
- $a: V \times V \to \mathbb{R}$ bilinear, continuous, and V-elliptic with $\alpha > 0$ as ellipticity constant; $a(\cdot, \cdot)$ is possibly nonsymmetric;
- $\phi: \mathbb{R} \to \mathbb{R}$, $\phi \in C^0(\mathbb{R})$, nondecreasing with $\phi(0) = 0$.

We then consider the following nonlinear elliptic equation (P) defined by: *Find $u \in V$ such that*

$$a(u, v) + \langle \phi(u), v \rangle = L(v), \qquad \forall\, v \in V,$$
$$\phi(u) \in L^1(\Omega) \cap H^{-1}(\Omega). \tag{P}$$

From the Riesz Representation Theorem it follows that there exists

$$A \in \mathscr{L}(V, V^*)$$

such that $a(u, v) = \langle Au, v\rangle, \forall\, u, v \in V$. Therefore (P) is equivalent to

$$Au + \phi(u) = f, \qquad u \in V, \quad \phi(u) \in L^1(\Omega) \cap H^{-1}(\Omega). \tag{2.1}$$

EXAMPLE 2.1. Let us consider a function $a_0 \in L^\infty(\Omega)$ such that

$$a_0(x) \geq \alpha > 0 \text{ a.e. in } \Omega. \tag{2.2}$$

Define $a(\cdot, \cdot)$ by

$$a(u, v) = \int_\Omega a_0(x)\nabla u \cdot \nabla v\, dx + \int_\Omega \beta \cdot \nabla u\, v dx, \tag{2.3}$$

where β is a constant vector in $\mathbb{R}^N$.

From the definition of $a_0(\cdot)$, and using the fact that $\int_\Omega \beta \cdot \nabla v\, v dx = 0$, $\forall\, v \in H_0^1(\Omega)$, we clearly have

$$a(v, v) \geq \alpha \|v\|_V^2. \tag{2.4}$$

From (2.3) we obtain

$$Au = -\nabla \cdot (a_0 \nabla u) + \beta \cdot \nabla u. \tag{2.5}$$

Hence, in this particular case, (2.1) becomes

$$-\nabla \cdot (a_0 \nabla u) + \beta \cdot \nabla u + \phi(u) = f, \qquad u \in V, \quad \phi(u) \in L^1(\Omega). \tag{2.6}$$

Remark 2.1. If $N = 1$, we have $H_0^1(\Omega) \subset C^0(\bar{\Omega})$. Because of this inclusion there is no great difficulty in the study of one-dimensional problems of type (P). If $N \geq 2$, the main difficulty is precisely related to the fact that $H_0^1(\Omega)$ is not contained in $C^0(\bar{\Omega})$.

Remark 2.2. The analysis given below may be extended to problems in which either $V = H^1(\Omega)$ or V is a convenient closed subspace of $H^1(\Omega)$.

2.2. A variational inequality related to (P)

2.2.1. *Definition of the variational inequality*

Let

$$\Phi(t) = \int_0^t \phi(\tau)\, d\tau, \tag{2.7}$$

$$D(\Phi) = \{v \in V, \Phi(v) \in L^1(\Omega)\}. \tag{2.8}$$

The functional $j: L^2(\Omega) \to \bar{\mathbb{R}}$ is defined by

$$j(v) = \int_\Omega \Phi(v)\, dx \quad \text{if } \Phi(v) \in L^1(\Omega), \qquad j(v) = +\infty \quad \text{if } \Phi(v) \notin L^1(\Omega). \tag{2.9}$$

Instead of studying the problem (P) directly, it is natural to associate with (P) the following EVI of the second kind:

$$a(u, v - u) + j(v) - j(u) \geq L(v - u), \qquad \forall\, v \in V, \quad u \in V. \tag{π}$$

If $a(\cdot, \cdot)$ is symmetrical, a standard method for studying (P) is to consider it to be the *formal Euler equation* of the following minimization problem encountered in the *calculus of variations*:

$$J(u) \leq J(v), \qquad \forall\, v \in V, \quad u \in V, \tag{2.10}$$

where

$$J(v) = \tfrac{1}{2}a(v, v) + \int_\Omega \Phi(v)\, dx - L(v). \tag{2.11}$$

EXERCISE 2.1. Prove that $D(\Phi)$ is a convex nonempty subset of V.

2.2.2. *Properties of* $j(\cdot)$

Since $\phi\colon \mathbb{R} \to \mathbb{R}$ is nondecreasing and continuous with $\phi(0) = 0$, we have

$$\Phi \in C^1(\mathbb{R}), \quad \Phi \text{ convex}, \quad \Phi(0) = 0; \qquad \Phi(t) \geq 0, \quad \forall\, t \in \mathbb{R}. \tag{2.12}$$

The properties of $j(\cdot)$ are given by the following:

Lemma 2.1. *The functional* $j(\cdot)$ *is convex, proper, and l.s.c. over* $L^2(\Omega)$.

PROOF. Since $j(v) \geq 0$, $\forall\, v \in L^2(\Omega)$, it follows that $j(\cdot)$ is proper. The convexity of $j(\cdot)$ is obvious from the fact that Φ is convex.

Let us prove that $j(\cdot)$ is l.s.c.. Let $\{v_n\}_n$, $v_n \in L^2(\Omega)$, be such that

$$\lim_{n \to \infty} v_n = v \text{ strongly in } L^2(\Omega).$$

Then we have to prove that

$$\liminf_{n \to \infty} j(v_n) \geq j(v). \tag{2.13}$$

If $\liminf_{n \to \infty} j(v_n) = +\infty$, the property is proved. Therefore assume that

$$\liminf_{n \to \infty} j(v_n) = l < +\infty.$$

Hence we can extract a subsequence $\{v_{n_k}\}_{n_k}$ such that

$$\lim_{k \to \infty} j(v_{n_k}) = l, \tag{2.14}$$

$$v_{n_k} \to v \text{ a.e. in } \Omega. \tag{2.15}$$

Since $\Phi \in C^1(\mathbb{R})$, (2.15) implies

$$\lim_{k \to \infty} \Phi(v_{n_k}) = \Phi(v) \text{ a.e.}. \tag{2.16}$$

Moreover, $\Phi(v) \geq 0$ a.e. and (2.14) imply that

$$\{\Phi(v_{n_k})\}_k \text{ is bounded in } L^1(\Omega). \tag{2.17}$$

Hence, by Fatou's Lemma, from (2.16) and (2.17), we have

$$\Phi(v) \in L^1(\Omega),$$

$$\liminf_{k\to\infty} \int_\Omega \Phi(v_{n_k})\,dx \geq \int_\Omega \Phi(v)\,dx. \tag{2.18}$$

From (2.14) and (2.18) we obtain (2.13).

This proves the lemma. □

Corollary 2.1. *The functional $j(\cdot)$ restricted to V is convex, proper, and l.s.c.*.

2.2.3. *Existence and uniqueness results for* (π)

Theorem 2.1. *Under the above hypotheses on V, $a(\cdot,\cdot)$, $L(\cdot)$, and $\phi(\cdot)$, problem (π) has a unique solution in $V \cap D(\Phi)$.*

PROOF. Since V, $a(\cdot,\cdot)$, $L(\cdot)$, and $j(\cdot)$ have the properties (cf. Corollary 2.1) required to apply Theorem 4.1 of Chapter I, Sec. 4, the EVI of the second kind, (π), has a unique solution u in V.

Let us show that $u \in D(\Phi)$. Taking $v = 0$ in (π), we obtain

$$a(u, u) + j(u) \leq L(u) \leq \|f\|_*\|u\|_V. \tag{2.19}$$

Since $j(u) \geq 0$, using the ellipticity of $a(\cdot,\cdot)$, we obtain

$$\|u\|_V \leq \frac{\|f\|_*}{\alpha}, \tag{2.20}$$

which implies

$$j(u) \leq \frac{\|f\|_*^2}{\alpha}. \tag{2.21}$$

This implies $u \in D(\Phi)$. □

Remark 2.3. If $a(\cdot,\cdot)$ is symmetric, (π) is equivalent to (2.10).

2.3. Equivalence between (P) and (π)

In this section we shall prove that (P) and (π) are equivalent. First we prove that the unique solution of (π) is also a solution of (P). In order to prove this result, we need to prove that $\phi(u)$ and $u\phi(u)$ belong to $L^1(\Omega)$.

Proposition 2.1. *Let u be the solution of (π). Then $u\phi(u)$ and $\phi(u)$ belong to $L^1(\Omega)$.*

PROOF. Here we use a *truncation* technique. Let n be a positive integer. Define

$$K_n = \{v \in V, |v(x)| \leq n \text{ a.e.}\}.$$

Since K_n is a closed convex nonempty subset of V, the variational inequality

$$a(u_n, v - u_n) + j(v) - j(u_n) \geq L(v - u_n), \qquad \forall\, v \in K_n, \quad u_n \in K_n, \tag{π_n}$$

has a unique solution (in order to apply Theorem 4.1 of Chapter I, we need to replace j by $j + I_{K_n}$, where I_{K_n} is the *indicator functional* of K_n).

Now we prove that $\lim_{n \to \infty} u_n = u$ weakly in V, where u is the solution of (π). Since $0 \in K_n$, taking $v = 0$ in (π_n), we obtain (as in Theorem 2.1 of this chapter)

$$\|u_n\|_V \leq \frac{\|f\|_*}{\alpha}, \tag{2.22}$$

$$j(u_n) \leq \frac{\|f\|_*^2}{\alpha}. \tag{2.23}$$

From (2.22) it follows that there exists a subsequence $\{u_{n_k}\}_{n_k}$ of $\{u_n\}_n$ and $u^* \in V$ such that

$$\lim_{k \to \infty} u_{n_k} = u^* \text{ weakly in } V. \tag{2.24}$$

Moreover, from the compactness of the canonical injection from $H_0^1(\Omega)$ to $L^2(\Omega)$ and from (2.24), it follows that

$$\lim_{k \to \infty} u_{n_k} = u^* \text{ strongly in } L^2(\Omega). \tag{2.25}$$

Relation (2.25) implies that we can extract a subsequence, still denoted by $(u_{n_k})_{n_k}$, such that

$$\lim_{k \to \infty} u_{n_k} = u^* \text{ a.e. in } \Omega. \tag{2.26}$$

Now let $v \in V \cap L^\infty(\Omega)$; then for large k, we have $v \in K_{n_k}$ and

$$a(u_{n_k}, u_{n_k}) + j(u_{n_k}) \leq a(u_{n_k}, v) + j(v) - L(v - u_{n_k}). \tag{2.27}$$

Since $\liminf_{k \to \infty} a(u_{n_k}, u_{n_k}) \geq a(u^*, u^*)$ and $\liminf_{k \to \infty} j(u_{n_k}) \geq j(u^*)$, it follows from (2.24) and (2.27) that

$$a(u^*, u^*) + j(u^*) \leq a(u^*, v) + j(v) - L(v - u^*), \qquad \forall\, v \in L^\infty(\Omega) \cap V, \quad u^* \in V,$$

which can also be written as

$$a(u^*, v - u^*) + j(v) - j(u^*) \geq L(v - u^*), \qquad \forall\, v \in V \cap L^\infty(\Omega), \quad u^* \in V. \tag{2.28}$$

For $n > 0$, define $\tau_n: V \to \mathrm{K}_n$ by

$$\tau_n v = \mathrm{Inf}(n, \mathrm{Sup}(-n, v)) \tag{2.29}$$

(see Fig. 2.1).

Then, from Corollary 2.1 of Chapter II, Sec. 2.2, we have

$$\begin{aligned} &\lim_{n \to \infty} \tau_n v = v \text{ strongly in } V, \\ &\lim_{n \to \infty} \tau_n v = v \text{ a.e. in } \Omega. \end{aligned} \tag{2.30}$$

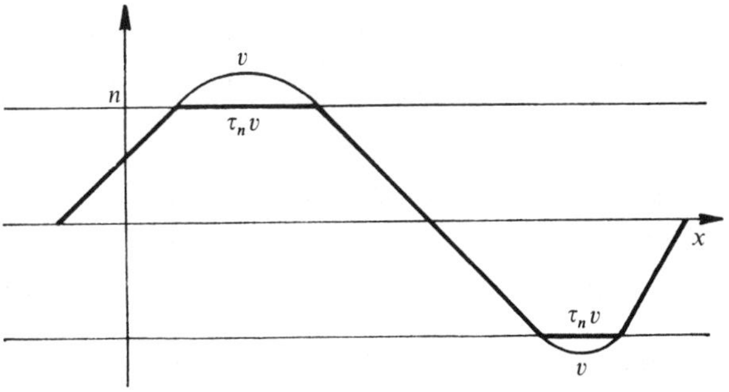

Figure 2.1

Moreover, we obviously have

$$|\tau_n v(x)| \leq |v(x)| \text{ a.e.,} \tag{2.31}$$

$$v(x)\tau_n v(x) \geq 0 \text{ a.e..} \tag{2.32}$$

From (2.30)–(2.32) and from the various properties of Φ, it then follows that

$$0 \leq \Phi(\tau_n v) \leq \Phi(v) \text{ a.e.,} \tag{2.33}$$

$$\lim_{n\to\infty} \Phi(\tau_n v) = \Phi(v) \text{ a.e..} \tag{2.34}$$

Since $\tau_n v \in L^\infty(\Omega) \cap V$, it follows from (2.28) that

$$a(u^*, \tau_n v - u^*) + j(\tau_n v) - j(u^*) \geq L(\tau_n v - u^*), \qquad \forall\, v \in V, \quad u^* \in V. \tag{2.35}$$

If $v \notin D(\Phi)$, then by Fatou's lemma,

$$\lim_{n\to\infty} j(\tau_n v) = +\infty.$$

If $v \in D(\Phi)$, by applying Lebesgue's dominated convergence theorem, it follows from (2.33) and (2.34) that

$$\lim_{n\to\infty} j(\tau_n v) = j(v).$$

From these convergence properties and from (2.30), and by taking the limit in (2.35), it follows that

$$a(u^*, v - u^*) + j(v) - j(u^*) \geq L(v - u^*), \qquad \forall\, v \in V, \quad u^* \in V. \tag{2.36}$$

Then u^* is a solution of (π), and from the uniqueness property, we have $u^* = u$. This proves that $\lim_{n\to\infty} u_n = u$ weakly in V.

Let us show that $\phi(u), u\phi(u) \in L^1(\Omega)$. Let $v \in K_n$. Then $u_n + t(v - u_n) \in K_n, \forall\, t \in]0, 1]$. Replacing v by $u_n + t(v - u_n)$ in (π_n) and dividing both sides of the inequality by t, we obtain

$$a(u_n, v - u_n) + \int_\Omega \frac{\Phi(u_n + t(v - u_n)) - \Phi(u_n)}{t}\, dx \geq L(v - u_n), \qquad \forall\, v \in K_n. \tag{2.37}$$

Since $\Phi \in C^1(\mathbb{R})$ and $\Phi' = \phi$, we have

$$\lim_{\substack{t \to 0 \\ t > 0}} \frac{\Phi(u_n + t(v - u_n)) - \Phi(u_n)}{t} = \phi(u_n)(v - u_n) \text{ a.e..} \tag{2.38}$$

Moreover, since Φ is convex, we also have, $\forall\, t \in]0, 1]$,

$$\phi(u_n)(v - u_n) \leq \frac{\Phi(u_n + t(v - u_n)) - \Phi(u_n)}{t} \leq \Phi(v) - \Phi(u_n) \text{ a.e..} \tag{2.39}$$

From (2.38), (2.39), and using Lebesgue's dominated convergence theorem in (2.37), we obtain

$$a(u_n, v - u_n) + \int_\Omega \phi(u_n)(v - u_n)\, dx \geq L(v - u_n), \qquad \forall\, v \in K_n. \tag{2.40}$$

Then, taking $v = 0$ in (2.40), we have

$$a(u_n, u_n) + \int_\Omega \phi(u_n) u_n\, dx \leq L(u_n),$$

which, using (2.22), implies

$$\int_\Omega \phi(u_n) u_n\, dx \leq \frac{\|f\|_*^2}{\alpha}. \tag{2.41}$$

But $\phi(v)v \geq 0, \forall\, v \in V$. Hence $\phi(u_n)u_n$ is bounded in $L^1(\Omega)$. Moreover, for some subsequence $\{u_{n_k}\}_{n_k}$ of $\{u_n\}_n$, we have

$$\phi(u_{n_k})u_{n_k} \to \phi(u)u \text{ a.e. in } \Omega.$$

Then by Fatou's lemma we obtain $u\phi(u) \in L^1(\Omega)$, and this completes the proof of the proposition, since $u\phi(u) \in L^1(\Omega)$ obviously implies that $\phi(u) \in L^1(\Omega)$. □

Incidentally, when proving the convergence of $(u_n)_n$ to u, we have proved the following useful lemma:

Lemma 2.2. *The solution u of* (π) *is characterized by*

$$a(u, v - u) + j(v) - j(u) \geq L(v - u),$$
$$\forall\, v \in V \cap L^\infty(\Omega), \qquad u \in V, \quad \Phi(u) \in L^1(\Omega). \tag{2.42}$$

In view of proving that (π) implies (P), we also need the following two lemmas:

Lemma 2.3. *The solution u of* (π) *is characterized by*

$$a(u, v - u) + \int_\Omega \phi(u)(v - u)\, dx \geq L(v - u),$$
$$\forall\, v \in L^\infty(\Omega) \cap V, \quad u \in V, \quad u\phi(u) \in L^1(\Omega). \tag{2.43}$$

PROOF. (1) (π) implies (2.43).

Let $v \in L^\infty(\Omega) \cap V$. Then $v \in D(\Phi)$, and since $D(\Phi)$ is convex, we have

$$u + t(v - u) \in D(\Phi), \qquad \forall\, t \in]0, 1].$$

Replacing v by $u + t(v - u)$ in (π) and dividing by t, we obtain $\forall\, t \in]0, 1]$,

$$a(u, v - u) + \int_\Omega \frac{\Phi(u + t(v - u)) - \Phi(u)}{t}\, dx \geq L(v - u), \qquad \forall\, v \in L^\infty(\Omega) \cap V. \tag{2.44}$$

Since $\Phi \in C^1$ and is convex, we have

$$\lim_{\substack{t \to 0 \\ t > 0}} \frac{\Phi(u + t(v - u)) - \Phi(u)}{t} = \phi(u)(v - u) \text{ a.e.}, \tag{2.45}$$

$$\phi(u)(v - u) \leq \frac{\Phi(u + t(v - u)) - \Phi(u)}{t} \leq \Phi(v) - \Phi(u). \tag{2.46}$$

By Proposition 2.1, we have $\phi(u)$, $u\phi(u) \in L^1(\Omega)$. Hence $\phi(u)(v - u) \in L^1(\Omega)$ and $\Phi(v) - \Phi(u) \in L^1(\Omega)$, $\forall v \in L^\infty(\Omega) \cap V$. Then, using the Lebesgue dominated convergence theorem, it follows from (2.45) and (2.46) that

$$\lim_{t \to 0} \int_\Omega \frac{\Phi(u + t(v - u)) - \Phi(u)}{t}\, dx = \int_\Omega \phi(u)(v - u)\, dx.$$

Using the above relation and (2.44), we obtain (2.43). This proves that $(\pi) \Rightarrow$ (2.43).

(2) We will now prove that (2.43) $\Rightarrow (\pi)$.

Let u be a solution of (2.43). Since Φ is convex, it follows that

$$-\Phi(u) = \Phi(0) - \Phi(u) \geq \phi(u)(0 - u) = -\phi(u)u.$$

This implies $0 \leq \Phi(u) \leq u\phi(u)$ and $\Phi(u) \in L^1(\Omega)$. Let $v \in L^\infty(\Omega) \cap V$. Then, from the inequality

$$\phi(u)(v - u) \leq \Phi(v) - \Phi(u) \text{ a.e. in } \Omega,$$

by integration we obtain

$$\int_\Omega \phi(u)(v - u)\, dx \leq j(v) - j(u), \qquad \forall\, v \in V \cap L^\infty(\Omega),$$

which combined with (2.43) and $\Phi(u) \in L^1(\Omega)$ implies (2.42). Hence from Lemma 2.2 we obtain that (2.43) implies (π). □

Lemma 2.4. *Let u be the solution of (π). Then u is characterized by*

$$a(u, v) + \int_\Omega \phi(u)v\, dx = L(v), \qquad \forall\, v \in L^\infty(\Omega) \cap V, \quad u \in V, \quad \phi(u) \in L^1(\Omega). \tag{2.47}$$

PROOF. (1) (π) implies (2.47).

Let $v \in V \cap L^\infty(\Omega)$. If u is the solution of (π), then u is also the unique solution of (2.43).

Let τ_n be defined by (2.29). Then $\tau_n u \in V \cap L^\infty(\Omega)$. Replacing v by $\tau_n u + v$ in (2.43), we obtain

$$a(u, v) + \int_\Omega \phi(u)v\,dx + a(u, \tau_n u - u) + \int_\Omega \phi(u)(\tau_n u - u)\,dx \geq L(v) + L(\tau_n u - u),$$
$$\forall\, v \in V \cap L^\infty(\Omega). \tag{2.48}$$

From (2.29)–(2.32) it follows that

$$\lim_{n\to\infty} a(u, \tau_n u - u) = 0,$$
$$\lim_{n\to\infty} L(\tau_n u - u) = 0, \tag{2.49}$$

$$\lim_{n\to\infty} \phi(u)(\tau_n u - u) = 0 \text{ a.e.,} \tag{2.50}$$

$$0 \leq \phi(u)(u - \tau_n u) \leq u\phi(u) \text{ a.e..} \tag{2.51}$$

Then, by the Lebesgue dominated convergence theorem and (2.50), (2.51), we obtain

$$\lim_{n\to\infty} \phi(u)(\tau_n u - u) = 0 \text{ strongly in } L^1(\Omega). \tag{2.52}$$

Then (2.48), (2.49), and (2.52) imply

$$a(u, v) + \int_\Omega \phi(u)v\,dx \geq L(v), \qquad \forall\, v \in V \cap L^\infty(\Omega).$$

Since the above relation also holds for $-v$, we have

$$a(u, v) + \int_\Omega \phi(u)v\,dx = L(v), \qquad \forall\, v \in V \cap L^\infty(\Omega). \tag{2.53}$$

By Proposition 2.1 we have $\phi(u) \in L^1(\Omega)$; combining this with (2.53) we obtain (2.47). This proves that $(\pi) \Rightarrow (2.47)$.

(2) (2.47) implies (π).
We have

$$a(u, v) + \int_\Omega \phi(u)v\,dx = L(v), \qquad \forall\, v \in V \cap L^\infty(\Omega).$$

Then

$$a(u, \tau_n u) + \int_\Omega \phi(u)\tau_n u\,dx = L(\tau_n u), \qquad \forall\, n. \tag{2.54}$$

Since $\tau_n u \to u$ strongly in V, $\{\int_\Omega \phi(u)\tau_n u\,dx\}_n$ is bounded. But $\phi(u)\tau_n u \geq 0$ a.e.. Hence we find that $\phi(u)\tau_n u$ is bounded in $L^1(\Omega)$. We also have $\lim_{n\to\infty} \tau_n u\phi(u) = u\phi(u)$ a.e.; hence, by Fatou's lemma, we have

$$u\phi(u) \in L^1(\Omega). \tag{2.55}$$

But now we observe that

$$0 \leq \phi(u)\tau_n u \leq u\phi(u).$$

Hence, by the Lebesgue dominated convergence theorem,

$$\lim_{n\to\infty} \int_\Omega \phi(u)\tau_n u \, dx = \int_\Omega \phi(u)u \, dx,$$

which along with (2.54) yields

$$a(u, u) + \int_\Omega \phi(u)u \, dx = L(u). \tag{2.56}$$

Then by substracting (2.56) from (2.47) we obtain

$$a(u, v - u) + \int_\Omega \phi(u)(v - u) \, dx = L(v - u),$$

$$\forall\, v \in V \cap L^\infty(\Omega), \quad u \in V, \quad u\phi(u) \in L^1(\Omega), \tag{2.57}$$

and obviously (2.57) implies (2.43). This completes the proof of the lemma. □

Corollary 2.2. *If u is the solution of* (π), *then u is also a solution of* (P).

PROOF. We recall that $V^* = H^{-1}(\Omega) \subset \mathscr{D}'(\Omega)$ and that $a(u, v) = \langle Au, v\rangle$, $\forall\, u, v \in V$ and $L(v) = \langle f, v\rangle$.

Let u be a solution of (π). Then u is characterized by (2.47), and since $\mathscr{D}(\Omega) \subset V$, we obtain

$$\langle Au, v\rangle + \int_\Omega \phi(u)v \, dx = \langle f, v\rangle, \qquad v \in \mathscr{D}(\Omega). \tag{2.58}$$

From (2.58) it follows that

$$Au + \phi(u) = f \text{ in } \mathscr{D}'(\Omega); \tag{2.59}$$

since Au and $f \in V^*$, we have $\phi(u) \in V^*$. Hence $\phi(u) \in L^1(\Omega) \cap H^{-1}(\Omega)$, and from (2.59), we obtain that u is a solution of (P). □

If we try to summarize what we have proved until now, we observe that the unique solution of (π) is also a solution of (P). Now we prove the reciprocal property; that is, every solution of (P) is a solution of (π) and hence (P) has a unique solution.

In order to prove this, we shall use the following density lemma:

Lemma 2.5. *$\mathscr{D}(\Omega)$ is dense in $V \cap L^\infty(\Omega)$, $V \cap L^\infty(\Omega)$ being equipped with the strong topology of V and the weak $*$ topology of $L^\infty(\Omega)$.*

PROOF. Let $v \in V \cap L^\infty(\Omega)$. Since $\overline{\mathscr{D}(\Omega)}^{H^1(\Omega)} = V$, there exists a sequence $\{v_n\}_n$, $v_n \in \mathscr{D}(\Omega)$, such that

$$\lim_{n\to\infty} v_n = v \text{ strongly in } V. \tag{2.60}$$

Let us define w_n by (see Fig. 2.2)

$$w_n = \min(v^+, v_n^+) - \min(v^-, v_n^-). \tag{2.61}$$

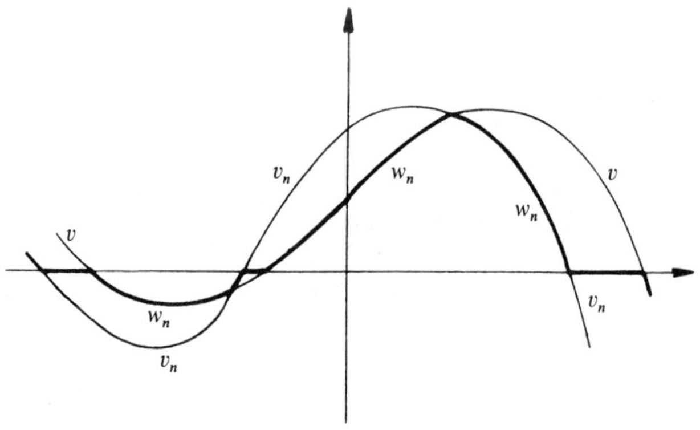

Figure 2.2

Then

$$w_n \text{ has a compact support in } \Omega, \tag{2.62}$$

$$\|w_n\|_{L^\infty(\Omega)} \leq \|v\|_{L^\infty(\Omega)}, \tag{2.63}$$

and it follows from Chapter II, Corollary 2.1 that

$$\lim_{n\to\infty} w_n = v \text{ strongly in } V. \tag{2.64}$$

From (2.63) and (2.64) we obtain $\lim_{n\to\infty} w_n = v$ for the weak$_*$ topology of $L^\infty(\Omega)$. Thus we have proved that

$$\mathscr{U} = \{v \in L^\infty(\Omega) \cap V, v \text{ has compact support in } \Omega\}$$

is dense in $L^\infty(\Omega) \cap V$ for the topology given in the statement of the lemma.

Let $v \in \mathscr{U}$ and $(\rho_n)_n$ be a mollifying sequence (see Chapter II, Lemma 2.4). Define v_n by

$$\tilde{v}(x) = \begin{cases} v(x) & \text{if } x \in \Omega, \\ 0, & \text{if } x \notin \Omega, \end{cases} \tag{2.65}$$

$$\tilde{v}_n = \rho_n * \tilde{v}. \tag{2.66}$$

Then

$$\tilde{v}_n \in \mathscr{D}(\mathbb{R}^N), \qquad \lim_{n\to\infty} \tilde{v}_n = \tilde{v} \text{ strongly in } H^1(\mathbb{R}^N), \tag{2.67}$$

$$\tilde{v}_n \text{ has a compact support in } \Omega \text{ for } n \text{ large enough.} \tag{2.68}$$

Let $v_n = \tilde{v}_n|_\Omega$; then for n large enough, $v_n \in \mathscr{D}(\Omega)$, and $\lim_{n\to\infty} v_n = v$ strongly in V. Since $\|\tilde{v}\|_{L^\infty(\mathbb{R}^N)} = \|v\|_{L^\infty(\Omega)}$, it follows from (2.66) that

$$|v_n(x)| \leq \int_{\mathbb{R}^N} \rho_n(x-y)|\tilde{v}(y)|\,dy \leq \|v\|_{L^\infty(\Omega)}. \tag{2.69}$$

From this it follows that

$$\|v_n\|_{L^\infty(\Omega)} \leq \|v\|_{L^\infty(\Omega)}. \tag{2.70}$$

Summarizing the above information, we have proved that $\forall\, v \in L^\infty(\Omega) \cap V$, there exists a sequence $(v_n)_n$, $v_n \in \mathscr{D}(\Omega)$, $\forall\, n$, such that

$$\lim_{n\to\infty} v_n = v \text{ strongly in } V, \tag{2.71}$$

$$\|v_n\|_{L^\infty(\Omega)} \le \|v\|_{L^\infty(\Omega)}, \qquad \forall\, n. \tag{2.72}$$

Hence from (2.71) and (2.72) we find that $v_n \to v$ in $L^\infty(\Omega)$ weak $*$. This completes the proof of the lemma. □

Theorem 2.2. *Under the above hypothesis on V, $a(\cdot,\cdot)$, $L(\cdot)$, and $\phi(\cdot)$, problems (π) and (P) are equivalent.*

PROOF. We have already proved that (π) implies (P). We need only prove that (P) implies (π).

From the definition of (P) we have

$$a(u, v) + \langle \phi(u), v\rangle = L(v), \qquad \forall\, v \in V, \quad u \in V, \quad \phi(u) \in H^{-1}(\Omega) \cap L^1(\Omega). \tag{2.73}$$

From (2.73) it follows that .

$$a(u, v) + \int_\Omega \phi(u) v \, dx = L(v), \qquad \forall\, v \in \mathscr{D}(\Omega). \tag{2.74}$$

If $v \in V \cap L^\infty(\Omega)$, from Lemma 2.5 we know that there exists a sequence $(v_n)_n$, $v_n \in \mathscr{D}(\Omega)$, such that

$$\lim_{n\to\infty} v_n = v \text{ strongly in } V, \tag{2.75}$$

$$\lim_{n\to\infty} v_n = v \text{ in } L^\infty(\Omega) \text{ weak} *. \tag{2.76}$$

Since $v_n \in \mathscr{D}(\Omega)$, from (2.74) we have

$$a(u, v_n) + \int_\Omega \phi(u) v_n \, dx = L(v_n). \tag{2.77}$$

From (2.77) it follows that $\lim_{n\to\infty} a(u, v_n) = a(u, v)$, $\lim_{n\to\infty} L(v_n) = L(v)$, and since $\phi(u) \in L^1(\Omega)$, (2.76) implies

$$\lim_{n\to\infty} \int_\Omega \phi(u) v_n \, dx = \int_\Omega \phi(u) v \, dx.$$

Thus, taking the limit in (2.77), we obtain

$$a(u, v) + \int_\Omega \phi(u) v \, dx = L(v), \qquad \forall\, v \in V \cap L^\infty(\Omega).$$

Therefore (P) implies (2.47) which in turn implies (π). This completes the proof of the theorem. □

EXERCISE 2.2. In $\mathbb{R}^2$, find a function v such that $v \in H^{-1}(\Omega) \cap L^1(\Omega)$, $v \notin L^p(\Omega)$, $\forall\, p > 1$, where Ω is some bounded open set in $\mathbb{R}^2$.

EXERCISE 2.3. Prove that if $u \geq 0$ a.e., then $\phi(u)v \in L^1(\Omega)$, $\forall\, v \in V$, where u is the solution of the problem (P).

2.4. Some comments on the continuous problem

We have studied (P) and (π) with rather weak hypotheses, namely, $\phi \in C^0(\mathbb{R})$ and nondecreasing, and $f \in V^*$. The proof we have given for the equivalence between (P) and (π) can be shortened by using more sophisticated tools of convex analysis and the theory of monotone operators (see Lions [1] and the bibliography therein). However, our proof is very elementary, and some of the lemmas we have obtained will be useful in the numerical analysis of the problem (P). Regularity results for problems a little more complicated than (P) and (π) are given in Brezis, Crandall, and Pazy [1]; in particular, for $f \in L^2(\Omega)$ and with convenient smoothness assumptions on A, the $H^2(\Omega)$-regularity of u is proved.

2.5. Finite element approximation of (π) and (P)

2.5.1. *Definition of the approximate problem*

Let Ω be a bounded polygonal domain of $\mathbb{R}^2$ and let $\mathscr{T}_h$ be a triangulation of Ω satisfying (2.21)–(2.23) of Chapter II. We approximate V by

$$V_h = \{v_h \in C^0(\overline{\Omega}),\ v_h|_\Gamma = 0,\ v_h|_T \in P_1, \qquad \forall\, T \in \mathscr{T}_h\}.$$

Then it is natural to approximate (P) and (π), respectively, by

$$a(u_h, v_h) + \int_\Omega \phi(v_h)v_h\,dx = L(v_h), \qquad \forall\, v_h \in V_h, \quad u_h \in V_h \tag{P$_h^*$}$$

and

$$a(u_h, v_h - u_h) + j(v_h) - j(u_h) \geq L(v_h - u_h), \qquad \forall\, v_h \in V_h, \quad u_h \in V_h \tag{π_h^*}$$

with

$$j(v_h) = \int_\Omega \Phi(v_h)\,dx.$$

Obviously (P_h^*) and (π_h^*) are equivalent. From a computational point of view, we cannot generally use (P$_h^*$) and (π_h^*) directly, since they involve the computation of integrals which cannot be done exactly. For this reason we shall have to modify (π_h^*) and (P$_h^*$) by using some numerical integration procedures. Actually we shall have to approximate $a(\cdot, \cdot)$, $L(\cdot)$, and $j(\cdot)$. Since the approximation of $(\cdot, \cdot)$ and $L(\cdot)$ is studied in Ciarlet [1, Chapter 8], we shall assume that we still work with $a(\cdot, \cdot)$ and $L(\cdot)$, but we shall approximate $j(\cdot)$.

Figure 2.3

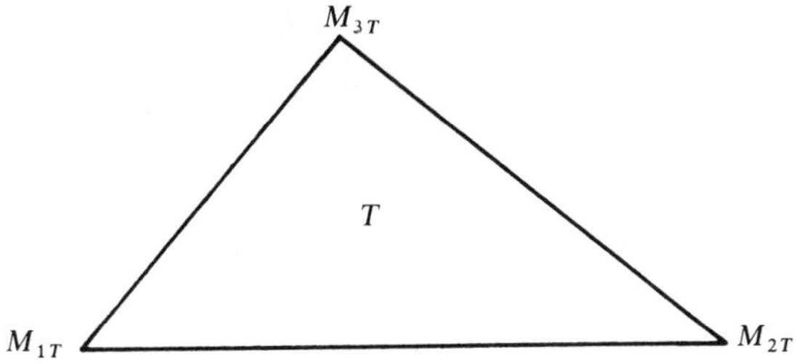

To approximate $j(\cdot)$ we shall use the *two-dimensional trapezoidal method.* Hence, using the notation of Fig. 2.3, we approximate $j(\cdot)$ by

$$j_h(v_h) = \sum_{T \in \mathcal{T}_h} \frac{\text{meas}(T)}{3} \sum_{i=1}^{3} \Phi(v_h(M_{iT})), \qquad \forall\, v_h \in V_h. \tag{2.78}$$

Actually $j_h(v_h)$ may be viewed as the exact integral of some piecewise constant function.

Using the notation of Chapter II, Sec. 2.5, assume that the set Σ_h of the nodes of $\mathcal{T}_h$ has been ordered by $i = 1, 2, \ldots, N_h$, where $N_h = \text{Card}(\Sigma_h)$. Let $M_i \in \Sigma_h$. We define a domain Ω_i by joining, as in Fig. 2.4, the centroids of the triangles, admitting M_i as a common vertex, to the midpoint of the edges admitting M_i as a common extremity (if M_i is a boundary point, the modification of Fig. 2.4 is trivial).

Let us define the space of piecewise constant functions

$$L_h = \left\{\mu_h \,|\, \mu_h = \sum_{i=1}^{N_h} \mu_i \chi_i,\ \mu_i \in \mathbb{R},\ i = 1, 2, \ldots, N_h\right\}, \tag{2.79}$$

where χ_i is the characteristic function of Ω_i.

We then define $q_h\colon C^0(\overline{\Omega}) \cap H_0^1(\Omega) \to L_h$ by

$$q_h v = \sum_{i=1}^{N_h} v(M_i)\chi_i. \tag{2.80}$$

Then it follows from (2.79) and (2.80) that

$$j_h(v_h) = \int_\Omega \Phi(q_h v_h)\, dx, \qquad \forall\, v_h \in V_h, \tag{2.81}$$

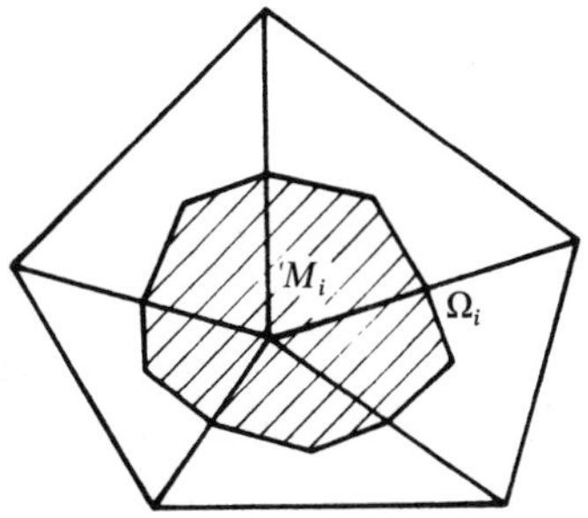

Figure 2.4

implying that

$$j_h(v_h) = j(q_h v_h), \qquad \forall\, v_h \in V_h. \tag{2.82}$$

Then we approximate (P) and (π) by

$$a(u_h, v_h) + \int_\Omega \phi(q_h u_h) q_h v_h\, dx = L(v_h), \qquad \forall\, v_h \in V_h, \quad u_h \in V_h \tag{P_h}$$

and

$$a(u_h, v_h - u_h) + j_h(v_h) - j_h(u_h) \ge L(v_h - u_h), \qquad \forall\, v_h \in V_h, \quad u_h \in V_h. \tag{π_h}$$

Then:

Theorem 2.3. *Problems* (P_h) *and* (π_h) *are equivalent and have a unique solution.*

EXERCISE 2.4. Prove Theorem 2.3.

2.5.2. *Convergence of the approximate solutions*

Theorem 2.4. *If, as* $h \to 0$, *the angles of* $\mathscr{T}_h$ *are uniformly bounded below by* $\theta_0 > 0$, *then*

$$\lim_{h \to 0} \|u_h - u\|_V = 0,$$

where u *and* u_h *are, respectively, the solution of* (P) *and* (P_h).

PROOF. Since $j(\cdot)$ is *not continuous* on V, the result of Chapter I, Sec. 6 on the approximation of the EVI of the second kind cannot be applied directly. However, the proof of convergence follows the same lines as in Theorem 6.3 of Chapter I.

(1) *A priori estimates for* u_h. Taking $v_h = 0$ in (π_h), we obtain

$$\|u_h\|_V \le \frac{\|f\|_*}{\alpha}, \tag{2.83}$$

$$0 \le \int_\Omega \Phi(q_h u_h)\, dx \le \frac{\|f\|_*^2}{\alpha}. \tag{2.84}$$

(2) *Weak convergence of* u_h. From (2.83) and from the compactness of the injection of V in $L^2(\Omega)$, it follows that we can extract from $(u_h)_h$ a subsequence, still denoted by $(u_h)_h$, such that

$$u_h \to u^* \text{ weakly in } V, \tag{2.85}$$

$$u_h \to u^* \text{ strongly in } L^2(\Omega), \tag{2.86}$$

$$u_h \to u^* \text{ a.e. in } \Omega. \tag{2.87}$$

Admitting, for the moment, the following inequality:

$$\|q_h v_h - v_h\|_{L^p(\Omega)} \le \frac{2h}{3} \|\nabla v\|_{L^p(\Omega) \times L^p(\Omega)}, \qquad \forall\, v_h \in V_h, \quad \forall\, p \text{ with } 1 \le p \le \infty, \tag{2.88}$$

it follows from (2.83) and (2.86) that

$$q_h u_h \to u^* \text{ strongly in } L^2(\Omega). \tag{2.89}$$

Then, modulo another extraction of a subsequence, we have

$$q_h u_h \to u^* \text{ a.e. in } \Omega, \tag{2.90}$$

from which it follows that

$$\Phi(q_h u_h) \to \Phi(u^*) \text{ a.e. in } \Omega. \tag{2.91}$$

Then taking $v \in \mathscr{D}(\Omega)$, it follows from Ciarlet [1], [2] and Strang and Fix [1] that under the assumptions on $\mathscr{T}_h$ of the statement of the theorem, we have

$$\|r_h v - v\|_{W^{1,\infty}(\Omega)} \le Ch\|v\|_{W^{2,\infty}(\Omega)}, \qquad \forall\, v \in \mathscr{D}(\Omega), \tag{2.92}$$

$$\|r_h v - v\|_{L^\infty(\Omega)} \le Ch^2\|v\|_{W^{2,\infty}(\Omega)}, \qquad \forall\, v \in \mathscr{D}(\Omega), \tag{2.93}$$

where C is a constant independent of v and h and r_h is the usual linear interpolation operator over $\mathscr{T}_h$. Moreover, (2.88) with $p = +\infty$, (2.92), and (2.93) imply

$$\lim_{h\to 0} \|q_h r_h v - v\|_{L^\infty(\Omega)} = 0, \qquad \forall\, v \in \mathscr{D}(\Omega). \tag{2.94}$$

Taking $v_h = r_h v$ in (π_h), we obtain

$$a(u_h, u_h) + \int_\Omega \Phi(q_h u_h)\, dx \le a(u_h, r_h v) + \int_\Omega \Phi(q_h r_h v)\, dx - L(r_h v - u_h), \qquad \forall\, v \in \mathscr{D}(\Omega). \tag{2.95}$$

From (2.85), (2.89), and Lemma 2.1 we have

$$a(u^*, u^*) + \int_\Omega \Phi(u^*)\, dx \le \liminf\left(a(u_h, u_h) + \int_\Omega \Phi(q_h u_h)\, dx\right).$$

Moreover,

$$\lim_{h\to 0} \int_\Omega \Phi(q_h r_h v)\, dx = \int_\Omega \Phi(v)\, dx = j(v), \qquad \forall\, v \in \mathscr{D}(\Omega).$$

Then in the limit in (2.95) we obtain

$$a(u^*, u^*) + j(u^*) \le a(u^*, v) + j(v) - L(v - u^*), \qquad \forall\, v \in \mathscr{D}(\Omega). \tag{2.96}$$

From Fatou's lemma applied to (2.84) and (2.91) we obtain

$$\Phi(u^*) \in L^1(\Omega). \tag{2.97}$$

Then from (2.96) and (2.97) it follows that u^* satisfies

$$a(u^*, v - u^*) + j(v) - j(u^*) \ge L(v - u^*), \qquad \forall\, v \in \mathscr{D}(\Omega), \quad u^* \in V, \quad \Phi(u^*) \in L^1(\Omega). \tag{2.98}$$

We now take $v \in V \cap L^\infty(\Omega)$; from Lemma 2.5 it follows that there exists a sequence $(v_n)_n$, $v_n \in \mathscr{D}(\Omega)$, such that

$$\lim_{n\to\infty} v_n = v \text{ strongly in } V, \tag{2.99}$$

$$\lim_{n\to\infty} v_n = v \text{ in } L^\infty(\Omega) \text{ weak}*. \tag{2.100}$$

From (2.98) we have

$$a(u^*, v_n - u^*) + j(v_n) - j(u^*) \geq L(v_n - u^*), \qquad \forall\, n, \quad u^* \in V, \quad \Phi(u^*) \in L^1(\Omega). \quad (2.101)$$

From (2.99) we obviously have

$$\lim_{n\to\infty} a(u^*, v_n - u^*) = a(u^*, v - u^*),$$

$$\lim_{n\to\infty} L(v_n - u^*) = L(v - u).$$

Since $v_n \to v$ in the weak $*$ topology of $L^\infty(\Omega)$, we have a constant C such that

$$\|v_n\|_{L^\infty(\Omega)} \leq C, \qquad \forall\, n. \quad (2.102)$$

Moreover, for some subsequence, (2.99) implies

$$\lim_{n\to\infty} v_n = v \text{ a.e. in } \Omega. \quad (2.103)$$

From (2.103) we obtain

$$\Phi(v_n) \to \Phi(v) \text{ a.e. in } \Omega.$$

From (2.102) and (2.103) one can easily see that the Lebesgue dominated convergence theorem can be applied to $(\Phi(u_n))_n$. Hence we obtain

$$\lim_{n\to\infty} j(v_n) = \lim_{n\to\infty} \int_\Omega \Phi(v_n)\, dx = \int_\Omega \Phi(v)\, dx = j(v).$$

Therefore, taking the limit in (2.101), we obtain

$$a(u^*, v - u^*) + j(v) - j(u^*) \geq L(v - u)^*, \qquad \forall\, v \in V \cap L^\infty(\Omega),$$
$$u^* \in V, \quad \Phi(u^*) \in L^1(\Omega). \quad (2.104)$$

Since from Lemma 2.2 we know that (2.104) is equivalent to (π), we have proved that $u^* = u$, where u is the solution of (π). From the uniqueness of the solution of (π), it follows that the whole sequence $(u_h)_h$ converges to u.

(3) *Strong convergence of* $(u_h)_h$. From the V-ellipticity of $a(\cdot,\cdot)$ and from the variational inequality satisfied by u_h, we obtain

$$\alpha\|u_h - u\|_V^2 + j_h(u_h) \leq a(u_h - u, u_h - u) + j_h(u_h) \leq -a(u_h, u) + a(u, u) + a(u_h, u_h)$$
$$- a(u, u_h) + j_h(u_h) \leq -a(u_h, u) + a(u, u) + a(u_h, r_h v) + j_h(r_h v) - L(r_h v - u_h)$$
$$- a(u, u_h), \qquad \forall\, v \in \mathscr{D}(\Omega). \quad (2.105)$$

Using the various convergence results of part (2), from (2.105) we obtain

$$j(u) \leq \liminf j_h(u_h) \leq \liminf(\alpha\|u_h - u\|_V^2 + j_h(u_h))$$
$$\leq \limsup(\alpha\|u_h - u\|_V^2 + j_h(u_h)) \leq a(u, v - u) + j(v) - L(v - u), \qquad \forall\, v \in \mathscr{D}(\Omega). \quad (2.106)$$

As in part (2), using the density of $\mathscr{D}(\Omega)$ in $L^\infty(\Omega) \cap V$ (for the strong topology of V and the weak $*$ topology of $L^\infty(\Omega)$), we find that (2.106) also holds for all $v \in V \cap L^\infty(\Omega)$. Then

$$j(u) \leq \liminf j_h(u_h) \leq \liminf(\alpha\|u_h - u\|_V^2 + j_h(u_h))$$
$$\leq \limsup(\alpha\|u_h - u\|_V^2 + j_h(u_h)) \leq a(u, \tau_n v - u) + j(\tau_n v) - L(\tau_n v - v), \qquad \forall\, v \in V. \quad (2.107)$$

Using the properties of $\tau_n v$, it then follows from (2.107), by taking the limit as $n \to \infty$, that (2.106) also holds for all $v \in V$. Hence, by taking $v = u$, we obtain

$$\begin{aligned} j(u) &\le \liminf j_h(u_h) \le \liminf(\alpha\|u_h - u\|_V^2 + j_h(u_h)) \\ &\le \limsup(\alpha\|u_h - u\|_V^2 + j_h(u_h)) \le j(u), \end{aligned}$$

which implies

$$\lim_{h\to 0} j_h(u_h) = j(u),$$

$$\lim_{h\to 0} \|u_h - u\|_V = 0.$$

This proves the theorem, modulo the proof of (2.88). □

We now prove (2.88).

Lemma 2.6. *We have* $\forall\, p,\ 1 \le p \le \infty,\ \|q_h v_h - v_h\|_{L^p(\Omega)} \le \frac{2}{3}h\|\nabla v_h\|_{L^p(\Omega)\times L^p(\Omega)}$ *where* q_h, v_h *are as before.*

PROOF. We use the notation of Sec. 2.5.1. We have (see Fig. 2.5)

$$|v_h(M) - q_h v_h(M)| = |v_h(M_i) - v_h(M)|, \qquad \forall\, M \in \Omega_i \cap T. \tag{2.108}$$

But since $v_h|_T \in P_1$, we have

$$v_h(M) = v_h(M_i) + \overrightarrow{M_iM}\cdot\nabla v_h, \qquad \forall\, M \in \Omega_i \cap T,$$

from which, combined with (2.108), it follows that

$$|q_h v_h(M) - v_h(M)| \le |\overrightarrow{M_iM}||\nabla v_h|, \qquad \forall\, M \in \Omega_i \cap T.$$

But from the definition of h, we have $|\overrightarrow{M_iM}| \le \frac{2}{3}h$, $\forall\, T$, so that we finally have

$$|q_h v_h(x) - v_h(x)| \le \tfrac{2}{3}h|\nabla v_h(x)|$$

a.e. in Ω, $\forall\, v_h \in V_h$. This implies

$$\|q_h v_h - v_h\|_{L^p(\Omega)} \le \tfrac{2}{3}h\|\nabla v_h\|_{L^p(\Omega)\times L^p(\Omega)}.$$

This proves the lemma. □

Remark 2.4. We have not considered the problem of error estimates. This problem is discussed in Glowinski [5].

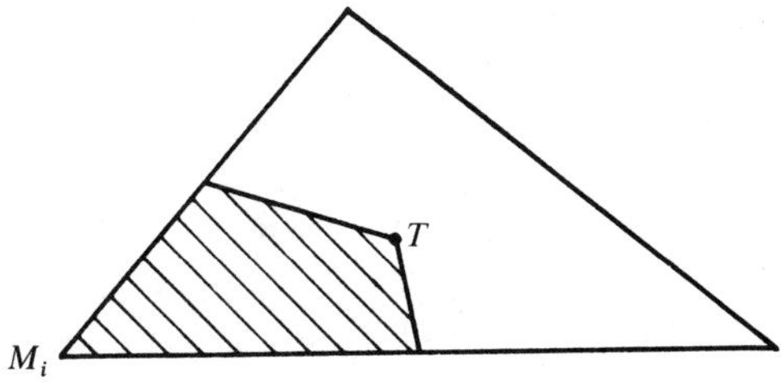

Figure 2.5

Remark 2.5. The numerical analysis of problems like (P), but with much stronger hypotheses on $a(\cdot, \cdot)$, ϕ, and f, is considered in Ciarlet, Schultz, and Varga [1].

2.6. Iterative methods for solving the discrete problem

2.6.1. *Introduction*

In this section we briefly describe some iterative methods which may be useful for computing the solution of (P_h) (and (π_h)). Actually most of these methods can be extended to other nonlinear problems. Many of the methods to be described here can be found in Ortega and Rheinboldt [1]. A method based on duality techniques will be described in Chapter VI.

2.6.2. *Formulation of the discrete problem*

Here we are using the notation of the continuous problem. Taking as unknowns the values of u_h at the interior nodes of $\mathscr{T}_h$, the problem (P_h) reduces to the finite-dimensional nonlinear problem

$$\mathbf{A}\mathbf{u} + \mathbf{D}\phi(\mathbf{u}) = \mathbf{f}, \tag{2.109}$$

where $\mathbf{A}$ is a $N \times N$ positive definite matrix, $\mathbf{D}$ is a diagonal matrix with positive diagonal elements d_i and where $\mathbf{u} = \{u_1, \ldots, u_N\} \in \mathbb{R}^N$, $\mathbf{f} \in \mathbb{R}^N$, $\phi(\mathbf{u}) \in \mathbb{R}^N$ with $(\phi(\mathbf{u}))_i = \phi(u_i)$. Clearly, from the properties of $\mathbf{A}, \mathbf{D}, \phi, \mathbf{f}$, we can see that problem (2.109) has a unique solution.

2.6.3. *Gradient Methods*

The basic algorithm with constant step (see Cea [1]) is given by

$$\mathbf{u}^0 \in \mathbb{R}^N \text{ given}, \tag{2.110}$$

$$\mathbf{u}^{n+1} = \mathbf{u}^n - \rho\mathbf{S}^{-1}(\mathbf{A}\mathbf{u}^n + \mathbf{D}\phi(\mathbf{u}^n) - \mathbf{f}), \qquad \rho > 0. \tag{2.111}$$

In (2.111), $\mathbf{S}$ is a symmetric positive-definite matrix; a canonical choice is $\mathbf{S} =$ identity. But in most problems it will yield a slow speed of convergence. If $\mathbf{A}$ is symmetrical, the natural choice is $\mathbf{S} = \mathbf{A}$, and if $\mathbf{A} \neq \mathbf{A}^*$, we can take $\mathbf{S} = (\mathbf{A} + \mathbf{A}^*)/2$.

For the convergence of $\mathbf{u}^n$ to $\mathbf{u}$ (where $\mathbf{u}$ is the solution of (2.109)), it is sufficient to have ϕ smooth enough (for example, ϕ locally Lipschitz continuous). Then $\lim_{n\to\infty} \mathbf{u}^n = \mathbf{u}$ if ρ is sufficiently small. Obviously the closer $\mathbf{u}^0$ is to $\mathbf{u}$, the faster is the convergence.

Remark 2.6. If $\mathbf{A} = \mathbf{A}^*$, then $\mathbf{A}\mathbf{v} + \mathbf{D}\phi(\mathbf{v}) - \mathbf{f}$ is the gradient at $\mathbf{v}$ of the functional $J(\mathbf{v}) = \frac{1}{2}(\mathbf{A}\mathbf{v}, \mathbf{v}) + \sum_{i=1}^N d_i\Phi(v_i) - (\mathbf{f}, \mathbf{v})$, where $(\cdot, \cdot)$ denotes the usual inner product of $\mathbb{R}^N$ and $\Phi(t) = \int_0^t \phi(\tau)\, d\tau$.

Remark 2.7. In each specific case, ρ has to be determined; this can be done theoretically, experimentally, or by using an automatic adjustment procedure which will not be described here.

Remark 2.8. Let us define $\mathbf{g}^n$ by

$$\mathbf{g}^n = \mathbf{A}\mathbf{u}^n + \mathbf{D}\phi(\mathbf{u}^n) - \mathbf{f}.$$

Instead of using a constant parameter ρ, we can use a family $(\rho_n)_n$ of positive parameters in (2.111). Therefore (2.111) can be written as

$$\mathbf{u}^{n+1} = \mathbf{u}^n - \rho_n \mathbf{S}^{-1}\mathbf{g}^n. \tag{2.112}$$

Suppose $\mathbf{A} = \mathbf{A}^*$; then if we use (2.110), (2.112) with ρ_n defined by

$$J(\mathbf{u}^n - \rho_n \mathbf{S}^{-1}\mathbf{g}^n) \leq J(\mathbf{u}^n - \rho\mathbf{S}^{-1}\mathbf{g}^n), \qquad \forall\, \rho \in \mathbb{R}, \quad \rho_n \in \mathbb{R}, \tag{2.113}$$

the resulting algorithm is, for obvious reasons, called the *steepest descent method*. This algorithm is convergent for $\phi \in C^0(\mathbb{R})$. We observe that at each iteration the determination of ρ_n requires the solution of a one-dimensional problem; for the solution of such one-dimensional problems, see Householder [1], Polak [1], and Brent [1].

Remark 2.9. At each iteration of (2.110), (2.111) or (2.110), (2.112), we have to solve a linear system related to $\mathbf{S}$. Since $\mathbf{S}$ is symmetric and positive definite, this system can be solved using the Cholesky method, provided the $\mathbf{S} = \mathbf{L}\mathbf{L}^*$ factorization has been done. From a practical point of view, it is obvious that the factorization of $\mathbf{S}$ will be made in the beginning once and for all. Then at each iteration we only have to solve two triangular systems, which is a trivial operation.

2.6.4. *Newton's method*

The Newton's algorithm is given by (for sufficient conditions of convergence, see Ortega and Rheinboldt [1])

$$\mathbf{u}^0 \in \mathbb{R}^N \text{ given}, \tag{2.114}$$

$$\mathbf{u}^{n+1} = (\mathbf{A} + \mathbf{D}\phi'(\mathbf{u}^n))^{-1}(\mathbf{D}\phi'(\mathbf{u}^n)\mathbf{u}^n - \mathbf{D}\phi(\mathbf{u}^n) + \mathbf{f}), \tag{2.115}$$

where $\phi'(\mathbf{v})$ denotes the diagonal matrix

$$\phi'(\mathbf{v}) = \begin{pmatrix} \phi'(v_1) & & 0 \\ & \ddots & \\ 0 & & \phi'(v_n) \end{pmatrix}.$$

Since ϕ is nondecreasing, $\phi' \geq 0$. This implies that $\mathbf{A} + \mathbf{D}\phi'(\mathbf{v})$ is positive definite, $\forall\, \mathbf{v} \in \mathbb{R}^N$.

Remark 2.10. At each iteration we have to solve a linear system. Since the matrix $\mathbf{A} + \mathbf{D}\phi'(\mathbf{u}^n)$ depends on n, this method may not be convenient for large N; for this reason a large number of variants of the Newton's method which avoid this drawback have been designed. In this regard, let us mention, among others, Broyden [1], [2], Dennis and More [1], Crisfield [1], Matthies and Strang [1], O'Leary [1], and also the references therein; most of the above references deal with the so-called *quasi-Newton's methods.*

Remark 2.11. The choice of $\mathbf{u}^0$ is very important when using Newton's method.

2.6.5. *Relaxation and over-relaxation methods.*

In this section we shall discuss the application of relaxation methods for solving the specific problem (2.109); the algorithms to be described below are, in fact, particular cases of a large family of algorithms to be discussed in Chapter V.

In this section we use the following notation:

$$\mathbf{A} = (a_{ij})_{1 \leq i, j \leq N},$$
$$\mathbf{f} = \{f_1, f_2, \ldots, f_N\}.$$

Since $\mathbf{A}$ is positive definite, we have $a_{ii} > 0$, $\forall\, i = 1, 2, \ldots, N$. Here we will describe three algorithms.

ALGORITHM 1

$$\mathbf{u}^0 \in \mathbb{R}^N \text{ given}; \tag{2.116}$$

then with $\mathbf{u}^n$ *known, we compute* $\mathbf{u}^{n+1}$, *component by component, using*

$$a_{ii}\bar{u}_i^{n+1} + d_i\phi(\bar{u}_i^{n+1}) = f_i - \sum_{j<i} a_{ij}u_j^{n+1} - \sum_{j>i} a_{ij}u_j^n, \tag{2.117}$$

$$u_i^{n+1} = u_i^n + \omega(\bar{u}_i^{n+1} - u_i^n), \qquad i = 1, 2, \ldots, N. \tag{2.118}$$

Since $a_{ii} > 0$, $\phi \in C^0(\mathbb{R})$ and ϕ is a nondecreasing function, we can always solve (2.117), and the solution is unique.

If $\omega = 1$, we recover an *ordinary relaxation method* (see Cea [1]). If $\mathbf{A} = \mathbf{A}^*$ and since ϕ is C^0 and nondecreasing, the solution $\mathbf{u}^n$ of (2.116)–(2.118) converges to the solution $\mathbf{u}$ of (2.109).

If, in (2.109), $\mathbf{A}$ is nonsymmetrical and $\omega \neq 1$, there are certain sufficient conditions for the convergence of $\mathbf{u}^n$ to $\mathbf{u}$, where $\mathbf{u}^n$ is given by (2.116)–(2.118) and where $\mathbf{u}$ is the solution of (2.109) (see Ortega and Rheinboldt [1] and S. Schechter [1], [2], [3]).

ALGORITHM 2. This algorithm is the variant of (2.116)–(2.118) obtained by replacing (2.117) and (2.118) by

$$a_{ii}u_i^{n+1} + d_i\phi(u_i^{n+1}) = (1 - \omega)(a_{ii}u_i^n + d_i\phi(u_i^n))$$
$$+ \omega\left(f_i - \sum_{j<i} a_{ij}u_j^{n+1} - \sum_{j>i} a_{ij}u_j^n\right) \quad \text{for } i = 1, 2, \ldots, N. \tag{2.119}$$

Remark 2.12. If $\omega = 1$ and/or ϕ is linear, the two algorithms coincide. In the general case the convergence of (2.116), (2.119) seems to be an open question. However, from numerical experiments it seems that the second algorithm is more "robust" that the first, perhaps because it is more implicit. Furthermore, it can be used even if ϕ is defined only on a bounded or semibounded interval $]\alpha, \beta[$ of $\mathbb{R}$ such that $\phi(\alpha) = -\infty$, $\phi(\beta) = +\infty$; in such a case, if $\phi \in C^0(]\alpha, \beta[)$ and ϕ is increasing, (2.109) has still a unique solution, but the use of (2.116)–(2.118) with $\omega > 1$ may be dangerous.

Remark 2.13. If $\phi \in C^1(\mathbb{R})$, an efficient method for computing $\bar{u}_i^{n+1}$ in (2.117) and u_i^{n+1} in (2.119) is the *one-dimensional Newton's method.*

Let $g \in C^1(\mathbb{R})$. In this case, Newton's algorithm for solving the equation $g(x) = 0$ is

$$x^0 \in \mathbb{R} \text{ given,} \tag{2.120}$$

$$x^{n+1} = x^n - \frac{g(x^n)}{g'(x^n)}. \tag{2.121}$$

If in the computation of $\bar{u}_i^{n+1}$ and u_i^{n+1}, we use only one iteration of Newton's method, starting from u_i^n, then the resulting algorithms are identical and we obtain:

$$\mathbf{u}^0 \in \mathbb{R}^N \text{ given,} \tag{2.122}$$

$$u_i^{n+1} = u_i^n - \omega \frac{\sum_{j<i} a_{ij} u_j^{n+1} + \sum_{j \geq i} a_{ij} u_j^n + d_i \phi(u_i^n) - f_i}{a_{ii} + d_i \phi'(u_i^n)}, \qquad i = 1, 2, \ldots, N. \tag{2.123}$$

In S. Schechter [1], [2], [3], sufficient conditions for the convergence of (2.122), (2.123) are given.

Remark 2.14. If $\omega > 1$ (resp., $\omega = 1$, $\omega < 1$), the previous algorithms are over-relaxation (resp., relaxation, under relaxation) algorithms.

Remark 2.15. In Glowinski and Marrocco [1], [2], we can find applications of relaxation methods for solving the nonlinear elliptic equations modeling the magnetic state of electrical machines.

2.6.6. *Alternating-direction methods.*

In this section we take $\rho > 0$. Here we will give two numerical methods for solving (2.109).

First method

$$\mathbf{u}^0 \in \mathbb{R}^N \textit{ given;} \tag{2.124}$$

knowing $\mathbf{u}^n$, *we compute* $\mathbf{u}^{n+1/2}$ *by*

$$\rho\mathbf{u}^{n+1/2} + \mathbf{A}\mathbf{u}^{n+1/2} = \rho\mathbf{u}^n - \mathbf{D}\phi(\mathbf{u}^n) + \mathbf{f}; \tag{2.125}$$

then we calculate $\mathbf{u}^{n+1}$ *by*

$$\rho\mathbf{u}^{n+1} + \mathbf{D}\phi(\mathbf{u}^{n+1}) = \rho\mathbf{u}^{n+1/2} - \mathbf{A}\mathbf{u}^{n+1/2} + \mathbf{f}. \tag{2.126}$$

For the convergence of (2.124)–(2.126), see R. B. Kellog [1].

Second method

$$\mathbf{u}^0 \in \mathbb{R}^N \text{ given}; \tag{2.127}$$

knowing $\mathbf{u}^n$, *we compute* $\mathbf{u}^{n+1/2}$ *by*

$$\rho\mathbf{u}^{n+1/2} + \mathbf{A}\mathbf{u}^{n+1/2} = \rho\mathbf{u}^n - \mathbf{D}\phi(\mathbf{u}^n) + \mathbf{f}; \tag{2.128}$$

then we calculate $\mathbf{u}^{n+1}$ *by*

$$\rho\mathbf{u}^{n+1} + \mathbf{D}(\mathbf{u}^{n+1}) = \rho\mathbf{u}^n - \mathbf{A}\mathbf{u}^{n+1/2} + \mathbf{f}. \tag{2.129}$$

Using the results of Lieutaud [1], we can prove that, for all $\rho > 0$, $\mathbf{u}^{n+1/2}$ and $\mathbf{u}^n$ converge to $\mathbf{u}$ if we suppose that $\mathbf{A}$ and ϕ satisfy the hypotheses given in Sec. 2.6.2; the same convergence result holds if we exchange the role of $\mathbf{A}$ and $\mathbf{D}\phi(\cdot)$ in (2.128), (2.129).

Remark 2.16. At each iteration we have to solve a linear system whose matrix is constant, since we use a constant step ρ. This is an advantage from the computational point of view (cf. Remark 2.9).

We also have to solve a nonlinear system of N equations, but in fact these equations are independent of each other and reduce to N nonlinear equations in one variable, which can be solved easily.

2.6.7. *Conjugate gradient methods.*

In this section we assume $\mathbf{A} = \mathbf{A}^*$ (if $\mathbf{A} \neq \mathbf{A}^*$, we can also use methods of conjugate gradient type). For a detailed study of these methods, we refer to Polak [1], Daniel [1], and Concus and Golub [1]. If the functional J defined in Remark 2.6 is not quadratic (i.e., ϕ is nonlinear), several conjugate gradient methods can be used. Let us describe two of them, whose convergence is studied in Polak [1]. Let J be given by

$$J(\mathbf{v}) = \tfrac{1}{2}(\mathbf{A}\mathbf{v}, \mathbf{v}) + \sum_{i=1}^{N} d_i \Phi(v_i) - (\mathbf{f}, \mathbf{v}),$$

where $\Phi(t) = \int_0^t \phi(\tau)\, d\tau$, ϕ being a nondecreasing continuous function on $\mathbb{R}$ with $\phi(0) = 0$.

Let S be a $N \times N$ symmetric positive-definite matrix.

First method (Fletcher–Reeves)

$$\mathbf{u}^0 \in \mathbb{R}^N \text{ given;} \tag{2.130}$$

$$\mathbf{g}^0 = \mathbf{S}^{-1}(\mathbf{A}\mathbf{u}^0 + \mathbf{D}\phi(\mathbf{u}^0) - \mathbf{f}), \tag{2.131}$$

$$\mathbf{w}^0 = \mathbf{g}^0. \tag{2.132}$$

Then assuming that $\mathbf{u}^n$ *and* $\mathbf{w}^n$ *are known, we compute* $\mathbf{u}^{n+1}$ *by*

$$\mathbf{u}^{n+1} = \mathbf{u}^n - \rho_n \mathbf{w}^n, \tag{2.133}$$

where ρ_n *is the solution of the one-dimensional minimization problem*

$$J(\mathbf{u}^n - \rho_n \mathbf{w}^n) \le J(\mathbf{u}^n - \rho \mathbf{w}^n), \qquad \forall \rho \in \mathbb{R}, \quad \rho_n \in \mathbb{R}. \tag{2.134}$$

Then

$$\mathbf{g}^{n+1} = \mathbf{S}^{-1}(\mathbf{A}\mathbf{u}^{n+1} + \mathbf{D}\phi(\mathbf{u}^{n+1}) - \mathbf{f}), \tag{2.135}$$

and compute $\mathbf{w}^{n+1}$ *by*

$$\mathbf{w}^{n+1} = \mathbf{g}^{n+1} + \lambda_n \mathbf{w}^n, \tag{2.136}$$

where

$$\lambda_n = \frac{(\mathbf{S}\mathbf{g}^{n+1}, \mathbf{g}^{n+1})}{(\mathbf{S}\mathbf{g}^n, \mathbf{g}^n)}. \tag{2.137}$$

Second method (Polak–Ribière). This method is like the previous method, except that (2.137) is replaced by

$$\lambda_n = \frac{(\mathbf{S}\mathbf{g}^{n+1}, \mathbf{g}^{n+1} - \mathbf{g}^n)}{(\mathbf{S}\mathbf{g}^n, \mathbf{g}^n)}. \tag{2.138}$$

Remark 2.17. For the computation of ρ_n in (2.134), see Remark 2.8 (and also Shanno [1]).

Remark 2.18. From Polak [1] it follows that if ϕ is sufficiently smooth, then the convergence of the above algorithms is superlinear, i.e., faster than the convergence of any geometric sequence.

Remark 2.19. The above algorithms are very sensitive to roundoff errors; hence double precision may be required for some problems. Moreover, it may be convenient to periodically take $\mathbf{w}^n = \mathbf{g}^n$ (see Powell [2] for this *restarting procedure* problem).

Remark 2.20. At each iteration we have to solve a linear system related to $\mathbf{S}$; Remark 2.9 still applies to this problem.

2.6.8. *Comments.*

The methods of this section may be applied to more general nonlinear systems than (2.109). They can be applied, of course, to finite-dimensional systems obtained by discretization of elliptic problems like

$$-\nabla\cdot(a_0(x)\nabla u) + \beta\cdot\nabla u + \phi(x, u) = f \text{ in } \Omega, \quad \text{plus suitable boundary conditions}$$

where, for fixed x, the function $t \to \phi(x, t)$ is continuous and nondecreasing on $\mathbb{R}$.

3. A Subsonic Flow Problem

3.1. Formulation of the continuous problem

Let Ω be a domain of $\mathbb{R}^N$ (in applications, we have $N = 1, 2, 3$) with a sufficiently smooth boundary Γ. Then the flow of a perfect compressible irrotational fluid (i.e., $\nabla \times \mathbf{v} = \mathbf{0}$, where $\mathbf{v}$ is the velocity vector of the flow) is described by

$$-\nabla \cdot (\rho(\phi)\nabla\phi) = 0 \text{ in } \Omega, \tag{3.1}$$

$$\rho(\phi) = \rho_0\left(1 - \frac{|\nabla\phi|^2}{[(\gamma + 1)/(\gamma - 1)]C_*^2}\right)^{1/(\gamma - 1)}, \tag{3.2}$$

with suitable boundary conditions. Here:

- ϕ is a potential and $\nabla\phi$ is the velocity of the flow;
- $\rho(\phi)$ is the density of the flow;
- ρ_0 is the density at $\nabla\phi = 0$; in the sequel we take $\rho_0 = 1$;
- γ is the ratio of specific heats ($\gamma = 1.4$ in air);
- C_* is the critical velocity.

The flow under consideration is subsonic if

$$|\nabla\phi| < C_* \text{ everywhere in } \Omega. \tag{3.3}$$

If $|\nabla\phi| \geq C_*$ in some part of Ω, then the flow is *transonic* or *supersonic*, and this leads to much more complicated problems (see Chapter VII for an introduction to the study of such flows).

Remark 3.1. In the case of a subsonic flow past a convex symmetrical airfoil and assuming (see Fig. 3.1) that $\mathbf{v}_\infty$ is parallel to the x-axis (Ω is the complement of the airfoil in $\mathbb{R}^2$ and $\partial\phi/\partial n|_\Gamma = 0$), H. Brezis and Stampacchia [1] have proved that the subsonic problem is equivalent to an EVI of the first kind in the hodograph plane (see Bers [1] and Landau and Lifschitz [1] for the hodograph transform). This EVI is related to a linear operator, and the corresponding convex set is the cone of non-negative functions.

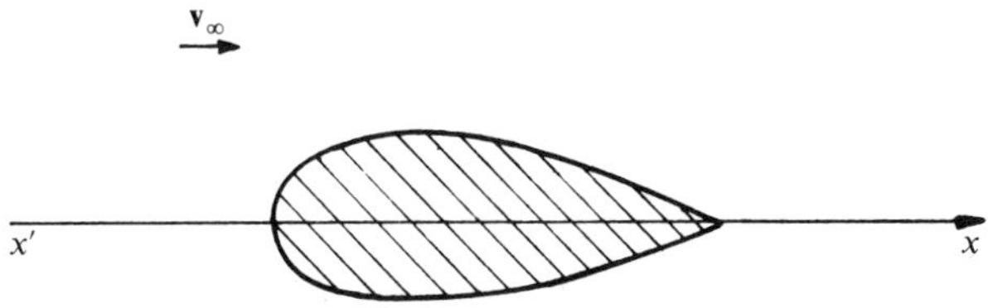

Figure 3.1

In the remainder of Sec. 3 (and also in Chapter VII), we shall work only in the physical plane, since it seems more convenient for the computation of nonsymmetric and/or transonic flows.

For the reader who is interested by the mathematical aspects of the flow mentioned above, see Bers [1] and Brezis and Stampacchia [1]. For the physical and mechanical aspects, see Landau and Lifschitz [1]. Additional references are given in Chapter VII.

3.2. Variational formulation of the subsonic problem

Preliminary Remark.

In the case of a nonsymmetric flow past an airfoil (see Fig. 3.2) the velocity potential has to be *discontinuous*, and a *circulation condition* is required to ensure the uniqueness (modulo a constant) of the solution of (3.1). If the airfoil has corners (as in Fig. 3.1), then the circulation condition is related to the so-called *Kutta–Joukowsky condition* from which it follows that for a physical flow, the velocity field is continuous at the corners (like 0 in Fig. 3.2). For more information about the Kutta–Joukowsky condition, see Landau and Lifschitz [1] (see also Chapter VII).

For the sake of simplicity, we shall assume in the sequel that either Ω is simply connected, as is the case for the nozzle of Fig. 3.3, or, if Ω is multiply connected, we shall assume (as in Fig. 3.1) that the flow is physically and geometrically symmetric, since in this case the Kutta–Joukowsky condition is automatically satisfied.

In the sequel we assume that the boundary conditions associated with (3.1), (3.2) are the following:

$$\phi = g_0 \textit{ over } \Gamma_0, \; \rho \frac{\partial \phi}{\partial n}\bigg|_{\Gamma_1} = g_1 \tag{3.4}$$

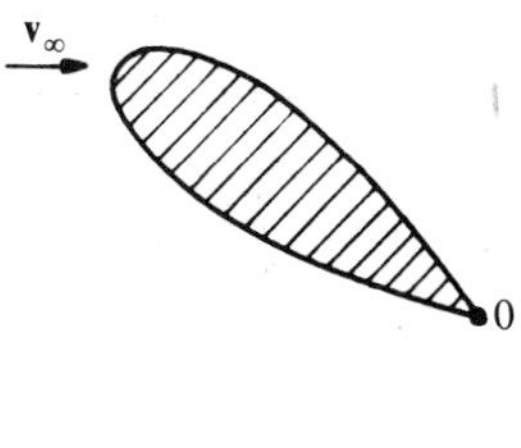

Figure 3.2

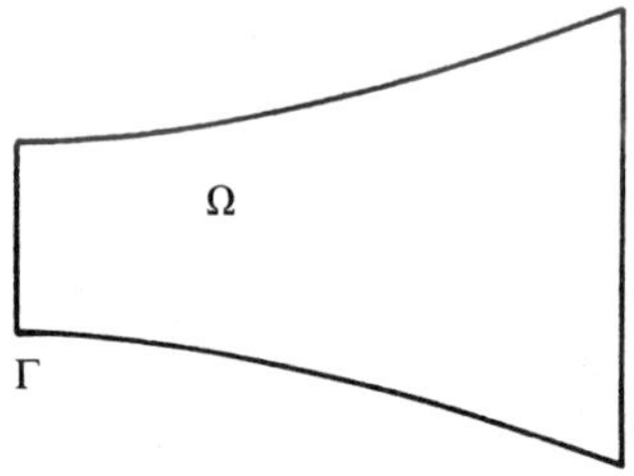

Figure 3.3

where $\Gamma_0, \Gamma_1 \subset \Gamma$ with $\Gamma_0 \cap \Gamma_1 = \varnothing$, $\Gamma_0 \cup \Gamma_1 = \Gamma$. Then the variational formulation for the flow problem (3.1), (3.2), (3.3), (3.4) is

$$\int_\Omega \rho(\phi)\nabla\phi \cdot \nabla v \, dx = \int_{\Gamma_1} g_1 v \, d\Gamma, \qquad \forall\, v \in V_0, \quad \phi \in V_{g_0}, \tag{3.5}$$

where

$$V_0 = \{v \in H^1(\Omega), v|_{\Gamma_0} = 0\}, \tag{3.6}$$

$$V_{g_0} = \{v \in H^1(\Omega), v|_{\Gamma_0} = g_0\}. \tag{3.7}$$

If g_0, g_1 are small enough, we can prove that (3.5) has a solution such that

$$|\nabla\phi| \leq M < C_* \text{ a.e.}.$$

When solving a practical flow problem, we may not know *a priori* whether the flow will be purely subsonic or not. Therefore, instead of solving (3.5), it may be convenient to consider (and solve) the following problem:

$$\int_\Omega \rho(\phi)\nabla\phi \cdot \nabla(v - \phi) \, dx \geq \int_{\Gamma_1} g_1(v - \phi) \, d\Gamma, \qquad \forall\, v \in K_\delta, \quad \phi \in K_\delta, \tag{3.8}$$

where

$$K_\delta = \{v \in V_{g_0}, |\nabla v| \leq \delta < C_* \text{ a.e.}\}. \tag{3.9}$$

The variational problem (3.8), (3.9) is an EVI of the first kind, but we have to observe that unlike the EVI's of Chapters I and II, it involves a nonlinear partial differential operator, namely, A defined by

$$A(\phi) = -\nabla \cdot (\rho(\phi)\nabla\phi).$$

Remark 3.2. In practical applications we shall take δ as close as possible to C_*.

Remark 3.3. Problem (3.8), (3.9) appears as a variant of the elasto-plastic torsion problem of Chapter II, Sec. 3.

3.3. Existence and uniqueness properties for problem (3.8)

In this section we shall assume that $\text{meas}(\Gamma_0) > 0$. To prove that (3.8) is well posed, we will use the following:

Lemma 3.1. *The function*

$$\xi \to -\left(1 - \frac{\xi^2}{[(\gamma - 1)/(\gamma - 1)]C_*^2}\right)^{\gamma/(\gamma-1)}$$

is convex if $\xi \in [0, C_*]$, *concave if*

$$\xi \in \left[C_*, \sqrt{\frac{\gamma + 1}{\gamma - 1}}\, C_*\right],$$

and strictly convex if $\xi \in [0, C_*[$.

EXERCISE 3.1. Prove Lemma 3.1.

We can now prove:

Theorem 3.1. *Assume that* Ω *is bounded and that* g_0, g_1 *are sufficiently smooth and small. Then* (3.8) *has a unique solution.*

PROOF. Since Ω is bounded, and if g_0 is sufficiently smooth and small, we observe that K_δ is a closed, convex, and nonempty bounded subset of $H^1(\Omega)$ (consisting of uniformly Lipschitz continuous functions).

Define $J(\cdot)$ by

$$J(v) = -\frac{\gamma + 1}{2\gamma} C_*^2 \int_\Omega \left(1 - \frac{|\nabla v|^2}{[(\gamma + 1)/(\gamma - 1)]C_*^2}\right)^{\gamma/(\gamma-1)} dx - \int_{\Gamma_1} g_1 v \, d\Gamma. \tag{3.10}$$

From Lemma 3.1 it follows that $J(\cdot)$ is strictly convex over K_δ. It is easy to check that $J(\cdot)$ is continuous and Gateaux-differentiable over K_δ with

$$(J'(v), w) = \int_\Omega \rho(v)\nabla v \cdot \nabla w \, dx - \int_{\Gamma_1} g_1 w \, d\Gamma. \tag{3.11}$$

Since K_δ is a closed convex nonempty subset of $H^1(\Omega)$ and $J(\cdot)$ is continuous and strictly convex over K_δ, from standard optimization theory in Hilbert spaces (see, e.g., Cea [1], [2]), it follows that the minimization problem

$$J(u) \le J(v), \qquad \forall\, v \in K_\delta, \quad u \in K_\delta, \tag{3.12}$$

has a unique solution.

Moreover, since $J(\cdot)$ is differentiable, the unique solution of (3.12) is characterized (see Cea [1], [2]) by

$$(J'(u), v - u) \ge 0, \qquad \forall\, v \in K_\delta, \quad u \in K_\delta;$$

from (3.11), this completes the proof of the theorem.

Remark 3.4. Suppose that $\Gamma_0 = \varnothing$. Then defining K_δ by

$$K_\delta = \{v \in H^1(\Omega), |\nabla v| \leq \delta < C_* \text{ a.e., } v(x_0) = v_0\}$$

with $x_0 \in \overline{\Omega}$ and v_0 arbitrarily given, we can prove that if Ω is bounded and g_1 is sufficiently smooth, then

$$\int_\Omega \rho(\phi)\nabla\phi \cdot \nabla(v - \phi)\, dx \geq \int_\Gamma g_1(v - \phi)\, d\Gamma, \qquad \forall\, v \in K_\delta, \quad \phi \in K_\delta, \tag{3.13}$$

has a unique solution (if ϕ is a solution of (3.13); then $\phi + C$ is the unique solution of the similar problem obtained by replacing v_0 by $v_0 + C$).

EXERCISE 3.2. Prove the statement of Remark 3.4.

Remark 3.5. In all the above arguments we assumed that Ω is bounded. We refer to Ciavaldini, Pogu, and Tournemine [1] which contains a careful study of the approximation of subsonic flow problems on an unbounded domain Ω_∞ by problems on a family $(\Omega_n)_n$ of bounded domains converging to Ω_∞ (actually they have obtained estimates for $\phi_\infty - \phi_n$).

The above EVI's will have a practical interest if we can prove that in the cases where a purely subsonic solution exists, then for δ large enough it is the solution of (3.8); actually this property is true and follows from:

Theorem 3.2. *Assuming the same hypotheses on Ω, g_0, g_1 as in Theorem 3.1, and that (3.1), (3.2), (3.4) has a unique solution in $H^1(\Omega)$ with*

$$|\nabla\phi| \leq \delta_0 < C_* \text{ a.e.,} \tag{3.14}$$

then ϕ is a solution of (3.8), (3.9), $\forall\, \delta \in [\delta_0, C_*[$. Conversely, if the solution of (3.8), (3.9) is such that $|\nabla\phi| \leq \delta_0 < \delta < C_*$ a.e., then ϕ is a solution of (3.1), (3.2), (3.4).

PROOF. (1) Let $\phi \in H^1(\Omega)$ satisfy (3.1), (3.2), (3.4), and (3.14). If $v \in V_0$, then using Green's formula, it follows from (3.1), (3.2), (3.4) that

$$\int_\Omega \rho(\phi)\nabla\phi \cdot \nabla v\, dx = \int_{\Gamma_1} g_1 v\, d\Gamma, \qquad \forall\, v \in V_0. \tag{3.15}$$

From (3.4), (3.15) and from the definition of V_{g_0} it follows that

$$\int_\Omega \rho(\phi)\nabla\phi \cdot \nabla(v - \phi)\, dx = \int_{\Gamma_1} g_1(v - \phi)\, d\Gamma, \qquad \forall\, v \in V_{g_0}. \tag{3.16}$$

Since $\phi \in K_\delta \subset V_{g_0}$, $\forall\, \delta \in [\delta_0, C_*[$, it follows from (3.16) that

$$\int_\Omega \rho(\phi)\nabla\phi \cdot \nabla(v - \phi)\, dx \geq \int_{\Gamma_1} g_1(v - \phi)\, d\Gamma, \qquad \forall\, v \in K_\delta, \quad \phi \in K_\delta$$

if $\delta \in [\delta_0, C_*[$; therefore ϕ is the solution of the EVI (3.8), (3.9), $\forall\, \delta \in [\delta_0, C_*[$.

(2) Define $U \subset V_0$ by

$$U = \{v \in C^\infty(\bar{\Omega}), v = 0 \text{ in a neighborhood of } \Gamma_0\}.$$

Then, if we suppose that Γ is sufficiently smooth, we find that the *closure* $\bar{U}^{H^1(\Omega)}$ of U in $H^1(\Omega)$ obeys

$$\bar{U}^{H^1(\Omega)} = V_0. \tag{3.17}$$

We assume that for $\delta < C_*$, (3.8) has a solution such that

$$|\nabla\phi| \leq \delta_0 < \delta \text{ a.e..} \tag{3.18}$$

Then $\forall\, v \in U$, and for $t > 0$ sufficiently small, $\phi + tv \in K_\delta$. Then replacing v by $\phi + tv$ in (3.8) and dividing by t, we obtain

$$\int_\Omega \rho(\phi)\nabla\phi \cdot \nabla v \, dx \geq \int_{\Gamma_1} g_1 v \, d\Gamma, \qquad \forall\, v \in U,$$

which implies

$$\int_\Omega \rho(\phi)\nabla\phi \cdot \nabla v \, dx = \int_{\Gamma_1} g_1 v \, d\Gamma, \qquad \forall\, v \in U. \tag{3.19}$$

Since $\mathscr{D}(\Omega) \subset U$, it follows from (3.19) that

$$\int_\Omega \rho(\phi)\nabla\phi \cdot \nabla v \, dx = 0, \qquad \forall\, v \in \mathscr{D}(\Omega), \tag{3.20}$$

i.e.,

$$-\nabla \cdot (\rho(\phi)\nabla\phi) = 0,$$

which proves (3.1).

Assuming (3.1) and using Green's formula, we obtain

$$\int_\Omega \rho(\phi)\nabla\phi \cdot \nabla v \, dx = \int_{\Gamma_1} \rho \frac{\partial\phi}{\partial n} v \, d\Gamma, \qquad \forall\, v \in U. \tag{3.21}$$

Using (3.17) and comparing with (3.19), we obtain

$$\rho \left.\frac{\partial\phi}{\partial n}\right|_{\Gamma_1} = g_1,$$

i.e., (3.4), which completes the proof of the theorem.

Remark 3.6. A similar theorem can be proved for the problem mentioned in Remark 3.4.

3.4. Comments

The solution of subsonic flow problems via EVI's like (3.8) or (3.13) is considered in Ciavaldini, Pogu, and Tournemine [2] (using a stream function approach) and in Fortin, Glowinski, and Marrocco [1].

Iterative methods for solving these EVI's may be found in the above references and also in Chapter VI of this book.

CHAPTER V

Relaxation Methods and Applications[1]

1. Generalities

The key idea of *relaxation methods* is to reduce, using some iterative process, the solution of some problems posed in a product space $V = \prod_{i=1}^{N} V_i$ (minimization of functionals, solution of systems of equations and/or inequalities, etc.) to the solution of a sequence of subproblems of the same kind, but simpler, since they are posed in the V_i.

A typical example of such methods is given by the classical *point* or *block Gauss–Seidel methods* and their variants (S.O.R., S.S.O.R., etc.). For the solution of finite-dimensional linear systems by methods of this type, we refer to Varga [1], Forsythe and Wasow [1], D. Young [1] and the bibliographies therein. For the solution of systems of nonlinear equations, we refer to Ortega and Rheinboldt [1], Miellou [1], [2], and the bibliographies therein.

For the minimization of convex functionals by methods of this kind, let us mention S. Schecter [1], [2], [3], Cea [1], [2], A. Auslender [1], Cryer [1], [2], Cea and Glowinski [1], Glowinski [6], and the bibliographies therein. The above list is far from complete.

The basic knowledge of convex analysis required for a good understanding of this chapter may be found in Cea [1], Rockafellar [1], and Ekeland and Temam [1].

2. Some Basic Results of Convex Analysis

In this section we shall give, without proof, some classical results on the existence, uniqueness, and characterization of the solution of convex minimization problems. Let

(i) V be a real reflexive Banach space, V^* its dual space,
(ii) K be a nonempty closed convex subset of V,

---

[1] In this chapter we follow Cea and Glowinski [1] and Glowinski [6].

(iii) $J_0: V \to \mathbb{R}$, be a convex functional Frechet or Gateaux differentiable[2] on V,

(iv) $J_1: V \to \overline{\mathbb{R}}$, be[3] a proper l.s.c. convex functional.

We assume that $K \cap \mathrm{Dom}(J_1) \neq \emptyset$, where

$$\mathrm{Dom}(J_1) = \{v \mid v \in V, J_1(v) \in \mathbb{R}\}.$$

We define $J: V \to \overline{\mathbb{R}}$ by $J = J_0 + J_1$ and assume that

$$\lim_{\substack{\|v\| \to +\infty \\ v \in K}} J(v) = +\infty. \tag{2.1}$$

Under the above assumptions on K and J, we have the following fundamental theorem.

Theorem 2.1. *The minimization problem*

$$J(u) \leq J(v), \qquad \forall\, v \in K, \quad u \in K, \tag{2.2}$$

has a solution characterized by

$$\langle J_0'(u), v - u\rangle + J_1(v) - J_1(u) \geq 0, \qquad \forall\, v \in K, \quad u \in K. \tag{2.3}$$

This solution is unique if J is strictly convex.[4]

Remark 2.1. If K is bounded, then (2.1) may be omitted.

Remark 2.2. Problem (2.3) is a *variational inequality* (see Chapter I of this book).

Let us now recall some definitions about monotone operators.

Definition 2.1. Let $A: V \to V^*$. The operator A is said to be *monotone* if

$$\langle A(v) - A(u), v - u\rangle \geq 0, \qquad \forall\, u, v \in V,$$

and *strictly monotone* if it is monotone and

$$\langle A(v) - A(u), v - u\rangle > 0, \qquad \forall\, u, v \in V, \quad u \neq v.$$

---

[2] Let $F: V \to \mathbb{R}$; the Gateaux-differentiability property means that

$$\lim_{\substack{t \to 0 \\ t \neq 0}} \frac{F(v + tw) - F(v)}{t} = \langle F'(v), w\rangle, \qquad \forall\, v, w \in V,$$

where $\langle \cdot, \cdot \rangle$ denotes the duality between V^* and V and $F'(v) \in V^*$; $F'(v)$ is said to be the Gateaux-derivative (or G-derivative) of F at v. Actually, we shall very often use the term gradient when referring to F'.

[3] $\overline{\mathbb{R}} = \mathbb{R} \cup \{+\infty\} \cup \{-\infty\}$.

[4] i.e., $J(tv + (1 - t)w) < tJ(v) + (1 - t)J(w)$, $\forall\, t \in]0, 1[$, $\forall\, v, w \in \mathrm{Dom}(J)$, $v \neq w$.

We shall introduce the following proposition which will be very useful in the sequel of this chapter.

Proposition 2.1. *Let $F: V \to \mathbb{R}$ be G-differentiable. Then there is equivalence between the convexity of F (resp., the strict convexity of F) and the monotonicity (resp., the strict monotonicity) of F'.*

To prove Theorem 2.1 and Proposition 2.1, we should use the following:

Proposition 2.2. *If F is G-differentiable, then F is convex if and only if*

$$F(w) - F(v) \geq \langle F'(v), w - v \rangle, \qquad \forall\, v, w \in V. \tag{2.4}$$

3. Relaxation Methods for Convex Functionals: Finite-Dimensional Case

3.1. Statement of the minimization problem. Notations

With respect to Sec. 2, we assume that $V = \mathbb{R}^N$, with

$$v = \{v_1, \ldots, v_N\}, \qquad v_i \in \mathbb{R}, \quad 1 \leq i \leq N.$$

The following notation will be used in the sequel:

$$(u, v) = \sum_{i=1}^{N} u_i v_i, \qquad \|v\| = \sqrt{(v, v)}.$$

We also assume that

$$K = \{v \mid v \in \mathbb{R}^N, v_i \in K_i = [a_i, b_i], a_i \leq b_i, 1 \leq i \leq N\}, \tag{3.1}$$

where the a_i (resp., the b_i) may take the value $-\infty$ (resp., $+\infty$); K is obviously a nonempty closed convex subset of $\mathbb{R}^N$. Furthermore, we assume that

$$J(v) = J_0(v) + \sum_{i=1}^{N} j_i(v_i), \tag{3.2}$$

where $J_0 \in C^1(\mathbb{R}^N)$ and is strictly convex and where the $j_i \in C^0(\mathbb{R})$ and are convex, $1 \leq i \leq N$. We note that we are not assuming the differentiability of the j_i. Finally, we assume that

$$\lim_{\|v\| \to +\infty} J(v) = +\infty. \tag{3.3}$$

Then, under the above assumptions and from Theorem 2.1, it follows that the problem

$$J(u) \leq J(v) \qquad \forall\, v \in K, \quad u \in K, \tag{3.4}$$

has a unique solution characterized by

$$(J_0'(u), v - u) + \sum_{i=1}^{N} (j_i(v_i) - j_i(u_i)) \geq 0, \qquad \forall\, v \in K, \quad u \in K, \tag{3.5}$$

where $J_0'(v)$ denotes the gradient of J_0 at v.

3.2. Description of the relaxation algorithm

To solve (3.4) we can use the following relaxation algorithm:

$$u^0 \in K \text{ arbitrarily given}; \tag{3.6}$$

then, u^n being known, we compute u^{n+1}, component by component, by

$$\begin{aligned} &J(u_1^{n+1}, \ldots, u_{i-1}^{n+1}, u_i^{n+1}, u_{i+1}^n, \ldots) \\ &\quad \leq J(u_1^{n+1}, \ldots, u_{i-1}^{n+1}, v_i, u_{i+1}^n, \ldots), \qquad \forall\, v_i \in K_i, \quad u_i^{n+1} \in K_i, \end{aligned} \tag{3.7}$$

where $1 \leq i \leq N$.

From the above assumptions on K and J, and from Theorem 2.1, we obtain:

Proposition 3.1. *Each subproblem (3.7) has a unique solution which is characterized by*

$$\frac{\partial J_0}{\partial v_i}(u_1^{n+1}, \ldots, u_i^{n+1}, u_{i+1}^n, \ldots)(v_i - u_i^{n+1}) + j_i(v_i) - j_i(u_i^{n+1}) \geq 0,$$
$$\forall\, v \in K_i, \quad u_i^{n+1} \in K_i. \tag{3.8}$$

Remark 3.1. To compute u_i^{n+1}, we can proceed as follows. First we compute $\bar{u}_i^{n+1}$ by solving

$$J(u_1^{n+1}, \ldots, u_{i-1}^{n+1}, \bar{u}_i^{n+1}, u_{i+1}^n, \ldots) \leq J(u_1^{n+1}, \ldots, u_{i-1}^{n+1}, v_i, u_{i+1}^n, \ldots),$$
$$\forall\, v_i \in \mathbb{R}, \quad \bar{u}_i^{n+1} \in \mathbb{R}. \tag{3.9}$$

Then we project $\bar{u}_i^{n+1}$ on K_i to obtain u_i^{n+1}; so

$$u_i^{n+1} = P_{K_i}(\bar{u}_i^{n+1}) = \operatorname{Max}(a_i, \operatorname{Min}(b_i, \bar{u}_i^{n+1})). \tag{3.10}$$

If j_i is differentiable, then $\bar{u}_i^{n+1}$ is the solution of the single-variable equation

$$\frac{\partial J_0}{\partial v_i}(u_1^{n+1}, \ldots, u_{i-1}^{n+1}, \bar{u}_i^{n+1}, u_{i+1}^n, \ldots) + \frac{dj_i}{dv_i}(\bar{u}_i^{n+1}) = 0. \tag{3.11}$$

Equation (3.11) can be solved by various methods (see, for example, Householder [1]).

3.3. A lemma on the monotonicity of the gradient of strictly convex functionals

To prove the convergence of algorithm (3.6), (3.7), we shall use the following:

Lemma 3.1. *Assume that $F \in C^1(\mathbb{R}^N)$ and is strictly convex. Then F is uniformly convex on the bounded sets of $\mathbb{R}^N$, i.e., $\forall\, M > 0$, $\exists\, \delta_M: [0, 2M] \to \mathbb{R}_+$, continuous, strictly increasing, and such that*

$$\delta_M(0) = 0, \tag{3.12}$$

$$(F'(v) - F'(u), v - u) \geq \delta_M(\|v - u\|)$$
$$\forall\, u, v \in \mathbb{R}^N, \quad \|u\| \leq M, \quad \|v\| \leq M, \tag{3.13}$$

and

$$F(v) \geq F(u) + (F'(u), v - u) + \tfrac{1}{2}\delta_M(\|v - u\|),$$
$$\forall\, u, v \in \mathbb{R}^N, \quad \|u\| \leq M, \quad \|v\| \leq M. \tag{3.14}$$

PROOF. Let $B_M = \{v \mid v \in \mathbb{R}^N, \|v\| \leq M\}$. For $\tau \in [0, 2M]$, we define δ_M^0 by

$$\delta_M^0(\tau) = \inf_{\substack{\|v-u\| = \tau \\ u, v \in B_M}} (F'(v) - F'(u), v - u). \tag{3.15}$$

From the definition of δ_M^0 it follows that

$$\delta_M^0(0) = 0 \tag{3.16}$$

and

$$(F'(v) - F'(u), v - u) \geq \delta_M^0(\|v - u\|), \qquad \forall\, u, v \in B_M. \tag{3.17}$$

Let $\tau_2 \in]0, 2M]$. From the continuity of $\{u, v\} \to (F'(v) - F'(u), v - u)$ and from the compactness of $B_M \times B_M$, it follows that there exists at least one pair $\{u_2, v_2\}$ realizing the minimum in (3.15). Then

$$\delta_M^0(\tau_2) = (F'(v_2) - F'(u_2), v_2 - u_2),$$

and

$$\delta_M^0(\tau_2) > 0$$

from the strict monotonicity of F' (cf. Sec. 2, Proposition 2.1). Let $\tau_1 \in]0, \tau_2[$. We define $w \in]u_2, v_2[$ by

$$w = u_2 + \frac{\tau_1}{\tau_2}(v_2 - u_2).$$

Since $0 < \tau_1/\tau_2 < 1$, from the strict monotonicity of F' it follows that

$$(F'(v_2) - F'(u_2), v_2 - u_2) > \left(F'\left(u_2 + \frac{\tau_1}{\tau_2}(v_2 - u_2)\right) - F'(u_2), v_2 - u_2\right).$$

This implies

$$(F'(v_2) - F'(u_2), v_2 - u_2) > \frac{\tau_2}{\tau_1}(F'(w) - F'(u_2), w - u_2) > (F'(w) - F'(u_2), w - u_2). \tag{3.18}$$

Since $\|w - u_2\| = \tau_1$, (3.18) in turn implies

$$\delta_M^0(\tau_2) > \delta_M^0(\tau_1).$$

Applying (3.17) to $\{u + t(v - u), u\}$, it follows that

$$(F'(u + t(v - u)), v - u) \geq (F'(u), v - u) + \frac{1}{t}\delta_M^0(t\|v - u\|), \qquad \forall\, t \in\,]0, 1]. \tag{3.19}$$

From the continuity of F', it easily follows that

$$\lim_{\tau \to 0+} \frac{1}{\tau}\delta_M^0(\tau) = 0. \tag{3.20}$$

Then from (3.20) it follows that (3.19) could be extended at $t = 0$. Integrating (3.19) on $[0, 1]$, it follows that

$$F(v) - F(u) \geq (F'(u), v - u) + \int_0^1 \delta_M^0(t\|v - u\|)\frac{dt}{t}. \tag{3.21}$$

We also have

$$F(u) - F(v) \geq (F'(v), u - v) + \int_0^1 \delta_M^0(t\|v - u\|)\frac{dt}{t}. \tag{3.22}$$

Then, by summation of (3.21), (3.22), we obtain

$$\begin{aligned}(F'(v) - F'(u), v - u) &\geq 2\int_0^1 \delta_M^0(t\|v - u\|)\frac{dt}{t} \\ &= 2\int_0^{\|v-u\|} \delta_M^0(s)\frac{ds}{s}.\end{aligned} \tag{3.23}$$

Therefore the function δ_M defined by

$$\delta_M(\tau) = 2\int_0^\tau \delta_M^0(s)\frac{ds}{s} \tag{3.24}$$

has the required properties. Furthermore, (3.14) follows from (3.21) and from the definition of δ_M. □

Remark 3.2. The term *forcing function* is frequently used for functions such as δ_M (see Ortega and Rheinboldt [1]).

3.4. Convergence of algorithm (3.6), (3.7)

We have:

Theorem 3.1. *Under the above assumptions on K and J, the sequence $(u^n)_n$ defined* (3.6), (3.7) *converges, $\forall\, u^0 \in K$, to the solution u of* (3.4).

PROOF. For the sake of simplicity, we have split the proof into several steps.

Step 1. We shall prove that the sequence $J(u^n)$ is decreasing. We have

$$J(u^n) - J(u^{n+1}) = \sum_{i=1}^{N} (J(u_1^{n+1}, \ldots, u_{i-1}^{n+1}, u_i^n, \ldots) - J(u_1^{n+1}, \ldots, u_i^{n+1}, u_{i+1}^n, \ldots)). \tag{3.25}$$

Since $u_i^n \in K_i$, it follows from (2.4), (3.8) that, $\forall\, i = 1, \ldots, N$, we have

$$\begin{aligned} &J(u_1^{n+1}, \ldots, u_{i-1}^{n+1}, u_i^n, \ldots) - J(u_1^{n+1}, \ldots, u_i^{n+1}, u_{i+1}^n, \ldots) \\ &= J_0(u_1^{n+1}, \ldots, u_{i-1}^{n+1}, u_i^n, \ldots) - J_0(u_1^{n+1}, \ldots, u_i^{n+1}, u_{i+1}^n, \ldots) + j_i(u_i^n) - j_i(u_i^{n+1}) \qquad (3.26) \\ &\geq \frac{\partial J_0}{\partial v_i}(u_1^{n+1}, \ldots, u_i^{n+1}, u_{i+1}^n, \ldots)(u_i^n - u_i^{n+1}) + j_i(u_i^n) - j_i(u_i^{n+1}) \geq 0. \end{aligned}$$

Then (3.26) combined with (3.25) implies

$$J(u^n) \geq J(u^{n+1}), \qquad \forall\, n \geq 0. \tag{3.27}$$

Moreover, since J satisfies (3.3), it follows from (3.27) that there exists a constant M such that

$$\begin{aligned} &\|u\| \leq M, \; \|u^n\| \leq M, \qquad \forall\, n, \\ &\|\{u_1^{n+1}, \ldots, u_i^{n+1}, u_{i+1}^n, \ldots\}\| \leq M, \qquad \forall\, i = 1, \ldots, N, \quad \forall\, n. \end{aligned} \tag{3.28}$$

Step 2. From (3.8), (3.14), (3.25), (3.26), and (3.28), it follows that

$$J(u^n) - J(u^{n+1}) \geq \frac{1}{2} \sum_{i=1}^{N} \delta_M(|u_i^{n+1} - u_i^n|). \tag{3.29}$$

The sequence $J(u^n)$ is decreasing and bounded below by $J(u)$, where u is the solution of (3.4). Therefore the sequence $J(u^n)$ is convergent, and this implies

$$\lim_{n \to +\infty} (J(u^n) - J(u^{n+1})) = 0. \tag{3.30}$$

From (3.29), (3.30), and the properties of δ_M, it follows that

$$\lim_{n \to +\infty} (u^n - u^{n+1}) = 0. \tag{3.31}$$

Step 3. Let u be the solution of (3.4). Then it follows from (3.15), (3.28) that

$$(J_0'(u^{n+1}) - J_0'(u), u^{n+1} - u) \geq \delta_M(\|u^{n+1} - u\|),$$

which implies

$$(J_0'(u^{n+1}) - J_0'(u), u^{n+1} - u) + J_1(u^{n+1}) - J_1(u) \geq J_1(u^{n+1}) - J_1(u) + \delta_M(\|u^{n+1} - u\|). \tag{3.32}$$

Since u is the solution of (3.4) and $u^{n+1} \in K$, we have (cf. (3.5))

$$(J_0'(u), u^{n+1} - u) + J_1(u^{n+1}) - J_1(u) \geq 0,$$

which, combined with (3.32), implies

$$(J_0'(u^{n+1}), u^{n+1} - u) + J_1(u^{n+1}) - J_1(u) \geq \delta_M(\|u^{n+1} - u\|). \tag{3.33}$$

Relation (3.33) implies

$$\begin{aligned}
&(J_0'(u^{n+1}), u^{n+1} - u) + J_1(u^{n+1}) - J_1(u) \\
&= \sum_{i=1}^{N} \left(\frac{\partial J_0}{\partial v_i}(u^{n+1}) - \frac{\partial J_0}{\partial v_i}(\hat{u}_i^{n+1}) \right)(u_i^{n+1} - u_i) \\
&\quad + \sum_{i=1}^{N} \left(\frac{\partial J_0}{\partial v_i}(\hat{u}_i^{n+1})(u_i^{n+1} - u_i) + j_i(u_i^{n+1}) - j_i(u_i) \right) \\
&\geq \delta_M(\|u^{n+1} - u\|),
\end{aligned} \tag{3.34}$$

where $\hat{u}_i^{n+1} = \{u_1^{n+1}, \ldots, u_i^{n+1}, u_{i+1}^n, \ldots\}$. Since $u_i \in K_i$, it follows from (3.8) that, $\forall\, i = 1, \ldots, N$,

$$\frac{\partial J_0}{\partial v_i}(\hat{u}_i^{n+1})(u_i^{n+1} - u_i) + j_i(u_i^{n+1}) - j_i(u_i) \leq 0. \tag{3.35}$$

Therefore (3.34) and (3.35) show that

$$\sum_{i=1}^{N} \left(\frac{\partial J_0}{\partial v_i}(u^{n+1}) - \frac{\partial J_0}{\partial v_i}(\hat{u}_i^{n+1}) \right)(u_i^{n+1} - u_i) \geq \delta_M(\|u^{n+1} - u\|). \tag{3.36}$$

Since $\|u^{n+1} - \hat{u}_i^{n+1}\| \leq \|u^{n+1} - u^n\|$, it follows from (3.31) that $\forall\, i = 1, \ldots, N$, we have

$$\lim_{n \to +\infty} (u^{n+1} - \hat{u}_i^{n+1}) = 0. \tag{3.37}$$

Since $J_0' \in C^0(\mathbb{R}^N)$, J_0' is uniformly continuous on the bounded subsets of $\mathbb{R}^N$. This property, combined with (3.37), implies, $\forall\, i = 1, \ldots, N$,

$$\lim_{n \to +\infty} \left\| \frac{\partial J_0}{\partial v_i}(u^{n+1}) - \frac{\partial J_0}{\partial v_i}(\hat{u}_i^{n+1}) \right\| = 0. \tag{3.38}$$

Therefore, from (3.28), (3.36), (3.38), and the properties of δ_M, it follows that

$$\lim_{n \to \infty} \|u^n - u\| = 0,$$

which completes the proof of the theorem. □

3.5. Various remarks

Remark 3.3. We assume that $K = \mathbb{R}^N$ and that $J \equiv J_0$ (i.e., $J_1 \equiv 0$), where

$J_0(v) = \frac{1}{2}(Av, v) - (b, v)$, where $b \in \mathbb{R}^N$ and

A is an $N \times N$ symmetrical positive-definite matrix.

The problem (3.4) associated with this choice of J and K obviously has an unique solution characterized (cf. (3.5)) by

$$Au = b. \tag{3.39}$$

If we apply the algorithm (3.6), (3.7) to this particular case, we obtain

$$u^0 \in \mathbb{R}^N, \text{ arbitrarily given}; \tag{3.40}$$

$$u_i^{n+1} = \frac{1}{a_{ii}}\left(b_i - \sum_{j<i} a_{ij}u_j^{n+1} - \sum_{j>i} a_{ij}u_j^n\right), \quad 1 \le i \le N. \tag{3.41}$$

The algorithm (3.40), (3.41) is known as the *Gauss–Seidel method* for solving (3.39) (see, e.g., Varga [1] and D. Young [1]). Therefore, when A is symmetric and positive definite, optimization theory yields another proof of the convergence of the Gauss–Seidel method through Theorem 3.1.

Remark 3.4. From the above remark it follows that the introduction of over- or under-relaxation parameters could be effective for increasing the speed of convergence. This possibility will be discussed in the sequel of this chapter.

Let $F: V \to \overline{\mathbb{R}}$. We define

$$D(F) = \{v \mid v \in V, |F(v)| < +\infty\}. \tag{3.42}$$

If F is convex and proper, then $D(F)$ is a nonempty convex subset of V.

Remark 3.5. If in Sec. 3.1 we replace the conditions $j_i \in C^0(\mathbb{R})$ and j_i convex, $\forall\, i = 1, \ldots, N$, by

$$j_i: \mathbb{R} \to \overline{\mathbb{R}} \text{ is convex, proper, and l.s.c.}$$

and we assume $K_i \cap D(j_i) \neq \emptyset$, $\forall\, i = 1, \ldots, N$, then the other assumptions being the same, (3.4) is still a well-posed problem and (3.5) still holds. Moreover, the algorithm (3.6), (3.7) could be used to solve (3.4), and the convergence result given by Theorem 3.1 would still hold.

Remark 3.6. We can complete Remark 3.5 in the following way. We take j_i as in Remark 3.5 and assume

$$J_0 \text{ strictly convex, proper, and l.s.c.},$$

$D(J_0)$ is an open set of $\mathbb{R}^N$ and $J_0 \in C^1(D(J_0))$. Then, if $D(J) \cap K \neq \emptyset$ and if $\lim_{\|v\| \to +\infty} J(v) = +\infty$, problem (3.4) is well posed and (3.5) still holds. Moreover, algorithm (3.6), (3.7) could be used to solve (3.4).

Remark 3.7. A typical situation in which algorithm (3.6), (3.7) could be used is $K = \mathbb{R}^N$, J_0 as in Remark 3.3 and $J_1(v) = \sum_{i=1}^N \alpha_i |v_i|$, $\alpha_i \ge 0$, $\forall\, i = 1, \ldots, N$.

3.6. Some dangerous generalizations

In this section we would like to discuss some of the limitations of the relaxation methods.

3.6.1. *Relaxation and nondifferentiable functionals*

We consider $K = \mathbb{R}^2$ and $J \in C^0(\mathbb{R}^2)$, strictly convex, defined by

$$J(v) = \tfrac{1}{2}(v_1^2 + v_2^2) + |v_1 - v_2| - (v_1 + v_2);$$

J is nondifferentiable on the line $v_1 = v_2$. The unique solution of

$$J(u) = \operatorname*{Min}_{v \in \mathbb{R}^2} J(v) \tag{3.43}$$

is obviously $u = \{1, 1\}$.

Now, starting from $u^0 = \{0, 0\}$, let us apply (3.6), (3.7) to (3.43). Since u_1^1 is the solution of

$$\operatorname*{Min}_{v_1 \in \mathbb{R}} (\tfrac{1}{2}v_1^2 + |v_1| - v_1).$$

we have $u_1^1 = 0$. In a similar way we have $u_2^1 = 0$, so that

$$u^n = \{0, 0\}, \qquad \forall\, n.$$

From the above result it follows that the algorithm (3.6), (3.7) *does not converge to u.*

3.6.2. *Relaxation and nonfactorable convex sets*

In this section we assume that K is a closed convex subset of $\mathbb{R}^N$, such that

$$K \neq \prod_{i=1}^{N} K_i.$$

If all the other assumptions of Sec. 3.1 hold, we can generalize algorithm (3.6), (3.7) as follows:

$$u^0 \in K, \tag*{(3.44)$_1$}$$

$$\begin{gathered} J(u_1^{n+1}, \ldots, u_{i-1}^{n+1}, u_i^{n+1}, u_{i+1}^n, \ldots) \leq J(u_1^{n+1}, \ldots, u_{i-1}^{n+1}, v_i, u_{i+1}^n, \ldots), \\ \forall\, \{u_1^{n+1}, \ldots, u_{i-1}^{n+1}, v_i, u_{i+1}^n, \ldots\} \in K, \\ \{u_1^{n+1}, \ldots, u_i^{n+1}, u_{i+1}^n, \ldots\} \in K; \qquad i = 1, \ldots, N. \end{gathered} \tag*{(3.44)$_2$}$$

Since $u^0 \in K$, it is very easy to prove that (3.44) is a well-posed problem, $\forall\, n \geq 0$. A very simple example will show that $(3.44)_1$, $(3.44)_2$ generally does not converge if K is nonfactorable. We take $J = J_0$ defined by $J(v) = v_1^2 + v_2^2$ and K defined (see Fig. 3.1) by

$$K = \{v \mid v \in \mathbb{R}^2, v_1 \geq 0, v_2 \geq 0, v_1 + v_2 \geq 1\}.$$

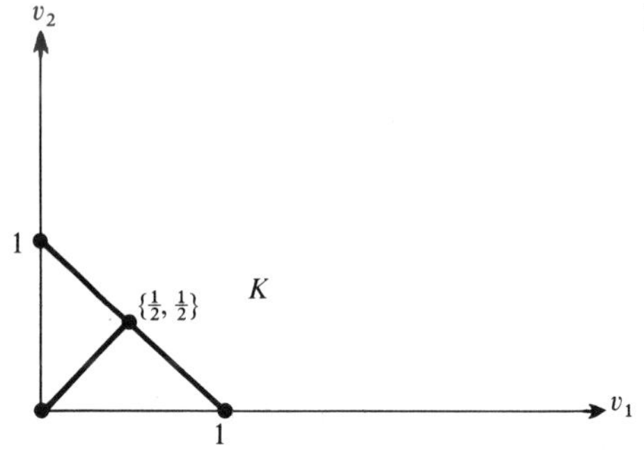

Figure 3.1

The solution of

$$J(u) = \operatorname*{Min}_{v \in K} J(v) \tag{3.45}$$

is obviously $u = \{\frac{1}{2}, \frac{1}{2}\}$. Then, starting from $u^0 = \{0, 1\}$, let us apply the algorithm $(3.44)_1, (3.44)_2$ to (3.45). It is easy to see that $u^n = \{0, 1\} \neq u, \forall\, n \geq 0$. Therefore we do not have convergence to the solution.

Remark 3.8. Let's consider the "one-dimensional" torsion problem (3.46) of Chapter II, Sec. 3.7.1. In the sequel we assume that $f \in C^0[0, 1]$. Using the notation $v_h(x_i) = v_i$, the above problem is approximated by the following variant of (3.41) (Chapter II, Sec. 3.7.1):

$$\int_0^1 \frac{du_h}{dx}\left(\frac{dv_h}{dx} - \frac{du_h}{dx}\right) dx \geq h \sum_{i=1}^{N-1} f_i(v_i - u_i), \qquad \forall\, v_h \in K_h, \quad u_h \in K_h, \tag{3.46}$$

where $f_i = f(x_i)$, $i = 1, \ldots, N$.

Problem (3.46) is equivalent to the minimization problem

$$J_h(\hat{u}_h) \leq J_h(\hat{v}_h), \qquad \forall\, \hat{v}_h \in \hat{K}_h, \quad \hat{u}_h \in \hat{K}_h, \tag{3.47}$$

where

$$\hat{v}_h = \{v_0, \ldots, v_N\}, \tag{3.48}$$

$$J_h(\hat{v}_h) = \frac{h}{2} \sum_{i=0}^{N-1} \left|\frac{v_{i+1} - v_i}{h}\right|^2 - h \sum_{i=1}^{N-1} f_i v_i, \tag{3.49}$$

$$\hat{K}_h = \left\{\hat{v}_h \,|\, \hat{v}_h \in \mathbb{R}^{N+1}, v_0 = v_N = 0, \left|\frac{v_{i+1} - v_i}{h}\right| \leq 1, \forall\, i = 0, \ldots, N - 1\right\}; \tag{3.50}$$

$\hat{K}_h$ is a nonfactorable closed convex subset of $\mathbb{R}^{N+1}$. However, it is proved in G.L.T. [1, Chapter 3], [3, Chapter 3] that $(3.44)_1$, $(3.44)_2$ converges to the solution $\hat{u}_h$ of (3.47) if $f \geq 0$ on [0, 1] and if we start with

$$\hat{u}_h^0 \leq \hat{u}_h \quad \text{(i.e., } u_i^0 \leq u_i, \forall\, i = 1, \ldots, N - 1\text{).}$$

Since $f \geq 0$ implies $u_i \geq 0$, $\forall\, i = 1, \ldots, N-1$, an obvious choice for $\hat{u}_h^0$ is $\hat{u}_h^0 = 0$.

4. Block Relaxation Methods

In this section we take $V = \mathbb{R}^N$ such that

$$V = \prod_{i=1}^{q} V_i \quad \text{with } V_i = \mathbb{R}^{N_i}, \tag{4.1}$$

where

$$N_i \geq 1 \quad \text{and} \quad \sum_{i=1}^{q} N_i = N. \tag{4.2}$$

If $v \in V$, then $v = \{v_1, \ldots, v_q\}$, $v_i \in V_i$. We assume that $K = \prod_{i=1}^{q} K_i$, where

$$K_i \text{ is a closed convex subset of } V_i. \tag{4.3}$$

Then (4.3) implies that K is a closed convex subset of V. Finally we consider a convex functional $J: V \to \overline{\mathbb{R}}$ such that

$$D(J) \cap K \neq \emptyset, \qquad \lim_{\|v\| \to +\infty} J(v) = +\infty, \quad J = J_0 + J_1, \tag{4.4}$$

where

$$J_0 \in C^1(V), \quad J_0 \text{ strictly convex} \tag{4.5}$$

and

$$J_1(v) = \sum_{i=1}^{q} j_i(v_i), \tag{4.6}$$

where the j_i are convex, proper, and l.s.c. on V_i.

From Theorem 2.1 of Sec. 2 it follows that

$$J(u) \leq J(v), \qquad \forall\, v \in K, \quad u \in K, \tag{4.7}$$

has a unique solution (in $D(J) \cap K$) characterized by

$$(J_0'(u), v - u) + \sum_{i=1}^{q} (j_i(v_i) - j_i(u_i)) \geq 0, \qquad \forall\, v \in K, \quad u \in K. \tag{4.8}$$

The generalization of the point relaxation algorithm (3.6), (3.7) is obviously

$$u^0 \in K, \tag{4.9}$$

$$J(u_1^{n+1}, \ldots, u_{i-1}^{n+1}, u_i^{n+1}, u_{i+1}^n, \ldots) \leq J(u_1^{n+1}, \ldots, u_{i-1}^{n+1}, v_i, u_{i+1}^n, \ldots),$$
$$\forall\, v_i \in K_i, \quad u_i^{n+1} \in K_i; \quad i = 1, \ldots, q. \tag{4.10}$$

Algorithm (4.9), (4.10) is a block-relaxation algorithm. Using a variant of the proof of Theorem 3.1, it may easily be proved that the sequence u^n defined by (4.9), (4.10) converges to the unique solution of (4.2).

Remark 4.1. Most of the remarks we have made on algorithm (3.6), (3.7) still apply to (4.9), (4.10). Unfortunately, Remark 3.1 does not apply in general.

Remark 4.2. In Cea and Glowinski [1] and Glowinski [6], it is proved that (4.9), (4.10) could also be used in infinite-dimensional situations.[5] In this case the assumptions $J_0 \in C^1$ and strictly convex are not sufficient to insure the convergence of the relaxation algorithm. The basic reason for this is that closed bounded sets are generally noncompact in Banach spaces of infinite dimension. Using the same notation, in the two references cited above, it is proved that sufficient convergence conditions are

$$J_0 \text{ is } C^1 \text{ with } J_0' \text{ Lipschitz continuous on the bounded sets of } V; \tag{4.11}$$

$$J_0' \text{ locally uniformly convex, i.e.,}$$

$$\langle J_0'(v) - J_0'(u), v - u \rangle \geq \delta_M(\|v - u\|), \qquad \forall\, u, v, \quad \|u\| \leq M, \quad \|v\| \leq M, \tag{4.12}$$

where δ_M is the same as in Sec. 3.3.

5. Constrained Minimization of Quadratic Functionals in Hilbert Spaces by Under and Over-Relaxation Methods: Application

We follow Cea and Glowinski [1, Sec. 3].

5.1. Statement of the minimization problem

Let $V = \prod_{i=1}^N V_i$, where V_i is a real Hilbert space, $\forall\, i = 1, \ldots, N$. Norm and scalar product on V_i are, respectively, denoted by $\|\cdot\|_i$ and $((\cdot, \cdot))_i$. If $v \in V$, then

$$v = \{v_1, \ldots, v_N\} \quad \text{with } v_i \in V_i, \quad i = 1, \ldots, N.$$

On V we define a scalar product and a norm by

$$((u, v)) = \sum_{i=1}^N ((u_i, v_i))_i, \tag{5.1}$$

$$\|v\| = \left(\sum_{i=1}^N \|v_i\|_i^2 \right)^{1/2}; \tag{5.2}$$

[5] More precisely, in reflexive Banach spaces.

V is a Hilbert space for $((\cdot, \cdot))$ and $\|\cdot\|$. Let K be a closed convex subset of V such that

$$K = \prod_{i=1}^{N} K_i, \; K_i \neq \varnothing, \qquad \forall\, i = 1, \ldots, N,$$

where K_i is closed and convex in V_i.

Let $J\colon V \to \mathbb{R}$ be defined by

$$J(v) = \tfrac{1}{2}a(v, v) - ((f, v)), \tag{5.3}$$

where the bilinear form $a(\cdot, \cdot)$ on $V \times V$, is continuous, symmetrical, and V-elliptic, i.e.,

$$\exists\, \alpha > 0 \text{ such that } a(v, v) \geq \alpha\|v\|^2, \qquad \forall\, v \in V,$$

and where $f \in V$.

Under the assumptions on V, K, and J, the optimization problem

$$J(u) \leq J(v), \qquad \forall\, v \in K, \quad u \in K \tag{5.4}$$

has a unique solution. This solution is characterized by

$$a(u, v - u) - ((f, v - u)) \geq 0, \qquad \forall\, v \in K, \quad u \in K. \tag{5.5}$$

5.2. Some preliminary results

From the properties of V it follows that

$$J(v) = J(v_1, \ldots, v_N) = \frac{1}{2} \sum_{1 \leq i, j \leq N} a_{ij}(v_i, v_j) - \sum_{i=1}^{N} ((f_i, v_i))_i, \tag{5.6}$$

where the a_{ij} are bilinear and continuous on $V_i \times V_j$ with $a_{ij} = a_{ji}^*$. The forms a_{ii} are V_i-elliptic (with the same constant α). Using the Riesz representation theorem, it is easily proved that there exists $A_{ij} \in \mathscr{L}(V_j, V_i)$ such that

$$a_{ij}(v_i, v_j) = ((v_i, A_{ij}v_j))_i, \tag{5.7}$$

$$A_{ij} = A_{ji}^*. \tag{5.8}$$

Moreover, the A_{ii} are self-adjoint and are isomorphisms from V_i to V_i. In the sequel it will be convenient to use the norm defined by a_{ii} on V_i, i.e.,

$$|||v_i|||_i^2 = a_{ii}(v_i, v_i) = ((A_{ii}v_i, v_i))_i, \qquad i = 1, \ldots, N. \tag{5.9}$$

The norms $\|\cdot\|_i$ and $|||\cdot|||_i$ are equivalent. The projection from V_i to K_i in the $|||\cdot|||_i$ norm will be denoted by P_i. Before giving the description of the iterative method, we shall prove some basic results on projections, useful in the sequel. Let:

(i) H be a real Hilbert space, with scalar product and norm denoted by $(\cdot, \cdot)$ and $\|\cdot\|$, respectively.

(ii) $b(\cdot, \cdot)$ be a bilinear form on H, continuous, symmetric, and H-elliptic (i.e., $\exists\, \beta > 0$ such that $b(v, v) \geq \beta\|v\|^2$, $\forall\, v \in H$).

Then from the Riesz representation theorem follows the existence of an isomorphism $B: H \to H$ such that

$$(Bu, v) = b(u, v), \qquad \forall\, u, v \in H,$$
$$B = B^*. \tag{5.10}$$

We denote by $[\cdot, \cdot]$ and $|\cdot|$ the scalar product on H and the norm on H, respectively, defined by

$$[u, v] = b(u, v), \qquad \forall\, u, v \in H, \tag{5.11}$$

$$|v|^2 = b(v, v), \qquad \forall\, v \in H. \tag{5.12}$$

The norms $|\cdot|$ and $\|\cdot\|$ are equivalent. Let

(iii) $C \neq \varnothing$ be a closed convex subset of H and π be the projector from $H \to C$ in the $|\cdot|$ norm.

(iv) $j: H \to \mathbb{R}$ be the functional defined by

$$j(v) = \tfrac{1}{2}b(v, v) - (g, v), \qquad \forall\, v \in H, \tag{5.13}$$

where $g \in H$.

Under the above assumptions, we have the following lemmas.

Lemma 5.1. *If u is the unique solution of*

$$j(u) \le j(v), \qquad \forall\, v \in C, \quad u \in C, \tag{5.14}$$

then

$$u = \pi(B^{-1}g). \tag{5.15}$$

PROOF. The solution u of (5.14) is characterized by

$$(Bu - g, v - u) \ge 0, \qquad \forall\, v \in C, \quad u \in C. \tag{5.16}$$

From (5.16) it follows that

$$(B(v - u), B^{-1}g - u) = b(v - u, B^{-1}g - u) \le 0, \qquad \forall\, v \in C, \quad u \in C, \tag{5.17}$$

and (5.17) characterizes u as the projection on C of $B^{-1}g$ in the norm $|\cdot|$. □

Lemma 5.2. *Let $u_0 \in C$ and let u_1 be defined by*

$$u_1 = \pi(u_0 + \omega(B^{-1}g - u_0)), \qquad \omega > 0. \tag{5.18}$$

Then

$$j(u_0) - j(u_1) \ge \frac{2 - \omega}{2\omega} |u_0 - u_1|^2. \tag{5.19}$$

PROOF. We have, $\forall\, v_1, v_2 \in H$,

$$j(v_1) - j(v_2) = \tfrac{1}{2}(|v_1 - B^{-1}g|^2 - |v_2 - B^{-1}g|^2). \tag{5.20}$$

Since $u_1 - B^{-1}g = u_0 - B^{-1}g + u_1 - u_0$, we have

$$|u_0 - B^{-1}g|^2 = |u_1 - B^{-1}g|^2 + 2[u_0 - B^{-1}g, u_0 - u_1] - |u_0 - u_1|^2. \tag{5.21}$$

From (5.18), and since $u_0 \in C$, it follows that

$$[u_0 + \omega(B^{-1}g - u_0) - u_1, u_0 - u_1] \leq 0.$$

This implies

$$|u_0 - u_1|^2 \leq \omega[u_0 - B^{-1}g, u_0 - u_1]. \tag{5.22}$$

Then (5.19) clearly follows from (5.20)–(5.22).

5.3. Description of the algorithm

For $i = 1, \ldots, N$, let the ω_i be positive numbers. Now let us consider the following algorithm:

$$u^0 = \{u_1^0, \ldots, u_N^0\} \textit{ arbitrarily given in } K; \tag{5.23}$$

u^n being known, we compute u^{n+1} by

$$u_i^{n+1} = P_i\left(u_i^n - \omega_i A_{ii}^{-1}\left(\sum_{j<i} A_{ij}u_j^{n+1} + \sum_{j\geq i} A_{ij}u_j^n - f_i\right)\right), \qquad i = 1, \ldots, N. \tag{5.24}$$

Remark 5.1. It follows from (5.24) that the computation of u_i^{n+1} can be achieved in three steps:

Step 1. On V_i we minimize the functional

$$v_i \to J(u_1^{n+1}, \ldots, u_{i-1}^{n+1}, v_i, u_{i+1}^n, \ldots, u_N^n).$$

Hence we obtain a solution

$$u_i^{n+1/3} = A_{ii}^{-1}\left(f_i - \sum_{j<i} A_{ij}u_j^{n+1} - \sum_{j>i} A_{ij}u_j^n\right).$$

Step 2. We compute $u_i^{n+2/3}$ by

$$u_i^{n+2/3} = u_i^n + \omega_i(u_i^{n+1/3} - u_i^n).$$

Step 3. At last we obtain u_i^{n+1} by

$$u_i^{n+1} = P_i(u_i^{n+2/3}).$$

We remark that if $\omega_i = 1$, $\forall\, i = 1, \ldots, N$, then algorithm (5.23), (5.24) reduces to the block algorithm (4.9), (4.10) (see Remark 4.2).

5.4. Convergence of algorithm (5.23), (5.24)

Proposition 5.1. *We have*

$$J(u^n) - J(u^{n+1}) \geq \sum_{i=1}^{N} \frac{2-\omega_i}{2\omega_i} |||u_i^{n+1} - u_i^n|||_i^2. \tag{5.25}$$

PROOF. We have

$$J(u^n) - J(u^{n+1}) = \sum_{i=1}^{N} (J(u_1^{n+1}, \ldots, u_{i-1}^{n+1}, u_i^n \ldots) - J(u_1^{n+1}, \ldots, u_i^{n+1}, u_{i+1}^n, \ldots)). \tag{5.26}$$

Then (5.25) clearly follows from (5.24), and from the application of Lemma 5.2 at each of the differences of the right-hand side of (5.26).

Proposition 5.2. *If* $0 < \omega_i < 2, \forall\, i = 1, \ldots, N$, *we have*

$$J(u^n) \geq J(u^{n+1}), \qquad \forall\, n$$

and

$$\lim_{n\to+\infty} (u^n - u^{n+1}) = 0 \text{ strongly in } V. \tag{5.27}$$

PROOF. Since $0 < \omega_i < 2$ implies $(2 - \omega_i)/\omega_i > 0, \forall\, i = 1, \ldots, N$, it follows from (5.25) that $J(u^n)$ is a decreasing sequence. Since $J(u^n)$ is bounded below by $J(u)$, where u is the solution of (5.4), $J(u^n)$ is convergent. This implies

$$\lim_{n\to+\infty} (J(u^n) - J(u^{n+1})) = 0. \tag{5.28}$$

Then, from (5.25), (5.27) and from $(2 - \omega_i)/\omega_i > 0, \forall\, i = 1, \ldots, N$, it clearly follows that

$$\lim_{n\to+\infty} \|u_i^{n+1} - u_i^n\|_i = 0, \qquad \forall\, i = 1, \ldots, N.$$

This implies (5.27).

From these two propositions we deduce:

Theorem 5.1. *If* $0 < \omega_i < 2, \forall\, i = 1, \ldots, N$, *then the sequence* u^n *defined by* (5.23), (5.24) *satisfies*

$$\lim_{n\to+\infty} u^n = u,$$

where u *is the solution of* (5.4).

PROOF. The V-ellipticity of $a(\cdot, \cdot)$ implies

$$a(u^{n+1} - u, u^{n+1} - u) \geq \alpha\|u^{n+1} - u\|^2. \tag{5.29}$$

From (5.29) it follows that

$$a(u^{n+1}, u^{n+1} - u) - ((f, u^{n+1} - u)) \geq a(u, u^{n+1} - u) - ((f, u^{n+1} - u)) + \alpha\|u^{n+1} + u\|^2. \tag{5.30}$$

Since u is the solution of (5.4), and since $u^{n+1} \in K$, we have

$$a(u, u^{n+1} - u) - ((f, u^{n+1} - u)) \geq 0,$$

which, combined with (5.30), implies

$$a(u^{n+1}, u^{n+1} - u) - ((f, u^{n+1} - u)) \geq \alpha \|u^{n+1} - u\|^2. \tag{5.31}$$

The left-hand side of (5.31) could be written as follows:

$$a(u^{n+1}, u^{n+1} - u) - ((f, u^{n+1} - u)) = \sum_{i=1}^{N} \left(\left(\sum_{j=1}^{N} A_{ij} u_j^{n+1} - f_i, u_i^{n+1} - u_i\right)\right)_i. \tag{5.32}$$

Let $\bar{u}_i^{n+1}$ be the vector of K_i for which the functional

$$v_i \to J(u_1^{n+1}, \ldots, u_{i-1}^{n+1}, v_i, u_{i+1}^n, \ldots)$$

attains its minimum on K_i. From Lemma 5.1 it follows that

$$\bar{u}_i^{n+1} = P_i\left(A_{ii}^{-1}\left(f_i - \sum_{j<i} A_{ij} u_j^{n+1} - \sum_{j>i} A_{ij} u_j^n\right)\right). \tag{5.33}$$

Moreover, from the usual characterization of the minimum we have

$$\left(\left(A_{ii}\bar{u}_i^{n+1} + \sum_{j<i} A_{ij} u_j^{n+1} + \sum_{j>i} A_{ij} u_j^n - f_i, v_i - \bar{u}_i^{n+1}\right)\right)_i \geq 0, \qquad \forall\, v_i \in K_i. \tag{5.34}$$

It follows from (5.32) that

$$\begin{aligned} &a(u^{n+1}, u^{n+1} - u) - ((f, u^{n+1} - u)) \\ &\quad = \sum_{i=1}^{N} ((A_{ii}(u_i^{n+1} - \bar{u}_i^{n+1}), u_i^{n+1} - u_i))_i \\ &\quad\quad + \sum_{i=1}^{N} \left(\left(\sum_{j>i} A_{ij}\, (u_j^{n+1} - u_j^n), u_i^{n+1} - u_i\right)\right)_i \\ &\quad\quad + \sum_{i=1}^{N} \left(\left(\sum_{j<i} A_{ij} u_j^{n+1} + A_{ii}\bar{u}_i^{n+1} + \sum_{j>i} A_{ij} u_j^n - f_i, u_i^{n+1} - \bar{u}_i^{n+1}\right)\right)_i \\ &\quad\quad + \sum_{i=1}^{N} \left(\left(\sum_{j<i} A_{ij} u_j^{n+1} + A_{ii}\bar{u}_i^{n+1} + \sum_{j>i} A_{ij} u_j^n - f_i, \bar{u}_i^{n+1} - u_i\right)\right)_i. \end{aligned} \tag{5.35}$$

Since $u_i \in K_i$, (5.34) implies that the last term on the right-hand side of (5.35) is ≤ 0. Therefore (5.31), (5.35) imply

$$\begin{aligned} &\sum_{i=1}^{N} ((A_{ii}(u_i^{n+1} - \bar{u}_i^{n+1}), u_i^{n+1} - u_i))_i + \sum_{i=1}^{N} \left(\left(\sum_{j<i} A_{ij} u_j^{n+1} + A_{ii}\bar{u}_i^{n+1}\right.\right. \\ &\quad \left.\left. + \sum_{j>i} A_{ij} u_j^n - f_i, u_i^{n+1} - \bar{u}_i^{n+1}\right)\right)_i \\ &\quad + \sum_{i=1}^{N} \left(\left(\sum_{j>i} A_{ij}(u_j^{n+1} - u_j^n), u_i^{n+1} - u_i\right)\right)_i \geq \alpha \|u^{n+1} - u\|^2. \end{aligned} \tag{5.36}$$

From (5.36) it follows that to prove the strong convergence of u^n to u, it suffices to prove that the left-hand side of (5.36) converges to zero. Since the A_{ij} are linear and continuous, the $u_i^n, \bar{u}_i^n$ are bounded uniformly in i and n, and $\lim_{n\to+\infty} \|u^{n+1} - u^n\| = 0$, it follows from (5.36) that $\lim_{n\to+\infty} \|\bar{u}^n - u^n\| = 0$ (where $\bar{u}^n = \{\bar{u}_1^n, \ldots, \bar{u}_i^n, \ldots, \bar{u}_N^n\}$) will imply convergence. Let us prove the last property. Since ω_i is >0, from (5.34) it follows that

$$\bar{u}_i^{n+1} = P_i\left(\bar{u}_i^{n+1} - \omega_i A_{ii}^{-1}\left(\sum_{j<i} A_{ij}u_j^{n+1} + A_{ij}\bar{u}_i^{n+1} + \sum_{j>i} A_{ij}u_j^n - f_i\right)\right). \tag{5.37}$$

From (5.24), (5.37), and since P_i is a contraction, it follows that

$$|||\bar{u}_i^{n+1} - u_i^{n+1}|||_i \le |1 - \omega_i| \, |||\bar{u}_i^{n+1} - u_i^n|||_i, \qquad \forall\, i = 1, 2, \ldots, N. \tag{5.38}$$

Since $0 < \omega_i < 2$ implies $0 < |1 - \omega_i| < 1$, it follows from (5.38) that

$$|||\bar{u}_i^{n+1} - u_i^{n+1}|||_i \le |||\bar{u}_i^{n+1} - u_i^n|||_i, \qquad \forall\, i = 1, \ldots, N. \tag{5.39}$$

From the triangle inequality and (5.38), (5.39), it follows that

$$\begin{aligned} |||u_i^n - u_i^{n+1}|||_i &\ge |||u_i^n - \bar{u}_i^{n+1}|||_i - |||\bar{u}_i^{n+1} - u_i^{n+1}|||_i \\ &\ge (1 - |1 - \omega_i|)|||\bar{u}_i^{n+1} - u_i^n|||_i \\ &\ge (1 - |1 - \omega_i|)|||\bar{u}_i^{n+1} - u_i^{n+1}|||_i, \end{aligned} \tag{5.40}$$

where $0 < 1 - |1 - \omega_i| < 1$.

From Proposition 5.2 we have $\lim_{n\to+\infty} \|u^{n+1} - u^n\| = 0$; this combined with (5.40) implies $\lim_{n\to\infty} \|\bar{u}^n - u^n\| = 0$, which completes the proof of the theorem. □

Remark 5.2. The above theorem generalizes Cryer [2], and also generalizes a classical result in finite dimensions (without constraints) for which we refer to R. S. Varga [1] and D. M. Young [1].

Remark 5.3. If $\omega_i > 1$ (resp., $\omega_i = 1$, $\omega_i < 1$), $\forall\, i = 1, \ldots, N$, then algorithm (5.23), (5.24) is an over-relaxation algorithm (resp., relaxation, under-relaxation algorithm) with projection.

5.5. Application to an elastic-plastic torsion problem

5.5.1. *Statement of the continuous problem*

We consider the elastic-plastic torsion problem described in Chapter II, Sec. 3 (we use the notation of Chapter II, Sec. 3):

$$a(u, v - u) \ge C \int_\Omega (v - u)\, dx, \qquad \forall\, v \in K, \quad u \in K, \tag{5.41}$$

where $a(u, v) = \int_\Omega \nabla u \cdot \nabla v \, dx$, and where

$$K = \{v \mid v \in H_0^1(\Omega), |\nabla v(x)| \le 1 \text{ a.e.}\}.$$

From the equivalence result of Brezis and Sibony [1] mentioned in Chapter II, Sec. 3.4, it follows that the solution of (5.41) is also the solution of

$$a(u, v - u) \geq C \int_\Omega (v - u)\, dx, \qquad \forall\, v \in \hat{K},$$

$$u \in \hat{K} = \{v \mid v \in H_0^1(\Omega), |v(x)| \leq d(x, \Gamma) \text{ a.e.}\}. \tag{5.42}$$

In this section we consider only the numerical analysis of (5.42).

5.5.2. *A finite element approximation of* (5.42)

We consider piecewise linear approximations only. As mentioned in Chapter II, Sec. 3.4, (5.42) is a variant of the obstacle problem we considered in Chapter II, Sec. 2.

We assume that Ω is a polygonal domain and that $\mathscr{T}_h$ is a standard triangulation of Ω (see Chapter II). From Chapter II, Sec. 2 it follows that $H_0^1(\Omega)$ and $\hat{K}$ may be, approximated respectively, by[6]

$$V_h = \{v_h \mid v_h \in C^0(\bar{\Omega}),\, v_h = 0 \text{ on } \Gamma,\, v_h|_T \in P_1,\, \forall\, T \in \mathscr{T}_h\},$$

$$\hat{K}_h = \{v_h \mid v_h \in V_h,\, |v_h(x)| \leq d(x, \Gamma),\, \forall\, x \in \mathring{\Sigma}_h\}.$$

Then (5.42) is approximated by

$$a(u_h, v_h - u_h) \geq C \int_\Omega (v_h - u_h)\, dx, \qquad \forall\, v_h \in \hat{K}_h, \quad u_h \in \hat{K}_h. \tag{5.43}$$

From the results of Chapters I and II it easily follows that:

Proposition 5.3. *The approximate problem* (5.43) *has a unique solution. This solution is also the unique solution of the minimization problem*

$$J(u_h) \leq J(v_h), \qquad \forall\, v_h \in \hat{K}_h, \quad u_h \in \hat{K}_h, \tag{5.44}$$

where $J(v) = \frac{1}{2}a(v, v) - C \int_\Omega v\, dx$.

Moreover, using the methods of Chapter II, we may prove the following:

Theorem 5.2. *Suppose the angles of* $\mathscr{T}_h$ *are uniformly bounded below by* $\theta_0 > 0$ *as* $h \to 0$. *Then*

$$\lim_{h \to 0} \|u_h - u\|_{H_0^1(\Omega)} = 0,$$

where u *and* u_h *are, respectively, the solutions of* (5.42) *and* (5.43).

---

[6] $\mathring{\Sigma}_h$: set of the interior nodes of $\mathscr{T}_h$.

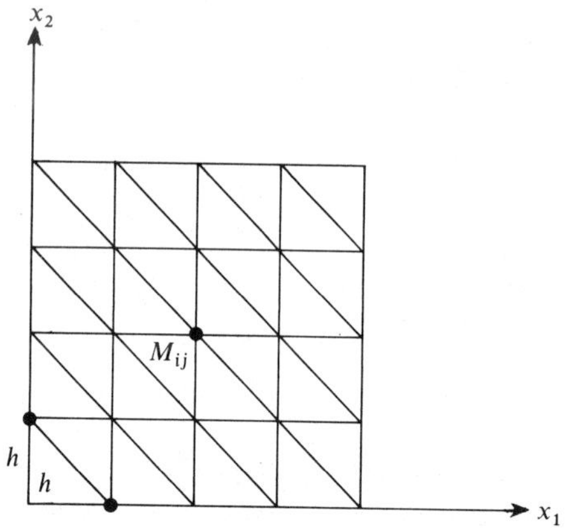

Figure 5.1

5.5.3. *Solution of the approximate problem*

The natural unknowns of the approximate problems (5.43) and (5.44) are $u_h(P)$, $P \in \mathring{\Sigma}_h$. Since $v_h \to J(v_h)$ is quadratic with respect to $v_h(P)$, it follows from the structure of $\hat{K}_h$ that algorithm (5.23), (5.24) could be used to solve (5.43), (5.44).

5.5.4. *Application to the case* $\Omega =]0, 1[\times]0, 1[$

When $\Omega =]0, 1[\times]0, 1[$, or, more generally, a rectangle, it is convenient to use a finite element approximation which is actually equivalent to a finite difference approximation. Let N be a positive integer and $h = 1/N$. On Ω we use the triangulation $\mathscr{T}_h$ of Fig. 5.1. We have

$$\Sigma_h = \{M_{ij} | M_{ij} = \{ih, jh\}, i, j \text{ integers}, 0 \le i, j \le N\},$$

$$\mathring{\Sigma}_h = \{M_{ij} | M_{ij} \in \Sigma_h, 1 \le i, j \le N - 1\}.$$

By analogy with the finite difference notation, it is convenient to denote $v_h(M_{ij})$ by v_{ij}. Then the approximate problems (5.43), (5.44) are easily expressed in function of v_{ij}, since

$$J(v_h) = \frac{h^2}{4} \sum_{0 \le i, j \le N} \left(\left(\frac{v_{i+1j} - v_{ij}}{h} \right)^2 + \left(\frac{v_{i-1j} - v_{ij}}{h} \right)^2 + \left(\frac{v_{ij+1} - v_{ij}}{h} \right)^2 + \left(\frac{v_{ij-1} - v_{ij}}{h} \right)^2 \right) - h^2 C \sum_{1 \le i, j \le N-1} v_{ij},$$

where $v_{kl} = 0$ if $M_{kl} \notin \mathring{\Sigma}_h$,

and since

$$\hat{K}_h = \{v_h | v_h \in V_h, |v_{ij}| \le d_{ij}, 1 \le i, j \le N - 1\},$$

where $d_{ij} = d(M_{ij}, \Gamma)$.

Algorithm (5.23), (5.24) has been used with $\omega_{ij} = \omega$, $\forall\, 1 \le i, j \le N - 1$. Therefore the explicit form of (5.23), (5.24) is

$$u_h^0 \in \hat{K}_h \text{ arbitrary given,} \tag{5.45}$$

and for $1 \le i, j \le N - 1$,

$$u_{ij}^{n+1/2} = (1 - \omega)u_{ij}^n + \frac{\omega}{4}(u_{i+1j}^n + u_{ij+1}^n + u_{i-1j}^{n+1} + u_{ij-1}^{n+1} + h^2 C) \tag{5.46}$$

$$u_{ij}^{n+1} = \max(-d_{ij}, \min(u_{ij}^{n+1/2}, d_{ij})), \tag{5.47}$$

where $u_{kl}^m = 0$ if $M_{kl} \notin \mathring{\Sigma}_h$.

From Theorem 5.1 it follows that u_h converges to the solution of (5.43), (5.44) if $0 < \omega < 2$.

Numerical experiments have been made with $C = 10$, $h = \frac{1}{40}$, $u_h^0 = 0$. The *stopping criterion* used was

$$\sum_{1 \le i, j \le N-1} |u_{ij}^{n+1} - u_{ij}^n| < 10^{-4}. \tag{5.48}$$

The number of iterations necessary to obtain the convergence is given below as a function of ω:

ω	1	1.4	1.5	1.6	1.7	1.8	1.9	2
Number of Iterations	290	138	118	98	93	108	188	

The C.P.U. time on an IBM 360/91 was 7.33 sec for $\omega = 1$ and 2.51 sec for $\omega = 1.7$.

Figure 5.2 shows the behavior of

$$R^n = \sum_{1 \le i, j \le N-1} |u_{ij}^{n+1} - u_{ij}^n|$$

as a function of n (when $\omega = 1.7$).

In the Fig. 5.3 we can see the elastic part of Ω (in white) and the plastic parts of Ω (striped area). In the plastic parts we have $|\nabla u| = 1$ and $|u(x)| = d(x, \Gamma)$. In the elastic part we have $-\Delta u = C$ and $|u(x)| < d(x, \Gamma)$.

Remark 5.4. The optimal value of ω is very close to the optimal value of ω corresponding to the solution, by S.O.R., of the Dirichlet problem $-\Delta u = C$ in the elastic part of Ω (with the same discretization step h). This follows directly from the fact that the plastic part of Ω[7] is very quickly obtained. From

[7] It is the part of Ω in which the constraints $|u| \le d(x, \Gamma)$ are active (i.e., $|u| = d(x, \Gamma)$).

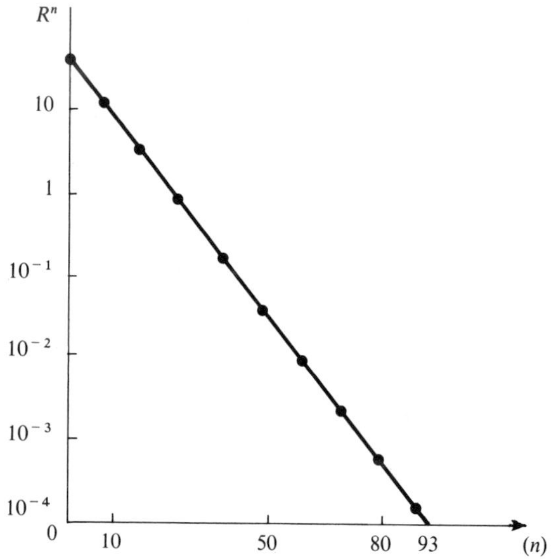

Figure 5.2

this fact it follows that the computation time is mainly used to solve $-\Delta u = C$ in the elastic part of Ω. Moreover, the plastic part increases with C; this fact explains that the optimal value of ω and the number of iterations needed to obtain the convergence are, for a given h, decreasing functions of C (as is the computational time).

Remark 5.5. In Cryer [1], [2] we can find a discussion on the choice of ω, when using point S.O.R. with projection to minimize quadratic functionals on a product of intervals.

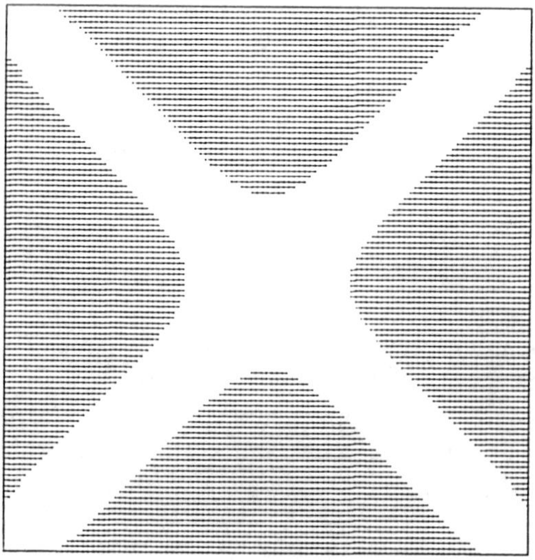

Figure 5.3

6. Solution of Systems of Nonlinear Equations by Relaxation Methods

6.1. Statement of the problem

In this section we very briefly describe some relaxation methods for solving systems of nonlinear equation such that

$$\begin{aligned} f_1(u_1, u_2, \ldots, u_N) &= 0, \\ &\vdots \\ f_i(u_1, u_2, \ldots, u_N) &= 0, \\ &\vdots \\ f_N(u_1, u_2, \ldots, u_N) &= 0, \end{aligned} \tag{6.1}$$

where $f_i\colon \mathbb{R}^N \to \mathbb{R}$. We define $F\colon \mathbb{R}^N \to \mathbb{R}^N$ by $F = \{f_1, \ldots, f_N\}$.

6.2. A first algorithm

We consider the following algorithm:

$$u^0 \textit{ given}; \tag{6.2}$$

u^n being known, we compute u^{n+1} by

$$\begin{gathered} f_i(u_1^{n+1}, \ldots, u_{i-1}^{n+1}, u_i^{n+1/2}, u_{i+1}^n, \ldots) = 0, \\ u_i^{n+1} = u_i^n + \omega(u_i^{n+1/2} - u_i^n), \\ 1 \leq i \leq N. \end{gathered} \tag{6.3}$$

If $F = \nabla J$, where $J\colon \mathbb{R}^N \to \mathbb{R}$ is a strictly convex C^1 function such that $\lim_{\|v\| \to +\infty} J(v) = +\infty$, then (6.3) has a unique solution (see Sec. 2). This solution is also the unique solution of

$$J(u) \leq J(v), \qquad \forall\, v \in \mathbb{R}^N, \quad u \in \mathbb{R}^N. \tag{6.4}$$

Moreover, if $\omega = 1$, it follows from Theorem 3.1 that algorithm (6.2), (6.3) converges to the solution u of (6.1), (6.4).

If $F = \nabla J$, $\omega \neq 1$, we refer to S. Schechter [1], [2], [3]. In these papers it is proved that under the hypothesis

(i) $J \in C^2(\mathbb{R}^N)$,
(ii) $(J'(w) - J'(v), w - v) \geq \alpha\|w - v\|^2, \forall\, v, w; \alpha > 0$,
(iii) $0 < \omega < \omega_M$,

algorithm (6.2), (6.3) converges to the solution of (6.1), (6.4). Moreover, estimates of ω_M and of the optimal value of ω are given.

For the convergence of (6.2), (6.3) when $F \neq \nabla J$, we refer to Ortega and Rheinboldt [1], Miellou [1], [2], and the bibliography therein.

6.3. A second algorithm

This algorithm is given by

$$u^0 \textit{ given}; \tag{6.5}$$

u^n being known, we compute u^{n+1} by

$$f_i(u_1^{n+1}, \ldots, u_i^{n+1}, u_{i+1}^n, \ldots) = (1-\omega)f_i(u_1^{n+1}, \ldots, u_{i-1}^{n+1}, u_i^n, \ldots),$$
$$1 \le i \le N. \tag{6.6}$$

To our knowledge the convergence of (6.5), (6.6) for $\omega \neq 1$ and F nonlinear has not yet been considered.

Remark 6.1. Algorithms (6.2), (6.3) and (6.5), (6.6) are identical if $\omega = 1$ and/or F is linear.

Remark 6.2. In many applications, from the numerical experiments it appears that (6.5), (6.6) is faster than (6.2), (6.3), ω having its (experimental) optimal value in both cases. Intuitively this seems to be related to the fact that (6.5), (6.6) is "more implicit" than (6.2), (6.3). For instance, (6.5), (6.6) could easily be used if F is only defined on a subset D on $\mathbb{R}^N$; in such a situation, when using (6.2), (6.3) with $\omega > 1$, it could happen that $\{u_1^{n+1}, \ldots, u_i^{n+1}, u_{i+1}^n, \ldots\} \notin D$.

6.4. A third algorithm

In this section we assume that $F \in C^1(\mathbb{R}^N)$. A natural method for computing $u_i^{n+1/2}$ in (6.3) or u_i^{n+1} in (6.6) is Newton's method. We recall that Newton's method applied to the solution of the single-variable equation

$$f(x) = 0$$

is basically:

$$x^0 \textit{ given}; \tag{6.7}$$

$$x^{m+1} = x^m - \frac{f(x^m)}{f'(x^m)}. \tag{6.8}$$

In the computation of $u_i^{n+1/2}$ in (6.3) or u_i^{n+1} in (6.6) by (6.7), (6.8), the obvious starting value is u_i^n. Then obvious variants of (6.2), (6.3) and (6.5), (6.6) are obtained if we run only one Newton iteration. Actually, in such a case, (6.2), (6.3) and (6.5), (6.6) reduce to the same algorithm, which is

$$u^0 \textit{ given}; \tag{6.9}$$

$$u_i^{n+1} = u_i^n - \omega \frac{f_i(u_1^{n+1}, \ldots, u_{i-1}^{n+1}, u_i^n, \ldots)}{(\partial f_i/\partial v_i)(u_1^{n+1}, \ldots, u_{i-1}^{n+1}, u_i^n, \ldots)}, \quad 1 \le i \le N. \tag{6.10}$$

In S. Schecter, *loc. cit.*, the convergence of (6.9), (6.10) is proved, if $F = \nabla J$, under the same assumptions as in Sec. 6.2 for algorithm (6.2), (6.3) (with a different ω_M in general).

Remark 6.3. In Glowinski and Marrocco [1], [2] and Concus [1], we can find comparisons between the above methods when applied to the numerical solution of the nonlinear elliptic equation modeling the magnetic state of ferromagnetic media (see also Winslow [1]). Applications of the first algorithm for solving minimal surface problems may be found in Jouron [1].

CHAPTER VI

Decomposition-Coordination Methods by Augmented Lagrangian: Applications[1]

1. Introduction

1.1. Motivation

A large number of problems in mathematics, physics, mechanics, economics, etc. may be formulated as

$$\operatorname*{Min}_{v \in V} \{F(Bv) + G(v)\}, \tag{P}$$

where

- V, H are topological vector spaces,
- $B \in \mathscr{L}(V, H)$,
- $F: H \to \overline{\mathbb{R}}$, $G: V \to \overline{\mathbb{R}}$ are convex proper l.s.c. functionals.

Let us give two examples taken from Chapter II.

EXAMPLE 1 This is the *Bingham flow problem* of Chapter II, Sec. 6; we recall that with Ω a bounded domain of $\mathbb{R}^2$, we consider the variational problem

$$\operatorname*{Min}_{v \in H_0^1(\Omega)} \left\{\frac{\nu}{2} \int_\Omega |\nabla v|^2\, dx + g \int_\Omega |\nabla v|\, dx - \int_\Omega fv\, dx\right\}, \tag{1.1}$$

where ν and g are positive constants. Then (1.1) is the particular problem (P) in which

- $V = H_0^1(\Omega)$,
- $H = L^2(\Omega) \times L^2(\Omega)$,
- $B = \nabla$,
- $F(q) = (\nu/2)\int_\Omega |q|^2\, dx + g\int_\Omega |q|\, dx \; (|q| = \sqrt{q_1^2 + q_2^2})$,
- $G(v) = -\int_\Omega fv\, dx$.

Actually, we can also take

- $F(q) = g\int_\Omega |q|\, dx$,
- $G(v) = (\nu/2)\int_\Omega |\nabla v|^2\, dx - \int_\Omega fv\, dx$.

[1] This chapter follows Fortin and Glowinski [2].

EXAMPLE 2 This is the *elastic-plastic torsion problem* of Chapter II, Sec. 3; with Ω still bounded in $\mathbb{R}^2$, we consider

$$\operatorname*{Min}_{v \in K} \left\{ \frac{1}{2} \int_\Omega |\nabla v|^2 \, dx - \int_\Omega fv \, dx \right\}, \tag{1.2}$$

where

$$K = \{v \in H_0^1(\Omega), |\nabla v| \leq 1 \text{ a.e.}\}.$$

Problem (1.2) is the particular problem (P) in which

- $V = H_0^1(\Omega)$, $H = L^2(\Omega) \times L^2(\Omega)$,
- $B = \nabla$,
- $F(q) = \frac{1}{2} \int_\Omega |q|^2 \, dx + I_{\hat{K}}(q)$,
- $G(v) = -\int_\Omega fv \, dx$,

where $I_{\hat{K}}$ is the *indicator functional* of the convex set

$$\hat{K} = \{q \in H, |q| \leq 1 \text{ a.e.}\},$$

i.e.,

$$I_{\hat{K}}(q) = \begin{cases} 0 & \text{if } q \in \hat{K} \\ +\infty & \text{if } q \notin \hat{K}. \end{cases}$$

We can also take

- $F(q) = I_{\hat{K}}$,
- $G(v) = \frac{1}{2} \int_\Omega |\nabla v|^2 - \int_\Omega fv \, dx$.

Synopsis. Problems of type (P) have a special structure, and in the sequel we shall introduce iterative methods of solution taking it into account.

1.2. Principle of the methods

The decomposition-coordination methods to follow are based on the following obvious equivalence result.

Theorem 1.1. (P) *is equivalent to*

$$\operatorname*{Min}_{\{v q\} \in W} \{F(q) + G(v)\}, \tag{π}$$

where

$$W = \{\{v, q\} \in V \times H, Bv - q = 0\}.$$

In the sequel we shall assume that V and H are *Hilbert spaces* with inner products and norms denoted by $((\cdot, \cdot))$, $\|\cdot\|$ and $(\cdot, \cdot)$, $|\cdot|$, respectively. We then define a Lagrangian functional $\mathscr{L}$ associated with (π), by

$$\mathscr{L}(v, q, \mu) = F(q) + G(v) + (\mu, Bv - q), \tag{1.3}$$

and for $r \geq 0$ an *Augmented Lagrangian* $\mathscr{L}_r$ is defined by

$$\mathscr{L}_r(v, q, \mu) = \mathscr{L}(v, q, \mu) + \frac{r}{2}|Bv - q|^2. \tag{1.4}$$

Remark 1.1. Augmented Lagrangian methods for solving general optimization problems have been introduced by Hestenes [1] and Powell [1]. Augmented Lagrangian methods for solving problems like (P) via (π) have been introduced by Glowinski and Marrocco [3] and also [4]–[6].

2. Properties of (P) and of the Saddle Points of $\mathscr{L}$ and $\mathscr{L}_r$

2.1. Existence and uniqueness properties for (P)

Let us define $J: V \to \overline{\mathbb{R}}$ by

$$J(v) = F(Bv) + G(v).$$

Then (P) can also be written

$$J(u) \leq J(v), \qquad \forall\, v \in V, \quad u \in V. \tag{2.1}$$

Let $j: X \to \overline{\mathbb{R}}$; we define the so-called *domain* of $j(\cdot)$ by

$$\operatorname{dom}(j) = \{x \in X, j(x) \in \mathbb{R}\}.$$

Then, if

$$\operatorname{dom}(F \circ B) \cap \operatorname{dom}(G) \neq \emptyset, \tag{2.2}$$

J is convex, proper, and l.s.c.. Therefore, sufficient conditions for (P) to have a unique solution are (cf. Cea [1], [2] and Ekeland and Temam [1]):

- $\lim_{\|v\| \to +\infty} J(v) = +\infty$,
- J strictly convex.

Remark 2.1. If B is an injection from V to H, with $R(B)$ (=range of B) closed in H, then $|Bv|$ is a norm on V equivalent to $\|v\|$.

2.2. Properties of the saddle points of $\mathscr{L}$ and $\mathscr{L}_r$

We have:

Theorem 2.1. *Let $\{u, p, \lambda\}$ be a saddle point of $\mathscr{L}$ on $V \times H \times H$; then $\{u, p, \lambda\}$ is also a saddle point of $\mathscr{L}_r$, $\forall\, r > 0$ and conversely. Moreover, u is a solution of* (P) *and $p = Bu$.*

PROOF. (i) $\{u, p, \lambda\}$ be a saddle point of $\mathscr{L}$; then $\mathscr{L}(u, p, \lambda) \in \mathbb{R}$ and

$$\mathscr{L}(u, p, \mu) \leq \mathscr{L}(u, p, \lambda) \leq \mathscr{L}(v, q, \lambda), \quad \forall \{v, q, \mu\} \in V \times H \times H, \quad \{u, p, \lambda\} \in V \times H \times H. \tag{2.3}$$

From the first inequality (2.3) and from (1.3) it follows that

$$(\mu, Bu - p) \leq (\lambda, Bu - p), \qquad \forall \mu \in H,$$

which obviously implies that

$$Bu = p. \tag{2.4}$$

From the second inequality (2.3) and from (1.3) and (2.4) it follows that

$$J(u) = \mathscr{L}(u, p, \lambda) \leq \mathscr{L}(v, q, \lambda), \qquad \forall \{v, q\} \in V \times H. \tag{2.5}$$

Taking $q = Bv$ in (2.5), it follows from (1.3) that

$$J(u) \leq \mathscr{L}(v, Bv, \lambda) = J(v), \qquad \forall v \in V; \tag{2.6}$$

hence u is a solution of (P). Since $p = Bu$, we have

$$\mathscr{L}_r(u, p, \mu) = \mathscr{L}(u, p, \mu) = J(u), \qquad \forall \mu \in H; \tag{2.7}$$

it then follows from (2.3), (2.7) that

$$\mathscr{L}_r(u, p, \mu) = \mathscr{L}_r(u, p, \lambda) \leq \mathscr{L}(v, q, \lambda), \qquad \forall \{v, q, \mu\} \in V \times H \times H. \tag{2.8}$$

Since $\mathscr{L}_r(v, q, \mu) = \mathscr{L}(v, q, \mu) + (r/2)|Bv - q|^2$, from (2.8) we obtain

$$\mathscr{L}_r(u, p, \mu) \leq \mathscr{L}_r(u, p, \lambda) \leq \mathscr{L}_r(v, q, \lambda), \qquad \forall \{v, q, \mu\} \in V \times H \times H, \tag{2.9}$$

which proves that $\{u, p, \lambda\}$ is also a saddle point of $\mathscr{L}_r$ on $V \times H \times H$. To conclude this part of the proof, we observe that from (2.3), $\{u, p\}$ is a solution of

$$\mathscr{L}(u, p, \lambda) \leq \mathscr{L}(v, q, \lambda), \qquad \forall \{v, q\} \in V \times H, \quad \{u, p\} \in V \times H, \tag{2.10}$$

from which it follows that $\{u, p\}$ is characterized (see Cea [1], [2], Ekeland and Temam [1], and also Chapter V, Sec. 2) by

$$F(q) - F(p) - (\lambda, q - p) \geq 0, \qquad \forall q \in H, \quad p \in H, \tag{2.11}$$

$$G(v) - G(u) + (\lambda, B(v - u)) \geq 0, \qquad \forall v \in V, \quad u \in V. \tag{2.12}$$

(ii) Let $\{u, p, \lambda\}$ be a saddle point of $\mathscr{L}_r$ with $r > 0$. Then, as in part (i), this implies that $p = Bu$ and that u is a solution of (P). Moreover, since $\{u, p, \lambda\}$ is solution of

$$\mathscr{L}_r(u, p, \lambda) \leq \mathscr{L}_r(v, q, \lambda), \qquad \forall \{v, q\} \in V \times H, \quad \{u, p\} \in V \times H, \tag{2.13}$$

it is characterized by

$$F(q) - F(p) + r(p - Bu, q - p) - (\lambda, q - p) \geq 0, \qquad \forall q \in H, \quad p \in H, \tag{2.14}$$

$$G(v) - G(u) + r(Bu - p, B(v - u)) + (\lambda, B(v - u)) \geq 0, \qquad \forall v \in V, \quad u \in V. \tag{2.15}$$

But since $Bu - p = 0$, (2.14), (2.15) reduce to (2.11), (2.12), and this fact implies that $\{u, p, \lambda\}$ satisfies (2.10). It then follows from (2.7) that $\{u, p, \lambda\}$ satisfies (2.3), and this completes the proof of the theorem. □

3. Description of the Algorithms

From Theorem 2.1 it follows that a method for solving (P) is to solve the saddle-point problem

$$\mathscr{L}_r(u, p, \mu) \leq \mathscr{L}_r(u, p, \lambda) \leq \mathscr{L}_r(v, q, \lambda),$$
$$\forall \{v, q, \mu\} \in V \times H \times H, \quad \{u, p, \lambda\} \in V \times H \times H. \quad (3.1)$$

To do this we shall use (see Cea [1], and G.L.T. [1, Chapter 2], [3, Chapter 2]) an algorithm of Uzawa type and a variant of it.

3.1. First algorithm

We denote by ALG 1 the following algorithm:

$$\lambda^0 \in H \textit{ given}; \quad (3.2)$$

then, λ^n known, we define u^n, p^n, λ^{n+1} by

$$\mathscr{L}_r(u^n, p^n, \lambda^n) \leq \mathscr{L}_r(v, q, \lambda^n), \qquad \forall \{v, q\} \in V \times H, \ \{u^n, p^n\} \in V \times H, \quad (3.3)$$
$$\lambda^{n+1} = \lambda^n + \rho_n(Bu^n - p^n), \qquad \rho_n > 0. \quad (3.4)$$

Problem (3.3) is in fact equivalent to the following system of *two coupled variational inequalities* (of the second kind):

$$G(v) - G(u^n) + (\lambda^n, B(v - u^n)) + r(Bu^n - p^n, B(v - u^n)) \geq 0,$$
$$\forall v \in V, \quad u^n \in V, \quad (3.5)$$
$$F(q) - F(p^n) - (\lambda^n, q - p^n) + r(p^n - Bu^n, q - p^n) \geq 0, \qquad \forall q \in H, \quad p^n \in H. \quad (3.6)$$

The convergence of ALG 1 will be studied in Sec. 4.

3.2. Second algorithm

The main drawback of ALG 1 is that it requires the solution of the coupled EVI's (3.5), (3.6) at each iteration. To overcome this difficulty, it is natural to consider the following variant of ALG 1 (denoted ALG 2 in the following):

$$\{p^0, \lambda^1\} \in H \times H \textit{ given}; \quad (3.7)$$

then, $\{p^{n-1}, \lambda^n\}$ known, we define $\{u^n, p^n, \lambda^{n+1}\}$ by

$$G(v) - G(u^n) + (\lambda^n, B(v - u^n)) + r(Bu^n - p^{n-1}, B(v - u^n)) \geq 0,$$
$$\forall v \in V, \quad u^n \in V, \quad (3.8)$$
$$F(q) - F(p^n) - (\lambda^n, q - p^n) + r(p^n - Bu^n, q - p^n) \geq 0, \qquad \forall q \in H, \quad p^n \in H, \quad (3.9)$$
$$\lambda^{n+1} = \lambda^n + \rho_n(Bu^n - p^n), \qquad \rho_n > 0. \quad (3.10)$$

The convergence of ALG 2 will be studied in Sec. 5.

4. Convergence of ALG 1

4.1. General case

In this subsection, V and H are possibly *infinite dimensional*; of course we assume that

$$\operatorname{dom}(F \circ B) \cap \operatorname{dom}(G) \neq \emptyset, \tag{4.1}$$

and also

$$B \text{ is an injection and } R(B) \text{ is closed in } H. \tag{4.2}$$

We also assume that

$$\lim_{|q| \to +\infty} \frac{F(q)}{|q|} = +\infty, \tag{4.3}$$

$$F = F_0 + F_1 \text{ with } F_0, F_1 \text{ convex, proper, and l.s.c.,} \tag{4.4}$$

$$F_0 \text{ is Gateaux-differentiable and uniformly convex on the bounded sets of } H. \tag{4.5}$$

By definition we say that F_0 is *uniformly convex* on the bounded sets of H if the following property holds:

$$\forall\, M > 0, \exists\, \delta_M : [0, 2M] \to \mathbb{R}, \text{ continuous, strictly increasing with } \delta_M(0) = 0, \text{ such that } \forall\, q, p \in H \text{ with } |p| \leq M, |q| \leq M \text{ we have } (F_0'(q) - F_0'(p), q - p) \geq \delta_M(|q - p|), \tag{4.6}$$

where $F_0' = \nabla F_0$ is the G-derivative of F_0. From the above properties, (P) has a unique solution u, and we define $p \in H$ by $p = Bu$.

EXERCISE 4.1. Prove that (P) is well posed if (4.1)–(4.5) hold.

On the convergence of ALG 1, we have:

Theorem 4.1. *We assume that $\mathscr{L}$ has a saddle point $\{u, p, \lambda\} \in V \times H \times H$. Then under the above assumptions on B, F, G and if*

$$0 < \alpha_0 \leq \rho_n \leq \alpha_1 < 2r, \tag{4.7}$$

the following convergence properties hold:

$$u^n \to u \text{ strongly in } V, \tag{4.8}$$

$$p^n \to p = Bu \text{ strongly in } H, \tag{4.9}$$

$$\lambda^{n+1} - \lambda^n \to 0 \text{ strongly in } H, \tag{4.10}$$

$$\lambda^n \text{ is bounded in } H. \tag{4.11}$$

Moreover, if λ^* is a weak cluster point of $\{\lambda^n\}_n$ in H, then $\{u, p, \lambda^*\}$ is a saddle point of $\mathscr{L}_r$ over $V \times H \times H$.

Proof. Since $\{u, p, \lambda\}$ is a saddle point of $\mathscr{L}_r$, we have

$$\mathscr{L}_r(u, p, \lambda) \le \mathscr{L}_r(v, q, \lambda), \qquad \forall \{v, q\} \in V \times H, \quad \{u, p\} \in V \times H. \tag{4.12}$$

Therefore $\{u, p\}$ is characterized by

$$G(v) - G(u) + (\lambda, B(v-u)) + r(Bu - p, B(v-u)) \ge 0, \qquad \forall v \in V, \quad u \in V, \tag{4.13}$$

$$(F_0'(p), q-p) + F_1(q) - F_1(p) - (\lambda, q-p) + r(p - Bu, q-p) \ge 0, \qquad \forall q \in H, \quad p \in H. \tag{4.14}$$

Moreover, from Theorem 2.1 we have $Bu = p$; therefore

$$\lambda = \lambda + \rho_n(Bu - p). \tag{4.15}$$

Let us define $\bar{u}^n, \bar{p}^n, \bar{\lambda}^n$ by

$$\bar{u}^n = u^n - u, \qquad \bar{p}^n = p^n - p, \qquad \bar{\lambda}^n = \lambda^n - \lambda.$$

From (3.4), (4.15) it then follows that

$$\bar{\lambda}^{n+1} = \bar{\lambda}^n + \rho_n(B\bar{u}^n - \bar{p}^n),$$

which implies

$$|\bar{\lambda}^{n+1}|^2 = |\bar{\lambda}^n|^2 + 2\rho_n(\bar{\lambda}^n, B\bar{u}^n - \bar{p}^n) + \rho_n^2|B\bar{u}^n - \bar{p}^n|^2$$

or, more conveniently,

$$|\bar{\lambda}^n|^2 - |\bar{\lambda}^{n+1}|^2 = -2\rho_n(\bar{\lambda}^n, B\bar{u}^n - \bar{p}^n) - \rho_n^2|B\bar{u}^n - \bar{p}^n|^2. \tag{4.16}$$

Since $\{u^n, p^n\}$ is a solution of (3.3), it is characterized by

$$G(v) - G(u^n) + (\lambda^n, B(v-u^n)) + r(Bu^n - p^n, B(v-u^n)) \ge 0, \qquad \forall v \in V, \quad u^n \in V, \tag{4.17}$$

$$(F_0'(p^n), q-p^n) + F_1(q) - F_1(p^n) - (\lambda^n, q-p^n) + r(p^n - Bu^n, q-p^n) \ge 0, \qquad \forall q \in H, \quad p^n \in H. \tag{4.18}$$

Taking $v = u$ (resp., $v = u^n$) in (4.17) (resp., (4.13)) and $q = p$ (resp., $q = p^n$) in (4.18) (resp., (4.14)), we obtain, by addition,

$$(\bar{\lambda}^n, B\bar{u}^n) + r(B\bar{u}^n - \bar{p}^n, B\bar{u}^n) \le 0, \tag{4.19}$$

$$(F_0'(p^n) - F_0'(p), \bar{p}^n) - (\bar{\lambda}^n, \bar{p}^n) + r(\bar{p}^n - B\bar{u}^n \cdot \bar{p}^n) \le 0, \tag{4.20}$$

which imply, also by addition,

$$(\bar{\lambda}^n, B\bar{u}^n - \bar{p}^n) + (F_0'(p^n) - F_0'(p), \bar{p}^n) + r|B\bar{u}^n - \bar{p}^n|^2 \le 0,$$

i.e.,

$$-(\bar{\lambda}^n, B\bar{u}^n - \bar{p}^n) \ge (F_0'(p^n) - F_0'(p), \bar{p}^n) + r|B\bar{u}^n - \bar{p}^n|^2. \tag{4.21}$$

Combining (4.16) and (4.21), we obtain

$$|\bar{\lambda}^n|^2 - |\bar{\lambda}^{n+1}|^2 \ge 2\rho_n(F_0'(p^n) - F_0'(p), \bar{p}^n) + \rho_n(2r - \rho_n)|B\bar{u}^n - \bar{p}^n|^2 \ge 0. \tag{4.22}$$

Assuming that (4.7) holds, it follows from (4.22) that

$$\lim_{n\to+\infty} |B\bar{u}^n - \bar{p}^n| = 0, \tag{4.23}$$

$$\lim_{n\to+\infty} (F_0'(p^n) - F_0'(p), p^n - p) = 0, \tag{4.24}$$

$$\lambda^n \text{ is bounded in } H. \tag{4.25}$$

Since $p = Bu$, it follows from (4.23) that

$$\lim_{n\to+\infty} |Bu^n - p^n| = 0. \tag{4.26}$$

Since F is proper, there exists $p_0 \in H$ such that $F(p_0) \in \mathbb{R}$; then, from the characterization (3.6), we have

$$F(p_0) - (\lambda^n, p_0) + r(p^n - Bu^n, p_0) \geq F(p^n) - (\lambda^n, p^n) + r(p^n - Bu^n, p^n). \tag{4.27}$$

Since λ^n and $p^n - Bu^n$ are bounded, (4.27) implies

$$\beta_0 \geq F(p^n) - \beta_1 |p^n|, \tag{4.28}$$

where β_0, β_1 are independent of n. From (4.3), (4.28) it then follows that

$$p^n \text{ is bounded in } H, \text{ i.e., } \exists\, M \text{ such that } |p^n| \leq M, \qquad \forall\, n. \tag{4.29}$$

Then, using the *uniform convexity* property (4.5), (4.6) of F_0, we obtain from (4.29) (assuming $M \geq |p|$) that

$$(F_0'(p^n) - F_0'(p), p^n - p) \geq \delta_M(|p^n - p|),$$

which combined with (4.24), implies,

$$\lim_{n\to+\infty} \delta_M(|p^n - p|) = 0 \Leftrightarrow \lim_{n\to+\infty} |p^n - p| = 0. \tag{4.30}$$

It then follows from (4.26), (4.30) that

$$\lim_{n\to+\infty} Bu^n = p = Bu \text{ strongly in } H. \tag{4.31}$$

Since B is an injection with $R(B)$ closed in H, then (4.31) implies that

$$\lim_{n\to+\infty} u^n = u \text{ strongly in } V. \tag{4.32}$$

The convergence result (4.10) clearly follows from (4.7), (4.26). Let λ^* be a weak cluster point of $\{\lambda^n\}_n$ in H. Then, passing to the limit in (3.5), (3.6) and using the l.s.c. property of F and G, we have

$$G(v) + (\lambda^*, B(v - u)) + r(Bu - p, B(v - u)) \geq \liminf G(u^n) \geq G(u), \qquad \forall\, v \in V, \quad u \in V,$$

$$F(q) - (\lambda^*, q - p) + r(p - Bu, q - p) \geq \liminf F(p^n) \geq F(p), \qquad \forall\, q \in H, \quad p \in H,$$

i.e.,

$$G(v) - G(u) + (\lambda^*, B(v - u)) + r(Bu - p, B(y - u)) \geq 0, \qquad \forall\, v \in V, \quad u \in V, \tag{4.33}$$

$$F(q) - F(p) - (\lambda^*, q - p) + r(p - Bu, q - p) \geq 0, \qquad \forall\, q \in H, \quad p \in H. \tag{4.34}$$

As noted before (see (2.13)–(2.15)), (4.33), (4.34) is equivalent to

$$\mathscr{L}_r(u, p, \lambda^*) \leq \mathscr{L}_r(v, q, \lambda^*), \qquad \forall \{v, q\} \in V \times H, \quad \{u, p\} \in V \times H. \tag{4.35}$$

Since, from $p = Bu$, we have

$$\mathscr{L}_r(u, p, \mu) = \mathscr{L}(u, p, \mu) = J(u), \qquad \forall \mu \in H,$$

we obtain

$$\mathscr{L}_r(u, p, \mu) = \mathscr{L}_r(u, p, \lambda^*), \qquad \forall \mu \in H. \tag{4.36}$$

It clearly follows from (4.35), (4.36) that $\{u, p, \lambda^*\}$ is a saddle point of $\mathscr{L}_r$;[2] this completes the proof of the theorem. □

4.2. Finite dimensional case

If V and H are finite dimensional, we have convergence of ALG 1 with weaker assumptions on F, B, and G than in Sec. 4.1. The reasons for this are the following:

(1) Since the constraint $Bv - q = 0$ is linear, if (P) has a solution, then $\mathscr{L}$ and $\mathscr{L}_r$ have a saddle point (see Rockafellar [1] and Cea [1], [2]).
(2) $R(B)$ is *always closed*.
(3) It follows from Cea and Glowinski [1] (and from Chapter V, Sec. 3.3) that F_0 satisfies the *uniform convexity* property (4.5), (4.6) if F_0 is C^1 and *strictly convex*.
(4) If F_0 is C^1 and strictly convex, then F_0' is C^0 and *strictly monotone*, i.e.,

$$(F_0'(q_2) - F_0'(q_1), q_2 - q_1) > 0, \qquad \forall q_1, q_2 \in H, \quad q_1 \neq q_2.$$

Then if (P) has a solution, the property

$$\lim_{|q| \to +\infty} \frac{F(q)}{|q|} = +\infty$$

is not necessary. This is related to the following:

Lemma 4.1. *Let H be finite dimensional and $A: H \to H$ be continuous and strictly monotone. Let $\{p^n\}_{n \geq 0}$, $p^n \in H, \forall n$, and $p \in H$ be such that*

$$\lim_{n \to +\infty} (A(p^n) - A(p), p^n - p) = 0; \tag{4.37}$$

then

$$\lim_{n \to +\infty} p^n = p. \tag{4.38}$$

---

[2] Concerning the convergence of $\{\lambda^n\}_n$, we have, in fact, a *weak convergence* of the *whole* sequence to a λ^* such that $\{u, p, \lambda^*\}$ is a saddle point of $\mathscr{L}_r$ (see Sec. 7).

Figure 4.1

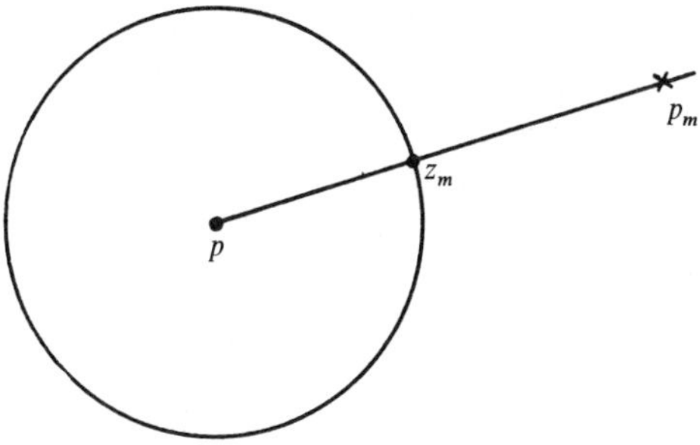

PROOF. Assume that (4.38) does not hold. Then there exist $\delta > 0$ and a subsequence, denoted $\{p^m\}_m$, such that

$$|p^m - p| \geq \delta, \qquad \forall\, m. \tag{4.39}$$

Let $S(p; \delta/2) = \{q \in H, |q - p| = \delta/2\}$. We define $z^m \in S(p; \delta/2)$ by

$$z^m = p + \frac{\delta}{2}\frac{p^m - p}{|p^m - p|}; \tag{4.40}$$

then $z^m \in]p, p^m[\subset H$ (see Fig. 4.1).

We introduce $t^m = \delta/2|p^m - p|$; then

$$z^m = p + t^m(p^m - p), \tag{4.41}$$

and from (4.39),

$$0 < t^m \leq \tfrac{1}{2}. \tag{4.42}$$

Since A is strictly monotone, we have

$$(A(p^m) - A(p), p^m - p) > (A(p + t(p^m - p)) - A(p), p^m - p), \qquad \forall\, t \in]0, 1[; \tag{4.43}$$

then, taking $t = t^m$ in (4.43), we obtain

$$(A(p^m) - A(p), p^m - p) > (A(z^m) - A(p), p^m - p) > 0. \tag{4.44}$$

From (4.41), (4.42), and (4.44), it then follows that

$$(A(p^m) - A(p), p^m - p) > \frac{1}{t^m}(A(z^m) - A(p), z^m - p) \geq 2(A(z^m) - A(p), z^m - p)$$

$$> (A(z^m) - A(p), z^m - p) > 0. \tag{4.45}$$

Since $S(p, \delta/2)$ is compact, we can extract from $\{z^m\}_m$ a subsequence—still denoted $\{z^m\}_m$—such that

$$\lim_{m \to +\infty} z^m = z, \qquad z \in S\left(p, \frac{\delta}{2}\right). \tag{4.46}$$

Since A is continuous, it follows from (4.37), (4.45), and (4.46) that

$$(A(z) - A(p), z - p) = 0. \tag{4.47}$$

The strict monotonicity of A and (4.47) imply that $z = p$, which is impossible since $|p - z| = \delta/2$. Therefore (4.39) cannot hold $\Rightarrow \lim_{n\to+\infty} p^n = p$. □

From the above properties we can easily prove the following:

Theorem 4.2. *Assume that V and H are finite dimensional and that* (P) *has a solution u. We suppose that:*

- *B is an injection,*
- *G is convex, proper, and l.s.c.,*
- *$F = F_0 + F_1$ with F_1 convex, proper, and l.s.c. over H and F_0 strictly convex and C^1 over H.*

Then (P) *has a unique solution, and if*

$$0 < \alpha_0 \leq \rho_n \leq \alpha_1 < 2r$$

holds, we have, for ALG 1, *the following convergence properties*:

$$\lim_{n\to+\infty} u^n = u,$$

$$\lim_{n\to+\infty} p^n = Bu,$$

$$\lim_{n\to+\infty} \lambda^{n+1} - \lambda^n = 0,$$

$$\lambda^n \text{ is bounded in } H.$$

Moreover, if λ is a cluster point of $\{\lambda^n\}_n$, then $\{u, p, \lambda\}$ is a saddle point of $\mathscr{L}_r$ over $V \times H \times H$.

4.3. Comments on the use of ALG 1. Further remarks

Assume that r is fixed and that we use a fixed value ρ for ρ_n. Then, from our computational experience, it appears that the best convergence is obtained for $\rho = r$. Concerning the choice of r, it can be theoretically proved that the larger the r, the *faster* the convergence; practically, the situation is not so simple, for the following reasons.

The larger the r, the worse the conditioning of the optimization problem (3.3) (or of the equivalent system (3.5), (3.6)). Then, since (3.3) is numerically (and not exactly) solved, at each iteration an error is made in the determination of $\{u^n, p^n\}$. The analysis of this error and the effect of it on the global behavior of ALG 1 is a very complicated problem since we have to take into account the conditioning of (3.3), the stopping criterion of the algorithms (usually iterative) solving (3.3), roundoff errors, etc.

Fortunately, it seems that the combined effect of all these factors is an algorithm which is not very sensitive to the choice of r (see Glowinski and Marrocco [5] and Fortin and Glowinski [1] for more details).

From a numerical point of view, the only nontrivial part in the use of ALG 1 is the solution at each iteration of problem (3.3). Taking into account the particular structure of (3.3), it follows from Chapter V, Cea and Glowinski [1], and Cea [2] that a method very well suited to the solution of (3.3) is the *block-relaxation* method described below:

All the problems (3.3) are of the following type:

$$\mathcal{L}_r(u, p, \mu) \leq \mathcal{L}_r(v, q, \mu), \qquad \forall \{v, q\} \in V \times H, \quad \{u, p\} \in V \times H, \tag{4.48}$$

where μ is given. The minimization problem (4.48) is equivalent to the system

$$G(v) - G(u) + (\mu, B(v - u)) + r(Bu - p, B(v - u)) \geq 0, \qquad \forall\, v \in V, u \in V, \tag{4.49}$$

$$F(q) - F(p) - (\mu, q - p) + r(p - Bu, q - p) \geq 0, \qquad \forall\, q \in H, \quad p \in H. \tag{4.50}$$

Then a block-relaxation method for solving (4.49), (4.50) is

$$\{u^0, p^0\} \ given; \tag{4.51}$$

then, $\{u^m, p^m\}$ *known, we obtain* $\{u^{m+1}, p^{m+1}\}$ *from*

$$\begin{aligned} G(v) - G(u^{m+1}) &+ (\mu, B(v - u^{m+1})) \\ &+ r(Bu^{m+1} - p^m, B(v - u^{m+1})) \geq 0, \qquad \forall\, v \in V, \quad u^{m+1} \in V, \end{aligned} \tag{4.52}$$

$$\begin{aligned} F(q) - F(p^{m+1}) - (\mu, q - p^{m+1}) + r(p^{m+1} - Bu^{m+1}, q - p^{m+1}) \geq 0, \\ \forall\, q \in H, \quad p^{m+1} \in H. \end{aligned} \tag{4.53}$$

Sufficient conditions for the convergence of (4.51)–(4.53) may be found in Chapter V, Cea and Glowinski, and Cea, *loc cit.*

In practice, when using (4.51)–(4.53), a stopping test of the following type will be used:

$$\text{Max}(\|u^{m+1} - u^m\|, |p^{m+1} - p^m|) \leq \varepsilon.$$

Another possibility is to stop after a fixed number of iterations. If, for instance, we stop after only one iteration of (4.51)–(4.53), and if at iteration n of ALG 1 we initialize with $\{u^{n-1}, p^{n-1}\}$ the computation of $\{u^p, p^n\}$ by (4.51)–(4.53), then we recover ALG 2.

Remark 4.1. Other relaxation methods can also be used; moreover, it can be worthwhile to introduce *over-relaxation parameters* to increase the speed of convergence of (4.51)–(4.53).

Remark 4.2. The choice $\rho = r$ may be motivated by the following proposition.

Proposition 4.1. *Suppose that* $F(q) = \frac{1}{2}|q|^2$ *and that* G *is linear. Then* $\forall\ \lambda^0 \in H$, *we have, for the sequence* $\{u^n\}_n$ *of* ALG 1, *convergence to the solution* u *of* (P) *in less than three iterations if we use* $\rho_n = \rho = r$, r *given.*

Preliminary Remark. In the above situation we have (P) equivalent to

$$B^t B u = f, \tag{4.54}$$

where $G(v) = ((f, v)),\ \forall\ v \in V$. Therefore, using ALG 1 for solving (P) has no practical interest. But even in this trivial case we shall observe that the behavior of ALG 1 is "interesting" since the convergence of u^n in a finite number of iterations does not imply a similar convergence property for p^n and λ^n.

Proof of Proposition 4.1 From (4.17), (4.18) it follows that in the particular case we are considering, ALG 1 reduces to

$$\lambda^0 \text{ given in } H; \tag{4.55}$$

$$rB^t B u^n = rB^t p^n - B^t \lambda^n + f, \tag{4.56}$$

$$p^n = \lambda^n + r(Bu^n - p^n), \tag{4.57}$$

$$\lambda^{n+1} = \lambda^n + r(Bu^n - p^n). \tag{4.58}$$

We can easily prove that the unique saddle point of $\mathscr{L}_r$ over $V \times H \times H$ is $\{u, Bu, Bu\}$, i.e., $p = Bu$, $\lambda = Bu$; using the notation $\bar{u}^n = u^n - u$, $\bar{p}^n = p^n - p$, $\bar{\lambda}^n = \lambda^n - \lambda$, it follows from (4.56)–(4.58) that

$$B^t \bar{\lambda}^n + rB^t(B\bar{u}^n - \bar{p}^n) = 0, \qquad \forall\ n \geq 0, \tag{4.59}$$

$$\lambda^{n+1} = p^n \Rightarrow \bar{\lambda}^{n+1} = \bar{p}^n, \qquad \forall\ n \geq 0, \tag{4.60}$$

$$\bar{p}^n = \bar{\lambda}^n + r(B\bar{u}^n - \bar{p}^n), \qquad \forall\ n \geq 0. \tag{4.61}$$

Multiplying (4.61) by B^t and comparing with (4.59), we obtain

$$B^t \bar{p}^n = 0, \qquad \forall\ n \geq 0. \tag{4.62}$$

Since (4.60), (4.61) imply

$$\bar{p}^{n+1} = \bar{p}^n + r(B\bar{u}^{n+1} - \bar{p}^{n+1}), \qquad \forall\ n \geq 0, \tag{4.63}$$

multiplying by B^t and taking (4.62) into account, we obtain

$$B^t B\bar{u}^{n+1} = 0, \qquad \forall\ n \geq 0 \Rightarrow B\bar{u}^{n+1} = 0, \qquad \forall\ n \geq 0. \tag{4.64}$$

Since $|Bv|$ is a norm on V, (4.64) implies that $u^n = u$, $\forall\ n \geq 1$. Hence the convergence of u^n to u requires two iterations at most. Using (4.63), (4.64), we have

$$\bar{p}^{n+1} = \frac{1}{1 + r}\bar{p}^n, \qquad \forall\ n \geq 0. \tag{4.65}$$

From (4.65) it follows that the larger the r, the faster p^n converges to $p = Bu$; for more details on the convergence of p^n, see Fortin and Glowinski [2]. □

5. Convergence of ALG 2

5.1. Synopsis

In this section we shall prove that under fairly general assumptions on F and G, we have convergence of ALG 2 if $0 < \rho_n = \rho < [(1 + \sqrt{5})/2]r$. We do not know if this result is optimal, since in some cases (G linear, for example) the upper bound of the interval of convergence is $2r$. Actually this question is rather academic, since in the various experiments we have done with ALG 2, the optimal choice seems to be $\rho = r$.

5.2. General case

We study the convergence of ALG 2 with the same hypotheses on B, F, G as in Sec. 4.1. We then have:

Theorem 5.1. *We suppose that $\mathscr{L}_r$ has a saddle point $\{u, p, \lambda\}$ over $V \times H \times H$. Then, if the assumptions on B, F, G are those of Sec. 4.1 and if*

$$0 < \rho_n = \rho < \frac{1 + \sqrt{5}}{2} r, \tag{5.1}$$

we have the following convergence properties:

$$u^n \to u \text{ strongly in } V, \tag{5.2}$$

$$p^n \to p \text{ strongly in } H, \tag{5.3}$$

$$\lambda^{n+1} - \lambda^n \to 0 \text{ strongly in } H, \tag{5.4}$$

$$\lambda^n \text{ is bounded in } H. \tag{5.5}$$

Moreover, if λ^* is a weak cluster point of $\{\lambda^n\}_n$, then $\{u, p, \lambda^*\}$ is a saddle point of $\mathscr{L}_r$ over $V \times H \times H$.

PROOF. Let us again define $\bar{u}^n, \bar{p}^n, \bar{\lambda}^n$ by

$$\bar{u}^n = u^n - u, \qquad \bar{p}^n = p^n - p, \qquad \bar{\lambda}^n = \lambda^n - \lambda.$$

Since $\{u, p, \lambda\}$ is a saddle point of $\mathscr{L}_r$ over $V \times H \times H$, we have

$$G(v) - G(u) + (\lambda, B(v - u)) + r(Bu - p, B(v - u)) \geq 0, \qquad \forall\, v \in V, \tag{5.6}$$

$$(F_0'(p), q - p) + F_1(q) - F_1(p) - (\lambda, q - p) + r(p - Bu, q - p) \geq 0 \qquad \forall\, q \in H, \tag{5.7}$$

$$\lambda = \lambda + \rho(Bu - p). \tag{5.8}$$

Moreover, (3.8)–(3.10) imply

$$G(v) - G(u^n) + (\lambda^n, B(v - u^n)) + r(Bu^n - p^{n-1}, B(v - u^n)) \geq 0, \qquad \forall\, v \in V, \tag{5.9}$$

$$\begin{aligned}(F_0'(p^n), q - p^n) + F_1(q) - F_1(p^n) - (\lambda^n, q - p^n)\\ + r(p^n - Bu^n, q - p^n) \geq 0, \qquad \forall\, q \in H,\end{aligned} \tag{5.10}$$

$$\lambda^{n+1} = \lambda^n + \rho(Bu^n - p^n). \tag{5.11}$$

Taking $v = u^n$ (resp. $v = u$) in (5.6) (resp., (5.9)) and $q = p^n$ (resp., $q = p$) in (5.7) (resp., (5.10)),

we obtain, by addition,

$$r(B\bar{u}^n - \bar{p}^{n-1}, B\bar{u}^n) + (\bar{\lambda}^n, B\bar{u}^n) \leq 0, \tag{5.12}$$

$$(F_0'(p^n) - F_0'(p), \bar{p}^n) + r(\bar{p}^n - B\bar{u}^n, \bar{p}^n) - (\bar{\lambda}^n, \bar{p}^n) \leq 0. \tag{5.13}$$

By addition of (5.12) and (5.13), it follows that

$$(F_0'(p^n) - F_0'(p), p^n - p) + r|B\bar{u}^n - \bar{p}^n|^2 + (\bar{\lambda}^n, B\bar{u}^n - \bar{p}^n) + r(\bar{p}^n - \bar{p}^{n-1}, B\bar{u}^n) \leq 0. \tag{5.14}$$

By subtracting (5.8) from (5.11), we obtain

$$|\bar{\lambda}^n|^2 - |\bar{\lambda}^{n+1}|^2 = -2\rho(B\bar{u}^n - \bar{p}^n, \bar{\lambda}^n) - \rho^2|B\bar{u}^n - \bar{p}^n|^2. \tag{5.15}$$

From (5.14), (5.15) it then follows that

$$|\bar{\lambda}^n|^2 - |\bar{\lambda}^{n+1}|^2 \geq 2\rho(F_0'(p^n) - F_0'(p), \bar{p}^n) + \rho(2r - \rho)|B\bar{u}^n - \bar{p}^n|^2 + 2\rho r(\bar{p}^n - \bar{p}^{n-1}, B\bar{u}^n). \tag{5.16}$$

Starting from

$$B\bar{u}^n = (B\bar{u}^n - B\bar{u}^{n-1}) + (B\bar{u}^{n-1} - \bar{p}^{n-1}) + \bar{p}^{n-1},$$

we obtain

$$\begin{aligned}(B\bar{u}^n, \bar{p}^n - \bar{p}^{n-1}) = (B\bar{u}^n - B\bar{u}^{n-1}, \bar{p}^n - \bar{p}^{n-1})\\ + (B\bar{u}^{n-1} - \bar{p}^{n-1}, \bar{p}^n - \bar{p}^{n-1}) + (\bar{p}^{n-1}, \bar{p}^n - \bar{p}^{n-1}).\end{aligned} \tag{5.17}$$

Since

$$(\bar{p}^{n-1}, \bar{p}^n - \bar{p}^{n-1}) = \tfrac{1}{2}(|\bar{p}^n|^2 - |\bar{p}^{n-1}|^2 - |\bar{p}^n - \bar{p}^{n-1}|^2),$$

it follows from (5.17) that

$$\begin{aligned}2\rho r(B\bar{u}^n, \bar{p}^n - \bar{p}^{n-1}) = 2\rho r(B\bar{u}^n - B\bar{u}^{n-1}, \bar{p}^n - \bar{p}^{n-1}) + 2\rho r(B\bar{u}^{n-1} - \bar{p}^{n-1}, \bar{p}^n - \bar{p}^{n-1})\\ + \rho r(|\bar{p}^n|^2 - |\bar{p}^{n-1}|^2 - |\bar{p}^n - \bar{p}^{n-1}|^2).\end{aligned} \tag{5.18}$$

Taking (5.10) at $n - 1$ instead of n, we have

$$\begin{aligned}(F_0'(p^{n-1}), q - p^{n-1}) + F_1(q) - F_1(p^{n-1}) - (\lambda^{n-1}, q - p^{n-1})\\ + r(p^{n-1} - Bu^{n-1}, q - p^{n-1}) \geq 0.\end{aligned} \tag{5.19}$$

Taking $q = p^{n-1}$ in (5.10) and $q = p^n$ in (5.19), we obtain, by addition,

$$\begin{aligned}(F_0'(p^n) - F_0'(p^{n-1}), p^n - p^{n-1}) - (\bar{\lambda}^n - \bar{\lambda}^{n-1}, \bar{p}^n - \bar{p}^{n-1}) + r|\bar{p}^n - \bar{p}^{n-1}|^2\\ - r(B\bar{u}^n - B\bar{u}^{n-1}, \bar{p}^n - \bar{p}^{n-1}) \leq 0.\end{aligned} \tag{5.20}$$

But since F'_0 is monotone, it follows from (5.20) that

$$r|\bar{p}^n - \bar{p}^{n-1}|^2 - (\bar{\lambda}^n - \bar{\lambda}^{n-1}, \bar{p}^n - \bar{p}^{n-1}) - r(B\bar{u}^n - B\bar{u}^{n-1}, \bar{p}^n - \bar{p}^{n-1}) \leq 0. \tag{5.21}$$

We have (from (3.10))

$$\lambda^n = \lambda^{n-1} + \rho(Bu^{n-1} - p^{n-1}),$$

which implies that

$$\bar{\lambda}^n - \bar{\lambda}^{n-1} = \rho(B\bar{u}^{n-1} - \bar{p}^{n-1}). \tag{5.22}$$

From (5.21), (5.22) it then follows that

$$r|\bar{p}^n - \bar{p}^{n-1}|^2 - \rho(B\bar{u}^{n-1} - \bar{p}^{n-1}, \ \bar{p}^n - \bar{p}^{n-1}) - r(B\bar{u}^n - B\bar{u}^{n-1}, \bar{p}^n - \bar{p}^{n-1}) \leq 0,$$

i.e.,

$$r(B\bar{u}^n - B\bar{u}^{n-1}, \bar{p}^n - \bar{p}^{n-1}) \geq r|\bar{p}^n - \bar{p}^{n-1}|^2 - \rho(B\bar{u}^{n-1} - \bar{p}^{n-1}, \bar{p}^n - \bar{p}^{n-1}). \tag{5.23}$$

From (5.18), (5.23) it then follows that

$$\begin{aligned} 2\rho r(B\bar{u}^n, \bar{p}^n - \bar{p}^{n-1}) \geq \rho r(|\bar{p}^n|^2 - |\bar{p}^{n-1}|^2) + \rho r|\bar{p}^n - \bar{p}^{n-1}|^2 \\ + 2\rho(r - \rho)(B\bar{u}^{n-1} - \bar{p}^{n-1}, \bar{p}^n - \bar{p}^{n-1}). \end{aligned} \tag{5.24}$$

Finally, combining (5.16) and (5.24), we obtain

$$\begin{aligned} (|\bar{\lambda}^n|^2 + \rho r|\bar{p}^{n-1}|^2) - (|\bar{\lambda}^{n+1}|^2 + \rho r|\bar{p}^n|^2) \geq 2\rho(F'_0(p^n) - F'_0(p), \bar{p}^n) \\ + \rho(2r - \rho)|B\bar{u}^n - \bar{p}^n|^2 + \rho r|\bar{p}^n - \bar{p}^{n-1}|^2) \\ + 2\rho(r - \rho)(B\bar{u}^{n-1} - \bar{p}^{n-1}, \bar{p}^n - \bar{p}^{n-1}). \end{aligned} \tag{5.25}$$

Then using, the Schwarz inequality, it follows from (5.25) that, $\forall\ \alpha > 0$, we have

$$\begin{aligned} (|\bar{\lambda}^n|^2 + \rho r|\bar{p}^{n-1}|^2) - (|\bar{\lambda}^{n+1}|^2 + \rho r|\bar{p}^n|^2) \geq 2\rho(F'_0(p^n) - F'_0(p), \bar{p}^n) \\ + \rho(2r - \rho)|B\bar{u}^n - \bar{p}^n|^2 + \rho r|\bar{p}^n - \bar{p}^{n-1}|^2 \\ - \rho|r - \rho|\left(\frac{1}{\alpha}|B\bar{u}^{n-1} - \bar{p}^{n-1}|^2 + \alpha|\bar{p}^n - \bar{p}^{n-1}|^2\right). \end{aligned} \tag{5.26}$$

If $\rho = r$, it is clear that, using the same method as in the proof of Theorem 4.1, we have (5.2)–(5.5). If $0 < \rho < r$, taking $\alpha = 1$ and observing that $|r - \rho| = r - \rho$, from (5.26) we observe that

$$\begin{aligned} &(|\bar{\lambda}^n|^2 + \rho r|\bar{p}^{n-1}|^2 + \rho(r - \rho)|B\bar{u}^{n-1} - \bar{p}^{n-1}|^2) \\ &\quad - (|\bar{\lambda}^{n+1}|^2 + \rho r|\bar{p}^n|^2 + \rho(r - \rho)|B\bar{u}^n - \bar{p}^n|^2) \\ &\quad \geq 2\rho(F'_0(p^n) - F'_0(p), \bar{p}^n) + \rho r|B\bar{u}^n - \bar{p}^n|^2 + \rho^2|\bar{p}^n - \bar{p}^{n-1}|^2 \geq 0, \end{aligned}$$

which clearly implies (5.2)–(5.5).

If $\rho > r$, we have $|r - \rho| = \rho - r$, and then it follows from (5.26) that (5.2)–(5.5) holds if we have $\rho < \rho_M$, where

$$\begin{aligned} \rho_M(2r - \rho_M) &= \frac{1}{\alpha}\rho_M(\rho_M - r), \\ \rho_M r &= \alpha\rho_M(\rho_M - r). \end{aligned} \tag{5.27}$$

By elimination of α, it follows from (5.27) that

$$\rho_M^2 - r\rho_M - r^2 = 0,$$

i.e. (since $\rho_M > 0$),

$$\rho_M = \frac{1 + \sqrt{5}}{2} r.$$

Then, using basically the same method as in the proof of Theorem 4.1, we can easily prove, from (5.2)–(5.5), that $\{u, Bu, \lambda^*\}$ is a saddle point of $\mathscr{L}_r$ over $V \times H \times H$ if λ^* is a weak cluster point of $\{\lambda^n\}_n$. □

5.3. Finite dimensional case

Using a variant of the proof of Theorem 5.1, and Lemma 4.1, we can easily prove:

Theorem 5.2. *Assume that the assumptions on V, H, F, B, G are those of the statement of Theorem 4.2. Then if*

$$0 < \rho_n = \rho < \frac{1 + \sqrt{5}}{2} r,$$

the conclusions of the statement of Theorem 4.2 still hold.

5.4. Comments on the choice of ρ and r

5.4.1. *Some remarks*

Remark 5.1. If G is linear, it has been proved by Gabay and Mercier [1] that ALG 2 converges if

$$0 < \rho_n = \rho < 2r.$$

The proof of this result is rather technical, and an open question is to decide if it can be extended to the more general cases considered here.

Remark 5.2. If G is linear, we observe that the step (3.8) of ALG 2 is a *linear problem* related to the *self-adjoint operator* B^tB. Therefore, in the finite-dimensional case, assuming B injective, it will be convenient to *factorize* (by a Cholesky method, for example) the *symmetric positive-definite* matrix B^tB once and for all, before starting the iterations of ALG 2.

5.4.2. *On the choice of ρ and r*

If r is given, our computational experience seems to indicate that the best choice for ρ is $\rho = r$. The choice of r is not clear, and ALG 2 appears to be *more sensitive* to the choice of r than ALG 1. By the way, ALG 1 seems to be more robust on very stiff problems than ALG 2; this means that the choice of the parameter is less critical and that the computational time for ALG 1 may become much shorter than for ALG 2.

Remark 5.3. In Remark 4.2 we have seen that if $F(q) = \frac{1}{2}|q|^2$ and G is linear, the sequence $\{u^n\}_n$ related to ALG 1 converges in two iterations (at most) if we use $\rho = r$. If we use ALG 2 with the same hypotheses on F and G, then we have convergence of $\{u^n\}_n$ in two iterations at most only if $\rho = r = 1$ (for any choice of $\{p^0, \lambda^1\}$). This fact also confirms the greater robustness of ALG 1.

6. Applications

6.1. Bingham flow in a cylindrical pipe

This is the problem considered in Chapter II, Sec. 6 and also in Sec. 1.1 of this chapter (we recall that Ω is a bounded domain of $\mathbb{R}^2$):

$$\operatorname*{Min}_{v \in H_0^1(\Omega)} \left\{ \frac{\nu}{2} \int_\Omega |\nabla v|^2 \, dx + g \int_\Omega |\nabla v| \, dx - \int_\Omega fv \, dx \right\}. \tag{6.1}$$

Then (6.1) is a particular problem (P) corresponding to

$$V = H_0^1(\Omega), \quad H = L^2(\Omega) \times L^2(\Omega), \qquad B = \nabla, \tag{6.2}$$

$$F(q) = \frac{\nu}{2} \int_\Omega |q|^2 \, dx + g \int_\Omega |q| \, dx, \tag{6.3}$$

$$G(v) = -\int_\Omega fv \, dx. \tag{6.4}$$

Moreover, we have $F = F_0 + F_1$ with

$$F_0(q) = \frac{\nu}{2} \int_\Omega |q|^2 \, dx, \qquad F_0'(q) = \nu q, \tag{6.5}$$

$$F_1(q) = g \int_\Omega |q| \, dx. \tag{6.6}$$

It then follows from (6.2)–(6.6) that the various assumptions required to apply Theorems 4.1 and 5.1 are satisfied. Therefore we can solve (6.1) by ALG 1 and ALG 2. Moreover, since G is linear, the result of Gabay and Mercier [1] holds (see Remark 5.1) and ALG 2 converges if $0 < \rho_n = \rho < 2r$. The augmented Lagrangian $\mathscr{L}_r$ to be used in this case is given by

$$\mathscr{L}_r(v, q, \mu) = \frac{\nu}{2} \int_\Omega |q|^2 \, dx + g \int_\Omega |q| \, dx - \int_\Omega fv \, dx + \int_\Omega \mu \cdot (\nabla v - q) \, dx$$
$$+ \frac{r}{2} \int_\Omega |\nabla v - q|^2 \, dx. \tag{6.7}$$

Solution of (6.1) *by* ALG 1. When applying ALG 1 to the solution of (6.1), it follows from (3.2)–(3.4), (6.7) that we have

$$\lambda^0 \in L^2(\Omega) \times L^2(\Omega), \textit{ arbitrarily given}; \tag{6.8}$$

then for $n \geq 0$,

$$-r\Delta u^n = f + \nabla \cdot \lambda^n - r\nabla \cdot p^n \textit{ on } \Omega, \quad u^n|_\Gamma = 0, \tag{6.9}$$

$$p^n(x) = 0 \quad \textit{if } g \geq |\lambda^n(x) + r\nabla u^n(x)|,$$

$$p^n(x) = \frac{\lambda^n(x) + r\nabla u^n(x)}{\nu + r}\left(1 - \frac{g}{|\lambda^n(x) + r\nabla u^n(x)|}\right) \textit{ elsewhere}, \tag{6.10}$$

$$\lambda^{n+1} = \lambda^n + \rho_n(\nabla u^n - p^n). \tag{6.11}$$

Solution of (6.1) *by* ALG 2. We have to replace (6.8) by

$$\{p^0, \lambda^1\} \textit{ arbitrarily given in } (L^2(\Omega))^2 \times (L^2(\Omega))^2, \tag{6.12}$$

and (6.9) by

$$-r\Delta u^n = f + \nabla \cdot \lambda^n - r\nabla \cdot p^{n-1} \textit{ on } \Omega, \quad u^n|_\Gamma = 0. \tag{6.13}$$

Remark 6.1. In practice, (6.8)–(6.11) and (6.12), (6.13), (6.10), and (6.11) will be applied to finite element or finite difference approximations of (6.1). It then follows from (6.9), (6.13) that it is easy to use either ALG 1 (combined with the block-relaxation method of Sec. 4.3) or ALG 2 once we have at our disposal an efficient program for solving approximate Dirichlet problems for $-\Delta$.

Bibliographical Comments. Numerical solutions of (6.1) by ALG 1 and ALG 2 may be found in Gabay and Mercier [1] and Fortin, Glowinski, and Marrocco [1]; in Fortin [2] we can also find an iterative method of solution of (6.1) close to ALG 2 but obtained by a different approach.

6.2. Elasto-plastic torsion of a cylindrical bar

This is the problem of Chapter II, Sec. 3, also considered in Sec. 1.1 of this chapter (Ω is a bounded domain of $\mathbb{R}^2$ in the sequel):

$$\underset{v \in K}{\text{Min}}\left[\frac{1}{2}\int_\Omega |\nabla v|^2\,dx - \int_\Omega fv\,dx\right], \tag{6.14}$$

where $K = \{v \mid v \in H_0^1(\Omega), |\nabla v| \leq 1 \text{ a.e.}\}$;

(6.14) is a particular problem (P) corresponding to

$$V = H_0^1(\Omega), \qquad H = L^2(\Omega) \times L^2(\Omega), \qquad B = \nabla, \tag{6.15}$$

$$G(v) = -\int_\Omega fv\,dx, \tag{6.16}$$

$$F = F_0 + F_1, \tag{6.17}$$

where

$$F_0(q) = \frac{1}{2}\int_\Omega |q|^2\,dx \Rightarrow F_0'(q) = q, \tag{6.18}$$

$$F_1(q) = I_{\hat{K}}(q) \tag{6.19}$$

with $\hat{K} = \{q \in H, |q| \leq 1 \text{ a.e.}\}$ and $I_{\hat{K}}$ the *indicator functional* of $\hat{K}$, i.e.,

$$I_{\hat{K}}(q) = \begin{cases} 0 & \text{if } q \in \hat{K}, \\ +\infty & \text{if } q \notin \hat{K}. \end{cases} \tag{6.20}$$

Here, too, it follows from (6.15)–(6.20) that the various assumptions required to apply Theorems 4.1 and 5.1 are satisfied. Therefore we can solve (6.14) by ALG 1 and ALG 2. Moreover, from the linearity of G, we have the convergence of ALG 2 if $0 < \rho_n = \rho < 2r$. In the present case, $\mathscr{L}_r$ is given by

$$\mathscr{L}_r(v, q, \mu) = \frac{1}{2}\int_\Omega |q|^2\,dx + I_{\hat{K}}(q) - \int_\Omega fv\,dx + \int_\Omega \mu\cdot(\nabla v - q)\,dx$$
$$+ \frac{r}{2}\int_\Omega |\nabla v - q|^2\,dx. \tag{6.21}$$

Solution of (6.14) *by* ALG 1. From (3.2)–(3.4), (6.21) it follows that when applying ALG 1 to (6.14), we obtain

$$\lambda^0 \textit{ arbitrarily given in } (L^2(\Omega))^2; \tag{6.22}$$

then for $n \geq 0$,

$$-r\Delta u^n = f + \nabla\cdot\lambda^n - r\nabla\cdot p^n \textit{ on } \Omega,$$
$$u^n|_\Gamma = 0, \tag{6.23}$$

$$p_n = \frac{\lambda^n + r\nabla u^n}{\sup(1 + r, |\lambda^n + ru^n|)}, \tag{6.24}$$

$$\lambda^{n+1} = \lambda^n + \rho_n(\nabla u^n - p^n). \tag{6.25}$$

Solution of (6.14) *by* ALG 2. We have to replace (6.22) by (6.12) and (6.23) by (6.13). Remark 6.1 still applies to (6.14), and numerical solutions of (6.14) by ALG 1 and ALG 2 may be found in Fortin, Glowinski, and Marrocco [1] and Gabay and Mercier [1].

6.3. A nonlinear Dirichlet problem

In this section we follow Glowinski and Marrocco [5]; let us consider $1 < s < +\infty$ and

$$W_0^{1,s}(\Omega) = \overline{\mathscr{D}(\Omega)}^{W^{1,s}(\Omega)} = \{v \in W^{1,s}(\Omega), v|_\Gamma = 0\},$$

where Ω is a bounded domain of $\mathbb{R}^N$.

Then we consider on Ω the following nonlinear Dirichlet problem:

$$\begin{aligned} -\nabla \cdot (|\nabla u|^{s-2}\nabla u) &= f, \\ u|_\Gamma &= 0, \end{aligned} \tag{6.26}$$

where $f \in V' = W^{-1,s'}(\Omega)(1/s + 1/s' = 1 \Rightarrow s' = s/(s-1))$. It can be proved (see, for instance, Glowinski and Marrocco [5]) that (6.26) has a unique solution which is also the solution of

$$\operatorname*{Min}_{v \in W_0^{1,s}(\Omega)} \left[\frac{1}{s}\int_\Omega |\nabla v|^s\,dx - \langle f, v\rangle\right]. \tag{6.27}$$

We observe that $W_0^{1,s}(\Omega)$ *is not a Hilbert space* if $s \neq 2$; therefore we cannot apply Theorems 4.1 and 5.1 to the iterative solution of (6.27). Nevertheless, once (6.27) has been approximated by a convenient finite element or finite difference method, it is possible to apply the above theorems (or Theorems 4.2 and 5.2) to the iterative solution of the approximate problem. For the sake of simplicity, we shall confine our study to the continuous problem, since it has simpler notation. We have

$$V = W_0^{1,s}(\Omega), \qquad H = (L^s(\Omega))^N, \qquad H' = (L^{s'}(\Omega))^N, \qquad B = \nabla,$$

$$F(q) = F_0(q) = \frac{1}{s}\int_\Omega |q|^s\,dx,\ F'(q) = q|q|^{s-2},$$

$$G(v) = -\langle f, v\rangle.$$

We observe that

$$\lim_{|q|_s \to +\infty} \frac{F(q)}{|q|_s} = +\infty,$$

where

$$|q|_s = \left(\int_\Omega |q|^s\,dx\right)^{1/s} = \|q\|_{(L^s(\Omega))^N}.$$

We also have $\forall\, p, q \in H$:

$$(F'(q) - F'(p), q - p) \geq |q - p|_s^s \quad \text{if } s \geq 2, \tag{6.28}$$

$$(F'(q) - F'(p), q - p) \geq \alpha \frac{|q - p|_s^2}{(|p|_s + |q|_s)^{2-s}} \quad \text{if } 1 < s \leq 2, \tag{6.29}$$

$$|F'(q) - F'(p)|_{s'} \leq \beta(|p|_s + |q|_s)^{s-2}|q - p|_s \quad \text{if } s \geq 2, \tag{6.30}$$

$$|F'(q) - F'(p)|_{s'} \leq \beta|q - p|_s^{s-1} \quad \text{if } 1 < s \leq 2, \tag{6.31}$$

where α, β are independent of p, q and are strictly positive.

EXERCISE 6.1. Prove (6.28)–(6.31).

We refer to Glowinski and Marrocco [5] for a detailed analysis, including error estimates, of a finite element approximation of (6.26), (6.27) (see also Ciarlet [2]).

From our numerical experiments it appears that solving (6.26), (6.27) if s is close to 1 (say $1 < s < 1.3$) or large (say $s > 5$) is a very difficult task if one uses standard iterative methods; to our knowledge the only very efficient methods are ALG 1 and ALG 2 (or closely related algorithms; see Glowinski and Marrocco, *loc. cit.*, for more details). The augmented Lagrangian $\mathscr{L}_r$ to be used for solving (6.26), (6.27) is defined by

$$\mathscr{L}_r(v, q, \mu) = \frac{1}{s}\int_\Omega |q|^s\,dx - \langle f, v\rangle + \frac{r}{2}\int_\Omega |\nabla v - q|^2\,dx + \int_\Omega \mu\cdot(\nabla v - q)\,dx. \tag{6.32}$$

Solution of (6.26), (6.27) *by* ALG 1. From (3.2)–(3.4), (6.32) it follows that when applying ALG 1 to (6.26), (6.27), we obtain:

$$\lambda^0 \in (L^{s'}(\Omega))^N; \tag{6.33}$$

then for $n \geq 0$,

$$-r\Delta u^n = f + \nabla\cdot\lambda^n - r\nabla\cdot p^n \text{ in } \Omega, \quad u^n|_\Gamma = 0, \tag{6.34}$$

$$|p^n|^{s-2}p^n + rp^n = r\nabla u^n + \lambda^n, \tag{6.35}$$

$$\lambda^{n+1} = \lambda^n + \rho_n(\nabla u^n - p^n). \tag{6.36}$$

The nonlinear system (6.34), (6.35) can be solved by the block-relaxation method of Sec. 4.3, and we observe that if u^n and λ^n are known (or estimated) in (6.35), the computation of p^n is an easy task since $|p^n|$ is solution of the single-variable nonlinear equation

$$|p^n|^{s-1} + r|p^n| = |r\nabla u^n + \lambda^n|, \tag{6.37}$$

which can easily be solved by various methods; once $|p^n|$ is known, we obtain p^n by solving a trivial linear equation (in $(L^s(\Omega))^N$).

Solution of (6.26), (6.27) *by* ALG 2. We have to replace (6.33) by

$$\{p^0, \lambda^1\} \in H \times H' \tag{6.38}$$

and (6.34) by

$$-r\Delta u^n = f + \nabla\cdot\lambda^n - r\nabla\cdot p^{n-1}, \quad u^n|_\Gamma = 0. \tag{6.39}$$

Remark 6.1 still applies to (6.26), (6.27), and since G is linear, we can take $0 < \rho_n = \rho < 2r$ if we are using ALG 2. For more details and comparisons with other methods, see Glowinski and Marrocco [5], [7] and Fortin, Glowinski, and Marrocco [1].

Remark 6.2. ALG 1 and ALG 2 have also been successfully applied to the iterative solution of *magneto-static* problems (see Glowinski and Marrocco [6]). They have also been applied by Fortin, Glowinski, and Marrocco [1] to the solution of the *subsonic flow* problem described in Chapter IV, Sec. 3; in the latter case, using ALG 1 and ALG 2, we obtain easy variants of (6.33)–(6.36) and (6.38), (6.39), (6.35), and (6.36).

6.4. Application to the solution of mildly nonlinear systems

Let $\mathbf{A}$ be a $N \times N$ symmetric positive-definite matrix, let $\mathbf{D}$ be a diagonal positive-semidefinite matrix, and let $\mathbf{f} \in \mathbb{R}^N$. Let $\phi: \mathbb{R} \to \mathbb{R}$ be a C^0 and non-decreasing function (we can always suppose that $\phi(0) = 0$). Using the same notation as in Chapter IV, Sec. 2.6, we associate to $\mathbf{v} = \{v_1, v_2, \ldots, v_N\} \in \mathbb{R}^N$ the vector $\phi(\mathbf{v}) \in \mathbb{R}^N$ defined by

$$(\phi(\mathbf{v}))_i = \phi(v_i), \qquad \forall\, i = 1, \ldots, N. \tag{6.40}$$

Then we consider the nonlinear system

$$\mathbf{Au} + \mathbf{D}\phi(\mathbf{u}) = \mathbf{f}. \tag{6.41}$$

In Chapter IV, Sec. 2.6, various methods for solving (6.41) have been given, but in this section we would like to show that (6.41) can also be solved by ALG 1 and ALG 2 once a convenient augmented Lagrangian has been introduced.

Remark 6.3. The methods to be described later are easily generalized to the case where $\mathbf{A}$ is not symmetric but still positive definite.

Let us define

$$\Phi(t) = \int_0^t \phi(\tau)\, d\tau.$$

Since ϕ is C^0 and nondecreasing, we find that Φ is C^1 and convex. It then follows from the symmetry of $\mathbf{A}$ that solving (6.41) is equivalent to solving the minimization problem

$$J(\mathbf{u}) \le J(\mathbf{v}), \qquad \forall\, \mathbf{v} \in \mathbb{R}^N, \quad \mathbf{u} \in \mathbb{R}^N. \tag{6.42}$$

In (6.42) we have

$$J(\mathbf{v}) = \frac{1}{2}(\mathbf{Av}, \mathbf{v}) + \sum_{i=1}^{N} d_i \Phi(v_i) - (\mathbf{f}, \mathbf{v}), \tag{6.43}$$

where $(\cdot, \cdot)$ denotes the usual inner product of $\mathbb{R}^N$ and $\|\cdot\|$ denotes the corresponding norm and where

$$\mathbf{D} = \begin{pmatrix} d_1 & & & & 0 \\ & \ddots & & & \\ & & d_i & & \\ & & & \ddots & \\ 0 & & & & d_N \end{pmatrix}.$$

From the above properties of **A**, **D**, and Φ, it follows from, e.g., Cea [1], [2], that (6.41), (6.42) has a unique solution.

Remark 6.4. In fact (6.41) has a unique solution if **A** is positive definite and possibly nonsymmetric, the assumptions on ϕ and **D** remaining the same.

Problem (6.42) is a particular problem (P) corresponding to

$$V = H = \mathbb{R}^N, \quad B = I, \tag{6.44}$$

$$G(\mathbf{v}) = \sum_{i=1}^{N} d_i \Phi(v_i) - (\mathbf{f}, \mathbf{v}), \tag{6.45}$$

$$F(\mathbf{q}) = F_0(\mathbf{q}) = \tfrac{1}{2}(\mathbf{Aq}, \mathbf{q}) \Rightarrow F_0'(\mathbf{q}) = \mathbf{Aq}. \tag{6.46}$$

From these properties we can solve (6.41), (6.42) by using ALG 1 and ALG 2 (we observe that, unlike in the above examples, G is nonlinear).

Remark 6.5. Instead of using G and F defined by (6.44), (6.45), we can use

$$G(v) = \sum_{i=1}^{N} d_i \Phi(v_i),$$

$$F(q) = \tfrac{1}{2}(\mathbf{Aq}, \mathbf{q}) - (\mathbf{f}, \mathbf{q}).$$

The augmented Lagrangian to be associated with (6.44)–(6.46) is

$$\mathscr{L}_r(\mathbf{v}, \mathbf{q}, \boldsymbol{\mu}) = \frac{1}{2}(\mathbf{Aq}, \mathbf{q}) + \sum_{i=1}^{N} d_i \Phi(v_i) - (\mathbf{f}, \mathbf{v}) + \frac{r}{2}\|\mathbf{v} - \mathbf{q}\|^2 + (\boldsymbol{\mu}, \mathbf{v} - \mathbf{q}). \tag{6.47}$$

Since the constraint $\mathbf{v} - \mathbf{q} = 0$ is linear, we know that $\mathscr{L}_r$ has a saddle point over $\mathbb{R}^N \times \mathbb{R}^N \times \mathbb{R}^N$; actually this saddle point is unique and is equal to $\{\mathbf{u}, \mathbf{u}, \mathbf{Au}\}$.

Solution of (6.41), (6.42) *by* ALG 1. From (3.2)–(3.4), (6.47) it follows that when applying ALG 1 to (6.41), (6.42), we obtain:

$$\boldsymbol{\lambda}^0 \in \mathbb{R}^N; \tag{6.48}$$

then for $n \geq 0$,

$$r\mathbf{u}^n + \mathbf{D}\phi(\mathbf{u}^n) = \mathbf{f} + r\mathbf{p}^n - \boldsymbol{\lambda}^n, \tag{6.49}$$

$$(r\mathbf{I} + \mathbf{A})\mathbf{p}^n = r\mathbf{u}^n + \boldsymbol{\lambda}^n, \tag{6.50}$$

$$\boldsymbol{\lambda}^{n+1} = \boldsymbol{\lambda}^n + \rho_n(\mathbf{u}^n - \mathbf{p}^n). \tag{6.51}$$

The nonlinear system (6.49), (6.50) can be solved by the block-relaxation method of Sec. 4.3, and we observe that if $\mathbf{p}^n$ and $\boldsymbol{\lambda}^n$ are known (or estimated) in (6.49), the computation of $\mathbf{u}^n$ is easy since it is reduced to the solution of N independent single-variable nonlinear equations of the following type:

$$r\xi + d\phi(\xi) = b \quad (\text{with } d \geq 0). \tag{6.52}$$

Since $r > 0$ and ϕ is C^0 and nondecreasing, (6.52) has a unique solution which can be computed by various standard methods (see, e.g., Householder [1] and Brent [1]). Similarly, if $\mathbf{u}^n$ and $\boldsymbol{\lambda}^n$ are known in (6.50), we obtain $\mathbf{p}^n$ by solving a linear system whose matrix is $r\mathbf{I} + \mathbf{A}$. Since r is independent of n, it is very convenient to prefactorize $r\mathbf{I} + \mathbf{A}$ (by the Cholesky or Gauss method).

Solution of (6.41), (6.42) *by* ALG 2. We have to replace (6.48) by

$$\{\mathbf{p}^0, \boldsymbol{\lambda}^1\} \in \mathbb{R}^N \times \mathbb{R}^N \tag{6.53}$$

and (6.49) by

$$r\mathbf{u}^n + \mathbf{D}\phi(\mathbf{u}^n) = \mathbf{f} + r\mathbf{p}^{n-1} - \boldsymbol{\lambda}^n. \tag{6.54}$$

From Theorem 5.2 it follows that we have convergence of (6.53), (6.54), (6.50), and (6.51) if $0 < \rho_n = \rho < (1 + \sqrt{5})/2r$.

Remark 6.6. Suppose that $\rho_n = \rho = r$ in ALG 2; we then have

$$\begin{aligned} r\mathbf{u}^n + \mathbf{D}\phi(\mathbf{u}^n) &= \mathbf{f} + r\mathbf{p}^{n-1} - \boldsymbol{\lambda}^n, \\ r\mathbf{p}^n + \mathbf{A}\mathbf{p}^n &= r\mathbf{u}^n + \boldsymbol{\lambda}^n, \\ \boldsymbol{\lambda}^{n+1} &= \boldsymbol{\lambda}^n + r(\mathbf{u}^n - \mathbf{p}^n). \end{aligned} \tag{6.55}$$

From (6.55) it follows that

$$\boldsymbol{\lambda}^{n+1} = \mathbf{A}\mathbf{p}^n. \tag{6.56}$$

Then from (6.55), (6.56) we obtain

$$r\mathbf{u}^n + \mathbf{D}\phi(\mathbf{u}^n) + \mathbf{A}\mathbf{p}^{n-1} = \mathbf{f} + r\mathbf{p}^{n-1}, \tag{6.57}$$

$$r\mathbf{p}^n + \mathbf{A}\mathbf{p}^n + \mathbf{D}\phi(\mathbf{u}^n) = \mathbf{f} + r\mathbf{p}^{n-1}. \tag{6.58}$$

Therefore, if $\rho_n = \rho = r$, ALG 2 reduces (with different notation) to the *alternating-direction method* described in Chapter IV, Sec. 2.6.6. (for more details on the relation existing between *alternating-direction* methods and augmented Lagrangian methods, we refer to G.L.T. [3, Appendix 2], Gabay [1], Bourgat, Dumay, and Glowinski [1], and Bourgat, Glowinski, and Le Tallec [1]).

Remark 6.7. From the numerical experiment performed in Chan and Glowinski [1], ALG 1, combined with the block-relaxation method of Sec. 4.3, is much more robust than ALG 2; this is the case if, for instance, we solve a finite element (or finite difference) approximation of the mildly nonlinear elliptic problem

$$-\Delta u + u|u|^{s-2} = f \quad \text{on } \Omega,$$
$$u|_\Gamma = 0, \tag{6.59}$$

with $1 < s < 2$.

In Chan and Glowinski, *loc. cit.*, we can find various numerical results and also comparisons with other methods (see also Chan, Fortin, and Glowinski [1]).

6.5. Solution of elliptic variational inequalities on intersections of convex sets

6.5.1. *Formulation of the problem*

Let V be a real Hilbert space and $a: V \times V \to \mathbb{R}$ be a bilinear form, continuous, symmetric, and V-elliptic. Let K be a closed convex nonempty subset of V such that

$$K = \bigcap_{i=1}^{N} K_i, \tag{6.60}$$

where, $\forall\, i = 1, \ldots, N$, K_i is a closed convex subset of V. We then consider the EVI problem

$$a(u, v - u) \geq L(v - u), \qquad \forall\, v \in K, \quad u \in K, \tag{6.61}$$

where $L: V \to \mathbb{R}$ is linear and continuous. Since $a(\cdot, \cdot)$ is symmetric, we know from Chapter I that the unique solution of (6.61) is also the solution of

$$J(u) \leq J(v), \qquad \forall\, v \in K, \quad u \in K, \tag{6.62}$$

where

$$J(v) = \tfrac{1}{2}a(v, v) - L(v). \tag{6.63}$$

6.5.2. *Decomposition of* (6.61), (6.62)

Let us define (with $q = \{q_1, \ldots, q_N\}$)

$$W = \{\{v, q\} \in V \times V^N, v - q_i = 0, \qquad \forall\, i = 1, \ldots, N\} \tag{6.64}$$

and

$$\mathscr{K} = \{\{v, q\} \in W, q_i \in K_i, \quad \forall\, i = 1, \ldots, N\}. \tag{6.65}$$

It is clear that (6.62) is equivalent to

$$\underset{\{v, q\} \in \mathscr{K}}{\text{Min}}\ j(v, q), \tag{6.66}$$

where

$$j(v, q) = \frac{1}{2N} \sum_{i=1}^{N} a(q_i, q_i) - L(v). \tag{6.67}$$

Remark 6.8. We have to observe that many other decompositions are possible; for instance,

$$W = \{\{v, q\} \in V \times V^N, v - q_1 = 0, q_{i+1} - q_i = 0, \quad \forall\, i = 1, \ldots, N - 1\}$$

with j and $\mathscr{K}$ again defined by (6.67), (6.65). We can also use

$$W = \{\{v, q\} \in V \times V^{N-1}, v - q_i = 0, \quad \forall\, i = 1, \ldots, N - 1\}$$

with

$$\mathscr{K} = \{\{v, q\} \in W, v \in K_1, q_i \in K_{i+1}, \quad \forall\, i = 1, \ldots, N - 1\}$$

and

$$j(v, q) = \frac{1}{2N} a(v, v) - L(v) + \frac{1}{2N} \sum_{i=1}^{N-1} a(q_i, q_i).$$

We suppose that in the sequel we use the decomposition defined by (6.64)–(6.67); then (6.66) is a particular problem (P) corresponding to

$$H = V^N, \qquad Bv = \{v, \ldots, v\}, \tag{6.68}$$

$$G(v) = -L(v), \tag{6.69}$$

$$F_0(q) = \frac{1}{2N} \sum_{i=1}^{N} a(q_i, q_i), \tag{6.70}$$

$$F_1(q) = \sum_{i=1}^{N} I_{K_i}(q_i) \tag{6.71}$$

with

$$I_{K_i}\text{: indicator function of } K_i.$$

It is easily shown that from the properties of B, G, F we can apply ALG 1 and ALG 2 to solve (6.62), via (6.66), provided that the augmented Lagrangian

$$\mathscr{L}_r(v, q, \mu) = F(q) + G(v) + \frac{r}{2N} \sum_{i=1}^{N} a(v - q_i, v - q_i) + \frac{1}{N} \sum_{i=1}^{N} (\mu_i, v - q_i) \tag{6.72}$$

has a saddle point over $V \times V^N \times V^N$. Such a saddle point exists if H is finite dimensional, since the constraints $v - q_i = 0$ are linear.

6.5.3. *Solution of* (6.62) *by* ALG 1.

From (3.2)–(3.4), (6.72) it follows that when applying ALG 1 to (6.62), we obtain:

$$\lambda^0 \in V^N \textit{ given}; \tag{6.73}$$

then for $n \geq 0$,

$$ra(u^n, v) = ra\left(\frac{1}{N}\sum_{i=1}^{N} p_i^n, v\right) - \left(\frac{1}{N}\sum_{i=1}^{N} \lambda_i^n, v\right) + L(v), \quad \forall\, v \in V, \quad u^n \in V, \tag{6.74}$$

$$(1 + r)a(p_i^n, q_i - p_i^n) \geq ra(u^n, q_i - p_i^n) + (\lambda_i^n, q_i - p_i^n), \quad \forall\, q_i \in K_i, \quad p_i^n \in K_i \qquad (6.75)_i$$

for $i = 1, 2, \ldots, N$;

$$\lambda_i^{n+1} = \lambda_i^n + \rho_n(u^n - p_i^n) \qquad (6.76)_i$$

for $i = 1, \ldots, N$.

The system (6.74), (6.75) is, for λ^n given, a system of coupled EVI's; a very convenient method for solving it is the *block-over-relaxation method with projection* described in Chapter V, Sec. 5 and also in Cea and Glowinski [1] and Cea [2]. This method will reduce the solution of (6.62) to a sequence of EVI's on K_i, $i = 1, \ldots, N$.

6.5.4. *Solution of* (6.62) *by* ALG 2

From (3.7)–(3.10), (6.72) it follows that to solve (6.62) by ALG 2, we have to use the variant of (6.73)–(6.76) obtained by replacing (6.73), (6.74) with:

$$\{p^0, \lambda^1\} \in V^N \times V^N \textit{ given}; \tag{6.77}$$

$$ra(u^n, v) = ra\left(\frac{1}{N}\sum_{i=1}^{N} p_i^{n-1}, v\right) - \left(\frac{1}{N}\sum_{i=1}^{N} \lambda_i^n, v\right) + L(v), \quad \forall\, v \in V, \quad u^n \in V. \tag{6.78}$$

Remark 6.9. The two algorithms above are well suited for use in multiprocessor computers, since many operations may be done in parallel; this is particularly clear for algorithm (6.77), (6.78), (6.75), (6.76).

Remark 6.10. Using different augmented Lagrangians, other than $\mathscr{L}_r$ defined by (6.72), we can solve (6.62) by algorithms better suited to *sequential computing* than to parallel computing. We leave to the reader, as exercises, the task of describing such algorithms.

Remark 6.11. The two algorithms described above can be extended to EVI's where $a(\cdot, \cdot)$ is not symmetric. Moreover, they have the advantage of reducing the solution of (6.62) to the solution of a sequence of simpler EVI's of the same type, to be solved over K_i, $i = 1, \ldots, N$, instead of K.

7. General Comments

As mentioned several times before, the methods described in this chapter may be extended to variational problems which are not equivalent to optimization problems. These methods have been applied by Begis and Glowinski [1] to the solution of fourth-order nonlinear problems in fluid mechanics (see also Begis [2] and G.L.T. [3, Appendix 6]).

From a historical point of view, the use of augmented Lagrangians for solving—via ALG 1 and ALG 2—nonlinear variational problems of type (P) (see Sec. 1.1) seems to be due to Glowinski and Marrocco [3], [4], [5]. For more details and other applications, see Gabay and Mercier [1], Fortin and Glowinski [1], [2], Glowinski and Marrocco, *loc. cit.*, and also Bourgat, Dumay, and Glowinski [1], Glowinski and Le Tallec [1], [2], Le Tallec [1], Bourgat, Glowinski, and Le Tallec [1], and Glowinski, Le Tallec, and Ruas [1], where ALG 1 and ALG 2 have been successfully used for solving nonlinear nonconvex variational problems occurring in finite elasticity (particularly in inextensible and/or incompressible finite elasticity).

With regard to Sec. 3.2, D. Gabay [1] has recently introduced the following variant of ALG 2:

$$\{p^0, \lambda^1\} \in H \times H \textit{ given}; \tag{7.1}$$

then, $\{p^{n-1}, \lambda^n\}$ *known, we define* $\{u^n, \lambda^{n+1/2}, p^n, \lambda^{n+1}\}$ *by*

$$G(v) - G(u^n) + (\lambda^n, B(v - u^n)) + r(Bu^n - p^{n-1}, B(v - u^n)) \geq 0,$$
$$\forall\, v \in V, \quad u^n \in V, \tag{7.2}$$

$$\lambda^{n+1/2} = \lambda^n + \rho(Bu^n - p^{n-1}), \tag{7.3}$$

$$F(q) - F(p^n) - (\lambda^{n+1/2}, q - p^n) + r(p^n - Bu^n, q - p^n) \geq 0,$$
$$\forall\, q \in H, \quad p^n \in H, \tag{7.4}$$

$$\lambda^{n+1} = \lambda^{n+1/2} + \rho(Bu^n - p^n), \tag{7.5}$$

with $\rho > 0$ in (7.3), (7.5). For additional details and convergence properties, see Gabay, *loc. cit.* (and also Gabay [2]).

To conclude this chapter, we have to mention that, using some results due to Opial [1], we have, in fact, in Theorems 4.1 and 5.1 (resp., 4.2 and 5.2) the weak convergence (resp., the convergence) of the whole sequence $\{\lambda^n\}_n$ to a λ^* such that $\{u, p, \lambda^*\}$ is a saddle point of $\mathscr{L}$ (and $\mathscr{L}_r$) over $V \times H \times H$. We refer to Glowinski, Lions, and Tremolieres [3, Appendix 2] for a proof of the above results in a more general context (see also Gabay [2]).

CHAPTER VII

Least-Squares Solution of Nonlinear Problems: Application to Nonlinear Problems in Fluid Dynamics

1. Introduction: Synopsis

In this chapter we would like to discuss the solution of some nonlinear problems in fluid dynamics by a combination of least-squares, conjugate gradient, and finite element methods. In view of introducing the reader to this technical subject, we consider in Sec. 2 the solution of systems of nonlinear equations in $\mathbb{R}^N$ by least-squares methods; then, in Sec. 3, the solution of a nonlinear Dirichlet model problem; also in Sec. 3 we make some comments about the use of pseudo-arc-length-continuation methods for solving nonlinear problems.

In Sec. 4 we discuss the application of the above methods to the solution of the nonlinear equation modelling potential transonic flows of inviscid compressible fluids; finally in Sec. 5 we discuss the solution of the Navier–Stokes equations, for incompressible viscous Newtonian fluids, by similar techniques.

This chapter is closely related to Bristeau, Glowinski, Periaux, Perrier, and Pironneau [1] and Bristeau, Glowinski, Periaux, Perrier, Pironneau, and Poirier [1]; other references will be given in the sequel.

2. Least-Squares Solution of Finite-Dimensional Systems of Equations

2.1. Generalities

Replacing the solution of finite-dimensional systems of equations by the solution of minimization problems is a very old idea, and many papers dealing with this approach can be found in the literature. Since referring to all those papers would be an almost impossible task, we shall mention just some of them, referring to the bibliographies therein for more references. The methods most widely used have been the least-squares methods in which the solution of

$$\mathbf{F}(\mathbf{x}) = \mathbf{0}, \tag{2.1}$$

where $\mathbf{F}\colon \mathbb{R}^N \to \mathbb{R}^N$ with $\mathbf{F} = \{f_1, \ldots, f_N\}$, is replaced by:

Find $\mathbf{x} \in \mathbb{R}^N$ *such that*

$$\|\mathbf{F}(\mathbf{x})\| \leq \|\mathbf{F}(\mathbf{y})\|, \qquad \forall\, \mathbf{y} \in \mathbb{R}^N, \tag{2.2}$$

where in (2.2), $\|\cdot\|$ denotes some *Euclidean norm*. If N is not too large, a natural choice for $\|\cdot\|$ is (if $\mathbf{y} = \{y_1, \ldots, y_N\}$)

$$\|\mathbf{y}\| = \left(\sum_{i=1}^{N} y_i^2\right)^{1/2}. \tag{2.3}$$

Suppose, for example, that

$$\mathbf{F}(\mathbf{x}) = \mathbf{A}\mathbf{x} - \mathbf{b}, \tag{2.4}$$

where $\mathbf{A}$ is an $N \times N$ matrix and $\mathbf{b} \in \mathbb{R}^N$. If $\|\cdot\|$ is defined by (2.3), then the corresponding problem (2.2) is equivalent to the well-known *normal equation*

$$\mathbf{A}^t\mathbf{A}\mathbf{x} = \mathbf{A}^t\mathbf{b}, \tag{2.5}$$

where $\mathbf{A}^t$ is the transpose matrix of $\mathbf{A}$. This simple example shows the main advantage of the method, which is to replace the original problem

$$\mathbf{A}\mathbf{x} = \mathbf{b}, \tag{2.6}$$

whose matrix is possibly *nonsymmetric* and *indefinite*, by the problem (2.5) whose matrix is *symmetric* and *positive semidefinite* (or equivalently, by the *minimization* of a quadratic *convex* functional). This *convexification* property (which can only be local in nonlinear problems) is fundamental since it will insure the good behavior (locally, at least) of most minimization methods used to solve the least-squares problem (2.2) (once a proper $\|\cdot\|$ has been chosen; see below).

Also, from (2.5) it is clear that a main drawback of the method is the possible deterioration of the conditioning which, for example, may make the solution of (2.2) sensitive to roundoff errors. Actually in many problems this drawback can be easily overcome by the use of a more sophisticated Euclidean norm than (2.3). Indeed, if $\|\cdot\|$ is defined by

$$\|\mathbf{y}\| = (\mathbf{S}\mathbf{y}, \mathbf{y})_{\mathbb{R}^N}^{1/2} \tag{2.7}$$

(where $\mathbf{S}$ is an $N \times N$ positive-definite symmetric matrix and $(\mathbf{x}, \mathbf{y})_{\mathbb{R}^N} = \sum_{i=1}^{N} x_i y_i$) and if $\mathbf{F}$ is still defined by (2.4), then (2.5) is replaced by

$$\mathbf{A}^t\mathbf{S}\mathbf{A}\mathbf{x} = \mathbf{A}^t\mathbf{S}\mathbf{b}. \tag{2.8}$$

With a proper choice of $\mathbf{S}$ we can dramatically improve the conditioning of the matrix in the normal equation (2.8) and make its solution much easier. This matrix $\mathbf{S}$ can be viewed as a *scaling* (or *preconditioning*) matrix. This idea of preconditioning stiff problems will be systematically used in the sequel.

The standard reference for linear least-squares problems is Lawson and Hanson [1]; concerning nonlinear least-squares problems of finite dimension and their solution, we shall mention, among many others, Levenberg [1], Marquardt [1], Powell [3], [4], Fletcher [1], Golub and Pereyra [1], Golub and Plemmons [1], Osborne and Watson [1], and More [1] (see also the references therein).

2.2. Conjugate gradient solution of the least-squares problem (2.2)

Conjugate gradient methods have been considered in Chapter IV, Sec. 2.6.7; actually they can also be used for solving the least-squares problem (2.2). We suppose that in (2.2) the Euclidean norm $\|\cdot\|$ is defined by (2.7), with $\mathbf{S}$ replaced by $\mathbf{S}^{-1}$, and we use the notation

$$(\mathbf{x}, \mathbf{y}) = (\mathbf{x}, \mathbf{y})_{\mathbb{R}^N} \quad \left(= \sum_{i=1}^{N} x_i y_i\right).$$

Let us define $J\colon \mathbb{R}^N \to \mathbb{R}$ by

$$J(\mathbf{y}) = \tfrac{1}{2}(\mathbf{S}^{-1}\mathbf{F}(\mathbf{y}), \mathbf{F}(\mathbf{y})); \tag{2.9}$$

we clearly have equivalence between (2.2) and the following:

Find $\mathbf{x} \in \mathbb{R}^N$ *such that*

$$J(\mathbf{x}) \le J(\mathbf{y}), \qquad \forall\, \mathbf{y} \in \mathbb{R}^N. \tag{2.10}$$

In the following we denote by $\mathbf{F}'$ and J' the differentials of $\mathbf{F}$ and J, respectively, we can identify $\mathbf{F}'$ with the (Jacobian) matrix $(\partial f_i/\partial x_j)_{1 \le i, j \le N}$, and we have

$$(J'(\mathbf{y}), \mathbf{z}) = (\mathbf{S}^{-1}\mathbf{F}(\mathbf{y}), \mathbf{F}'(\mathbf{y})\mathbf{z}), \qquad \forall\, \mathbf{y}, \mathbf{z} \in \mathbb{R}^N \tag{2.11}$$

which implies

$$J'(\mathbf{y}) = (\mathbf{F}'(\mathbf{y}))^t \mathbf{S}^{-1}\mathbf{F}(\mathbf{y}). \tag{2.12}$$

To solve (2.2) (via (2.10)), we can use the following conjugate gradient algorithm in which $\mathbf{S}$ is used as a *scaling* (or *preconditioning*) matrix (most of the notation is the same as in Chapter IV, Sec. 2.6.7).

First algorithm (Fletcher–Reeves)

$$\mathbf{x}^0 \in \mathbb{R}^N \textit{ given}; \tag{2.13}$$

$$\mathbf{g}^0 = \mathbf{S}^{-1}J'(\mathbf{x}^0), \tag{2.14}$$

$$\mathbf{w}^0 = \mathbf{g}^0. \tag{2.15}$$

Then assuming that $\mathbf{x}^n$ *and* $\mathbf{w}^n$ *are known, we compute* $\mathbf{x}^{n+1}$ *by*

$$\mathbf{x}^{n+1} = \mathbf{x}^n - \rho_n \mathbf{w}^n, \tag{2.16}$$

where ρ_n *is the solution of the one-dimensional minimization problem*

$$J(\mathbf{x}^n - \rho_n \mathbf{w}^n) \le J(\mathbf{x}^n - \rho \mathbf{w}^n), \qquad \forall\, \rho \in \mathbb{R}, \quad \rho_n \in \mathbb{R}. \tag{2.17}$$

Then

$$\mathbf{g}^{n+1} = \mathbf{S}^{-1}J'(\mathbf{x}^{n+1}) \tag{2.18}$$

and compute $\mathbf{w}^{n+1}$ *by*

$$\mathbf{w}^{n+1} = \mathbf{g}^{n+1} + \lambda_n \mathbf{w}^n, \tag{2.19}$$

where

$$\lambda_n = \frac{(\mathbf{S}\mathbf{g}^{n+1}, \mathbf{g}^{n+1})}{(\mathbf{S}\mathbf{g}^n, \mathbf{g}^n)}. \tag{2.20}$$

Second algorithm (Polak–Ribière). This method is like the first algorithm except that (2.20) is replaced by

$$\lambda_n = \frac{(\mathbf{S}\mathbf{g}^{n+1}, \mathbf{g}^{n+1} - \mathbf{g}^n)}{(\mathbf{S}\mathbf{g}^n, \mathbf{g}^n)}. \tag{2.21}$$

Remarks 2.17, 2.19, and 2.20 of Chapter IV, Sec. 2.6.7 still hold for algorithms (2.13)–(2.20) and (2.13)–(2.19), (2.21).

As a stopping test for the above conjugate gradient algorithms, we may use, for example, either $J(x^n) \le \varepsilon$ or $\|g^n\| \le \varepsilon$ (where ε is a "small" positive number), but other tests are possible.

3. Least-Squares Solution of a Nonlinear Dirichlet Model Problem

In order to introduce the methods that we shall apply in Secs. 4 and 5 to the solution of fluid dynamics problems, we shall consider the solution of a simple nonlinear Dirichlet problem by least-squares and conjugate gradient methods after briefly describing (in Sec. 3.2) the solution of the model problem introduced in Sec. 3.1 by some more standard interative methods; in Sec. 3.5 we shall briefly discuss the use of *pseudo-arc-length-continuation methods* for solving nonlinear problems via least-squares and conjugate gradient algorithms.

3.1. Formulation of the model problem

Let $\Omega \subset \mathbb{R}^N$ be a bounded domain with a smooth boundary $\Gamma = \partial\Omega$; let T be a nonlinear operator from $V = H_0^1(\Omega)$ to $V^* = H^{-1}(\Omega)$ ($H^{-1}(\Omega)$: topological dual space of $H_0^1(\Omega)$). We consider the nonlinear Dirichlet problem:

Find $u \in H_0^1(\Omega)$ *such that*

$$-\Delta u - T(u) = 0 \text{ in } \Omega, \tag{3.1}$$

and we observe that $u \in H_0^1(\Omega)$ implies

$$u = 0 \text{ on } \Gamma.$$

Here we shall not discuss the existence and uniqueness properties of the solutions of (3.1), since we do not want to be very specific about the operator T.

3.2. Review of some standard iterative methods for solving the model problem

3.2.1. *Gradient methods*

The simplest algorithm that we can imagine for solving (3.1) is as follows:

$$u^0 \text{ given}; \tag{3.2}$$

then for $n \geq 0$, define u^{n+1} from u^n by

$$\begin{aligned} -\Delta u^{n+1} &= T(u^n) \text{ in } \Omega, \\ u^{n+1} &= 0 \text{ on } \Gamma. \end{aligned} \tag{3.3}$$

Algorithm (3.2), (3.3) has been extensively used (see, e.g., Norrie and De Vries [1] and Periaux [1]) for the numerical simulation of subsonic potential flows for compressible inviscid fluids like those considered in Chapter IV, Sec. 3. Unfortunately algorithm (3.2), (3.3) usually blows up in the case of transonic flows. Actually (3.2), (3.3) is a particular case of the following algorithm:

$$u^0 \text{ given}; \tag{3.4}$$

then for $n \geq 0$, define u^{n+1} from u^n by

$$\begin{aligned} -\Delta u^{n+1/2} &= T(u^n) \text{ in } \Omega, \\ u^{n+1/2} &= 0 \text{ on } \Gamma, \end{aligned} \tag{3.5}$$

$$u^{n+1} = u^n + \rho(u^{n+1/2} - u^n), \qquad \rho > 0. \tag{3.6}$$

If $\rho = 1$ in (3.4)–(3.6), we recover (3.2), (3.3). Since (3.5), (3.6) are equivalent to

$$u^{n+1} = u^n - \rho(-\Delta)^{-1}(-\Delta u^n - T(u^n)), \tag{3.7}$$

with $(-\Delta)^{-1}$ corresponding to Dirichlet boundary conditions, algorithm (3.4)–(3.6) is very close to a gradient method (and is rigorously a gradient algorithm if T is the derivative of some functional).

Let us define $A: H_0^1(\Omega) \to H^{-1}(\Omega)$ by

$$A(v) = -\Delta v - T(v). \tag{3.8}$$

We can easily prove the following:

Proposition 3.1. *Suppose that the following properties hold*
(i) *A is Lipschitz continuous on the bounded sets of $H_0^1(\Omega)$;*
(ii) *A is strongly elliptic, i.e., there exists $\alpha > 0$ such that*

$$\langle A(v_2) - A(v_1), v_2 - v_1 \rangle \geq \alpha \|v_2 - v_1\|^2_{H_0^1(\Omega)}, \qquad \forall\, v_1, v_2 \in H_0^1(\Omega) \tag{3.9}$$

(where $\langle \cdot, \cdot \rangle$ denotes the duality between $H^{-1}(\Omega)$ and $H_0^1(\Omega)$).

Then problem (3.1) *has a unique solution; moreover, for every $u^0 \in H_0^1(\Omega)$, there exists $\rho_M > 0$ (depending upon u^0 in general) such that*

$$0 < \rho < \rho_M \tag{3.10}$$

implies the strong convergence of (3.4)–(3.6) *to the solution u of* (3.1).

PROOF. See, e.g., Brezis and Sibony [2]. □

3.2.2. *Newton's methods*

Assuming that T is *differentiable*, one may try to solve (3.1) by a Newton's method. For this case, using a prime to denote differentiation, we obtain:

$$u^0 \text{ given}; \tag{3.11}$$

then for $n \geq 0$, *define* u^{n+1} *from* u^n *by*

$$-\Delta u^{n+1} - T'(u^n) \cdot u^{n+1} = T(u^n) - T'(u^n) \cdot u^n \text{ in } \Omega, \quad u^{n+1} = 0 \text{ on } \Gamma. \tag{3.12}$$

Algorithm (3.11), (3.12) is the particular case corresponding to $\rho = 1$ of the following:

$$u^0 \text{ given}; \tag{3.13}$$

then for $n \geq 0$, *define* u^{n+1} *from* u^n *by*

$$-\Delta u^{n+1/2} - T'(u^n) \cdot u^{n+1/2} = T(u^n) - T'(u^n) \cdot u^n \text{ in } \Omega, \quad u^{n+1/2} = 0 \text{ on } \Gamma, \tag{3.14}$$

$$u^{n+1} = u^n + \rho(u^{n+1/2} - u^n), \qquad \rho > 0. \tag{3.15}$$

The various comments made in Chapter IV, Sec. 2.6.4 about Newton's methods still hold for algorithms (3.11), (3.12) and (3.13)–(3.15).

3.2.3. *Time-dependent approach*

A well-known technique is the following: one associates with (3.1) the time-dependent problem

$$\frac{\partial u}{\partial t} - \Delta u - T(u) = 0 \text{ in } \Omega, \tag{3.16}$$

$$u = 0 \text{ on } \Gamma, \tag{3.17}$$

$$u(x, 0) = u_0(x) \quad (\textit{initial condition}). \tag{3.18}$$

Since $\lim_{t \to +\infty} u(t)$ are usually solutions of (3.1), a natural method for solving (3.1) is the following:

(i) Use a space approximation to replace (3.16)–(3.18) by a system of ordinary differential equations.

(ii) Use an efficient method for numerically integrating systems of ordinary differential equations.

(iii) Then integrate from 0 to $+\infty$ (in practice, to a large value of t).

In the case of a stiff problem, it may be necessary to use an implicit method to integrate the initial-value problem (3.16)–(3.18). Therefore each time step will require the solution of a problem like (3.1). If one uses the ordinary backward implicit scheme (see Chapter III, Sec. 3) one obtains

$$u^0 = u_0, \tag{3.19}$$

and for $n \geq 0$,

$$\frac{u^{n+1} - u^n}{k} - \Delta u^{n+1} - T(u^{n+1}) = 0 \text{ in } \Omega, \quad u^{n+1} = 0 \text{ on } \Gamma \tag{3.20}$$

(where k denotes the time-step size). At each step one has to solve

$$\frac{u^{n+1}}{k} - \Delta u^{n+1} - T(u^{n+1}) = \frac{u^n}{k} \text{ in } \Omega, \quad u^{n+1} = 0 \text{ on } \Gamma, \tag{3.21}$$

which is very close to (3.1) (but usually better conditioned); actually, in practice, instead of (3.21), we solve a finite-dimensional system obtained from (3.16)–(3.18) by a space discretization.

3.2.4. *Alternating-direction methods*

These methods have been considered in Chapter IV, Sec. 2.6.6; actually they are also closely related to the time-dependent approach as can be seen in, e.g., Lions and Mercier [1] to which we refer for further results and comments.

Two possible algorithms are the following:

First Algorithm. This is a nonlinear variant of the *Peaceman–Rachford* algorithm (see Peaceman and Rachford [1], Varga [1], Kellog [1], Lions and Mercier [1], and Gabay [1]) defined by:

$$u^0 \text{ given;} \tag{3.22}$$

then for $n \geq 0$, u^n being given, we compute $u^{n+1/2}$, u^{n+1} from u^n by

$$r_n u^{n+1/2} - T(u^{n+1/2}) = r_n u^n + \Delta u^n, \tag{3.23}$$

$$r_n u^{n+1} - \Delta u^{n+1} = r_n u^{n+1/2} + T(u^{n+1/2}). \tag{3.24}$$

Second Algorithm. This is a nonlinear variant of the *Douglas–Rachford* algorithm (see Douglas and Rachford [1], Lieutaud [1], Varga, Lions and Mercier, and Gabay, *loc. cit.*) defined by (3.22) and

$$r_n u^{n+1/2} - T(u^{n+1/2}) = r_n u^n + \Delta u^n, \tag{3.25}$$

$$r_n u^{n+1} - \Delta u^{n+1} = r_n u^n + T(u^{n+1/2}). \tag{3.26}$$

In both algorithms $\{r_n\}_{n \geq 0}$ is a sequence of positive parameters (usually a cyclic sequence). In the nonlinear case, the determination of optimal sequences $\{r_n\}_{n \geq 0}$ is a difficult problem.[1] We also have to observe that if in algorithm

[1] See, however, Doss and Miller [1] in which alternating-direction methods more sophisticated than (3.22)–(3.24) and (3.22), (3.25), (3.26) are also discussed and tested.

(3.22)–(3.24) operators T and Δ play the same role, this is no longer true in (3.22), (3.25), (3.26), and it is usually safer to have Δ as an "acting" operator in the second step (if we suppose that $-\Delta$ is "more" elliptic than $-T$).

3.3. Least squares formulations of the model problem (3.1)

3.3.1. *Generalities*

We shall consider least-squares formulations of the model problem (3.1). An obvious least-squares formulation consists of the statement that the required function u minimizes the left-hand side of (3.1) in a $L^2(\Omega)$-least-squares sense. That is,

$$\operatorname*{Min}_{v \in V} \int_\Omega |\Delta v + T(v)|^2 \, dx, \tag{3.27}$$

where V is a space of feasible functions. Let us introduce ξ by

$$\begin{aligned} -\Delta\xi &= T(v) \text{ in } \Omega, \\ \xi &= 0 \text{ on } \Gamma. \end{aligned} \tag{3.28}$$

Then (3.27) is equivalent to

$$\operatorname*{Min}_{v \in V} \int_\Omega |\Delta(v - \xi)|^2 \, dx, \tag{3.29}$$

where ξ is a (nonlinear) function of v, through (3.28). From Lions [4] and Cea [1], [2], for example, it is clear that (3.28), (3.29) has the structure of an *optimal control* problem where

(i) v is the *control vector*,
(ii) ξ is the *state vector*,
(iii) (3.28) is the *state equation*, and
(iv) the functional occurring in (3.29) is the *cost function*.

Another least-squares optimal control formulation is

$$\operatorname*{Min}_{v \in V} \int_\Omega |v - \xi|^2 \, dx, \tag{3.30}$$

where ξ again satisfies (3.28). This formulation has been used by Cea and Geymonat [1] to solve nonlinear partial differential problems (including the steady Navier–Stokes equations). Actually the two above least-squares formulations may lead to a slow convergence, since the norm occurring in the cost functions is not appropriate for the state equation. An alternate choice, very well suited to nonlinear second-order Dirichlet problems, will be discussed in the next section.

3.3.2. *A H^{-1}-least-squares formulation of* (3.1)

Let us recall some properties of $H^{-1}(\Omega)$, the topological dual space of $H_0^1(\Omega)$. If $L^2(\Omega)$ has been identified with its dual space, then

$$H_0^1(\Omega) \subset L^2(\Omega) \subset H^{-1}(\Omega);$$

moreover $\Delta\ (=\nabla^2)$ is an isomorphism from $H_0^1(\Omega)$ onto $H^{-1}(\Omega)$. In the sequel the duality pairing $\langle \cdot, \cdot \rangle$ between $H^{-1}(\Omega)$ and $H_0^1(\Omega)$ is chosen in such a way that

$$\langle f, v \rangle = \int_\Omega fv\, dx, \qquad \forall\, f \in L^2(\Omega), \quad \forall\, v \in H_0^1(\Omega). \tag{3.31}$$

The topology of $H^{-1}(\Omega)$ is defined by $\|\cdot\|_*$, where, $\forall\, f \in H^{-1}(\Omega)$,

$$\|f\|_* = \sup_{v \in H_0^1(\Omega)-\{0\}} \frac{|\langle f, v\rangle|}{\|v\|_{H_0^1(\Omega)}}. \tag{3.32}$$

A convenient[2] least-squares formulation for solving the model problem (3.1) seems to be

$$\operatorname*{Min}_{v \in H_0^1(\Omega)} \|\Delta v + T(v)\|_*. \tag{3.33}$$

It is clear that if (3.1) has a solution, then this solution will be a solution of (3.33) for which the cost function will vanish. Let us introduce $\xi \in H_0^1(\Omega)$ by

$$\begin{aligned} \Delta\xi &= \Delta v + T(v) \text{ in } \Omega, \\ \xi &= 0 \text{ on } \Gamma, \end{aligned} \tag{3.34}$$

so that (3.33) reduces to

$$\operatorname*{Min}_{v \in H_0^1(\Omega)} \|\Delta\xi\|_*, \tag{3.35}$$

where ξ is a function of v through (3.34).

Actually it can be proved that if $\|\cdot\|_*$ is defined by (3.32) with $\langle \cdot, \cdot \rangle$ obeying (3.31), then

$$\|\Delta v\|_* = \|v\|_{H_0^1(\Omega)} \quad \left(= \left(\int_\Omega |\nabla v|^2\, dx\right)^{1/2}\right), \qquad \forall\, v \in H_0^1(\Omega). \tag{3.36}$$

From (3.36) it then follows that (3.35) may also be formulated by

$$\operatorname*{Min}_{v \in H_0^1(\Omega)} \int_\Omega |\nabla\xi|^2\, dx \tag{3.37}$$

---

[2] Convenient because the space $H_0^1(\Omega)$ in (3.33) is also the space in which we want to solve (3.1) (from the properties of Δ and T).

where ξ is a function of v through (3.34); (3.37) also has the structure of an optimal control problem.

Remark 3.1. Nonlinear boundary-value problems have been treated by Lozi [1] using a formulation closely related to (3.34), (3.37).

3.4. Conjugate gradient solution of the least squares problem (3.34), (3.37)

Let us define $J\colon H_0^1(\Omega) \to \mathbb{R}$ by

$$J(v) = \frac{1}{2}\int_\Omega |\nabla \xi|^2\,dx, \tag{3.38}$$

where ξ is a function of v in accordance with (3.34); then (3.37) may also be written as follows:

Find $u \in H_0^1(\Omega)$ such that

$$J(u) \le J(v), \qquad \forall\, v \in H_0^1(\Omega). \tag{3.39}$$

To solve (3.39), we shall use a conjugate gradient algorithm. Among the possible conjugate gradient algorithms, we have selected the *Polak–Ribière* version (see Polak [1]), since this algorithm produced the best performances in the various experiments that we performed (the good performances of the Polak–Ribière algorithm are discussed in Powell [2]). Let us denote the differential of $J(\cdot)$ by $J'(\cdot)$; then the Polak–Ribière version of the conjugate gradient method applied to the solution of (3.39) is

Step 0: Initialization

$$u^0 \in H_0^1(\Omega) \textit{ given}; \tag{3.40}$$

then compute $g^0 \in H_0^1(\Omega)$ from

$$\begin{aligned} -\Delta g^0 &= J'(u^0) \text{ in } \Omega, \\ g^0 &= 0 \text{ on } \Gamma, \end{aligned} \tag{3.41}$$

and set

$$z^0 = g^0. \tag{3.42}$$

Then for $n \ge 0$, assuming u^n, g^n, z^n known, compute $u^{n+1}, g^{n+1}, z^{n+1}$ by:

Step 1: Descent

$$u^{n+1} = u^n - \lambda_n z^n, \tag{3.43}$$

where λ_n is the solution of the one-dimensional minimization problem

$$\lambda_n \in \mathbb{R}, \qquad J(u^n - \lambda_n z^n) \le J(u^n - \lambda z^n), \qquad \forall\, \lambda \in \mathbb{R}. \tag{3.44}$$

Step 2: Construction of the New Descent Direction. Define $g^{n+1} \in H_0^1(\Omega)$ by

$$-\Delta g^{n+1} = J'(u^{n+1}) \text{ in } \Omega, \qquad g^{n+1} = 0 \text{ on } \Gamma; \tag{3.45}$$

then

$$\gamma_n = \frac{\int_\Omega \nabla g^{n+1} \cdot \nabla(g^{n+1} - g^n)\, dx}{\int_\Omega |\nabla g^n|^2\, dx}, \tag{3.46}$$

$$z^{n+1} = g^{n+1} + \gamma_n z^n, \tag{3.47}$$

$$n = n + 1 \text{ go to (3.43).} \tag{3.48}$$

The two nontrivial steps of algorithm (3.40)–(3.48) are:

(i) The solution of the single-variable minimization problem (3.44); the corresponding *line search* can be achieved by *dichotomy* or *Fibonacci* methods (see, for example, Polak [1], Wilde and Beightler [1], and Brent [1]). We have to observe that each evaluation of $J(v)$ for a given argument v requires the solution of the linear Dirichlet problem (3.34) to obtain the corresponding ξ.

(ii) The calculation of g^{n+1} from u^{n+1} which requires the solution of two linear Dirichlet problems (namely (3.34) with $v = u^{n+1}$ and (3.45)).

Calculation of $J'(u^n)$ and g^n. Due to the importance of step (ii), let us describe the calculation of $J'(u^n)$ and g^n in detail. Let $w \in H_0^1(\Omega)$; then $J'(v)$ may be defined by

$$\langle J'(v), w\rangle = \lim_{\substack{t \to 0 \\ t \neq 0}} \frac{J(v + tw) - J(v)}{t}; \tag{3.49}$$

from (3.34), (3.38), (3.49) we obtain

$$\langle J'(v), w\rangle = \int_\Omega \nabla\xi \cdot \nabla\eta\, dx, \tag{3.50}$$

where $\eta \in H_0^1(\Omega)$ is the solution of

$$\Delta\eta = \Delta w + T'(v) \cdot w \text{ in } \Omega, \qquad \eta = 0 \text{ on } \Gamma; \tag{3.51}$$

(3.51) has the following variational formulation:

$$\int_\Omega \nabla\eta \cdot \nabla z\, dx = \int_\Omega \nabla w \cdot \nabla z\, dx - \langle T'(v) \cdot w, z\rangle, \qquad \forall\, z \in H_0^1(\Omega), \quad \eta \in H_0^1(\Omega). \tag{3.52}$$

Taking $z = \xi$ in (3.52), from (3.50) we obtain

$$\langle J'(v), w \rangle = \int_\Omega \nabla \xi \cdot \nabla w \, dx - \langle T'(v) \cdot w, \xi \rangle, \qquad \forall\, v, w \in H_0^1(\Omega). \tag{3.53}$$

Therefore $J'(v)$ $(\in H^{-1}(\Omega))$ may be identified with the linear functional on $H_0^1(\Omega)$ defined by

$$w \to \int_\Omega \nabla \xi \cdot \nabla w \, dx - \langle T'(v) \cdot w, \xi \rangle. \tag{3.54}$$

From (3.45), (3.53), (3.54) it then follows that g^n is the solution of the following linear variational problem:

Find $g^n \in H_0^1(\Omega)$ such that, $\forall\, w \in H_0^1(\Omega)$,

$$\int_\Omega \nabla g^n \cdot \nabla w \, dx = \int_\Omega \nabla \xi^n \cdot \nabla w \, dx - \langle T'(u^n) \cdot w, \xi^n \rangle, \tag{3.55}$$

where ξ^n is the solution of (3.34) corresponding to $v = u^n$.

Remark 3.2. It is clear from the above observations that an efficient *Poisson solver* will be a basic tool for solving (3.1) (in fact, a finite-dimensional approximation of it) by the conjugate gradient algorithm (3.40)–(3.48).

Remark 3.3. The fact that $J'(v)$ is known through (3.53) is not at all a drawback if a *Galerkin* or a *finite element* method is used to approximate (3.1). Indeed we only need to know the value of $\langle J'(v), w \rangle$ for w belonging to a basis of the finite dimensional subspace of $H_0^1(\Omega)$ corresponding to the Galerkin or finite element approximation under consideration.

EXERCISE 3.1. Extend the above least-squares–conjugate-gradient methodology to the solution in $H_0^1(\Omega) \times H_0^1(\Omega)$ of the nonlinear boundary-value system

$$\begin{aligned} -\Delta u_1 + u_1 \frac{\partial u_1}{\partial x_1} + u_2 \frac{\partial u_1}{\partial x_2} &= f_1 \text{ in } \Omega, \\ -\Delta u_2 + u_1 \frac{\partial u_2}{\partial x_1} + u_2 \frac{\partial u_2}{\partial x_2} &= f_2 \text{ in } \Omega, \\ u_1 = u_2 &= 0 \text{ on } \Gamma, \end{aligned} \tag{3.56}$$

where Ω is a bounded domain of $\mathbb{R}^2$ and $f_1, f_2 \in H^{-1}(\Omega)$.

3.5. A nonlinear least-squares approach to arc-length-continuation methods

3.5.1. *Generalities: Synopsis*

We would like to show that the above least-squares methodology can be (slightly) modified in order to solve nonlinear problems by arc-length-continuation methods, directly inspired by H. B. Keller [1], [2] (where the basic iterative methods are Newton's and quasi-Newton's, instead of conjugate gradient).

As a test problem we have chosen a variant of the nonlinear Dirichlet problem (3.1); let us consider the following family of nonlinear Dirichlet problems,[3] parametrized by $\lambda \in \mathbb{R}$,

$$\begin{aligned} -\Delta u &= \lambda T(u) \text{ in } \Omega, \\ u &= 0 \text{ on } \Gamma; \end{aligned} \tag{3.57}$$

(3.1) corresponds to $\lambda = 1$.

3.5.2. *Solution of* (3.57) *via arc-length-continuation methods*

Following H. B. Keller [1], [2] [for which we refer for justification (see also Percell [1])], we associate to (3.57) a "continuation" equation; we have chosen[4]

$$\int_\Omega |\nabla \dot{u}|^2 \, dx + \dot{\lambda}^2 = 1, \tag{3.58}$$

where $\dot{u} = \partial u/\partial s$, $\dot{\lambda} = d\lambda/ds$, and where the *curvilinear abscissa* s is defined by

$$\delta s = \dot{\lambda}\delta\lambda + \int_\Omega \nabla \dot{u} \cdot \nabla \delta u \, dx, \tag{3.59}$$

or equivalently by

$$(\delta s)^2 = (\delta\lambda)^2 + \int_\Omega \nabla \delta u \cdot \nabla \delta u \, dx. \tag{3.60}$$

In fact we are considering a path in $H_0^1(\Omega) \times \mathbb{R}$ whose arc length is defined by (3.58)–(3.60). Then, in order to solve (3.57), we consider the family (parametrized by s) of nonlinear systems (3.57), (3.58). In practice we shall approximate (3.57), (3.58) by the following *discrete family* of nonlinear systems, where Δs is an *arc-length step, positive or negative* (possibly varying with n) and $u^n \simeq u(n\Delta s)$:

$$\textit{Take } u^0 = 0, \lambda^0 = 0 \textit{ and suppose that } \dot{\lambda}(0), \dot{u}(0) \textit{ are given} \tag{3.61}$$

(initialization (3.61) is justified by the fact that $u = 0$ is the unique solution of (3.57) if $\lambda = 0$); *then for* $n \geq 0$, *assuming* u^{n-1}, λ^{n-1}, u^n, λ^n *known, we obtain* $\{u^{n+1}, \lambda^{n+1}\} \in H_0^1(\Omega) \times \mathbb{R}$ *by*

$$\begin{aligned} -\Delta u^{n+1} &= \lambda^{n+1} T(u^{n+1}) \textit{ in } \Omega, \\ u^{n+1} &= 0 \textit{ on } \Gamma, \end{aligned} \tag{3.62}$$

---

[3] To be solved in $H_0^1(\Omega)$

[4] Other choices are possible

and

$$\int_\Omega \nabla(u^1 - u^0)\cdot\nabla\dot{u}(0)\,dx + (\lambda^1 - \lambda^0)\dot{\lambda}(0) = \Delta s \quad \text{if } n = 0, \tag{3.63}$$

$$\int_\Omega \nabla(u^{n+1} - u^n)\cdot\nabla\left(\frac{u^n - u^{n-1}}{\Delta s}\right)dx + (\lambda^{n+1} - \lambda^n)\left(\frac{\lambda^{n+j} - \lambda^{n+j-1}}{\Delta s}\right) = \Delta s$$

$$\text{with } j = 0 \text{ or } 1, \quad \text{if } n \geq 1. \tag{3.64}$$

Obtaining $\dot{u}(0)$ and $\dot{\lambda}(0)$ is an easy task since we have (from (3.57))

$$\begin{aligned} -\Delta\dot{u}(0) &= \dot{\lambda}(0)T(0) \text{ in } \Omega, \\ \dot{u}(0) &= 0 \text{ on } \Gamma, \end{aligned} \tag{3.65}$$

and therefore

$$\dot{\lambda}^2(0)\left(1 + \int_\Omega |\nabla\hat{u}|^2\,dx\right) = 1, \tag{3.66}$$

where $\hat{u} \in H_0^1(\Omega)$ is the solution of

$$\begin{aligned} -\Delta\hat{u} &= T(0) \text{ in } \Omega, \\ \hat{u} &= 0 \text{ on } \Gamma \end{aligned} \tag{3.67}$$

(then we clearly have $\dot{u}(0) = \dot{\lambda}(0)\hat{u}$).

Relations (3.61)–(3.64) look like clearly a discretization scheme for solving the Cauchy problem for *first-order ordinary differential equations*; from this analogy we can derive many other discretization schemes for the approximation of (3.57), (3.58) (Runge–Kutta, multisteps, etc...) and also methods for the *automatic adjustment of* Δs.

3.5.3. *Nonlinear least squares and conjugate gradient solution of* (3.62)–(3.64).

Without going into details (for which we refer to Reinhart [1] and Glowinski, Keller, and Reinhart [1]) we can solve (3.62)–(3.64) by a variant of (3.40)–(3.48) defined on the Hilbert space $H_0^1(\Omega) \times \mathbb{R}$ equipped with the metric and inner-product corresponding to

$$\{v, \mu\} \to \int_\Omega |\nabla v|^2\,dx + \mu^2; \tag{3.68}$$

it is clear that many other norms than (3.68) are possible, however in all cases the scaling of a conjugate gradient algorithm using a discrete variant of[5]

$$\begin{pmatrix} -\Delta & 0 \\ 0 & 1 \end{pmatrix} (\textit{or similar operators}) \tag{3.69}$$

will require an efficient solver and the conclusions of Sec. 3.4 still hold.

[5] In (3.69), Δ corresponds to the homogeneous Dirichlet boundary conditions.

Remark 3.4. To initialize the conjugate gradient algorithm solving (3.62)–(3.64) we have used $\{2\lambda^n - \lambda^{n-1}, 2u^n - u^{n-1}\}$ as initial guess to compute $\{\lambda^{n+1}, u^{n+1}\}$; with such a choice we obtain a much faster convergence than by taking $\{\lambda^n, u^n\}$ as initial guess.

3.5.4. *Application*

3.5.4.1. *Formulation of the problem.* We shall apply the methods described in Secs. 3.5.2 and 3.5.3 to the solution of the following classical problem:

$$\begin{aligned} -\Delta u &= \lambda e^u \text{ in } \Omega, \\ u &= 0 \text{ on } \Gamma, \end{aligned} \tag{3.70}$$

where Ω is a bounded domain of $\mathbb{R}^N$. We have to observe that unless $N = 1$, the mapping T defined by

$$T(v) = e^v, \qquad v \in H_0^1(\Omega),$$

is not continuous from $H_0^1(\Omega)$ to V^* $(=H^{-1}(\Omega))$. We consider only the case where $\lambda > 0$, since if $\lambda \leq 0$, the operator $v \to -\Delta v - \lambda e^v$ is monotone and therefore the methods of Chapter IV, Sec. 2 can be applied (take $\phi(t) = -\lambda(e^t - 1), f = \lambda$), showing the existence of a unique solution of (3.70) (which is $u = 0$ if $\lambda = 0$).

With $\lambda > 0$, problem (3.70) has been considered by many authors (Henri Poincaré—with $\Omega = \mathbb{R}^N$—among them). With regard to recent publications, let us mention, among others, Crandall and Rabinowitz [1], [2], Amann [1], Mignot and Puel [1], and Mignot, Murat, Puel [1]. In particular, in Mignot, Murat, and Puel [1] we may find an interesting discussion showing the relationships between (3.70) and *combustion phenomena.*

From a numerical point of view, problem (3.70) has been investigated by, among others, Kikuchi [1] and Reinhart [1] to which we refer for more details and further references (see also Simpson [1], Moore and Spence [1], Glowinski, Keller, and Reinhart [1], and Chan and Keller [1]).

3.5.4.2. *Numerical implementation of the methods of Secs.* 3.5.2. and 3.5.3. We have chosen to solve the particular case of (3.70) where $\Omega =]0, 1[\times]0, 1[$.

The practical application of the methods of Secs. 3.5.2 and 3.5.3 requires the reduction of (3.70) to a finite-dimensional problem; to do this we have used the finite element method described in Chapter IV, Sec. 2.5, taking for $\mathcal{T}_h$ the triangulation consisting of 512 triangles indicated in Fig. 3.1.

The unknowns are the values taken by the approximate solution u_h at the interior nodes of $\mathcal{T}_h$; we have 225 such nodes.

Algorithm (3.61)–(3.64) has been applied with $\Delta s = 0.1$, and $j = 0$ in (3.64); we observe that $T(0) = 1$ in (3.67); algorithm (3.61)–(3.64) ran "nicely," since an accurate least-squares solution of the nonlinear system (3.62), (3.64) required basically no more than 3 or 4 conjugate gradient iterations, even close to the turning point.

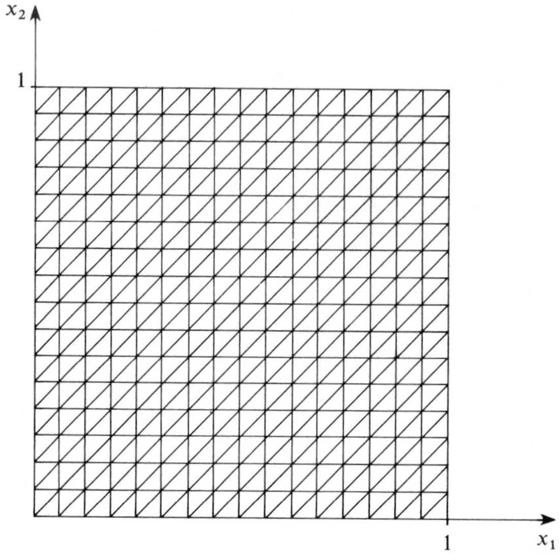

Figure 3.1

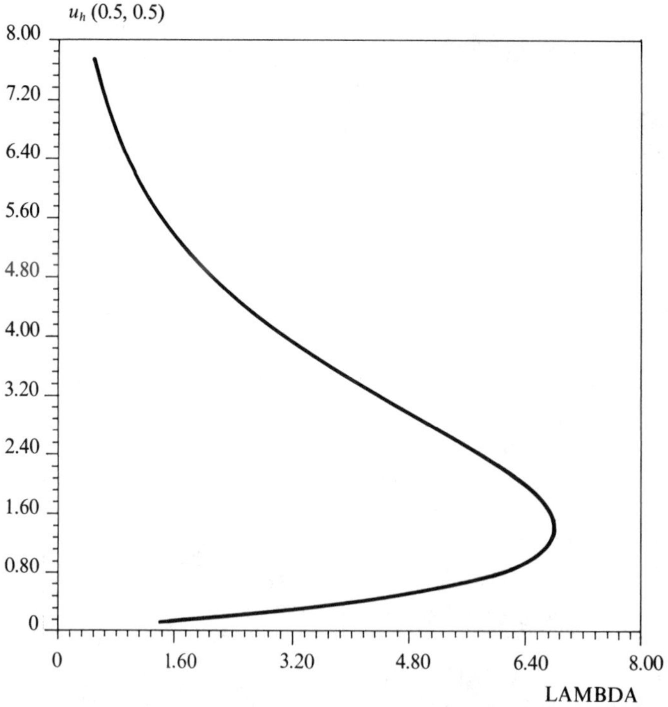

Figure 3.2

In Fig. 3.2 we show the maximal value (reached at $x_1 = x_2 = 0.5$) of the computed solution u_h as a function of λ; the computed turning point is at $\lambda = 6.8591\ldots$.

The initialization of the conjugate gradient algorithm used to solve system (3.62), (3.64), via least squares, was performed according to Remark 3.4.

3.5.5. *Further comments*

The least-squares conjugate gradient continuation method described in Secs. 3.5.2, 3.5.3, and 3.5.4 has been applied to the solution of nonlinear problems more complicated than (3.70); among them let us mention the Navier–Stokes equations for incompressible viscous fluids at high Reynold's number and also problems involving genuine bifurcation phenomena like the Von Karman equations[6] for plates. The details of these calculations can be found in Reinhart [1], [2] and Glowinski, Keller, and Reinhart [1].

4. Transonic Flow Calculations by Least-Squares and Finite Element Methods

4.1. Introduction

In Chapter IV, Sec. 3, we considered the nonlinear elliptic equation modelling the subsonic potential flows of an inviscid compressible fluid. In this section, which closely follows Bristeau, Glowinski, Periaux, Perrier, Pironneau, and Poirier[7] [2] (see also G.L.T. [3, Appendix 4]). we would like to show that the least-squares conjugate gradient methods of Sec. 2 and 3 can be applied (via convenient finite element approximations) to the computation of transonic flows for similar fluids.

Given the importance and complexity of the above problem, we would like to point out that the following considerations are just an introduction to a rather difficult subject. Many methods, using very different approaches, exist in the specialized literature, and we shall concentrate on a few of them only (see the following references for other methods). We would also like to mention that from a mathematical point of view, the methods to be described in the following sections are widely heuristical.

---

[6] For which we refer to the monograph by Ciarlet and Rabier [1] (and the references therein).

[7] B.G.4P. in the sequel

4.2. Generalities. The physical problem

The theoretical and numerical studies of transonic potential flows for inviscid compressible fluids have always been very important questions. But these problems have become even more important in recent years in relation to the design and development of large subsonic economical aircrafts.

From the theoretical point of view, many open questions still remain, with their counterparts in numerical methodology. The difficulties are quite considerable for the following reasons:

(1) The equations governing these flows are nonlinear and of changing type (elliptic in the subsonic part of the flow; hyperbolic in the supersonic part).
(2) Shocks may exist in these flows corresponding to discontinuities of velocity, pressure, and density.
(3) An entropy condition has to be included "somewhere" in order to eliminate rarefaction shocks, since they correspond to nonphysical situations.

Concerning the fluids and flows under consideration, we suppose that these fluids are compressible and inviscid (nonviscous) and that their flows are potential (and therefore quasi-isentropic) with weak shocks only; in fact this potential property is no longer true after a shock (cf. Landau and Lifchitz [1]).

In the case of flows past bodies, we shall suppose that these bodies are *sufficiently thin and parallel to the main flow in order not to create a wake in the outflow.*

4.3. Mathematical formulation of the transonic flow problem. References

4.3.1. *Governing equations*

If Ω is the region of the flow and Γ its boundary, it follows from Landau and Lifchitz [1] that the flow is governed by the so-called *full potential equation*:

$$\nabla \cdot \rho \mathbf{u} = 0 \text{ in } \Omega, \tag{4.1}$$

where

$$\rho = \rho_0 \left(1 - \frac{|\mathbf{u}|^2}{[(\gamma + 1)/(\gamma - 1)]C_*^2}\right)^{1/(\gamma-1)}, \tag{4.2}$$

$$\mathbf{u} = \nabla \phi, \tag{4.3}$$

and

(a) ϕ is the velocity potential,
(b) ρ is the density of the fluid,
(c) γ is the ratio of specific heats ($\gamma = 1.4$ in air),
(d) C_* is the critical velocity.

Figure 4.1

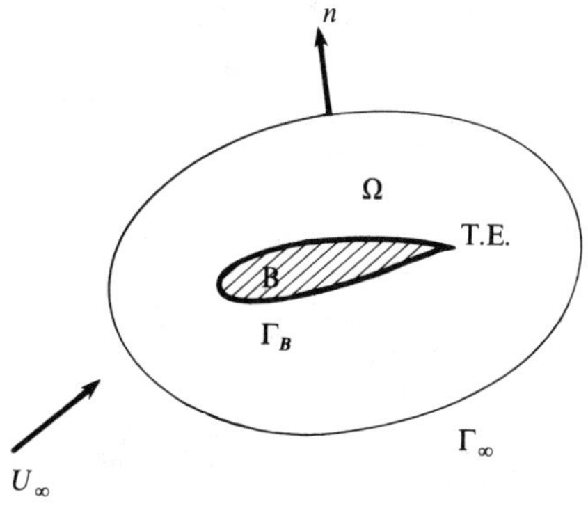

4.3.2. *Boundary conditions*

For an *airfoil* B (see Fig. 4.1), the flow is supposed to be uniform on Γ_∞ and tangential at Γ_B.

We then have

$$\frac{\partial \phi}{\partial n} = \mathbf{u}_\infty \cdot \mathbf{n} \text{ on } \Gamma_\infty, \tag{4.4}$$

$$\frac{\partial \phi}{\partial n} = 0 \text{ on } \Gamma_B. \tag{4.5}$$

Since only *Neumann boundary conditions* are involved in the above case, the potential ϕ is determined only to within an arbitrary constant. To remedy this, we can prescribe the value of ϕ at some point within $\Omega \cup \Gamma$ and, for example, we may conveniently use

$$\phi = 0 \text{ at the trailing edge T.E. of } B. \tag{4.6}$$

4.3.3. *Lifting airfoils and the Kutta–Joukowsky condition*

For two-dimensional flows a slit Σ (see Fig. 4.2) has to be introduced in Ω in order to render the potential ϕ single valued, and we choose ϕ to be discontinuous across Σ. The circulation β is another unknown depending on the

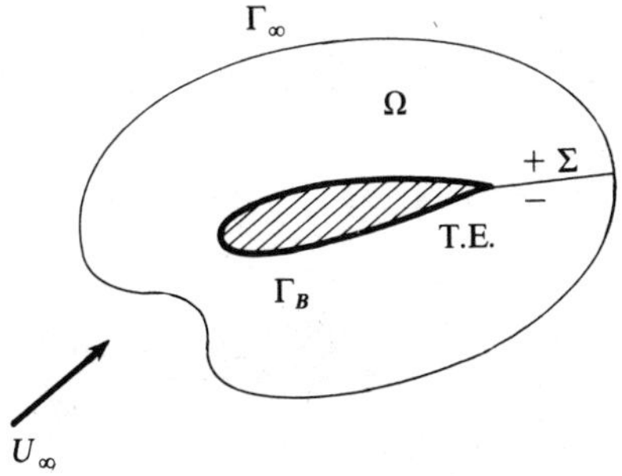

Figure 4.2

boundary conditions and the geometry of B. Along Σ, the following relation is required:

$$\phi^+ - \phi^- = \beta. \tag{4.7}$$

To find β one uses the *Kutta–Joukowsky condition*

$$p^+ = p^- \text{ at T.E.} \quad (p\text{: pressure}), \tag{4.8}$$

which, by applying the *Bernouilli law*, may also be written as

$$|\nabla\phi^+| = |\nabla\phi^-| \text{ at T.E.}. \tag{4.9}$$

Observe that (4.9) is a nonlinear relation.

Remark 4.1. In two-dimensional cases we can use linear formulations of the Kutta–Joukowsky condition. Furthermore, if there is no cusp at T.E., it has been proved in Ciavaldini, Pogu, and Tournemine [3], for strictly subsonic flows (i.e., $|\nabla\phi| < C_*$ everywhere), that the physical solution is characterized by

$$\nabla\phi = 0 \text{ at T.E.}, \tag{4.10}$$

which can be taken as the Kutta–Joukowsky condition. A similar result has not yet been proved for genuine transonic flows. Let us again emphasize that (4.10) is no longer true if there is a cusp at T.E. (as is the case, for example, for the celebrated *Korn's airfoil*).

The treatment of the Kutta–Joukowsky condition for three-dimensional flows is much more complicated; we refer to B.G. 4P. [1] and Bristeau, Glowinski, Perrier, Periaux, and Pironneau [1] for the practical implementation of the three-dimensional Kutta–Joukowsky condition.

4.3.4. *Shock conditions*

Across a shock the flow has to satisfy the *Rankine–Hugoniot conditions*

$$(\rho\mathbf{u}\cdot\mathbf{n})_+ = (\rho\mathbf{u}\cdot\mathbf{n})_- \tag{4.11}$$

(where $\mathbf{n}$ is normal at the shock line or surface);

the tangential component of the velocity is continuous. (4.12)

A suitable weak formulation of (4.1)–(4.3) will take (4.11), (4.12) into account automatically.

4.3.5. *Entropy condition*

This condition can be formulated as follows (see Landau and Lifchitz [1] for further details):

Following the flow we cannot have a positive variation of velocity through a shock, since this would imply a negative variation of entropy which is a nonphysical phenomenon. (4.13)

The numerical implementation of (4.13) will be discussed in Sec. 4.6.

4.3.6. *Some references: Synopsis*

The mathematical analysis of the above transonic flow problem is quite difficult. Some standard references are Bers [1] and Moravetz [1]–[5] (see also Landau and Lifchitz [1] and Courant and Friedrichs [1] for the physical aspects).

From the numerical point of view, the more commonly used finite difference methods have originated from Murman and Cole [1], and we shall mention, among many other references, Bauer and Garabedian, and Korn [1], Bauer, Garabedian, Korn, and Jameson [1], [2], Jameson [1], [2] ,[3], [4], Holst [1], and Osher [1], and the bibliographies therein (see also Hewitt and Hillingworth and co-editors [1]). The above numerical methods use the key idea of Murman and Cole, which consists of using a finite difference scheme (centered in the subsonic part of the flow) backward[8] (in the direction of the flow) in the supersonic part. The switching between these two schemes is automatically accomplished via a truncation operator active only in the supersonic part of the flow (see Jameson and Holst, *loc. cit.* for more details). Relaxation or alternating-direction methods (or a combination of both) are then used to solve the resulting nonlinear system.

These finite difference methods of solution have been extended to finite elements (of quadrilateral type) by Eberle [1], [2], Deconinck and Hirsh [1], and Amara, Joly, and Thomas [1].

The methods to be described in Secs. 4.4, 4.5, and 4.6 allow the use of triangular (or tetrahedral) elements and are well suited to a least-squares conjugate gradient solution.

4.4. Least-squares formulation of the continuous problem

In this section we do not consider the practical implementation of (4.13); we only discuss the variational formulation of (4.1)–(4.5), (4.11), (4.12) and of an associated nonlinear least-squares formulation.

4.4.1. *A variational formulation of the continuity equation*

For simplicity we consider the situation of Fig. 4.3 which shows a symmetric flow, subsonic at infinity, around a symmetric airfoil; thus the Kutta–Joukowsky condition is automatically satisfied.

For practicality (but other approaches are possible) we imbed the airfoil in a "large" bounded domain. Using the notation of Sec. 4.3, the continuity equation and the boundary conditions are

$$\nabla \cdot \rho(\phi)\nabla\phi = 0 \text{ in } \Omega, \tag{4.14}$$

[8] One also says *upwinded* or *one sided*.

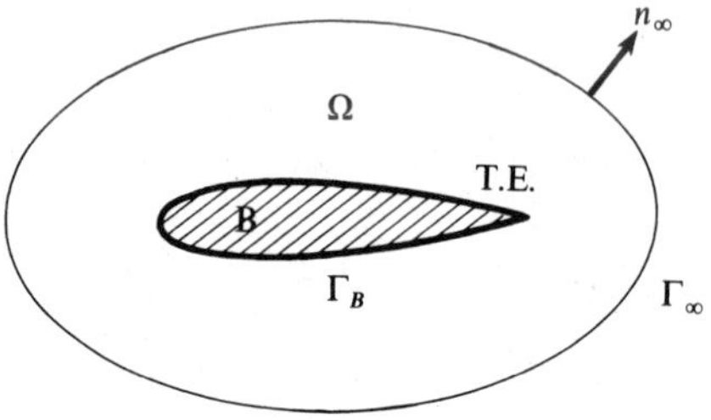

Figure 4.3

with

$$\rho(\phi) = \rho_0\left(1 - \frac{|\nabla\phi|^2}{[(\gamma + 1)/(\gamma - 1)]C_*^2}\right)^{1/(\gamma-1)}, \tag{4.15}$$

and

$$\rho\frac{\partial\phi}{\partial n} = 0 \text{ on } \Gamma_B, \qquad \rho\frac{\partial\phi}{\partial n} = \rho_\infty \mathbf{u}_\infty \cdot \mathbf{n}_\infty \text{ on } \Gamma_\infty; \tag{4.16}$$

on Γ $(=\Gamma_B \cup \Gamma_\infty)$, we define g by

$$g = 0 \text{ on } \Gamma_B, \qquad g = \rho_\infty \mathbf{u}_\infty \cdot \mathbf{n}_\infty \text{ on } \Gamma_\infty. \tag{4.17}$$

We clearly have

$$\rho\frac{\partial\phi}{\partial n} = g \text{ on } \Gamma \quad \text{and} \quad \int_\Gamma g\, d\Gamma = 0. \tag{4.18}$$

An equivalent variational formulation of (4.14), (4.18) is

$$\int_\Omega \rho(\phi)\nabla\phi \cdot \nabla v\, dx = \int_\Gamma gv\, d\Gamma, \qquad \forall\, v \in H^1(\Omega), \quad \phi \in W^{1,\infty}(\Omega)/\mathbb{R}, \tag{4.19}$$

where (cf. Adams [1] and Necas [1]), for $p \geq 1$, $W^{1,p}(\Omega)$ is the *Sobolev space* defined by

$$W^{1,p}(\Omega) = \left\{v \mid v \in L^p(\Omega), \frac{\partial v}{\partial x_i} \in L^p(\Omega), \forall\, i\right\}$$

(with $H^1(\Omega) = W^{1,2}(\Omega)$); the function ϕ is determinated only to within an arbitrary constant.

Remark 4.2. The space $W^{1,\infty}(\Omega)$ (space of *Lipschitz continuous* functions) is a natural choice for ϕ since physical flows require (among other properties) a positive density ρ; therefore, from (4.2), (4.3), ϕ must satisfy

$$|\nabla\phi| \leq \delta < \left(\frac{\gamma + 1}{\gamma - 1}\right)^{1/2} C_* \text{ a.e. on } \Omega. \tag{4.20}$$

4.4.2. *A least-squares formulation of* (4.19)

For a genuine transonic flow, problem (4.19) is not equivalent to a standard problem of the calculus of variations (as is the case for purely subsonic flows; see Chapter IV, Sec. 3). To remedy this situation, and—in some sense—*convexify* the problem under consideration, we consider, as in Sec. 3.3 of this chapter, a nonlinear least-squares formulation of the transonic flow problem (4.18), defined as follows:

Let X be a set of feasible transonic flow solutions; the least-squares problem is then

$$\operatorname*{Min}_{\xi \in X} J(\xi), \tag{4.21}$$

with

$$J(\xi) = \frac{1}{2}\int_\Omega |\nabla y(\xi)|^2\, dx, \tag{4.22}$$

where, in (4.22), $y(\xi)$ $(=y)$ *is the solution of the state equation*:

Find $y \in H^1(\Omega)/\mathbb{R}$ *such that*

$$\int_\Omega \nabla y \cdot \nabla v\, dx = \int_\Omega \rho(\xi)\nabla\xi \cdot \nabla v\, dx - \int_\Gamma gv\, d\Gamma, \qquad \forall\, v \in H^1(\Omega). \tag{4.23}$$

If the transonic flow problem has solutions, these solutions solve the least-squares problem and give the value zero to the cost function J.

4.5. Finite element approximation and least-squares-conjugate gradient solution of the approximate problems

We consider only two-dimensional problems; but the following methods can be (and have been) applied to three-dimensional problems.

4.5.1. *Finite element approximation of the nonlinear variational equation* (4.19)

We still consider the nonlifting situation of Sec. 4.4.1. Once the flow region has been imbedded in a large bounded domain Ω, we approximate this later domain by a polygonal domain Ω_h; with $\mathcal{T}_h$ a standard triangulation of Ω_h, we approximate $H^1(\Omega)$ (and in fact $W^{1,p}(\Omega)$, $\forall\, p \geq 1$) by

$$H_h^1 = \{v_h | v_h \in C^0(\overline{\Omega}_h),\ v_h|_T \in P_1,\ \forall\, T \in \mathcal{T}_h\}, \tag{4.24}$$

where, in (4.24), P_1 is the space of the polynomials in two variables of degree ≤ 1.

We prescribe the value zero for the potential at T.E.; this leads to

$$V_h = \{v_h \in H_h^1,\ v_h(\text{T.E.}) = 0\}. \tag{4.25}$$

We clearly have

$$\dim H_h^1 = \dim V_h + 1 = \text{number of vertices of } \mathcal{T}_h. \tag{4.26}$$

We then approximate the variational equation (4.19) by:

Find $\phi_h \in V_h$ *such that*

$$\int_{\Omega_h} \rho(\phi_h)\nabla\phi_h \cdot \nabla v_h \, dx = \int_{\Gamma_h} g_h v_h \, d\Gamma, \qquad \forall\, v_h \in V_h \tag{4.27}$$

where, in (4.27), g_h is an approximation of the function g of (4.17) (and $\Gamma_h = \partial\Omega_h$).

The above discrete variational formulation implies that $\rho\, \partial\phi/\partial n|_\Gamma = g$ is approximately satisfied automatically.

Let $\mathcal{B}_h = \{w_i\}_{i=1}^{N_h}$ be a vector basis of V_h (with $N_h = \dim V_h$); then (4.27) is equivalent to the nonlinear finite-dimensional system

$$\phi_h = \sum_{j=1}^{N_h} \phi_j w_j, \int_{\Omega_h} \rho(\phi_h)\nabla\phi_h \cdot \nabla w_i \, dx = \int_{\Gamma_h} g_h w_i \, d\Gamma, \qquad \forall i = 1, \ldots, N_h. \tag{4.28}$$

If $\{P_j\}_{j=1}^{N_h}$ is the set of the vertices of $\mathcal{T}_h$ different from T.E., we take (for $\mathcal{B}_h$) the set defined by

$$\begin{aligned} \mathcal{B}_h = \{w_j\}_{j=1}^{N_h} \quad &\text{with } w_j \in V_h, \qquad \forall\, j = 1, \ldots, N_n, \\ w_j(P_j) = 1, \qquad &w_j(P_k) = 0, \qquad \forall\, k \neq j; \end{aligned} \tag{4.29}$$

we then have $\phi_j = \phi_h(P_j)$.

From the above choice for H_h^1 and V_h, there is no problem of numerical integration since, in (4.27), (4.28), $\nabla\phi_h$, ∇v_h (and therefore $\rho(\phi_h)$) are piecewise constant functions.

4.5.2. *Least-squares formulation of the discrete problem* (4.27), (4.28)

For simplicity we set $\Omega_h = \Omega$, $\Gamma_h = \Gamma$. Combining the results of Secs. 4.4.2 and 4.5.1, we introduce the following least-squares formulation of the approximate problem (4.27), (4.28):

$$\operatorname*{Min}_{\xi_h \in X_h} J_h(\xi_h), \tag{4.30}$$

where, in (4.30), X_h is the set of the feasible discrete solutions and

$$J_h(\xi_h) = \frac{1}{2}\int_\Omega |\nabla y_h(\xi_h)|^2 \, dx, \tag{4.31}$$

with $y_h(\xi_h)$ $(= y_h)$ the solution of the discrete variational state equation:

Find $y_h \in V_h$ *such that*

$$\int_\Omega \nabla y_h \cdot \nabla v_h \, dx = \int_\Omega \rho(\xi_h)\nabla\xi_h \cdot \nabla v_h \, dx - \int_\Gamma g_h v_h \, d\Gamma, \qquad \forall\, v_h \in V_h. \tag{4.32}$$

4.5.3. *Conjugate gradient solution of the least-squares problem* (4.30)–(4.32)

We follow B.G. 4P. [1], [2], Bristeau, Glowinski, Periaux, Perrier, and Pironneau [1], and Periaux [2]; a preconditioned conjugate gradient algorithm for solving (4.30)–(4.32) (with $X_h = V_h$) is as follows.

Step 0: Initialization

$$\phi_h^0 \in V_h \text{ given}; \tag{4.33}$$

then compute g_h^0 from

$$g_h^0 \in V_h, \quad \int_\Omega \nabla g_h^0 \cdot \nabla v_h \, dx = \langle J_h'(\phi_h^0), v_h \rangle, \qquad \forall\, v_h \in V_h, \tag{4.34}$$

and set

$$z_h^0 = g_h^0. \tag{4.35}$$

Then for $n \geq 0$, assuming that ϕ_h^n, g_h^n, z_h^n are known, compute $\phi_h^{n+1}, g_h^{n+1}, z_h^{n+1}$ by the following.

Step 1: Descent. Compute

$$\lambda^n = \operatorname*{Arg\,min}_{\lambda \in \mathbb{R}} J_h(\phi_h^n - \lambda z_h^n), \tag{4.36}$$

$$\phi_h^{n+1} = \phi_h^n - \lambda^n z_h^n. \tag{4.37}$$

Step 2: Construction of the new descent direction. Define g_h^{n+1} by

$$g_h^{n+1} \in V_h, \quad \int_\Omega \nabla g_h^{n+1} \cdot \nabla v_h \, dx = \langle J_h'(\phi_h^{n+1}), v_h \rangle, \qquad \forall\, v_h \in V_h; \tag{4.38}$$

then

$$\gamma_{n+1} = \frac{\int_\Omega \nabla g_h^{n+1} \cdot \nabla (g_h^{n+1} - g_h^n) \, dx}{\int_\Omega |\nabla g_h^n|^2 \, dx}, \tag{4.39}$$

$$z_h^{n+1} = g_h^{n+1} + \gamma_{n+1} z_h^n, \tag{4.40}$$

$n = n + 1$ *go to* (4.36).

The two nontrivial steps of algorithm (4.33)–(4.40) are as follows:

(i) The solution of the single-variable minimization problem (4.36); the corresponding line search can be achieved by the methods mentioned in Sec. 3.4. We observe that each evaluation of $J_h(\xi_h)$, for a given argument ξ_h, requires the solution of the linear approximate Neumann's problem (4.32) to obtain the corresponding y_h.

(ii) The calculation of g_h^{n+1} from ϕ_h^{n+1} which requires the solution of two linear approximate Neumann's problems (namely (4.32) with $\xi_h = \phi_h^{n+1}$ and (4.38)).

Calculation of $J_h'(\phi_h^n)$ *and* g_h^n: Due to the importance of step (ii), let us describe the calculation of $J_h'(\phi_h^n)$ and g_h^n in detail (for simplicity we suppose that $\rho_0 = 1$).

By differentiation we have

$$\langle J_h'(\xi_h), \delta\xi_h \rangle = \int_\Omega \nabla y_h \cdot \nabla \delta y_h \, dx, \tag{4.41}$$

where δy_h is from (4.32), the solution of:

$\delta y_h \in V_h$ *and* $\forall\, v_h \in V_h$, *we have*

$$\int_\Omega \nabla \delta y_h \cdot \nabla v_h \, dx = \int_\Omega \rho(\xi_h) \nabla \delta\xi_h \cdot \nabla v_h \, dx + \int_\Omega \delta\rho(\xi_h) \nabla \xi_h \cdot \nabla v_h \, dx. \tag{4.42}$$

Since $\rho(\xi_h) = (1 - K|\nabla\xi_h|^2)^\alpha$, with $K = (1/C_*^2)[(\gamma - 1)/(\gamma + 1)]$ and $\alpha = 1/(\gamma - 1)$, we have

$$\delta\rho(\xi_h) = -2K\alpha(1 - K|\nabla\xi_h|^2)^{\alpha-1} \nabla\xi_h \cdot \nabla\delta\xi_h. \tag{4.43}$$

From (4.41)–(4.43) it follows that

$$\int_\Omega \nabla y_h \cdot \nabla \delta y_h \, dx = \int_\Omega \rho(\xi_h) \nabla y_h \cdot \nabla \delta\xi_h \, dx$$
$$- 2K\alpha \int_\Omega (\rho(\xi_h))^{2-\gamma} \nabla\xi_h \cdot \nabla y_h \nabla\xi_h \cdot \nabla\delta\xi_h \, dx. \tag{4.44}$$

Finally $J_h'(\xi_h)$ can be identified with the linear functional on V_h defined by

$$\eta_h \to \int_\Omega \rho(\xi_h) \nabla y_h \cdot \nabla \eta_h \, dx - 2K\alpha \int_\Omega (\rho(\xi_h))^{2-\gamma} \nabla\xi_h \cdot \nabla y_h \nabla\xi_h \cdot \nabla\eta_h \, dx. \tag{4.45}$$

It is then quite easy to obtain g_h^{n+1} from ϕ_h^{n+1} using (4.38), (4.45).

Remark 4.3. An efficient discrete Poisson's solver will be a basic tool if one uses the above conjugate gradient algorithm (4.33)–(4.40).

EXERCISE 4.1. Show that algorithm (4.33)–(4.40) is a particular case of algorithm (2.13)–(2.19), (2.21) of Sec. 2.2. Identify the matrix **S**.

4.5.4. *Numerical solution of a test problem.*

We apply the above iterative and approximation methods to the numerical simulation of a flow around a disk; we suppose that the flow is *subsonic and uniform at infinity*. From the symmetry, the Kutta–Joukowsky condition is automatically satisfied. If $|\mathbf{u}_\infty|$ is sufficiently small, the flow is purely subsonic, and a very good solution is obtained in very few iterations of algorithm (4.33)–(4.40) ($\simeq 5$ iterations). For greater values of $|\mathbf{u}_\infty|$, a supersonic pocket appears, and if we plot the computed[9] Mach distribution on the skin of the

[9] We still have a very fast convergence ($\simeq 10$ iterations).

Figure 4.4

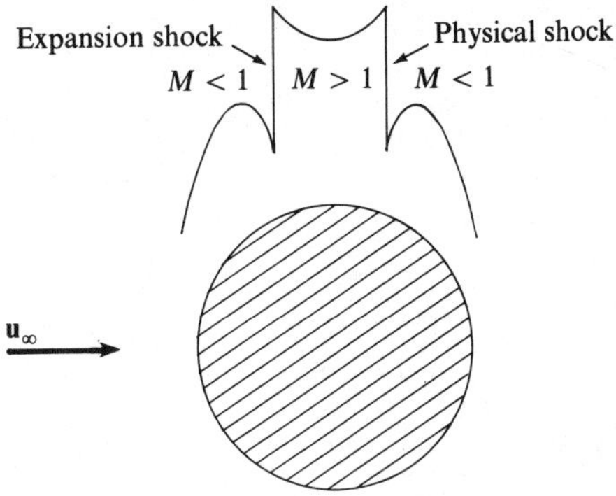

Figure 4.5

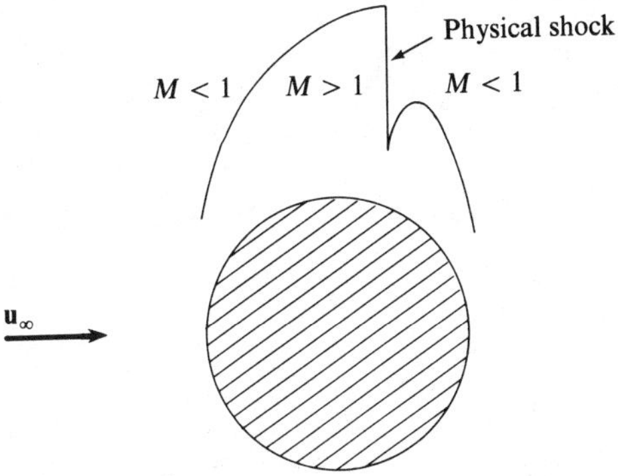

body, we observe the distribution of Fig. 4.4, showing a *rarefaction* (or *expansion*) shock which is a nonphysical phenomenon. The correct Mach distribution is shown on Fig. 4.5.

From the above numerical experiments, it appears that in our numerical process we have to include a device (some dissipation, for example) which eliminates those nonphysical shocks violating the entropy condition (4.13).

The numerical implementation of (4.13) is discussed in Sec. 4.6.

4.6. Numerical implementation of the Entropy Condition

4.6.1. *Generalities: Synopsis*

Several methods based on *penalty* and/or *artificial viscosity* have been discussed and numerically tested in B.G.4P. [1], Bristeau [2], Periaux [2], Bristeau, Glowinski, Periaux, Perrier, and Pironneau [1], and Poirier [1].

In Sec. 4.6.2. we describe (following B. G. 4P. [2]) an *interior penalty method*, and in Sec. 4.6.3 we describe a method based on *upwinding of the density* (as in Eberle [1], [2], Deconinck and Hirsch [1], and Holst [1]).

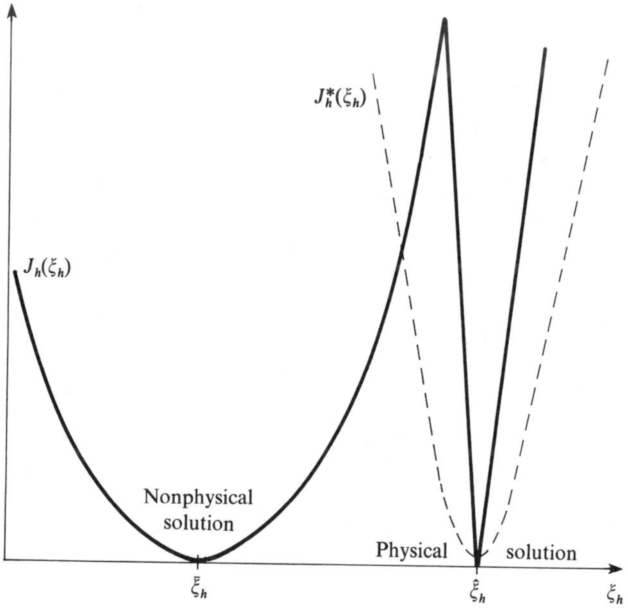

Figure 4.6

If we go back to the example of Sec. 6.5.4 and try to represent the corresponding least-squares functional $\xi_h \rightarrow J_h(\xi_h)$, it is very likely that it looks like the graph of Fig. 4.6 (if we suppose $|\mathbf{u}_\infty|$ sufficiently large).

Our feeling about Fig. 4.6 is motivated by the fact that $\bar{\xi}_h$ is a very powerful attractor for almost any iterative method, particularly for algorithm (4.33)–(4.40); on the other hand, the physical solution lacks these attractor properties (except possibly in a very small neighborhood of this solution). From these observations a reasonable idea is to replace the continuous curve in Fig. 4.6 with the discontinuous one, corresponding to a functional $\xi_h \rightarrow J_h^*(\xi_h)$, taking a very large value at the nonphysical solution $\bar{\xi}_h$, and taking its minimal value close to $\hat{\xi}_h$.

4.6.2. *An interior penalty method with truncation*

4.6.2.1. *The physical motivation.* From Landau and Lifchitz [1] it is known that in the case of a weak shock, we have, following the stream lines,

$$[\mathscr{E}] = O(-[u]^3), \tag{4.46}$$

where in (4.46), $[\mathscr{E}]$ (resp., $[u]$) denotes the *jump of entropy* (resp., *the jump of velocity*).

We find that, in a physical shock,

$$[\mathscr{E}] \geq 0; \tag{4.47}$$

Figure 4.7

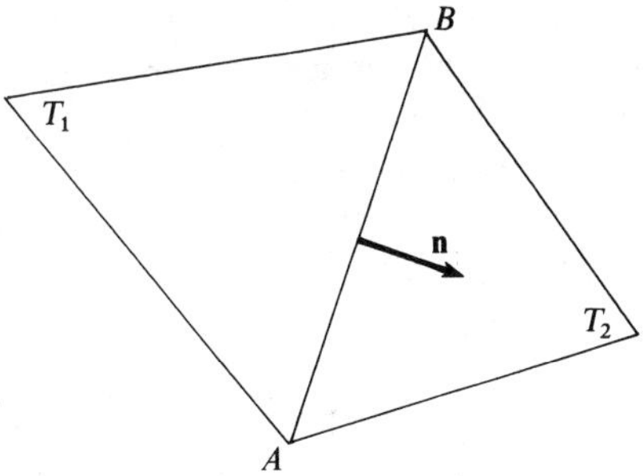

combining (4.46), (4.47), we shall use the entropy condition

$$[u]^3 \leq 0. \tag{4.48}$$

This choice is motivated by the fact that our finite element method gives us direct access to velocity variables.

4.6.2.2. *Finite element realization of* (4.48) *by an interior penalty method.* Let us consider two adjacent triangles, as in Fig. 4.7 [actually the following method can be (and has been) applied to three-dimensional problems]; we denote by $\mathbf{n}$ the unit vector of the normal at their common edge AB, oriented from T_1 to T_2. Since we use a piecewise linear approximation, we have

$$\mathbf{u}_{h|T_j} = \nabla\phi_{h|T_j}, \qquad j = 1, 2 \tag{4.49}$$

with $\nabla\phi_h$ piecewise constant.

Moreover, since we use C^0-conforming elements, the component along AB of the velocity $\mathbf{u}_h = \nabla\phi_h$ is continuous, unlike the normal component which is clearly discontinuous; from these considerations the discontinuities of $\mathbf{u}_h$ are supported by the local normal components of the velocity.

From these observations and from (4.49), we introduce the functional E_h defined as follows:

$$E_h(\phi_h) = \sum_i \sigma_i\left(\left[\frac{\partial\phi_h}{\partial n_i}\right]^{3+}\right)^2 \tag{4.50}$$

where the summation is done with respect to the common edges of the adjacent triangles of $\mathscr{T}_h$, where σ_i is a coefficient which takes the local size of the triangulation into account, and where

$$t^+ = \max(0, t), \qquad \forall\, t \in \mathbb{R}$$

($[\partial\phi_h/\partial n_i]$ denotes the jump of normal derivative in the direction of $\mathbf{n}_i$).

Remark 4.4. We observe that in fact (4.50) is very close to the integral over Ω of the sixth power of a truncated second-order derivative; for homogeneity this suggests that we take

$$\sigma_i = d_i^\alpha, \tag{4.51}$$

where d_i is the length of the common edge and $\alpha = -4$ (we should take $\alpha = -5$ for one-dimensional problems and $\alpha = -3$ for three-dimensional problems).

Combining the functional E_h and the approximate problem (4.30)–(4.32) of Sec. 4.5.2, we obtain the following discrete problem, containing a dissipative device in order to avoid nonphysical shocks:

$$\operatorname*{Min}_{\xi_h \in X_h} J_{rh}(\xi_h), \tag{4.52}$$

where X_h is the set of the feasible discrete solutions and where

$$J_{rh}(\xi_h) = \frac{1}{2}\int_\Omega |\nabla y_h(\xi_h)|^2\, dx + rE_h(\xi_h), \tag{4.53}$$

with $r > 0$ and $y_h(\xi_h)\ (= y_h)$, the solution of the discrete variational state equation:

Find $y_h \in V_h$ such that

$$\int_\Omega \nabla y_h \cdot \nabla v_h\, dx = \int_\Omega \rho(\xi_h)\nabla \xi_h \cdot \nabla v_h\, dx - \int_\Gamma g_h v_h\, d\Gamma, \qquad \forall\, v_h \in V_h. \tag{4.54}$$

To solve (4.52)–(4.54), we can use a variant of the conjugate gradient algorithm (4.33)–(4.40) (with J_h replaced by J_{rh}).

The numerical results obtained by the above methods are quite good since the sonic transition (without shock) from the subsonic to the supersonic region is very well approximated. Furthermore, the shocks are well located and are very sharp. We refer to Periaux [2] and also to Sec. 4.7 where numerical results obtained by the above methods are presented and discussed.

A practical and very important problem is the proper choice of r, since for a given airfoil the *optimal* r seems to be a complicated and sensitive function of $|\mathbf{u}_\infty|$ and of the angle of attack; in Sec. 4.6.2.3 we shall discuss a technique which seems to be effective for the removal (or at least the reduction) of sensitivity to the choice of r.

Remark 4.5. We complete Remark 4.4; since $E_h(\xi_h)$ appears as the integral over Ω of the sixth power of a discrete truncated second-order derivative of ξ_h, we can associate to E_h—by differentiation—a discrete fourth-order nonlinear operator. Since E_h is a convex functional, its differential is a monotone operator. From these properties the addition of rE_h to J_h may be interpreted as a nonlinear fourth-order elliptic regularization process.

Remark 4.6. We justify the title of Sec. 4.6.2 by observing that in fact the functional E_h introduced to regularize our problem is a variant (more complicated) of the interior penalty functionals discussed in Douglas and Dupont [1] and Wheeler [1].

Remark 4.7. We have performed computations with E_h of (4.50) using exponents different from 2 and 3; actually these two exponents seem to be the optimal combination in view of the quality of the computed solutions.

EXERCISE 4.2. Compute the differential of the functional J_{rh} defined by (4.53).

4.6.2.3. *A nonlinearly weighted interior penalty method.* From the comments in Sec. 4.6.2.2 about the sensitivity to the choice of r in J_{rh}—which is clearly related to the strong nonlinearity of our problem—it is very tempting to introduce a local control of the regularization process associated with $([\partial\phi_h/\partial n]^+)^6$. To achieve such a goal, we have introduced, in the above functional E_h (defined by (4.50)), a nonlinear weight directly related to the local values of the density; more precisely, instead of using E_h, we use $\tilde{E}_h$ defined by

$$\tilde{E}_h(\phi_h) = \sum_i \frac{\sigma_i}{\rho_i^{2n(\gamma-1)}} \left(\left[\frac{\partial\phi_h}{\partial n_i}\right]^+\right)^6, \tag{4.55}$$

where $1/\rho_i^\beta$ is a symbolic notation defined by either

$$\frac{1}{\rho_i^\beta} = \frac{1}{2}\left(\frac{1}{\rho_{i1}^\beta} + \frac{1}{\rho_{i2}^\beta}\right), \tag{4.56$_1$}$$

or

$$\frac{1}{\rho_i^\beta} = \frac{1}{2^\beta}\left(\frac{1}{\rho_{i1}} + \frac{1}{\rho_{i2}}\right)^\beta \tag{4.56$_2$}$$

with

$$\rho_{ij} = \left(1 - \frac{|(\nabla\phi_h)_{ij}|^2}{[(\gamma+1)/(\gamma-1)]C_*^2}\right)^{1/(\gamma-1)}, \tag{4.57}$$

where

$$(\nabla\phi_h)_{ij} = \nabla\phi_{h|T_{ij}},$$

T_{i1}, T_{i2} being the adjacent triangles of $\mathscr{T}_h$ having the ith edge in common.

Remark 4.8. From our computer experiments, a "good" value for n in (4.55) is $n = 2$.

Replacing E_h by $\tilde{E}_h$, we obtain a variant of the approximate problem (4.52)–(4.54) which can be solved by the same type of iterative methods; computer experiments performed with $\tilde{E}_h$ have shown that for a NACA 0012 airfoil, the same r was optimal (or nearly optimal) in the conditions listed in Table 4.1. The corresponding results are described in Sec. 4.7.

Table 4.1

Angle of Attack (Degrees)	Mach Number at Infinity (M_∞)
6	0.6
1	0.78
0	0.8
0	0.85

4.6.2.4. *An interior penalty method using density jumps.* The interior penalty method discussed in Sec. 4.6.2.3 combines two effects:

(i) It penalizes rarefaction shocks via $[u]^+$.
(ii) Because of the nonlinear weight that we have introduced, the regularization effect is reinforced in regions at high Mach number, since in these regions ρ is "small".

It is then natural to look for a variant of (4.55), combining both effects more closely; we start from the observation that in a physical shock we have (still following the stream lines)

$$\left[\frac{1}{\rho}\right] \leq 0. \tag{4.58}$$

This jump condition (4.58) leads to the following variant of the entropy functionals E_h and $\tilde{E}_h$ of Secs. 4.6.2.2 and 4.6.2.3 (whose notation has been retained):

$$R_h = \sum_i \sigma_i \left(\left(\frac{w_i}{C_*} \left[\frac{1}{\rho^\beta}\right]_i \right)^+ \right)^{2n}. \tag{4.59}$$

In (4.59) the notation is self-explanatory, except for w_i which is defined by

$$w_i = \frac{1}{2} \sum_{j=1}^{2} \left(\frac{\partial \phi_h}{\partial n_i} \right)_j, \tag{4.60}$$

the indices $j = 1, 2$ corresponding to the two adjacent triangles of $\mathcal{T}_h$ having the ith edge in common. It is then quite easy to formulate an approximate problem in which E_h or $\tilde{E}_h$ is replaced by R_h; moreover, the same type of iterative methods will hold for this new type of approximate problem.

Numerical experiments have to be performed to check the validity of this new approach and also to make an optimal choice for the two exponents β and n in (4.59).

4.6.3. *Implementation of the entropy condition by upwinding of the density*

4.6.3.1. *Generalities: Synopsis.* Density upwinding has been extensively used (see Eberle [1], [2], Holst [1], and Deconinck and Hirsh [1] and the references

Figure 4.8

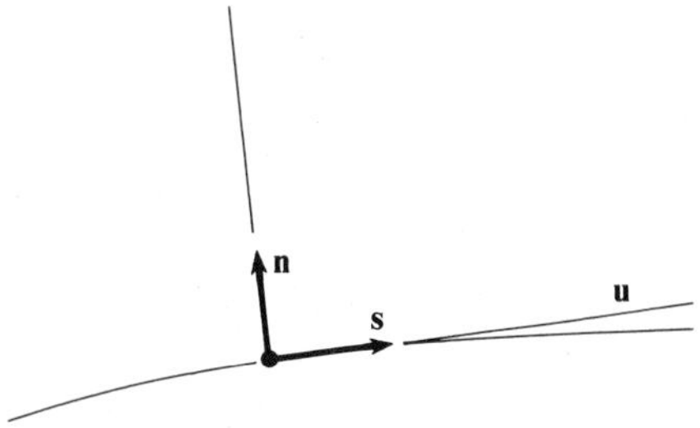

therein) in order to eliminate nonphysical shocks; these upwinding techniques have been very effective, coupled with alternating-direction methods (implicit or semi-implicit; see Holst [1] and Deconinck and Hirsch [1]) if the computational mesh is regular (finite differences or regular finite element grids). In particular, their application, combined with finite element techniques, have been limited (see Eberle [1], [2] and Deconinck and Hirsh [1]) to quadrilateral elements on quasiregular quadrangulations (fairly close to finite difference methods, in our opinion).

In Sec. 4.6.3.2 we would like to discuss a method (due to M. O. Bristeau) which also makes use of an upwinding of the density; this method can be used with simplicial[10] finite elements (triangles in two dimension, tetrahedra in three dimensions) and has been very effective for computing flows at high Mach numbers and around complicated two- and three-dimensional geometries.

4.6.3.2. *A modified discrete continuity equation by upwinding of the density in the flow direction.* Following Jameson [1]–[4] and Bristeau [2], [3], we may write the continuity equation (4.14) in a system of local coordinates $\{\mathbf{s}, \mathbf{n}\}$, where (see Fig. 4.8) for a two-dimensional flow, $\mathbf{s}$ is the unit vector of the stream direction (i.e., $\mathbf{s} = \mathbf{u}/|\mathbf{u}|$ if $\mathbf{u} \neq \mathbf{0}$) and $\mathbf{n}$ is the corresponding normal unit vector (conventionally oriented).

Using $\{\mathbf{s}, \mathbf{n}\}$ and setting[11]

$$k = \frac{\gamma - 1}{\gamma + 1}, \qquad \alpha = \frac{1}{\gamma - 1}, \qquad U = \frac{|\nabla\phi|}{C_*},$$

we obtain (from (4.14))

$$\rho \frac{\partial^2 \phi}{\partial n^2} + \rho \frac{1 - U^2}{1 - kU^2} \frac{\partial^2 \phi}{\partial s^2} = 0; \tag{4.61}$$

[10] We use here the terminology of Ciarlet [1], [2].

[11] U is the Mach number M_*.

the elliptic-hyperbolic aspect of the problem is clear from (4.61). Actually (4.61) can also be written

$$\rho \frac{\partial^2 \phi}{\partial n^2} + \frac{U^2 - 1}{2k\alpha U^2} \nabla\phi \cdot \nabla\rho = 0 \tag{4.62}$$

[we have $1/2k\alpha = (\gamma + 1)/2$].

EXERCISE 4.3. Prove (4.61), (4.62).

We use (4.62) to modify the discrete continuity equation (4.27) as follows: *Find $\phi_h \in V_h$ such that*

$$\int_\Omega \rho(\phi_h)\nabla\phi_h \cdot \nabla v_h\, dx + \frac{1}{2k\alpha}\int_\Omega \left(\frac{\partial}{\partial s} h_s(U_h^2 - 1)^+ \nabla\phi_h \cdot \nabla\rho_h\right)_h v_h\, dx$$

$$= \int_\Gamma g_h v_h\, d\Gamma, \qquad \forall\, v_h \in V_h. \tag{4.63}$$

The approximate problem (4.63) has been introduced by M. O. Bristeau and is a finite element variant of a finite difference scheme due to Holst [1]. In (4.63):

(i) h_s is a measure of the local size of the finite element mesh in the flow direction,
(ii) $(\partial/\partial s)_h$ is an approximation of $(\nabla\phi_h/|\nabla\phi_h|) \cdot \nabla$,
(iii) U_h, ρ_h are upwinded approximations of U and ρ, respectively.

More precisely, we write the second integral in the left-hand side of (4.63) as follows.

$$\int_\Omega \left(\frac{\partial}{\partial s} h_s(U_h^2 - 1)^+ \nabla\phi_h \cdot \nabla\rho_h\right)_h v_h\, dx$$

$$= (-1)\sum_{P_i \in \Sigma_h} v_h(P_i)\sum_{j=1}^{2} u_{ji} \sum_{T \in \mathscr{T}_i} m(T)h_s(T)(U_h^2 - 1)^+ \nabla\phi_h \cdot \nabla\rho_h \frac{\partial w_i}{\partial x_j}, \tag{4.64}$$

where

(a) $\Sigma_h = \{P_i\}_{i=0}^{N_h}$ is the set of the vertices of $\mathscr{T}_h$, with $P_0 =$ T.E.;
(b) w_i is the basis function of H_h^1 (cf. (4.24)) associated with P_i by

$$w_i \in H_h^1, \qquad \forall\, i = 0, \ldots, N_h,\ w_i(P_i) = 1, \qquad w_i(P_j) = 0, \qquad \forall\, j \neq i; \tag{4.65}$$

(c) $\mathscr{T}_i$ is the subset of $\mathscr{T}_h$ consisting of those triangles having P_i as a common vertex;
(d) $m(T) = \text{meas}(T)$;

Figure 4.9

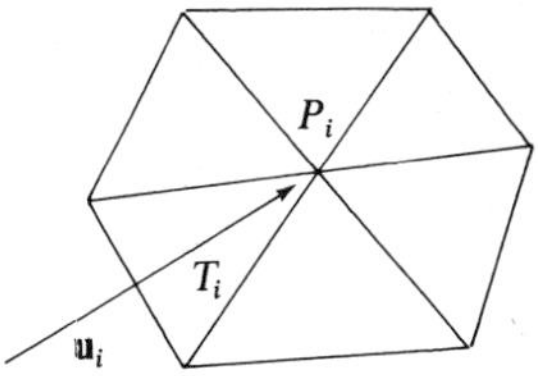

(e) $\{u_{ji}\}_{j=1}^2$ is an approximation of $\nabla\phi_h/|\nabla\phi_h|$ at vertex P_i obtained, for $j = 1, 2$, by the following averaging formulas:

$$\left(\frac{\partial\phi_h}{\partial x_j}\right)_i = \frac{\sum_{T\in\mathcal{T}_i} \frac{\partial\phi_h}{\partial x_j}\left|\frac{\partial w_i}{\partial x_j}\right|}{\sum_{T\in\mathcal{T}_i}\left|\frac{\partial w_i}{\partial x_j}\right|},$$

$$u_{ji} = \frac{\left(\frac{\partial\phi_h}{\partial x_j}\right)_i}{\left(\sum_{l=1}^{2}\left(\frac{\partial\phi_h}{\partial x_l}\right)_i^2\right)^{1/2}} \tag{4.66}$$

(others averaging methods are possible);

(f) $h_s(T)$ is the length of T in the **s** direction, i.e.,

$$h_s(T) = \frac{2}{\left(\sum_{k=1}^{3}\left|\frac{\partial w_{kT}}{\partial s}\right|\right)}, \tag{4.67}$$

where $w_{kT}, k = 1, 2, 3$ are the basis functions associated with the vertices P_{kT} of T; actually (4.67) can also be written

$$h_s(T) = \frac{2|\nabla\phi_h|_T}{\sum_{k=1}^{3}|\nabla\phi_h\cdot\nabla w_{kT}|_T}. \tag{4.68}$$

(g) We define U_h^2 as follows:
With each vertex P_i of $\mathcal{T}_h$ we associate an inflow triangle T_i which is the triangle of $\mathcal{T}_h$ having P_i as vertex and which is crossed by the vector $\mathbf{u}_i = \{u_{ij}\}_{j=1}^2$ pointing to P_i (as shown on Fig. 4.9); we then define U_i^2 as

$$U_i^2 = \left.\frac{|\nabla\phi_h|^2}{C_*^2}\right|_{T_i},$$

and for each triangle T of $\mathscr{T}_h$,

$$U_h^2 = \frac{\sum_{k=1}^{3} U_{kT}^2 \left|\frac{\partial w_{kT}}{\partial s}\right|_T}{\sum_{k=1}^{3} \left|\frac{\partial w_{kT}}{\partial s}\right|_T}. \tag{4.69}$$

(h) We finally obtain $\rho_h \in H_h^1$ as follows:
As for U_i^2, we define ρ_i by

$$\rho_i = \rho(\phi_h)|_{T_i},$$

and then

$$\rho_h = \sum_{i=0}^{N_h} \rho_i w_i. \tag{4.70}$$

4.6.3.3. *Some brief comments on the least-squares solution of* (4.63). For solving the discrete upwinded continuity equation (4.63), we can use the following least-squares formulation:

$$\operatorname*{Min}_{\xi_h \in V_h} J_h(\xi_h), \tag{4.71}$$

where

$$J_h(\xi_h) = \frac{1}{2}\int_\Omega |\nabla y_h(\xi_h)|^2\, dx,$$

with $y_h(\xi_h)$ $(= y_h)$, the solution of the following linear variational equation:
Find $y_h \in V_h$ *such that*

$$\int_\Omega \nabla y_h \cdot \nabla v_h\, dx = \int_\Omega \rho(\phi_h)\nabla\phi_h \cdot \nabla v_h\, dx - \int_\Gamma g_h v_h\, d\Gamma$$
$$+ \frac{1}{2k\alpha}\int_\Omega \left(\frac{\partial}{\partial s} h_s(U_h^2 - 1)^+ \nabla\phi_h \cdot \nabla\rho_h\right)_h v_h\, dx, \qquad \forall\, v_h \in V_h. \tag{4.72}$$

Since J_h is a nondifferentiable functional of ξ_h, to solve (4.71) we have used (instead of algorithm (4.33)–(4.40)) a generalization of the conjugate gradient method due to Lemarechal [1], [2] which also applies to the minimization of nondifferentiable functionals (actually good results are also obtained if one uses algorithm (4.33)–(4.40) with $J_h(\cdot)$ defined as in (4.71); however, more iterations are needed).

4.7. Numerical experiments

In this section we shall present some of the numerical results obtained using the above methods. The results of Sec. 4.7.1 are related to a NACA 0012 airfoil; those of Sec. 4.7.2 (resp., 4.7.3) to a Korn airfoil (resp., a two-piece airfoil).

4.7.1. *Simulation of flows around a* NACA 0012 *airfoil.*

As a first example we have considered flows around a NACA 0012 airfoil at various angles of attack and Mach numbers at infinity. The corresponding pressure distributions on the skin of the airfoil are shown on Figs. 4.10–4.17 in which the *isomach lines* (in the supersonic region only on Figs. 4.10–4.14) are also shown. The results shown in Figs. 4.10–4.14 have been obtained using the interior penalty method of Sec. 4.6.2; those of Fig. 4.15–4.17 have been obtained using the upwinding method of Sec. 4.6.3.

We observe that the physical shocks are quite neat and also that the transition (without shock) from the subsonic to the supersonic region is smoothly restituted, implying that the entropy condition has been satisfied. The above numerical results are very close to those obtained by various authors using finite difference methods (see, particularly, Jameson [1]).

4.7.2. *Flow around a Korn's airfoil*

In Fig. 4.18 we have represented the pressure distribution corresponding to the flow around a Korn's airfoil at $M_\infty = 0.75$ and $\alpha = 0.11$; the computation method is the interior penalty method of Sec. 4.6.2. The agreement with a finite difference solution is good, as indicated in Fig. 4.18.

4.7.3. *Flows around a two-piece airfoil*

The tested two-piece airfoil is shown on Figs. 4.19 and 4.20. Each piece is a NACA 0012 airfoil (the body No. 1 is the upper body). The pressure distribution and the isomach lines (computed by the interior penalty method of Sec. 4.6.2) are shown on Figs. 4.19 and 4.20. We observe that the region between the two airfoils acts as a nozzle; we also observe supersonic regions, in particular, between the two airfoils.

4.8. Transonic flow simulations on large bounded computational domains

Consider the situation depicted in Fig. 4.3; if the supersonic zone extends far from the airfoil, it is necessary to use a very large computational domain.

Let 0 be the origin of coordinates; it is then reasonable to take the circle $\{x \in \mathbb{R}^2, r = R_\infty\}$ for Γ_∞, with $r = \sqrt{x_1^2 + x_2^2}$. Now suppose that the disk of center 0 and radius R_0 is sufficiently large to contain the airfoil B in its interior; we then introduce

$$\Omega_1 = \{x \in \mathbb{R}^2, 0 \leq r < R_0, x \notin B\},$$

$$\Omega_2 = \{x \in \mathbb{R}^2, R_0 < r < R_\infty\}.$$

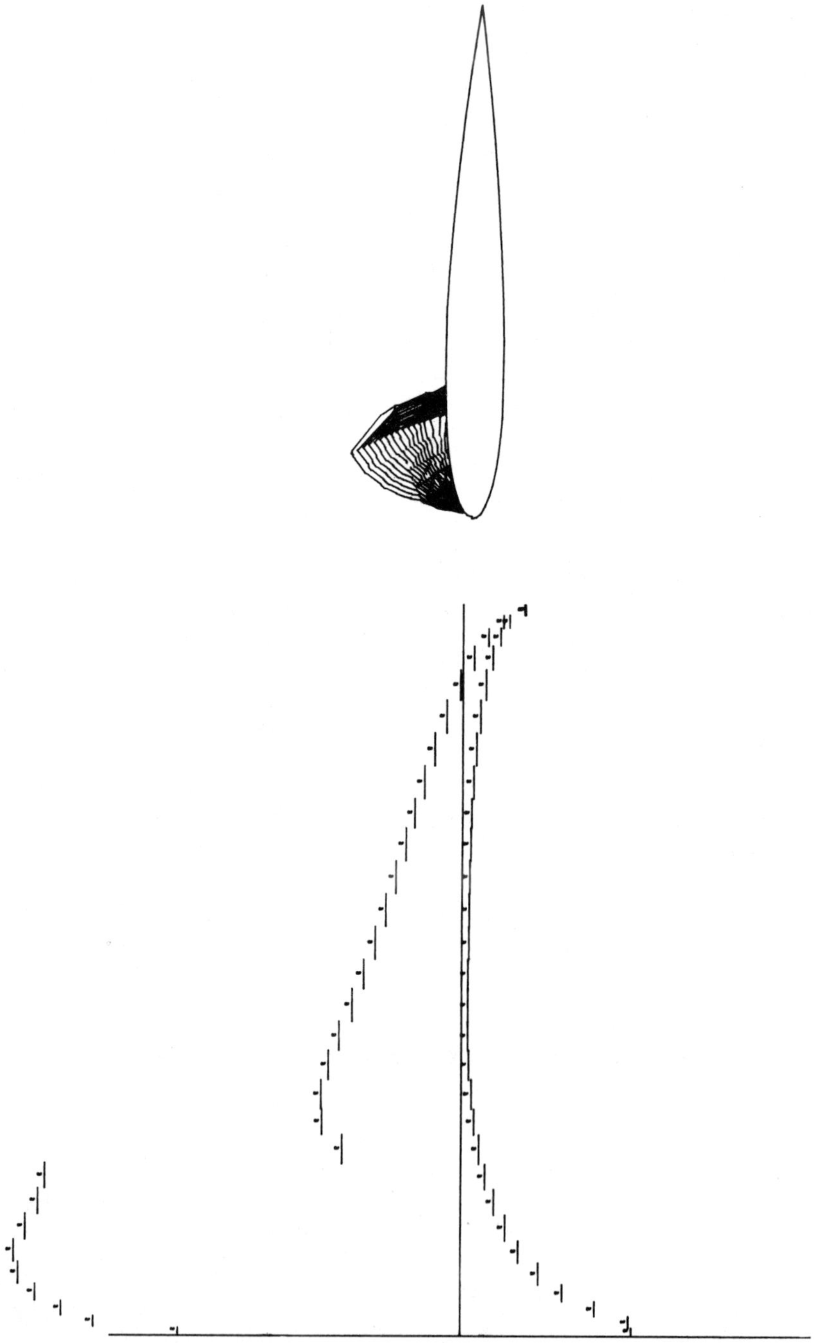

Figure 4.10. NACA 0012 airfoil. $M_\infty = 0.6$; $\alpha = 6$. (Interior penalty method.)

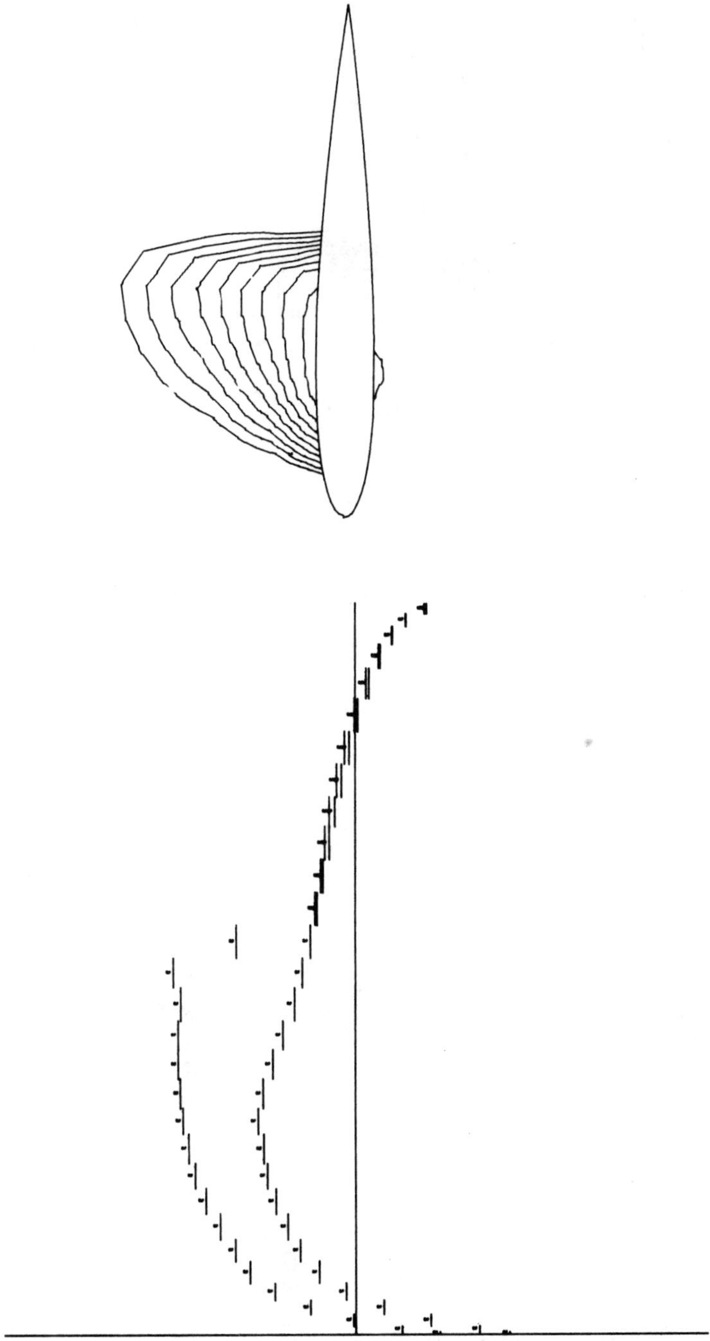

Figure 4.11. NACA 0012 airfoil. $M_\infty = 0.78$; $\alpha = 1$. (Interior penalty method.)

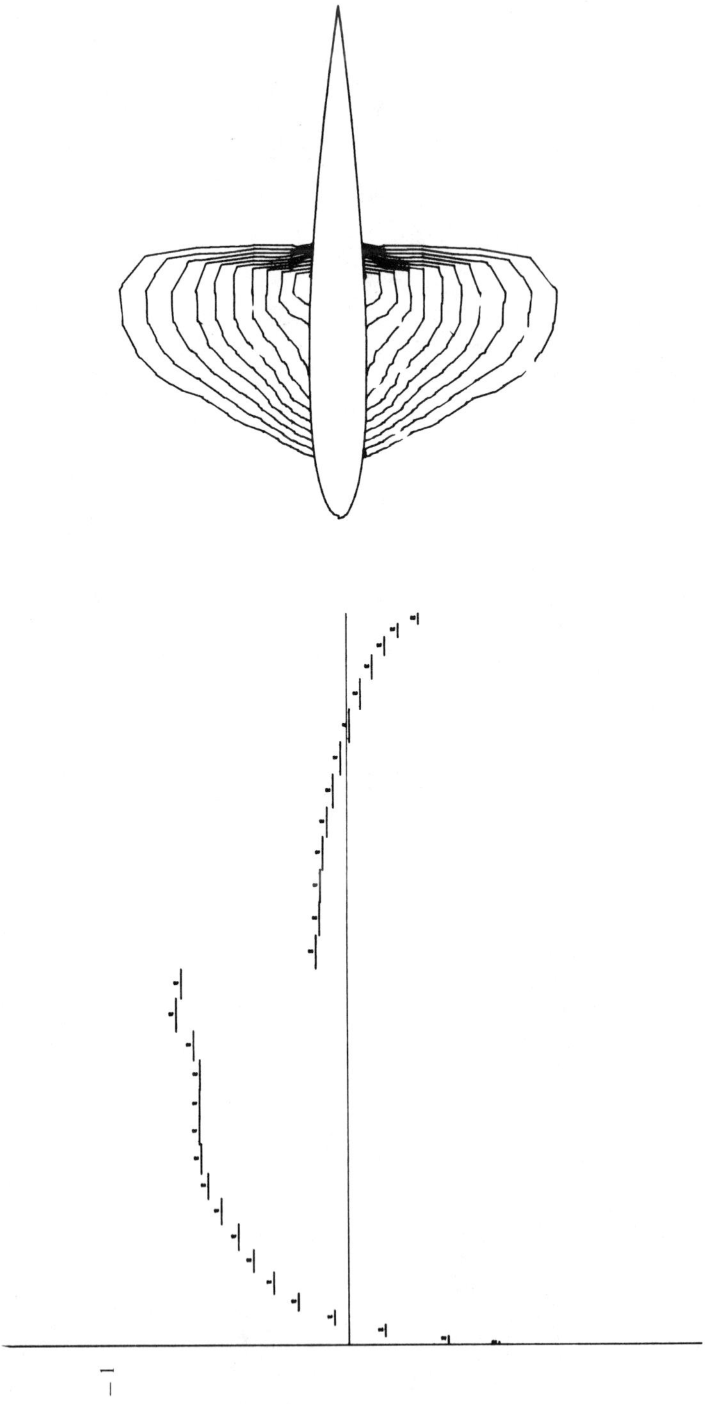

Figure 4.12. NACA 0012 airfoil. $M_\infty = 0.8$; $\alpha = 0$. (Interior penalty method.)

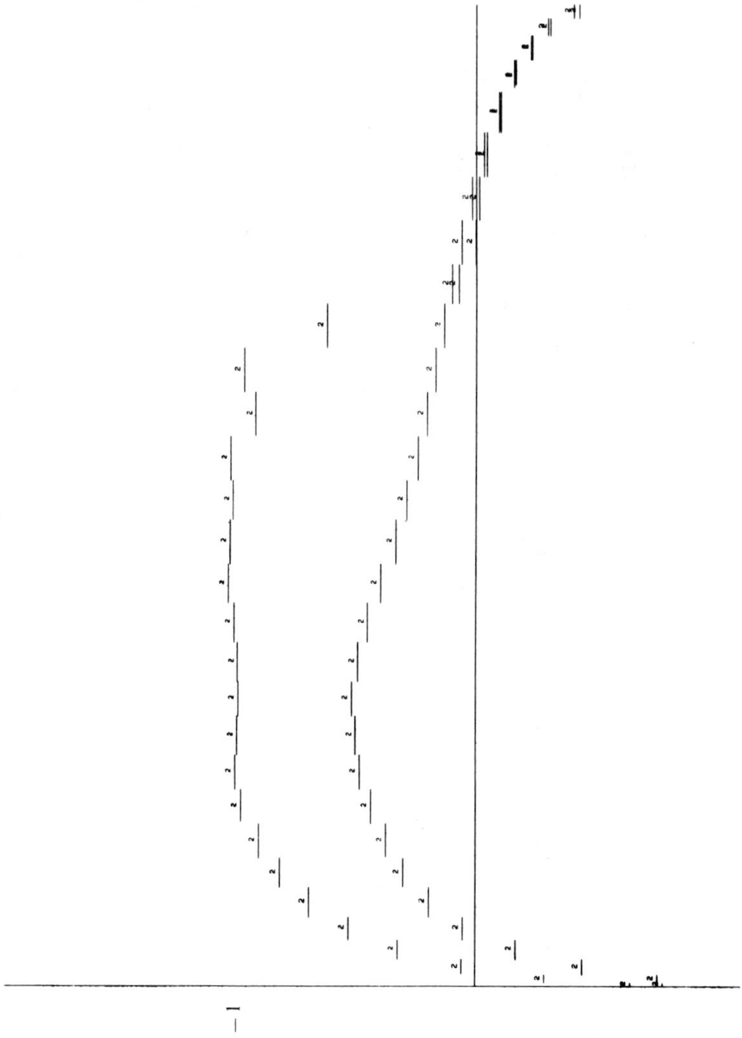

Figure 4.13. NACA 0012 airfoil. $M_{\infty} = 0.8$; $\alpha = 1.1$. (Interior penalty method.)

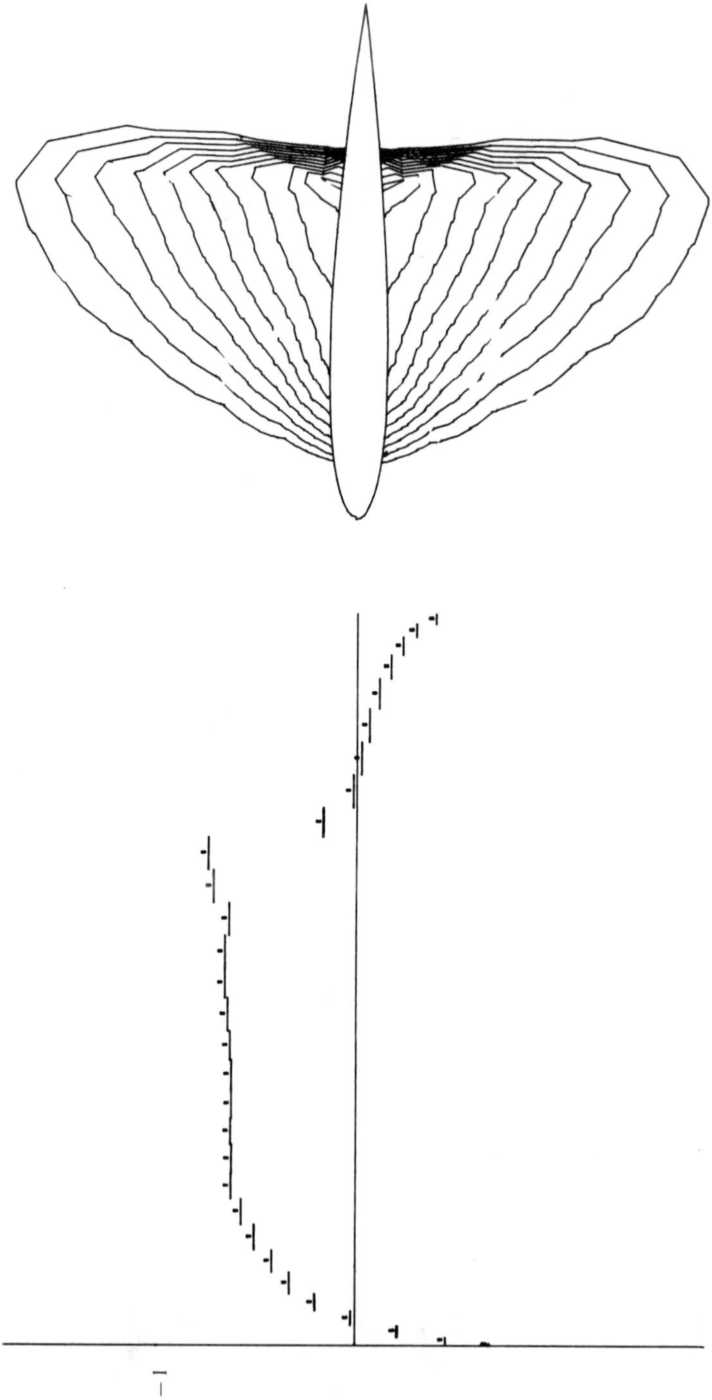

Figure 4.14. NACA 0012 airfoil. $M_\infty = 0.85$; $\alpha = 0$. (Interior penalty method.)

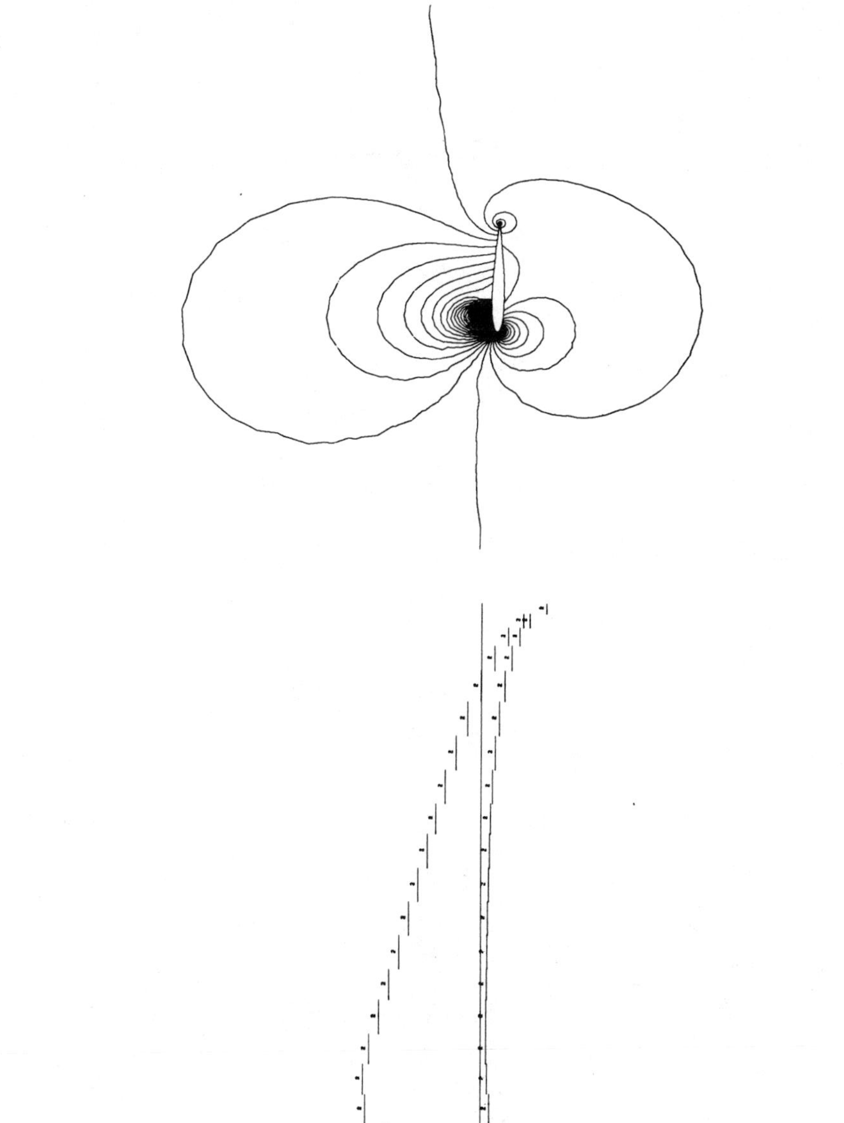

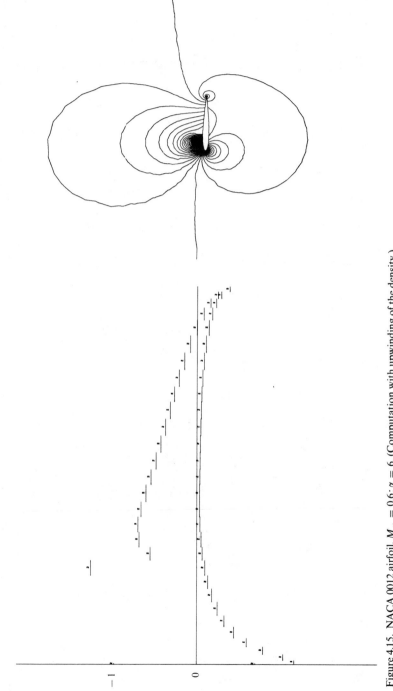

Figure 4.15. NACA 0012 airfoil. $M_\infty = 0.6$; $\alpha = 6$. (Computation with upwinding of the density.)

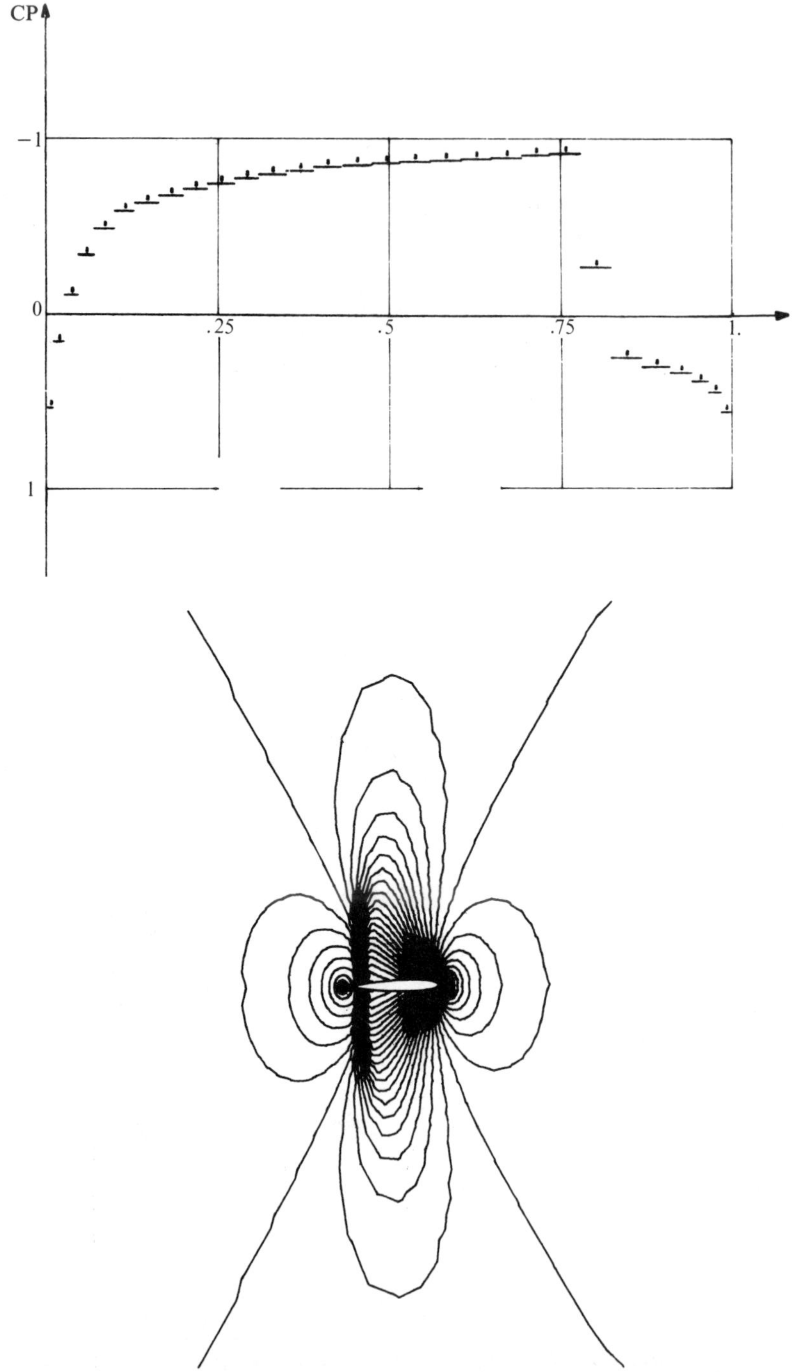

Figure 4.16. NACA 0012 airfoil. $M_\infty = 0.85; \alpha = 0$. (Computation with upwinding of the density.)

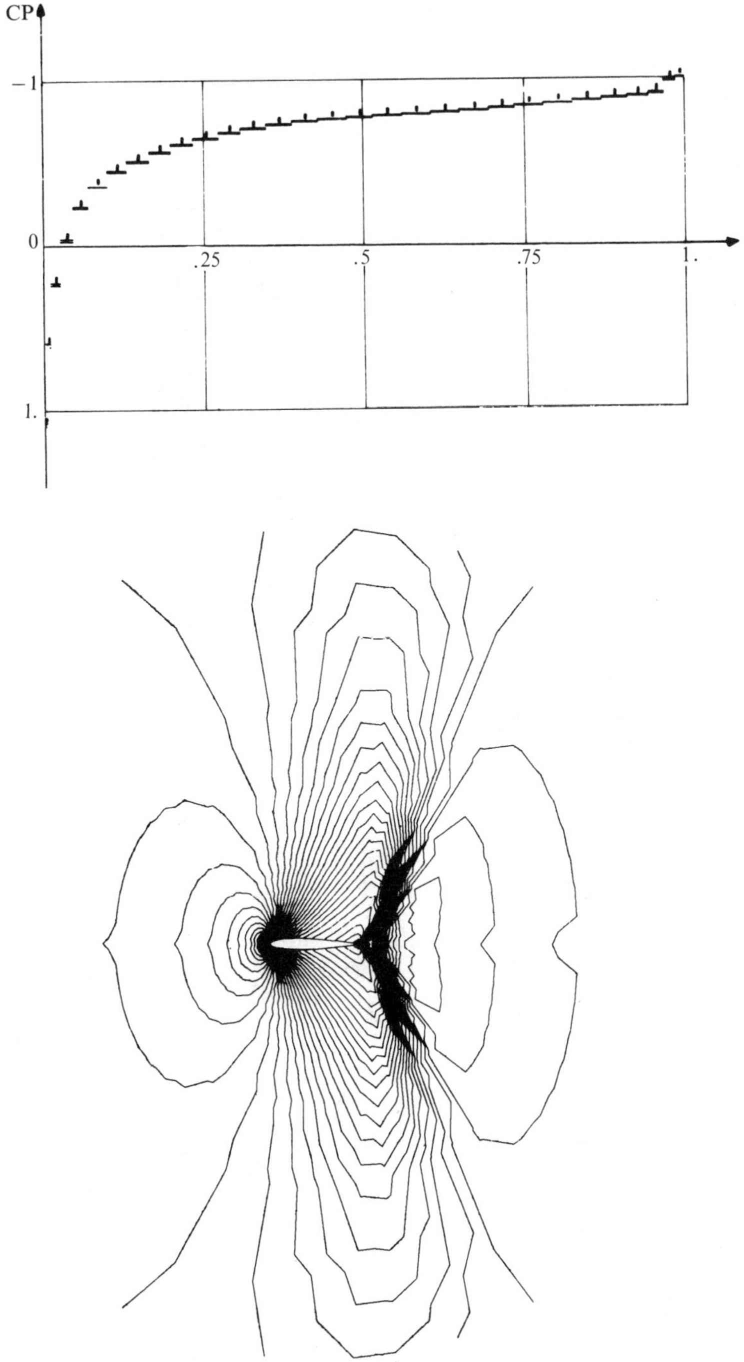

Figure 4.17. NACA 0012 airfoil. $M_\infty = 0.9$; $\alpha = 0$. (Computation with upwinding of the density.)

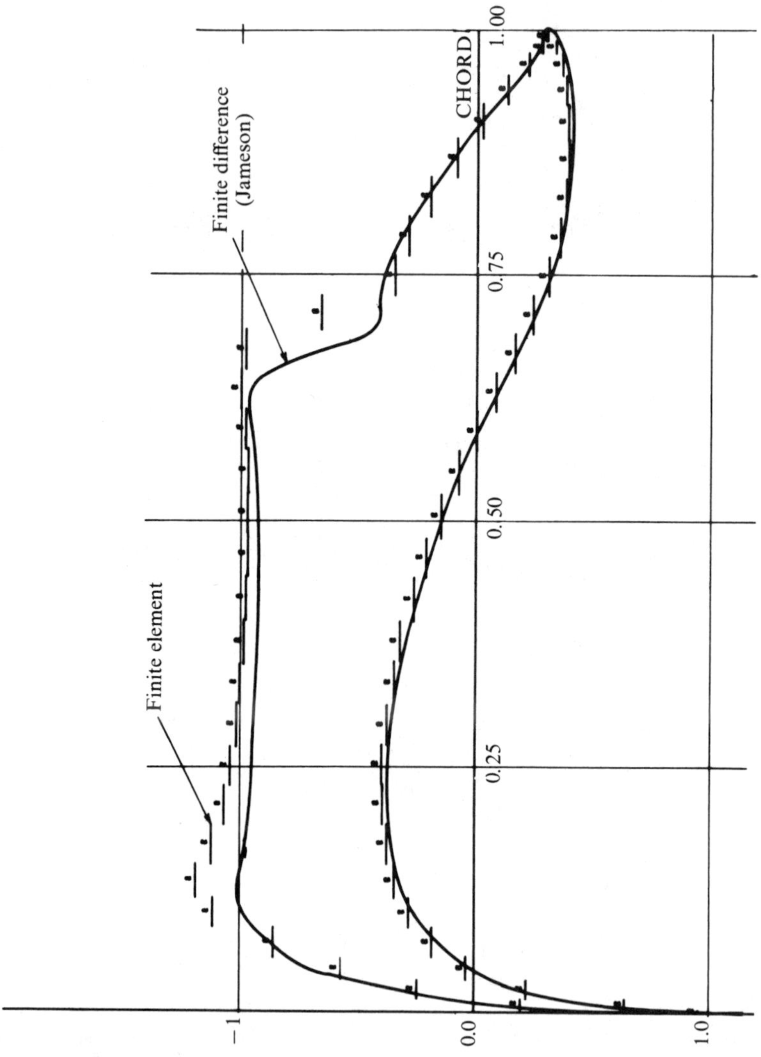

Figure 4.18. Korn airfoil. $M_\infty = 0.75$; $\alpha = 0.11$. (Interior penalty method.)

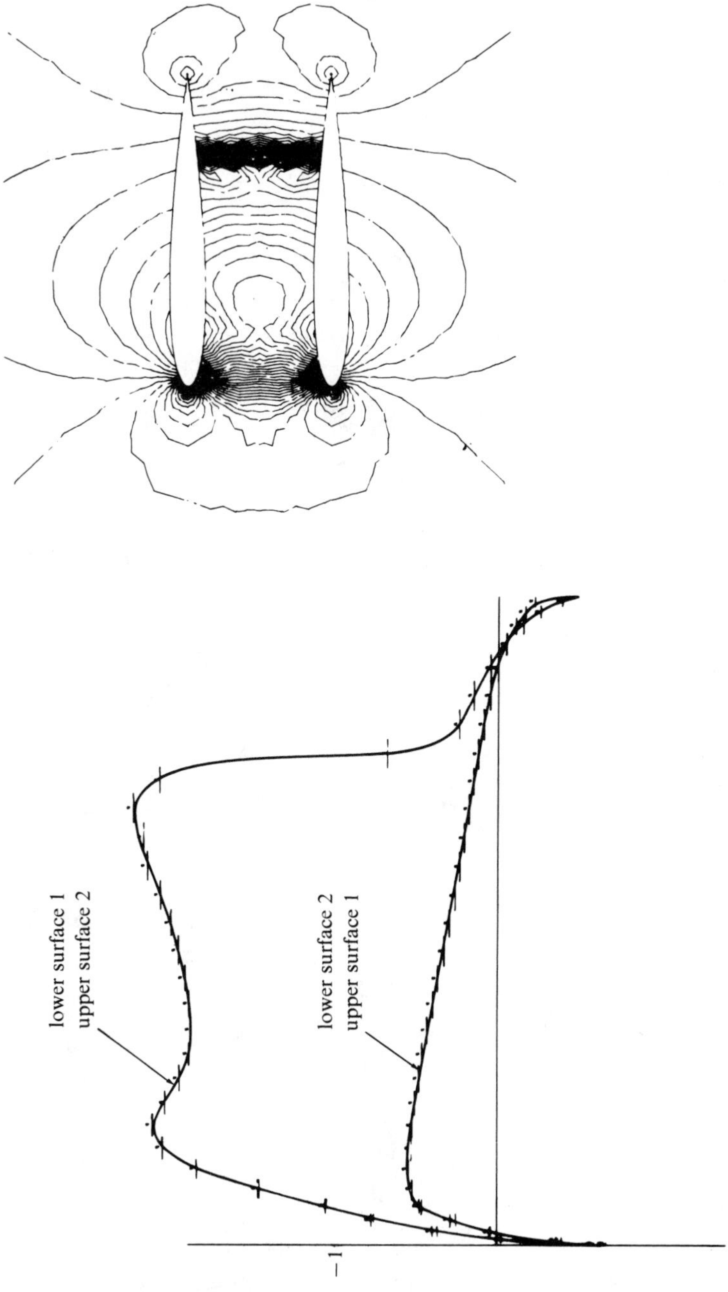

Figure 4.19. Bi NACA 0012 airfoil. $M_\infty = 0.6$; $\alpha = 0$. (Interior penalty method.)

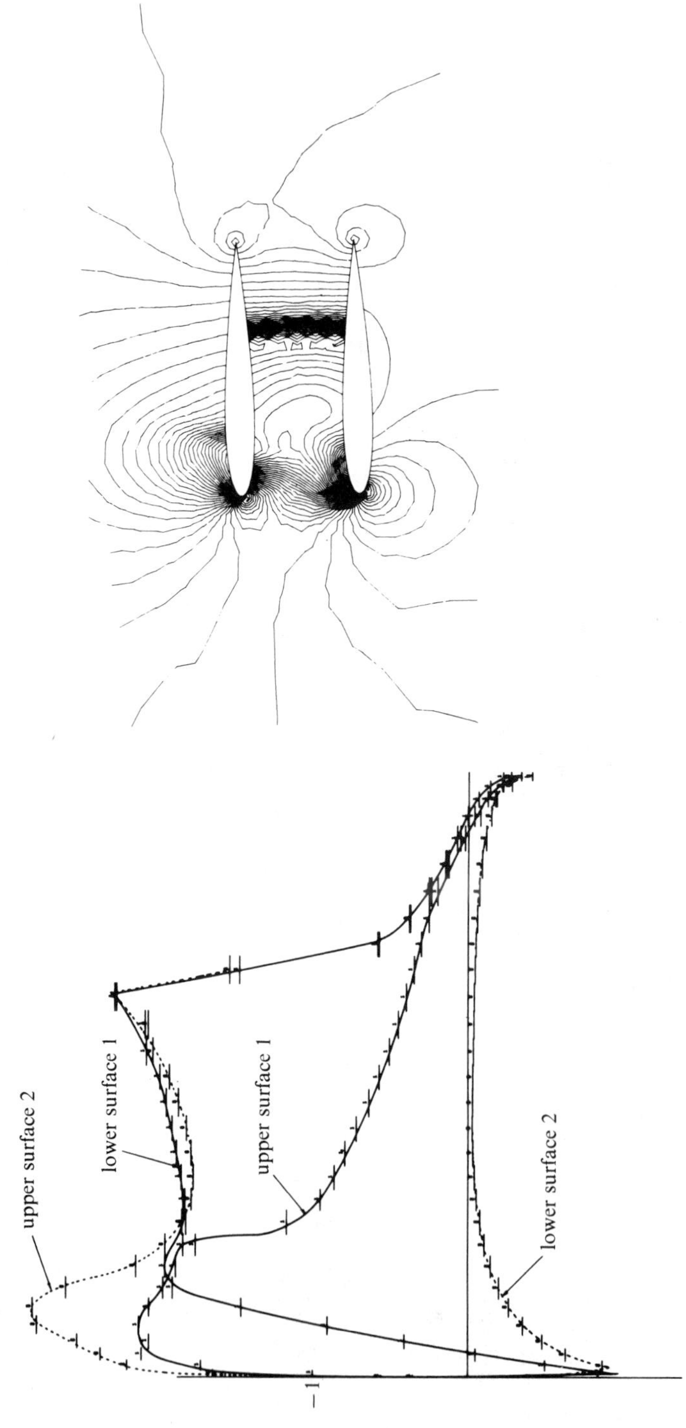

Figure 4.20. Bi NACA 0012 airfoil. $M_\infty = 0.6$; $\alpha = 6$. (Interior penalty method.)

With $\{r, \theta\}$ as a standard polar coordinate system associated with 0, we define the following new variables:

$$\begin{aligned} &\xi_1 = x_1, \xi_2 = x_2, \qquad \forall\, x = \{x_1, x_2\} \in \Omega_1 \\ &\xi_1 = R_0\left(1 + \operatorname{Log}\frac{r}{R_0}\right), \qquad \xi_2 = \theta, \qquad \forall \{r, \theta\} \in \Omega_2; \end{aligned} \tag{4.73}$$

we use the notation $\xi = \{\xi_1, \xi_2\}$ in the sequel.

We now introduce $\tilde{\Omega}_1 = \Omega_1$, and

$$\tilde{\Omega}_2 = \left\{\xi \in \mathbb{R}^2, R_0 < \xi_1 < R_0\left(1 + \operatorname{Log}\frac{R_\infty}{R_0}\right), 0 < \xi_2 < 2\pi\right\};$$

we then define

$$\begin{aligned} \tilde{H}^1 = \Big\{\tilde{v} = \{\tilde{v}_1, \tilde{v}_2\} \in H^1(\tilde{\Omega}_1) \times H^1(\tilde{\Omega}_2),\ &\tilde{v}_1(R_0 \cos \xi_2, R_0 \sin \xi_2) \\ = \tilde{v}_2(R_0, \xi_2),\ &\forall\, \xi_2 \in \,]0, 2\pi[,\ \tilde{v}_2(\xi_1, 0) = \tilde{v}_2(\xi_1, 2\pi), \\ &\forall\, \xi_1 \in \,]R_0, R_0\left(1 + \operatorname{Log}\frac{R_\infty}{R_0}\right)\Big\}. \end{aligned}$$

The space $\tilde{H}^1$ is clearly isomorphic to $H^1(\Omega)$ via the transformation $\tilde{v} \to v$: $\tilde{H}^1 \to H^1(\Omega)$:

$$v(x) = \begin{cases} \tilde{v}_1(x_1, x_2) & \text{if } x = \{x_1, x_2\} \in \Omega_1, \\ \tilde{v}_2\left(R_0\left(1 + \operatorname{Log}\frac{r}{R_0}\right), \theta\right) & \text{if } x = \{r, \theta\} \in \Omega_2. \end{cases} \tag{4.74}$$

Using the above properties, the continuity equation (4.19) is now formulated as follows:

Find $\tilde{\phi} \in \tilde{H}^1$ such that $\forall\, \tilde{v} \in \tilde{H}^1$

$$\begin{aligned} &\int_{\tilde{\Omega}_1} \rho(\tilde{\phi}_1)\nabla\tilde{\phi}_1 \cdot \nabla\tilde{v}_1\, d\xi + \int_{\tilde{\Omega}_2} \tilde{\rho}(\tilde{\phi}_2)\left(R_0 \frac{\partial\tilde{\phi}_2}{\partial\xi_1}\frac{\partial\tilde{v}_2}{\partial\xi_1} + \frac{1}{R_0}\frac{\partial\tilde{\phi}_2}{\partial\xi_2}\frac{\partial\tilde{v}_2}{\partial\xi_2}\right) d\xi \\ &\quad = R_\infty \rho_\infty \int_0^{2\pi} (u_{1\infty} \cos \xi_2 + u_{2\infty} \sin \xi_2)\tilde{v}_2\left(R_0\left(1 + \operatorname{Log}\frac{R_\infty}{R_0}\right), \xi_2\right) d\xi_2, \end{aligned} \tag{4.75}$$

with

$$\{u_{1\infty}, u_{2\infty}\} = \mathbf{u}_\infty$$

and

$$\tilde{\rho}(\tilde{\phi}_2) = \rho_0\left[1 - \frac{\gamma - 1}{\gamma + 1}\frac{1}{C_*^2}\left(\left|\frac{\partial\tilde{\phi}_2}{\partial\xi_1}\right|^2 + R_0^{-2}\left|\frac{\partial\tilde{\phi}_2}{\partial\xi_2}\right|^2\right)e^{2(R_0 - \xi_1)/R_0}\right]^{1/(\gamma-1)}. \tag{4.76}$$

The discretization of (4.75), (4.76) can be done as follows: On the one hand, one uses in $\tilde{\Omega}_1$ (which is part of the physical space) a standard finite element approximation taking into account the possible complexities of the geometry; on the other hand, since $\tilde{\Omega}_2$ is a rectangle, one may use a finite difference discretization which can in fact be obtained via a finite element approximation on a uniform triangle or quadrilateral grid (like the one in Fig. 3.1 of Sec. 3.5.4 of this chapter).

Remark 4.9. From the above decomposition of the computational domain (in $\tilde{\Omega}_1$ and $\tilde{\Omega}_2$), it is natural to solve the approximate problems by iterative methods, taking this decomposition into account, and also the fact that finite differences can be used in $\tilde{\Omega}_2$ (since finite differences allow special solvers on $\tilde{\Omega}_2$). We are presently working on such methods and also on their generalization to three-dimensional problems; the corresponding results will be presented in a forthcoming paper.

5. Numerical Solution of the Navier–Stokes Equations for Incompressible Viscous Fluids by Least-Squares and Finite Element Methods

5.1. Introduction. Synopsis

The numerical solution of the Navier–Stokes equations for incompressible viscous fluids has motivated so many authors that giving a complete bibliography has become an impossible task. Therefore, restricting our attention to very recent contributions making use of finite element approximations, we shall mention, among many others, B. G. 4P. [1], Bristeau, Glowinski, Periaux, Perrier, and Pironneau [1], Bristeau, Glowinski, Mantel, Periaux, Perrier, and Pironneau [1], Glowinski, Mantel, Periaux, Perrier, and Pironneau [1], Gartling and Becker [1], [2], Hughes, Liu, and Brooks [1], Temam [1], Bercovier and Engelman [1], Fortin and Thomasset [1], Girault and Raviart [1], Le Tallec [2], [3], Johnson [2], Glowinski and Pironneau [1], [2], Gresho, Lee, Chan, and Sani [1], Rannacher [1], Benque, Ibler, Keramsi, and Labadie [1], Thomasset [1], and Brooks and Hughes [1]; see also the references therein.

In this section (which very closely follows Bristeau, Glowinski, Mantel, Periaux, Perrier, and Pironneau [1]) we would like to discuss several methods for the effective solution of the above Navier–Stokes problems in the *steady*[12] and *unsteady*[13] cases. The basic ingredients of the methods to be described are the following:

(i) *Mixed finite element approximations* of a *pressure-velocity* formulation of the original problem.
(ii) *Time discretizations* of the unsteady problem by *finite differences*; several schemes will be presented.
(iii) *Iterative solution* of the approximate problems by least-squares conjugate gradient methods (possibly combined with an alternating-direction method).
(iv) *Efficient solvers* for the *discrete Stokes problems.*

The possibilities of the above methodology will be illustrated by the results of various numerical experiments concerning nontrivial two-dimensional flows.

5.2. Formulation of the steady and unsteady Navier–Stokes equations for incompressible viscous fluids

Let us consider a Newtonian, viscous, and incompressible fluid. If Ω and Γ denote the region of the flow[14] and its boundary, respectively, then this flow is governed by the Navier–Stokes equations

$$\frac{\partial \mathbf{u}}{\partial t} - \nu\Delta\mathbf{u} + (\mathbf{u}\cdot\nabla)\mathbf{u} + \nabla p = \mathbf{f} \text{ in } \Omega, \tag{5.1}$$

$$\nabla\cdot\mathbf{u} = 0 \text{ in } \Omega \text{ (\textit{incompressibility condition})}, \tag{5.2}$$

which in the steady case reduce to

$$-\nu\Delta\mathbf{u} + (\mathbf{u}\cdot\nabla)\mathbf{u} + \nabla p = \mathbf{f} \text{ in } \Omega, \tag{5.3}$$

$$\nabla\cdot\mathbf{u} = 0 \text{ in } \Omega. \tag{5.4}$$

In (5.1)–(5.4):

(i) $\mathbf{u} = \{u_i\}_{i=1}^N$ is the flow velocity,
(ii) p is the *pressure*,
(iii) ν is the viscosity of the fluid ($\nu = 1/\text{Re}$, where Re is the Reynold's number),
(iv) $\mathbf{f}$ is the density of external forces;

---

[12] One also says *stationary*.
[13] One also says *nonstationary*.
[14] $\Omega \subset \mathbb{R}^N$, $N = 2, 3$ in practice.

in (5.1), (5.3), $(\mathbf{u} \cdot \nabla)\mathbf{u}$ is a symbolic notation for the nonlinear (vector) term:

$$\left\{\sum_{j=1}^{N} u_j \frac{\partial u_i}{\partial x_j}\right\}_{i=1}^{N}.$$

Boundary conditions have to be added; for example, in the case of the airfoil, B of Fig. 4.1 of Sec. 4.3.2 of this chapter, we have (since the fluid is viscous) the following adherence condition:

$$\mathbf{u} = \mathbf{0} \text{ on } \partial B = \Gamma_B. \tag{5.5}$$

Typical conditions at infinity are

$$\mathbf{u} = \mathbf{u}_\infty, \tag{5.6}$$

where $\mathbf{u}_\infty$ is a constant vector (with regard to the space variables, at least).

Finally, for the time-dependent problem (5.1), (5.2), an initial condition such as

$$\mathbf{u}(x, 0) = \mathbf{u}_0(x) \text{ a.e. on } \Omega, \tag{5.7}$$

where $\mathbf{u}_0$ is given, is usually prescribed.

Other boundary and/or initial conditions may be prescribed (periodicity in space and/or time, pressure given on $\partial\Omega$ or on a part of it, etc.).

In two dimensions it may be convenient to formulate the Navier–Stokes equations using a *stream-function–vorticity* formulation (see, e.g., Bristeau, Glowinski, Periaux, Perrier, and Pironneau [1, Sec. 4], Fortin and Thomasset [1], Girault and Raviart [1], Glowinski and Pironneau [1], Reinhart [1], and Glowinski, Keller, and Reinhart [1]). To conclude this section, let us mention that a mathematical analysis of the Navier–Stokes equations for incompressible viscous fluids can be found in, e.g., Lions [1], Ladyshenskaya [1], Temam [1], and Tartar [1].

5.3. A mixed finite element method for the Stokes and Navier–Stokes problems

5.3.1. *Synopsis*

In this section we discuss a mixed finite element approximation of the Navier–Stokes problems which have been introduced in Sec. 5.2. For simplicity we shall begin our discussion with the approximation of the steady Stokes problem for incompressible viscous fluids, i.e.,

$$-\nu\Delta\mathbf{u} + \nabla p = \mathbf{f} \text{ in } \Omega,$$

$$\nabla \cdot \mathbf{u} = 0 \text{ in } \Omega; \tag{5.8}$$

as boundary conditions, we choose

$$\mathbf{u} = \mathbf{g} \text{ on } \Gamma \tag{5.9}$$

(with $\int_\Gamma \mathbf{g} \cdot \mathbf{n}\, d\Gamma = 0$, $\mathbf{n}$ being the unit vector of the outward normal at Γ), more general boundary conditions are discussed in Appendix 3.

Also, for simplicity, in the following we suppose that Ω is a bounded polygonal domain of $\mathbb{R}^2$; but the following methods are easily extended to domains with curved boundary in $\mathbb{R}^2$ and $\mathbb{R}^3$.

5.3.2. *A mixed variational formulation of the Stokes problem* (5.8), (5.9)

5.3.2.1. *Some functional spaces: Standard formulation of the Stokes problem.* The following (Sobolev) spaces play an important role in the sequel (in fact they have been extensively used in the other parts of this book):

$$H^1(\Omega) = \left\{\phi \in L^2(\Omega), \frac{\partial \phi}{\partial x_i} \in L^2(\Omega), \forall\, i = 1, \ldots, N\right\},$$

$$H_0^1(\Omega) = \overline{\mathscr{D}}^{H^1(\Omega)} = \{\phi \in H^1(\Omega), \phi = 0 \text{ on } \Gamma\},$$

$$H^2(\Omega) = \left\{\phi \in H^1(\Omega), \frac{\partial^2 \phi}{\partial x_i\, \partial x_j} \in L^2(\Omega), \forall\, i, j, 1 \le i, j \le N\right\},$$

with

$$\mathscr{D}(\Omega) = \{\phi \in C^\infty(\overline{\Omega}), \phi \text{ has a compact support in } \Omega\}.$$

From these spaces we also define

$$V_g = \{\mathbf{v} \mid \mathbf{v} \in (H^1(\Omega))^N, \nabla \cdot \mathbf{v} = 0 \text{ in } \Omega, \mathbf{v} = \mathbf{g} \text{ on } \Gamma\}.$$

Suppose that $\mathbf{f} \in (H^{-1}(\Omega))^N$, where $H^{-1}(\Omega) = (H_0^1(\Omega))^* =$ dual space of $H_0^1(\Omega)$ and that $\mathbf{g} \in (H^{1/2}(\Gamma))^N$, $\int_\Gamma \mathbf{g} \cdot \mathbf{n}\, d\Gamma = 0$; it then follows, from Lions, Temam, Ladyshenskaya, Tartar, *loc. cit.*, that (5.8), (5.9) has a unique solution $\{\mathbf{u}, p\} \in V_g \times (L^2(\Omega)/\mathbb{R})$. If $\mathbf{f} \in (L^2(\Omega))^N$ and $\mathbf{g} \in (H^{3/2}(\Gamma))^N$, then $\{\mathbf{u}, p\} \in (V_g \cap (H^2(\Omega))^N) \times (H^1(\Omega)/\mathbb{R})$ if Γ is sufficiently smooth. The above $\mathbf{u}$ is also the unique solution of the following variational problem (where V_0 is obtained by setting $\mathbf{g} = \mathbf{0}$ in the above definition of V_g):

Find $\mathbf{u} \in V_g$ *such that*

$$\nu \int_\Omega \nabla \mathbf{u} \cdot \nabla \mathbf{v}\, dx = \langle \mathbf{f}, \mathbf{v} \rangle, \qquad \forall\, \mathbf{v} \in V_0, \tag{5.10}$$

where $\langle \cdot, \cdot \rangle$ denotes the duality between $(H^{-1}(\Omega))^N$ and $(H_0^1(\Omega))^N$ and

$$\int_\Omega \nabla \mathbf{u} \cdot \nabla \mathbf{v}\, dx = \sum_{i=1}^N \int_\Omega \nabla u_i \cdot \nabla v_i\, dx.$$

5.3.2.2. *A new variational formulation of the Stokes problem* (5.8), (5.9). For simplicity we suppose that $\mathbf{f} \in (L^2(\Omega))^N$; we then define $W_g \in (H^1(\Omega))^N \times H_0^1(\Omega)$ by

$$W_g = \left\{\{\mathbf{v}, \phi\} \in (H^1(\Omega))^N \times H_0^1(\Omega), \mathbf{v}|_\Gamma = \mathbf{g},\right.$$

$$\left.\int_\Omega \nabla \phi \cdot \nabla w\, dx = \int_\Omega \nabla \cdot \mathbf{v} w\, dx, \forall\, w \in H^1(\Omega)\right\}. \tag{5.11}$$

We now consider the following variational problem[15]

Find $\{\mathbf{u}, \psi\} \in W_g$ *such that*

$$\nu \int_\Omega \nabla \mathbf{u} \cdot \nabla \mathbf{v} \, dx = \int_\Omega \mathbf{f} \cdot (\mathbf{v} + \nabla \phi) \, dx, \qquad \forall \{\mathbf{v}, \phi\} \in W_0. \tag{P}$$

In Glowinski and Pironneau [2] the following has been proved:

Theorem 5.1. *Problem* (P) *has a unique solution* $\{\mathbf{u}, \psi\}$ *such that*

$$\psi = 0, \tag{5.12}$$

$\mathbf{u}$ *is the solution of the Stokes problem* (5.8), (5.9) (*and* (5.10)). (5.13)

EXERCISE 5.1. Prove Theorem 5.1.

Remark 5.1. The potential ϕ introduced above is not at all mysterious. Indeed, the formulation (P) can be interpreted as follows: if $\mathbf{v} \in (H^1(\Omega))^N$ and if Γ is sufficiently smooth, there exist $\phi \in H^2(\Omega) \cap H_0^1(\Omega)$ and $\boldsymbol{\omega} \in (H^1(\Omega))^N$ with $\nabla \cdot \boldsymbol{\omega} = 0$, such that

$$\mathbf{v} = -\nabla \phi + \boldsymbol{\omega}, \tag{5.14}$$

and the decomposition (5.14) is unique.

In the formulation (P), instead of directly imposing $\nabla \cdot \mathbf{v} = 0$, we try to impose $\phi = 0$; these procedures are equivalent in the continuous case but not at all in the discrete case, as will be seen below.

5.3.3. *A mixed finite element approximation of the steady Stokes problem* (5.8), (5.9)

As mentioned before, Ω is a bounded polygonal domain of $\mathbb{R}^2$. In this section we follow B.G. 4P. [1] and Glowinski and Pironneau [2], [3].

5.3.3.1. *Triangulation of* Ω: *fundamental discrete spaces.* Let $\{\mathcal{T}_h\}_h$ be a family of triangulations of Ω such that $\overline{\Omega} = \bigcup_{T \in \mathcal{T}_h} T$. We set $h(T)$ = length of the greatest side of T, $h = \max_{T \in \mathcal{T}_h} h(T)$, and we suppose that

$$\frac{h}{\min_{T \in \mathcal{T}_h} h(T)} \leq \beta, \qquad \forall \mathcal{T}_h. \tag{5.15}$$

We then define the following finite element spaces:

$$H_h^1 = \{\phi_h \in C^0(\overline{\Omega}), \phi_{h|T} \in P_1, \forall T \in \mathcal{T}_h\}, \tag{5.16}$$

$$H_{0h}^1 = \{\phi_h \in H_h^1, \phi_h = 0 \text{ on } \Gamma\} = H_h^1 \cap H_0^1(\Omega), \tag{5.17}$$

$$V_h = \{\mathbf{v}_h \in (C^0(\overline{\Omega}))^2, \mathbf{v}_{h|T} \in P_2 \times P_2, \forall T \in \mathcal{T}_h\}, \tag{5.18}$$

$$V_{gh} = \{\mathbf{v}_h \in V_h, \mathbf{v}_h = \mathbf{g}_h \text{ on } \Gamma\} \tag{5.19}$$

[15] Where W_0 is obtained by setting $\mathbf{g} = \mathbf{0}$ in (5.11).

Figure 5.1

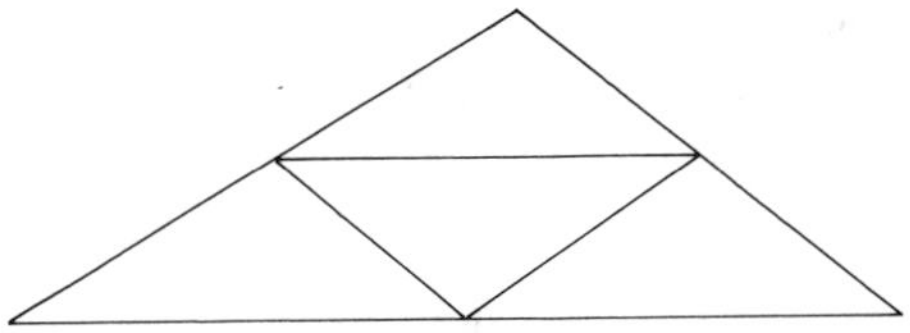

(where, in (5.19), $\mathbf{g}_h$ is a convenient approximation of $\mathbf{g}$ whose construction will be discussed in Appendix 3; if $\mathbf{g} = \mathbf{0}$, one takes $\mathbf{g}_h = \mathbf{0}$ in (5.19) to obtain V_{0h}),

$$W_{gh} = \left\{\{\mathbf{v}_h, \phi_h\} \in V_{gh} \times H_{0h}^1, \int_\Omega \nabla\phi_h \cdot \nabla w_h \, dx = \int_\Omega \nabla \cdot \mathbf{v}_h w_h \, dx, \forall\, w_h \in H_h^1\right\}. \tag{5.20}$$

We shall also use the variants of V_{gh}, W_{gh} obtained from

$$V_h = \{\mathbf{v}_h \in (C^0(\bar{\Omega}))^2, \mathbf{v}_{h|T} \in P_1 \times P_1, \forall\, T \in \tilde{\mathscr{T}}_h\}, \tag{5.21}$$

where $\tilde{\mathscr{T}}_h$ is the triangulation obtained from $\mathscr{T}_h$ by subdividing each triangle $T \in \mathscr{T}_h$ into four subtriangles (by joining the mid-sides; see Figure 5.1). In the above definitions, P_k denotes (as usual) the space of the polynomials in two variables of degree $\leq k$.

5.3.3.2. *Definition of the approximate problems. Characterization of the approximate solutions.* We approximate (P) of Sec. 5.3.2.2 (therefore, the Stokes problem (5.8), (5.9)) by:

Find $\{\mathbf{u}_h, \psi_h\} \in W_{gh}$ *such that*

$$\nu \int_\Omega \nabla\mathbf{u}_h \cdot \nabla\mathbf{v}_h \, dx = \int_\Omega \mathbf{f} \cdot (\mathbf{v}_h + \nabla\phi_h) \, dx, \qquad \forall\, \{\mathbf{v}_h, \phi_h\} \in W_{0h}. \tag{P_h}$$

In Glowinski and Pironneau [2] one may find the proof of the following:

Theorem 5.2. *Problem* (P_h) *has a unique solution* $\{\mathbf{u}_h, \psi_h\}$ *which is characterized by the existence of a discrete pressure* $p_h \in H_h^1$ *such that*

$$\int_\Omega \nabla p_h \cdot \nabla w_h \, dx = \int_\Omega \mathbf{f} \cdot \nabla w_h \, dx, \qquad \forall\, w_h \in H_{0h}^1, \tag{5.22}$$

$$\nu \int_\Omega \nabla\mathbf{u}_h \cdot \nabla\mathbf{v}_h \, dx = \int_\Omega (-\nabla p_h + \mathbf{f}) \cdot \mathbf{v}_h \, dx, \qquad \forall\, \mathbf{v}_h \in V_{0h}, \tag{5.23}$$

$$\{\mathbf{u}_h, \psi_h\} \in W_{gh} \; (\textit{with } \psi_h \neq 0 \textit{ in general}). \tag{5.24}$$

EXERCISE 5.2. Prove Theorem 5.2.

5.3.3.3. *Uniqueness of the discrete pressure. Convergence of the approximate solutions.* We introduce the notation

$$|q|_{1,\Omega} = \left(\int_\Omega |\nabla q|^2 \, dx\right)^{1/2}, \qquad |\mathbf{v}|_{0,\Omega} = \|\mathbf{v}\|_{L^2(\Omega) \times L^2(\Omega)}.$$

The following lemma (proved in Bercovier and Pironneau [1] and Glowinski and Pironneau [2]) plays a fundamental role in the study of the convergence.

Lemma 5.1. *Suppose that the angles of $\mathscr{T}_h$ are bounded from below by $\theta_0 > 0$, independent of h. Also suppose that $\forall\, T \in \mathscr{T}_h$, T has at most two vertices belonging to Γ. Then if H_h^1 and V_{0h} are defined by either* (5.16) *and* (5.18), (5.19), *or* (5.16) *and* (5.21), (5.19) (*with* $\mathbf{g}_h = \mathbf{0}$ *in* (5.19)), *there exists C_0 independent of h such that*

$$\forall\, q_h \in H_h^1, \quad |q_h|_{1,\Omega} \le C_0 \max_{\mathbf{v}_h \in V_{0h} - \{\mathbf{0}\}} \frac{\int_\Omega \mathbf{v}_h \cdot \nabla q_h \, dx}{|\mathbf{v}_h|_{0,\Omega}}. \tag{5.25}$$

A first consequence of Lemma 5.1 is given by the following:

Theorem 5.3. *Suppose that* (5.25) *holds. Then the discrete pressure p_h occurring in* (5.22)–(5.24) *is unique in $H_h^1/\mathbb{R}$ (i.e., up to an arbitrary constant).*

For a proof see Glowinski and Pironneau [2].

EXERCISE 5.3. Prove Theorem 5.3.

A fairly complete discussion of the convergence properties of the approximate solutions is done in Glowinski and Pironneau [2]; in the two following theorems we summarize the results which have been obtained there:

Theorem 5.4. *Suppose that Ω is a bounded polygonal convex domain of $\mathbb{R}^2$ and that $\{\mathscr{T}_h\}_h$ obeys* (5.15) *and the statement of Lemma* 5.1; *then the following error estimates hold (if* $\mathbf{g}$ *has been conveniently approximated by* $\mathbf{g}_h$*), where $\{\mathbf{u}_h, \psi_h\}$ is the solution of* (P_h) *and p_h is the corresponding discrete pressure*:

(i) *If H_h^1 and V_h are defined by* (5.16), (5.18), *and if* $\mathbf{u} \in (H^3(\Omega))^2$, $p \in H^2(\Omega)/\mathbb{R}$, *we have*

$$\|\mathbf{u}_h - \mathbf{u}\|_{(H^1(\Omega))^2} \le Ch^2(\|\mathbf{u}\|_{(H^3(\Omega))^2} + \|p\|_{H^2(\Omega)/\mathbb{R}}), \tag{5.26}$$

$$\|p_h - p\|_{H^1(\Omega)/\mathbb{R}} \le Ch(\|\mathbf{u}\|_{(H^3(\Omega))^2} + \|p\|_{H^2(\Omega)/\mathbb{R}}). \tag{5.27}$$

$$\|\psi_h\|_{H^1(\Omega)} \le Ch^2(\|\mathbf{u}\|_{(H^3(\Omega))^2} + \|p\|_{H^2(\Omega)/\mathbb{R}}). \tag{5.28}$$

(ii) *If H_h^1 and V_h are defined by either* (5.16), (5.18) *or* (5.16), (5.21) *and if* $\mathbf{u} \in (H^2(\Omega))^2$, $p \in H^1(\Omega)/\mathbb{R}$, *we have*

$$\|\mathbf{u}_h - \mathbf{u}\|_{(H^1(\Omega))^2} \le Ch(\|\mathbf{u}\|_{(H^2(\Omega))^2} + \|p\|_{H^1(\Omega)/\mathbb{R}}). \tag{5.29}$$

$$\|\psi_h\|_{H^1(\Omega)} \le Ch(\|\mathbf{u}\|_{(H^2(\Omega))^2} + \|p\|_{H^1(\Omega)/\mathbb{R}}). \tag{5.30}$$

In (5.26)–(5.30), *C denotes various quantities independent of* $\mathbf{u}$, *p, and h.*

Theorem 5.5. *If the hypotheses on Ω and $\{\mathscr{T}_h\}_h$ are those of Theorem 5.4, then the following error estimates hold*:

(i) *If* H_h^1 *and* V_h *are defined by* (5.16), (5.18), *and if* $\mathbf{u} \in (H^3(\Omega))^2$, $p \in H^2(\Omega)/\mathbb{R}$, *we have*

$$\|p_h - p\|_{L^2(\Omega)/\mathbb{R}} \leq Ch^2(\|\mathbf{u}\|_{(H^3(\Omega))^2} + \|p\|_{H^2(\Omega)/\mathbb{R}}), \tag{5.31}$$

$$\|(\mathbf{u}_h + \nabla\psi_h) - \mathbf{u}\|_{(L^2(\Omega))^2} \leq Ch^3(\|\mathbf{u}\|_{(H^3(\Omega))^2} + \|p\|_{H^2(\Omega)/\mathbb{R}}). \tag{5.32}$$

(ii) *If* H_h^1 *and* V_h *are defined by either* (5.16), (5.18) *or* (5.16), (5.21) *and if* $\mathbf{u} \in (H^2(\Omega))^2$, $p \in H^1(\Omega)/\mathbb{R}$, *we have*

$$\|p_h - p\|_{L^2(\Omega)/\mathbb{R}} \leq Ch(\|\mathbf{u}\|_{(H^2(\Omega))^2} + \|p\|_{H^1(\Omega)/\mathbb{R}}), \tag{5.33}$$

$$\|(\mathbf{u}_h + \nabla\psi_h) - \mathbf{u}\|_{(L^2(\Omega))^2} \leq Ch^2(\|\mathbf{u}\|_{(H^2(\Omega))^2} + \|p\|_{H^1(\Omega)/\mathbb{R}}). \tag{5.34}$$

In (5.31)–(5.34), *C denotes various quantities independent of* $\mathbf{u}$, *p, and h.*

We observe that the error estimates (5.26)–(5.34) are of optimal order. For more information concerning the convergence, see Glowinski and Pironneau [2], where the extension to finite element approximations of (5.8), (5.9) by rectangular elements is also considered (in Sec. 7).

Remark 5.2. Let X be a *Banach functional space containing the constant functions*; we then define $\|v\|_{X/\mathbb{R}}$, $\forall\, v \in X$, by

$$\|v\|_{X/\mathbb{R}} = \operatorname*{Inf}_{c \in \mathbb{R}} \|v + c\|_X.$$

5.3.4. *A mixed finite element approximation of the stationary Navier–Stokes equations*

We now consider the steady Navier–Stokes equations

$$-\nu\Delta\mathbf{u} + (\mathbf{u}\cdot\nabla)\mathbf{u} + \nabla p = \mathbf{f} \text{ in } \Omega, \tag{5.35}$$

$$\nabla\cdot\mathbf{u} = 0 \text{ in } \Omega, \tag{5.36}$$

$$\mathbf{u} = \mathbf{g} \text{ on } \Gamma. \tag{5.37}$$

Using the notation of Sec. 5.3.3, we approximate (5.35)–(5.37) by:

Find $\{\mathbf{u}_h, \psi_h\} \in W_{gh}$ *such that*, $\forall\, \{\mathbf{v}_h, \phi_h\} \in W_{0h}$,

$$\nu \int_\Omega \nabla\mathbf{u}_h \cdot \nabla\mathbf{v}_h \, dx + \int_\Omega (\mathbf{u}_h\cdot\nabla)\mathbf{u}_h \cdot (\mathbf{v}_h + \nabla\phi_h)\, dx = \int_\Omega \mathbf{f}\cdot(\mathbf{v}_h + \nabla\phi_h)\, dx. \tag{5.38$_1$}$$

Actually to every solution of $(5.38)_1$ we can associate a discrete pressure $p_h \in H_h^1$, uniquely defined in $H_h^1/\mathbb{R}$ if (5.25) holds, and such that $(5.38)_1$ is equivalent to

$$\int_\Omega \nabla p_h \cdot \nabla w_h \, dx + \int_\Omega (\mathbf{u}_h\cdot\nabla)\mathbf{u}_h \cdot \nabla w_h \, dx = \int_\Omega \mathbf{f}\cdot\nabla w_h \, dx, \qquad \forall\, w_h \in H_{0h}^1,$$

$$\nu \int_\Omega \nabla\mathbf{u}_h \cdot \nabla\mathbf{v}_h \, dx + \int_\Omega (\mathbf{u}_h\cdot\nabla)\mathbf{u}_h \cdot \mathbf{v}_h \, dx + \int_\Omega \nabla p_h \cdot \mathbf{v}_h \, dx = \int_\Omega \mathbf{f}\cdot\mathbf{v}_h \, dx,$$

$$\forall\, \mathbf{v}_h \in V_{0h}, \quad \{\mathbf{u}_h, \psi_h\} \in W_{gh}, \quad p_h \in H_h^1. \tag{5.38$_2$}$$

In Le Tallec [2], [3] it is proved that if some reasonable assumptions on the smoothness of $\mathbf{u}$ and p are satisfied, then the estimates (5.26)–(5.31), (5.33) still hold; it is very likely that (5.32), (5.34) also hold, but proving these estimates is still an open problem.

5.3.5. *A semidiscrete approximation of the time-dependent Navier–Stokes equations*

Using suitable finite element bases for the spaces defined in Sec. 5.3.3.1, it is possible to reduce the unsteady Navier–Stokes equations to a system of ordinary differential equations. Suppose that the time-dependent problem under consideration is

$$\frac{\partial \mathbf{u}}{\partial t} - \nu\Delta\mathbf{u} + (\mathbf{u}\cdot\nabla)\mathbf{u} + \nabla p = \mathbf{f} \text{ in } \Omega, \tag{5.39}$$

$$\nabla\cdot\mathbf{u} = 0 \text{ in } \Omega, \tag{5.40}$$

$$\mathbf{u} = \mathbf{g}\ (=\mathbf{g}(x, t)) \text{ on } \Gamma, \tag{5.41}$$

$$\mathbf{u}(x, 0) = \mathbf{u}_0(x) \text{ in } \Omega; \tag{5.42}$$

we then approximate (5.39)–(5.42) by:

Find $\{\mathbf{u}_h(t), \psi_h(t)\} \in W_{gh}(t)$ *a.e. in t such that*

$$\int_\Omega \frac{\partial \mathbf{u}_h}{\partial t}\cdot\mathbf{v}_h\,dx + \nu\int_\Omega \nabla\mathbf{u}_h\cdot\nabla\mathbf{v}_h\,dx + \int_\Omega (\mathbf{u}_h\cdot\nabla)\mathbf{u}_h\cdot(\mathbf{v}_h + \nabla\phi_h)\,dx$$
$$= \int_\Omega \mathbf{f}\cdot(\mathbf{v}_h + \nabla\phi_h)\,dx, \qquad \forall\ \{\mathbf{v}_h, \phi_h\} \in W_{0h}, \tag{5.43}$$

$$\mathbf{u}_h(0) = \mathbf{u}_{0h} \in V_h \textit{ given } (\mathbf{u}_{0h}\textit{: approximation of } \mathbf{u}_0).$$

In (5.43) the space $W_{gh}(t)$ is defined as follows:

$$W_{gh}(t) = \left\{\{\mathbf{v}_h, \phi_h\} \in V_{gh}(t) \times H^1_{0h}, \int_\Omega \nabla\phi_h\cdot\nabla w_h\,dx = \int_\Omega \nabla\cdot\mathbf{v}_h w_h\,dx, \forall\ w_h \in H^1_h\right\} \tag{5.44}$$

with

$$V_{gh}(t) = \{\mathbf{v}_h \in V_h, \mathbf{v}_{h|\Gamma} = \mathbf{g}_h(t)\}, \tag{5.45}$$

$\mathbf{g}_h(t)$ being a suitable approximation of $\mathbf{g}(t)$.

The approximate problem (5.43) is not directly suitable for computation, and for this purpose a convenient time discretization is required in order to obtain a fully discrete approximate problem; several such time discretization schemes are given in Sec. 5.4.

5.4. Time discretization of the time-dependent Navier–Stokes equations

We follow the presentation of B.G. 4P. [1] (see also Le Tallec [2] and Periaux [2]); in the sequel, $k = \Delta t$ is the time discretization step. We only consider fully implicit schemes, since the stability condition of the semi-implicit schemes that we have tested (on realistic problems) seems to be a severe limitation with regard to today's computers (a description of such semi-implicit schemes is given in B.G. 4P. [1], Le Tallec [2], and Periaux [2]).

The following considerations are, of course, closely related to Chapter III of this book.

5.4.1. *An ordinary implicit scheme*

This scheme is defined as follows ($\mathbf{u}_h^n$ approximates $\mathbf{u}_h(nk)$):

$$\mathbf{u}_h^0 = \mathbf{u}_{0h} \text{ (see (5.43))}; \tag{5.46}$$

then for $n \geq 0$, *we obtain* $\mathbf{u}_h^{n+1}$ *from* $\mathbf{u}_h^n$ *by solving*

$$\int_\Omega \frac{\mathbf{u}_h^{n+1} - \mathbf{u}_h^n}{k} \cdot \mathbf{v}_h \, dx + \nu \int_\Omega \nabla \mathbf{u}_h^{n+1} \cdot \nabla \mathbf{v}_h \, dx + \int_\Omega (\mathbf{u}_h^{n+1} \cdot \nabla)\mathbf{u}_h^{n+1} \cdot (\mathbf{v}_h + \nabla \phi_h) \, dx$$
$$= \int_\Omega \mathbf{f}^{n+1} \cdot (\mathbf{v}_h + \nabla \phi_h) \, dx, \qquad \forall \{\mathbf{v}_h, \phi_h\} \in W_{0h}, \quad \{\mathbf{u}_h^{n+1}, \psi_h^{n+1}\} \in W_{gh}^{n+1}, \tag{5.47}$$

where (see (5.44), (5.45)) $W_{gh}^m = W_{gh}(mk)$, *and* $f^m = f(mk)$.

To obtain $\mathbf{u}_h^{n+1}$ from $\mathbf{u}_h^n$ in (5.47), we have to solve a finite-dimensional nonlinear problem very close to the steady discrete Navier–Stokes problem (5.38). The above scheme has a time truncation error in $O(\Delta t)$ and appears to be unconditionally stable.

5.4.2. *A Crank–Nicholson implicit scheme*

This scheme is defined as follows:

$$\mathbf{u}_h^0 = \mathbf{u}_{0h}; \tag{5.48}$$

then for $n \geq 0$, *we obtain* $\mathbf{u}_h^{n+1}$ *from* $\mathbf{u}_h^n$ *by solving*

$$\int_\Omega \frac{\mathbf{u}_h^{n+1} - \mathbf{u}_h^n}{k} \cdot \mathbf{v}_h \, dx + \nu \int_\Omega \nabla \mathbf{u}_h^{n+1/2} \cdot \nabla \mathbf{v}_h \, dx$$
$$+ \int_\Omega (\mathbf{u}_h^{n+1/2} \cdot \nabla)\mathbf{u}_h^{n+1/2} \cdot (\mathbf{v}_h + \nabla \phi_h) \, dx$$
$$= \int_\Omega \mathbf{f}^{n+1/2} \cdot (\mathbf{v}_h + \nabla \phi_h) \, dx, \qquad \forall \{\mathbf{v}_h, \phi_h\} \in W_{0h},$$
$$\{\mathbf{u}_h^{n+1/2}, \psi_h^{n+1/2}\} \in W_{gh}^{n+1/2} \; (= W_{gh}((n + 1/2)k)), \tag{5.49}$$

where

$$\mathbf{u}_h^{n+1/2} = \tfrac{1}{2}(\mathbf{u}_h^n + \mathbf{u}_h^{n+1}). \tag{5.50}$$

Since $\mathbf{u}_h^{n+1} = \mathbf{u}_h^n + 2(\mathbf{u}_h^{n+1/2} - \mathbf{u}_h^n)$, we can eliminate $\mathbf{u}_h^{n+1}$ in (5.49) and therefore reduce this problem to a variant of the discrete Navier–Stokes problem (5.38). The above scheme has a time truncation error in $O(|\Delta t|^2)$ and appears to be unconditionally stable.

5.4.3. *A two-step implicit scheme*

The scheme is defined by

$$\mathbf{u}_h^0 = \mathbf{u}_{0h}, \mathbf{u}_h^1 \textit{ given}; \tag{5.51}$$

then for $n \geq 1$, *we obtain* $\mathbf{u}_h^{n+1}$ *from* $\mathbf{u}_h^n, \mathbf{u}_h^{n-1}$ *by solving*

$$\begin{aligned}
&\int_\Omega \frac{3\mathbf{u}_h^{n+1} - 4\mathbf{u}_h^n + \mathbf{u}_h^{n-1}}{2k} \cdot \mathbf{v}_h \, dx + \nu \int_\Omega \nabla \mathbf{u}_h^{n+1} \cdot \nabla \mathbf{v}_h \, dx \\
&\quad + \int_\Omega (\mathbf{u}_h^{n+1} \cdot \nabla)\mathbf{u}_h^{n+1} \cdot (\mathbf{v}_h + \nabla \phi_h) \, dx \\
&= \int_\Omega \mathbf{f}^{n+1} \cdot (\mathbf{v}_h + \nabla \phi_h) \, dx, \qquad \forall \{\mathbf{v}_h, \phi_h\} \in W_{0h}, \quad \{\mathbf{u}_h^{n+1}, \psi_h^{n+1}\} \in W_{gh}^{n+1}.
\end{aligned} \tag{5.52}$$

To obtain $\mathbf{u}_h^1$ from $\mathbf{u}_h^0$, we may use either one of the two schemes discussed in Secs. 5.4.1 and 5.4.2 or one of the semi-implicit schemes described in B.G. 4P. [1]; scheme (5.51), (5.52) appears to be unconditionally stable and its truncation error is in $O(|\Delta t|^2)$.

5.4.4. *Alternating-direction methods*

Previously we have used alternating-direction methods to solve various kinds of steady nonlinear problems (cf. Chapter IV, Sec. 2.6.6 and also Sec. 3.2.4 of this chapter); actually these methods are also very useful for solving time-dependent problems and most particularly the unsteady Navier–Stokes equations as indicated by the two methods described below (and the corresponding numerical experiments).

5.4.4.1. *A Peaceman–Rachford alternating-direction method for solving the unsteady Navier–Stokes equations.* The method (inspired by Peaceman and Rachford [1]) is defined, with $0 < \theta < 1$, by

$$\mathbf{u}_h^0 = \mathbf{u}_{0h}; \tag{5.53}$$

then for $n \geq 0$, we obtain $\mathbf{u}_h^{n+1/2}$, $\mathbf{u}_h^{n+1}$ from $\mathbf{u}_h^n$ by solving

$$\int_\Omega \frac{\mathbf{u}_h^{n+1/2} - \mathbf{u}_h^n}{k/2} \cdot \mathbf{v}_h \, dx + \theta\nu \int_\Omega \nabla \mathbf{u}_h^{n+1/2} \cdot \nabla \mathbf{v}_h \, dx + (1-\theta)\nu \int_\Omega \nabla \mathbf{u}_h^n \cdot \nabla \mathbf{v}_h \, dx$$

$$+ \int_\Omega (\mathbf{u}_h^n \cdot \nabla)\mathbf{u}_h^n \cdot (\mathbf{v}_h + \nabla\phi_h) \, dx = \int_\Omega \mathbf{f}^{n+1/2} \cdot (\mathbf{v}_h + \nabla\phi_h) \, dx, \qquad (5.54)_1$$

$$\forall \{\mathbf{v}_h, \phi_h\} \in W_{0h}, \quad \{\mathbf{u}_h^{n+1/2}, \psi_h^{n+1/2}\} \in W_{gh}^{n+1/2},$$

which is equivalent to

$$\int_\Omega \nabla p_h^{n+1/2} \cdot \nabla w_h \, dx + \int_\Omega (\mathbf{u}_h^n \cdot \nabla)\mathbf{u}_h^n \cdot \nabla w_h \, dx$$

$$= \int_\Omega \mathbf{f}^{n+1/2} \cdot \nabla w_h \, dx, \qquad \forall\, w_h \in H_{0h}^1, \quad \int_\Omega \frac{\mathbf{u}_h^{n+1/2} - \mathbf{u}_h^n}{(k/2)} \cdot \mathbf{v}_h \, dx$$

$$+ \theta\nu \int_\Omega \nabla \mathbf{u}_h^{n+1/2} \cdot \nabla \mathbf{v}_h \, dx + \int_\Omega \nabla p_h^{n+1/2} \cdot \mathbf{v}_h \, dx$$

$$+ (1-\theta)\nu \int_\Omega \nabla \mathbf{u}_h^n \cdot \nabla \mathbf{v}_h \, dx + \int_\Omega (\mathbf{u}_h^n \cdot \nabla)\mathbf{u}_h^n \cdot \mathbf{v}_h \, dx$$

$$= \int_\Omega \mathbf{f}^{n+1/2} \cdot \mathbf{v}_h \, dx, \qquad \forall\, \mathbf{v}_h \in V_{0h},$$

$$\{\mathbf{u}_h^{n+1/2}, \psi_h^{n+1/2}\} \in W_{gh}^{n+1/2}, \quad p_h^{n+1/2} \in H_h^1/\mathbb{R}, \qquad (5.54)_2$$

and then

$$\int_\Omega \frac{\mathbf{u}_h^{n+1} - \mathbf{u}_h^{n+1/2}}{(k/2)} \cdot \mathbf{v}_h \, dx + (1-\theta)\nu \int_\Omega \nabla \mathbf{u}_h^{n+1} \cdot \nabla \mathbf{v}_h \, dx$$

$$+ \int_\Omega (\mathbf{u}_h^{n+1} \cdot \nabla)\mathbf{u}_h^{n+1} \cdot \mathbf{v}_h \, dx + \theta\nu \int_\Omega \nabla \mathbf{u}_h^{n+1/2} \cdot \nabla \mathbf{v}_h \, dx$$

$$+ \int_\Omega \nabla p_h^{n+1/2} \cdot \mathbf{v}_h \, dx = \int_\Omega \mathbf{f}^{n+1} \cdot \mathbf{v}_h \, dx,$$

$$\forall\, \mathbf{v} \in V_{0h}, \quad \mathbf{u}_h^{n+1} \in V_{gh}^{n+1} \; (= V_{gh}((n+1)k)). \qquad (5.55)$$

The above scheme (whose practical implementation will be discussed in Sec. 5.4.4.3) has a time truncation error in $O(\Delta t)$ and appears to be unconditionally stable in practice.

The equivalence between $(5.54)_1$ and $(5.54)_2$ is left to the reader as an exercise (see also Sec. 5.7.2.3).

5.4.4.2. *A second alternating-direction method for solving the unsteady Navier–Stokes equations.* This is the variant [16] of (5.53)–(5.55) defined as follows (again with $0 < \theta < 1$):

$$\mathbf{u}_h^0 = \mathbf{u}_{0h}; \tag{5.56}$$

then for $n \geq 0$, we obtain $\mathbf{u}_h^{n+1/4}$, $\mathbf{u}_h^{n+3/4}$, $\mathbf{u}_h^{n+1}$ from $\mathbf{u}_h^n$ by solving

$$\begin{aligned}
&\int_\Omega \frac{\mathbf{u}_h^{n+1/4} - \mathbf{u}_h^n}{k/4} \cdot \mathbf{v}_h \, dx + (1-\theta)\nu \int_\Omega \nabla \mathbf{u}_h^{n+1/4} \cdot \nabla \mathbf{v}_h \, dx \\
&\qquad + \theta\nu \int_\Omega \nabla \mathbf{u}_h^n \cdot \nabla \mathbf{v}_h \, dx + \int_\Omega (\mathbf{u}_h^n \cdot \nabla)\mathbf{u}_h^n \cdot (\mathbf{v}_h + \nabla\phi_h) \, dx \\
&\qquad = \int_\Omega \mathbf{f}^{n+1/4} \cdot (\mathbf{v}_h + \nabla\phi_h) \, dx, \\
&\qquad \forall \{\mathbf{v}_h, \phi_h\} \in W_{0h}, \quad \{\mathbf{u}_h^{n+1/4}, \psi_h^{n+1/4}\} \in W_{gh}^{n+1/4} \; (= W_{gh}((n+1/4)k)),
\end{aligned} \tag{5.57$_1$}$$

which is equivalent to

$$\begin{aligned}
&\int_\Omega \nabla p_h^{n+1/4} \cdot \nabla w_h \, dx + \int_\Omega (\mathbf{u}_h^n \cdot \nabla)\mathbf{u}_h^n \cdot \nabla w_h \, dx \\
&\qquad = \int_\Omega \mathbf{f}^{n+1/4} \cdot \nabla w_h \, dx, \qquad \forall w_h \in H_{0h}^1, \; \int_\Omega \frac{\mathbf{u}_h^{n+1/4} - \mathbf{u}_h^n}{k/4} \cdot \mathbf{v}_h \, dx \\
&\qquad + (1-\theta)\nu \int_\Omega \nabla \mathbf{u}_h^{n+1/4} \cdot \nabla \mathbf{v}_h \, dx + \int_\Omega \nabla p_h^{n+1/4} \cdot \mathbf{v}_h \, dx \\
&\qquad + \theta\nu \int_\Omega \nabla \mathbf{u}_h^n \cdot \nabla \mathbf{v}_h \, dx + \int_\Omega (\mathbf{u}_h^n \cdot \nabla)\mathbf{u}_h^n \cdot \mathbf{v}_h \, dx \\
&\qquad = \int_\Omega \mathbf{f}^{n+1/4} \cdot \mathbf{v}_h \, dx, \qquad \forall \mathbf{v}_h \in V_{0h}, \\
&\qquad \{\mathbf{u}_h^{n+1/4}, \psi_h^{n+1/4}\} \in W_{gh}^{n+1/4}, \quad p_h^{n+1/4} \in H_h^1/\mathbb{R};
\end{aligned} \tag{5.57$_2$}$$

then

$$\begin{aligned}
&\int_\Omega \frac{\mathbf{u}_h^{n+3/4} - \mathbf{u}_h^{n+1/4}}{k/2} \cdot \mathbf{v}_h \, dx + \theta\nu \int_\Omega \nabla \mathbf{u}_h^{n+3/4} \cdot \nabla \mathbf{v}_h \, dx \\
&\quad + \int_\Omega (\mathbf{u}_h^{n+3/4} \cdot \nabla)\mathbf{u}_h^{n+3/4} \cdot \mathbf{v}_h \, dx + (1-\theta)\nu \int_\Omega \nabla \mathbf{u}_h^{n+1/4} \cdot \nabla \mathbf{v}_h \, dx \\
&\quad + \int_\Omega \nabla p_h^{n+1/4} \cdot \mathbf{v}_h \, dx = \int_\Omega \mathbf{f}^{n+3/4} \cdot \mathbf{v}_h \, dx, \qquad \forall \mathbf{v}_h \in V_{0h}, \quad \mathbf{u}_h^{n+3/4} \in V_{gh}^{n+3/4},
\end{aligned} \tag{5.58}$$

[16] Inspired to us by Leveque and Oliger [1] (see also Leveque [1]).

and finally

$$\int_\Omega \frac{\mathbf{u}_h^{n+1} - \mathbf{u}_h^{n+3/4}}{k/4} \cdot \mathbf{v}_h \, dx + (1 - \theta)\nu \int_\Omega \nabla \mathbf{u}_h^{n+1} \cdot \nabla \mathbf{v}_h \, dx$$

$$+ \theta\nu \int_\Omega \nabla \mathbf{u}_h^{n+3/4} \cdot \nabla \mathbf{v}_h \, dx + \int_\Omega (\mathbf{u}_h^{n+3/4} \cdot \nabla)\mathbf{u}_h^{n+3/4} \cdot (\mathbf{v}_h + \nabla \phi_h) \, dx$$

$$= \int_\Omega \mathbf{f}^{n+1} \cdot (\mathbf{v}_h + \nabla \phi_h) \, dx, \qquad \forall \{\mathbf{v}_h, \phi_h\} \in W_{0h}, \quad \{\mathbf{u}_h^{n+1}, \psi_h^{n+1}\} \in W_{gh}^{n+1}. \tag{5.59}$$

Scheme (5.56)–(5.59) has a time truncation error in $O(|\Delta t|^2)$ and appears to be unconditionally stable in practice; the practical implementation of (5.56)–(5.59) is discussed in Sec. 5.4.4.3.

5.4.4.3. *Practical implementation of* (5.53)–(5.55) *and* (5.56)–(5.59): *further remarks*. The main advantage of the alternating-direction methods (5.53)–(5.55) and (5.56)–(5.59), compared to the methods of Secs. 5.4.1, 5.4.2, and 5.4.3, is that they decouple the two main difficulties of the Navier–Stokes equations, which are nonlinearity and incompressibility.

In both methods, the nonlinear problems solved at each time step (namely (5.55) and (5.58), respectively) are in fact finite-dimensional variants of the nonlinear problem (3.56) (cf. Sec. 3.4, Exercise 3.1), and therefore they can be solved by least-squares conjugate gradient methods very close to those methods used in Sec. 3.4 for solving the model problem (3.1).

On the other hand, problems (5.54) and (5.57), (5.59) are closely related to the discrete Stokes problem (P_h) (cf. Sec. 5.3.3.2) whose numerical solution will be discussed in Sec. 5.7.

Remark 5.3. With respect to storage requirements, the optimal values[17] of θ are $\theta = \frac{1}{2}$ for (5.53)–(5.55) and $\theta = \frac{1}{3}$ for (5.56)–(5.59); these choices will be justified in Sec. 5.7 and are directly related to the properties of the methods used to solve the nonlinear problems (5.55), (5.58) and the discrete Stokes problems (5.54), (5.57), and (5.59).

Remark 5.4. From our computational experiments, (5.55) (resp., (5.58)) is the most costly[18] step (in term of computational time) of method (5.53)–(5.55) (resp., (5.56)–(5.59)); this observation justifies the choice made in (5.56)–(5.59) for the symmetrization process leading to an $O(|\Delta t|^2)$ time truncation error; indeed the cost of step (5.57)–(5.59) is practically the same as the cost of step (5.54), (5.55).

---

[17] Actually these values also lead to shorter computational times.

[18] About 5 to 10 times.

Remark 5.5. The alternating-direction method (5.53)–(5.55) is in fact a discrete variant of the following semidiscrete scheme for integrating the time-dependent Navier–Stokes equations (5.39)–(5.42):

$$\mathbf{u}^0 = \mathbf{u}_0; \tag{5.60}$$

then for $n \geq 0$, *we obtain* $\{\mathbf{u}^{n+1/2}, p^{n+1/2}\}$, $\mathbf{u}^{n+1}$ *from* $\mathbf{u}^n$ *by solving*

$$\frac{\mathbf{u}^{n+1/2} - \mathbf{u}^n}{k/2} - \theta\nu\Delta\mathbf{u}^{n+1/2} + \nabla p^{n+1/2} - (1-\theta)\nu\Delta\mathbf{u}^n + (\mathbf{u}^n \cdot \nabla)\mathbf{u}^n = \mathbf{f}^{n+1/2} \text{ in } \Omega,$$

$$\nabla \cdot \mathbf{u}^{n+1/2} = 0 \text{ in } \Omega, \quad \mathbf{u}^{n+1/2} = \mathbf{g}((n+1/2)k) \text{ on } \Gamma, \tag{5.61}$$

and

$$\frac{\mathbf{u}^{n+1} - \mathbf{u}^{n+1/2}}{k/2} - (1-\theta)\nu\Delta\mathbf{u}^{n+1} + (\mathbf{u}^{n+1} \cdot \nabla)\mathbf{u}^{n+1} - \theta\nu\Delta\mathbf{u}^{n+1/2} + \nabla p^{n+1/2}$$

$$= \mathbf{f}^{n+1} \text{ in } \Omega, \quad \mathbf{u}^{n+1} = \mathbf{g}((n+1)k) \text{ on } \Gamma. \tag{5.62}$$

Similarly, method (5.56)–(5.59) is the discrete variant of a semidiscrete scheme that the reader may formulate by himself.

Remark 5.6. Alternating-direction methods are closely related to fractional step methods (basic references for fractional step methods are Yanenko [1] and Marchouk [1]). Actually alternating-direction and fractional step methods have been extensively used for quite a long time for solving time-dependent partial differential equation problems; concentrating particularly on Navier–Stokes equations for incompressible viscous fluids, we shall mention Chorin [1]–[4], Fortin [3], and Temam [1] (see also the references therein).

Remark 5.7. A fractional step variant of the following alternating-direction method has been used by Benque, Ibler, Keramsi, and Labadie [1][19] to solve the unsteady Navier–Stokes equations; the alternating-direction method is defined by[20]:

$$\mathbf{u}^0 = \mathbf{u}_0; \tag{5.63}$$

then for $n \geq 0$, *we obtain* $\mathbf{u}^{n+1/2}$, $p^{n+1/2}$ *and* $\mathbf{u}^{n+1}$ *from* $\mathbf{u}^n$ *by solving*

$$\frac{\mathbf{u}^{n+1/2} - \mathbf{u}^n}{k/2} - \nu\Delta\mathbf{u}^{n+1/2} + \nabla p^{n+1/2} + (\mathbf{u}^n \cdot \nabla)\mathbf{u}^n = \mathbf{f}^{n+1/2} \text{ in } \Omega,$$

$$\nabla \cdot \mathbf{u}^{n+1/2} = 0 \text{ in } \Omega, \ \mathbf{u}^{n+1/2} = \mathbf{g} \text{ on } \Gamma, \tag{5.64}$$

[19] See also Pironneau [1].

[20] We suppose, for simplicity, that $\mathbf{g}\ (= \mathbf{u}|_\Gamma)$ is independent of t.

and then,

$$\frac{\mathbf{u}^{n+1} - \mathbf{u}^{n+1/2}}{k/2} + (\mathbf{u}^{n+1/2} \cdot \nabla)\mathbf{u}^{n+1} - \nu\Delta\mathbf{u}^{n+1/2} + \nabla p^{n+1/2}$$

$$= \mathbf{f}^{n+1} \text{ in } \Omega,\ \mathbf{u}^{n+1} = \mathbf{g} \text{ on } \Gamma_-, \quad (5.65)$$

where

$$\Gamma_- = \{x \mid x \in \Gamma, \mathbf{g}(x) \cdot \mathbf{n}(x) < 0\}$$

($\mathbf{n}(x)$ is the unit vector of the outward normal at Γ, at x).

Problem (5.65) is a linear first-order system which has been solved in Benque, Ibler, Keramsi, and Labadie [1] by a method of characteristics; on the other hand, problem (5.64), which is quite close to the Stokes problem (5.8), (5.9), has been solved in the above reference by the methods of Sec. 5.7 via the finite element approximations of Sec. 5.3.

5.5. Least-squares conjugate gradient solution of the stationary Navier–Stokes equations: (I) The continuous case

5.5.1. *Generalities: synopsis*

The various discrete Navier–Stokes problems that we have encountered in Secs. 5.3.4, 5.4.1, 5.4.2, and 5.4.3 are in fact finite-dimensional variants (obtained by finite element approximations) of the following nonlinear problem (with $\alpha \geq 0$)

$$\alpha\mathbf{u} - \nu\Delta\mathbf{u} + (\mathbf{u} \cdot \nabla)\mathbf{u} + \nabla p = \mathbf{f} \text{ in } \Omega,\ \nabla \cdot \mathbf{u} = 0 \text{ in } \Omega,$$

$$\mathbf{u} = \mathbf{g} \quad \text{on } \Gamma \left(\text{with } \int_\Gamma \mathbf{g} \cdot \mathbf{n}\, d\Gamma = 0\right), \quad (5.66)_1$$

which is clearly a generalization of the steady Navier–Stokes problem (5.35)–(5.37).

Problem $(5.66)_1$ can also be formulated as a nonlinear variational equation by:

Find $\mathbf{u} \in V_g$ *such that*

$$\alpha \int_\Omega \mathbf{u} \cdot \mathbf{v}\, dx + \int_\Omega \nabla\mathbf{u} \cdot \nabla\mathbf{v}\, dx + \int_\Omega \mathbf{v} \cdot (\mathbf{u} \cdot \nabla)\mathbf{u}\, dx = \int_\Omega \mathbf{f} \cdot \mathbf{v}\, dx, \qquad \forall\, \mathbf{v} \in V_0$$

$$(5.66)_2$$

(see Sec. 5.3.2.1 for the definition of V_0, V_g).

Following B.G. 4P. [1], Bristeau, Glowinski, Periaux, Perrier, and Pironneau [1], Bristeau, Glowinski, Mantel, Periaux, Perrier, and Pironneau [1], and Periaux [2], we shall describe, in Sec. 5.5.2, a least-squares formulation of

(5.66) and in Sec. 5.5.3 a conjugate gradient algorithm for solving the least-squares problem. The discrete variants of the methods described in Secs. 5.5.2 and 5.5.3 will be described in Sec. 5.6.

5.5.2. *A least-squares formulation of* (5.66)

A natural least squares formulation of (5.66) is

Find $\mathbf{u} \in V_g$ *such that*

$$J(\mathbf{u}) \leq J(\mathbf{v}), \tag{5.67}$$

with

$$J(\mathbf{v}) = \frac{\alpha}{2}\int_\Omega |\boldsymbol{\xi}(\mathbf{v})|^2\,dx + \frac{\nu}{2}\int_\Omega |\nabla\boldsymbol{\xi}(\mathbf{v})|^2\,dx, \tag{5.68}$$

where $\boldsymbol{\xi}(\mathbf{v})$ ($=\boldsymbol{\xi}$) is a function of $\mathbf{v}$ through the state equation

$$\alpha\boldsymbol{\xi} - \nu\Delta\boldsymbol{\xi} + \nabla\pi = \alpha\mathbf{v} - \nu\Delta\mathbf{v} + (\mathbf{v}\cdot\nabla)\mathbf{v} - \mathbf{f} \text{ in } \Omega,\ \nabla\cdot\boldsymbol{\xi} = 0 \text{ in } \Omega,\quad \boldsymbol{\xi} = \mathbf{0} \text{ on } \Gamma. \tag{5.69}$$

We observe that $\boldsymbol{\xi}$ is obtained from $\mathbf{v}$ through the solution of a Stokes problem, and also that the formulation (5.67)–(5.69) is a natural generalization of (3.37) in Sec. 3.3.2.

The above least-squares formulation is justified by the following obvious proposition.

Proposition 5.1. *Suppose that* $\{\mathbf{u}, p\} \in V_g \times (L^2(\Omega)/\mathbb{R})$ *is a solution of* (5.66); *then* $\mathbf{u}$ *is also a solution of the least-squares problem* (5.67), *and we have, for the corresponding pair* $\{\boldsymbol{\xi}, \pi\}$,

$$\boldsymbol{\xi} = \mathbf{0}, \qquad \pi = -p; \tag{5.70}$$

we also have $J(\mathbf{u}) = 0$.

5.5.3. *Conjugate gradient solution of the least-squares problem* (5.67)–(5.69)

5.5.3.1. *Description of the conjugate gradient algorithm.* A conjugate gradient algorithm for solving the least-squares problem (5.67)–(5.69) is defined as follows.

Step 0: Initialization

$$\mathbf{u}^0 \in V_g \textit{ given}; \tag{5.71}$$

then compute $\mathbf{z}^0$ *as the solution of the linear variational equation*

$$\alpha\int_\Omega \mathbf{z}^0\cdot\boldsymbol{\eta}\,dx + \nu\int_\Omega \nabla\mathbf{z}^0\cdot\nabla\boldsymbol{\eta}\,dx = \langle J'(\mathbf{u}^0), \boldsymbol{\eta}\rangle, \qquad \forall\,\boldsymbol{\eta}\in V_0,\quad \mathbf{z}_0 \in V_0, \tag{5.72}$$

and set

$$\mathbf{w}^0 = \mathbf{z}^0. \tag{5.73}$$

For $m \geq 0$, *assuming that* $\mathbf{u}^m$, $\mathbf{z}^m$, $\mathbf{w}^m$ *are known, compute* $\mathbf{u}^{m+1}$, $\mathbf{z}^{m+1}$, $\mathbf{w}^{m+1}$ *by the following.*

Step 1: Descent

$$\text{Compute } \lambda^m = \operatorname*{Arg\,min}_{\lambda \in \mathbb{R}} J(\mathbf{u}^m - \lambda \mathbf{w}^m), \tag{5.74}$$

$$\mathbf{u}^{m+1} = \mathbf{u}^m - \lambda^m \mathbf{w}^m. \tag{5.75}$$

Step 2: Construction of the new descent direction. Define $\mathbf{z}^{m+1}$ *as the solution of*

$$\mathbf{z}^{m+1} \in V_0, \quad \alpha \int_\Omega \mathbf{z}^{m+1} \cdot \boldsymbol{\eta}\, dx + \nu \int_\Omega \nabla \mathbf{z}^{m+1} \cdot \nabla \boldsymbol{\eta}\, dx = \langle J'(\mathbf{u}^{m+1}), \boldsymbol{\eta} \rangle, \qquad \forall\, \boldsymbol{\eta} \in V_0; \tag{5.76}$$

then (Polak–Ribière strategy)

$$\gamma_{m+1} = \frac{\alpha \int_\Omega \mathbf{z}^{m+1} \cdot (\mathbf{z}^{m+1} - \mathbf{z}^m)\, dx + \nu \int_\Omega \nabla \mathbf{z}^{m+1} \cdot \nabla(\mathbf{z}^{m+1} - \mathbf{z}^m)\, dx}{\alpha \int_\Omega |\mathbf{z}^m|^2\, dx + \nu \int_\Omega |\nabla \mathbf{z}^m|^2\, dx}, \tag{5.77}$$

and finally

$$\mathbf{w}^{m+1} = \mathbf{z}^{m+1} + \gamma_{m+1} \mathbf{w}^m, \tag{5.78}$$

$m = m + 1$ *go to* (5.74).

Remark 5.8. To obtain $\mathbf{z}^{m+1}$ from $\mathbf{u}^{m+1}$, via (5.76), we also have to solve a Stokes problem (written in a variational form).

5.5.3.2. *Calculation of* $J'(\mathbf{u}^{m+1})$ *and* $\mathbf{z}^{m+1}$. A most important step in order to use algorithm (5.71)–(5.78) is the calculation of $J'(\mathbf{u}^{m+1})$; due to the importance of this step, we shall show this calculation in detail. We have

$$\delta J = \langle J'(\mathbf{v}), \delta \mathbf{v} \rangle = \alpha \int_\Omega \boldsymbol{\xi} \cdot \delta \boldsymbol{\xi}\, dx + \nu \int_\Omega \nabla \boldsymbol{\xi} \cdot \nabla \delta \boldsymbol{\xi}\, dx; \tag{5.79}$$

to express $\delta \boldsymbol{\xi}$ as a function of $\delta \mathbf{v}$, we observe that (5.69) also has the following variational formulation:

$\boldsymbol{\xi} \in V_0$ *and* $\forall\, \boldsymbol{\eta} \in V_0$, *we have*

$$\alpha \int_\Omega \boldsymbol{\xi} \cdot \boldsymbol{\eta}\, dx + \nu \int_\Omega \nabla \boldsymbol{\xi} \cdot \nabla \boldsymbol{\eta}\, dx = \alpha \int_\Omega \mathbf{v} \cdot \boldsymbol{\eta}\, dx + \nu \int_\Omega \nabla \mathbf{v} \cdot \nabla \boldsymbol{\eta}\, dx + \int_\Omega \boldsymbol{\eta} \cdot (\mathbf{v} \cdot \nabla)\mathbf{v}\, dx - \int_\Omega \mathbf{f} \cdot \boldsymbol{\eta}\, dx, \tag{5.80}$$

which in turn implies: $\delta\boldsymbol{\xi} \in V_0$ *and* $\forall\, \boldsymbol{\eta} \in V_0$, *we have*

$$\alpha \int_\Omega \delta\boldsymbol{\xi} \cdot \boldsymbol{\eta}\, dx + \nu \int_\Omega \boldsymbol{\nabla}\delta\boldsymbol{\xi} \cdot \boldsymbol{\nabla}\boldsymbol{\eta}\, dx = \alpha \int_\Omega \delta\mathbf{v} \cdot \boldsymbol{\eta}\, dx + \nu \int_\Omega \boldsymbol{\nabla}\delta\mathbf{v} \cdot \boldsymbol{\nabla}\boldsymbol{\eta}\, dx$$
$$+ \int_\Omega \boldsymbol{\eta} \cdot (\delta\mathbf{v} \cdot \boldsymbol{\nabla})\mathbf{v}\, dx + \int_\Omega \boldsymbol{\eta} \cdot (\mathbf{v} \cdot \boldsymbol{\nabla})\delta\mathbf{v}\, dx. \quad (5.81)$$

Since $\boldsymbol{\xi} \in V_0$, we can take $\boldsymbol{\eta} = \boldsymbol{\xi}$ in (5.81); it then follows, from (5.79), that

$$\delta J = \langle J'(\mathbf{v}), \delta\mathbf{v}\rangle = \alpha \int_\Omega \boldsymbol{\xi} \cdot \delta\mathbf{v}\, dx + \nu \int_\Omega \boldsymbol{\nabla}\boldsymbol{\xi} \cdot \boldsymbol{\nabla}\delta\mathbf{v}\, dx$$
$$+ \int_\Omega \boldsymbol{\xi} \cdot (\delta\mathbf{v} \cdot \boldsymbol{\nabla})\mathbf{v}\, dx + \int_\Omega \boldsymbol{\xi} \cdot (\mathbf{v} \cdot \boldsymbol{\nabla})\delta\mathbf{v}\, dx. \quad (5.82)$$

Finally, from (5.82) it follows that $J'(\mathbf{v})$ can be identified with the linear functional on V_0 defined by

$$\boldsymbol{\eta} \to \alpha \int_\Omega \boldsymbol{\xi} \cdot \boldsymbol{\eta}\, dx + \nu \int_\Omega \boldsymbol{\nabla}\boldsymbol{\xi} \cdot \boldsymbol{\nabla}\boldsymbol{\eta}\, dx + \int_\Omega \boldsymbol{\xi} \cdot (\boldsymbol{\eta} \cdot \boldsymbol{\nabla})\mathbf{v}\, dx + \int_\Omega \boldsymbol{\xi} \cdot (\mathbf{v} \cdot \boldsymbol{\nabla})\boldsymbol{\eta}\, dx. \quad (5.83)$$

From (5.83) it then follows that to compute $\mathbf{z}^{m+1}$ from $\mathbf{u}^{m+1}$, we have to first solve the Stokes problem (5.69), (5.80) with $\mathbf{v} = \mathbf{u}^{m+1}$, which gives $\boldsymbol{\xi}^{m+1}$; then from (5.83) we have $\forall\, \boldsymbol{\eta} \in V_0$

$$\langle J'(\mathbf{u}^{m+1}), \boldsymbol{\eta}\rangle = \alpha \int_\Omega \boldsymbol{\xi}^{m+1} \cdot \boldsymbol{\eta}\, dx + \nu \int_\Omega \boldsymbol{\nabla}\boldsymbol{\xi}^{m+1} \cdot \boldsymbol{\nabla}\boldsymbol{\eta}\, dx$$
$$+ \int_\Omega \boldsymbol{\xi}^{m+1} \cdot (\boldsymbol{\eta} \cdot \boldsymbol{\nabla})\mathbf{u}^{m+1}\, dx + \int_\Omega \boldsymbol{\xi}^{m+1} \cdot (\mathbf{u}^{m+1} \cdot \boldsymbol{\nabla})\boldsymbol{\eta}\, dx. \quad (5.84)$$

Finally we obtain $\mathbf{z}^{m+1}$ from (5.76), (5.84).

To summarize, at each iteration of the conjugate gradient algorithm (5.71)–(5.78), we have to solve several Stokes problems, namely:

(i) the Stokes problem (5.69), (5.80) with $\mathbf{v} = \mathbf{u}^{m+1}$ to obtain $\boldsymbol{\xi}^{m+1}$ from $\mathbf{u}^{m+1}$;
(ii) the Stokes problem (5.76) to obtain (via (5.84)) $\mathbf{z}^{m+1}$ from $\mathbf{u}^{m+1}$, $\boldsymbol{\xi}^{m+1}$;
(iii) the solution of the one-dimensional problem (5.74) requires several calculations of J, i.e., several solutions of the state equation (5.69), (5.80) (which is a Stokes problem).

Thus an efficient "Stokes solver" will be an important tool for solving the Navier–Stokes equations by algorithm (5.71)–(5.78) via the least-squares formulation (5.67)–(5.69). The description of such solvers is given in Sec. 5.7.

5.6. Least-squares conjugate gradient solution of the stationary Navier–Stokes equations: (II) The discrete case

In this section we consider the discrete analogue of Sec. 5.5. We apologize in advance for the ponderous notation to be used (unfortunately this complicated notation seems to be inherent to mixed finite element methods).

5.6.1. *Formulation of the basic problem*

The discrete Navier–Stokes problems encountered in Secs. 5.3.4, 5.4.1, 5.4.2, and 5.4.3 have the general formulation:

Find $\{\mathbf{u}_h, \psi_h\} \in W_{gh}$ *such that*

$$\alpha \int_\Omega \mathbf{u}_h \cdot \mathbf{v}_h \, dx + \nu \int_\Omega \nabla \mathbf{u}_h \cdot \nabla \mathbf{v}_h \, dx + \int_\Omega (\mathbf{u}_h \cdot \nabla)\mathbf{u}_h \cdot (\mathbf{v}_h + \nabla \phi_h) \, dx$$

$$= \int_\Omega \mathbf{f}_{0h} \cdot \mathbf{v}_h \, dx + \int_\Omega \mathbf{f}_{1h} \cdot (\mathbf{v}_h + \nabla \phi_h) \, dx, \qquad \forall \{\mathbf{v}_h, \phi_h\} \in W_{0h}, \tag{5.85}$$

where $\alpha \geq 0$, $\nu > 0$.

5.6.2. *A least-squares formulation of* (5.85)

By analogy with Sec. 5.5.2, we shall consider the following least-squares formulation:

$$\operatorname*{Min}_{\{\mathbf{v}_h, \phi_h\} \in W_{gh}} J_h(\mathbf{v}_h, \phi_h) \tag{5.86}$$

with

$$J_h(\mathbf{v}_h, \phi_h) = \frac{\alpha}{2} \int_\Omega |\boldsymbol{\xi}_h|^2 \, dx + \frac{\nu}{2} \int_\Omega |\nabla \boldsymbol{\xi}_h|^2 \, dx, \tag{5.87}$$

where $\boldsymbol{\xi}_h$ is a function of $\{\mathbf{v}_h, \phi_h\}$ according to the state equation

$$\{\boldsymbol{\xi}_h, \chi_h\} \in W_{0h},$$

$$\alpha \int_\Omega \boldsymbol{\xi}_h \cdot \boldsymbol{\eta}_h \, dx + \nu \int_\Omega \nabla \boldsymbol{\xi}_h \cdot \nabla \boldsymbol{\eta}_h \, dx = \alpha \int_\Omega \mathbf{v}_h \cdot \boldsymbol{\eta}_h \, dx + \nu \int_\Omega \nabla \mathbf{v}_h \cdot \nabla \boldsymbol{\eta}_h \, dx$$

$$+ \int_\Omega (\mathbf{v}_h \cdot \nabla)\mathbf{v}_h \cdot (\boldsymbol{\eta}_h + \nabla \omega_h) \, dx - \int_\Omega \mathbf{f}_{0h} \cdot \boldsymbol{\eta}_h \, dx - \int_\Omega \mathbf{f}_{1h} \cdot (\boldsymbol{\eta}_h + \nabla \omega_h) \, dx,$$

$$\forall \{\boldsymbol{\eta}_h, \omega_h\} \in W_{0h}. \tag{5.88}$$

The least-squares formulation (5.86)–(5.88) is justified by the following:

Proposition 5.2. *Suppose that* $\{\mathbf{u}_h, \psi_h\} \in W_{gh}$ *is a solution of* (5.85); $\{\mathbf{u}_h, \psi_h\}$ *is also a solution of the least-squares problem* (5.86), *and we have, for the corresponding pair* $\{\boldsymbol{\xi}_h, \chi_h\}$,

$$\boldsymbol{\xi}_h = \mathbf{0}, \qquad \chi_h = 0;$$

we also have $J_h(\mathbf{u}_h, \psi_h) = 0$.

5.6.3. *Computation of the gradient* J'_h

From (5.87), (5.88) we have

$$\delta J_h = \alpha \int_\Omega \boldsymbol{\xi}_h \cdot \delta\boldsymbol{\xi}_h \, dx + \nu \int_\Omega \nabla\boldsymbol{\xi}_h \cdot \nabla\delta\boldsymbol{\xi}_h \, dx, \tag{5.89}$$

where $\{\delta\boldsymbol{\xi}_h, \delta\chi_h\}$ is the solution of:

$\{\delta\boldsymbol{\xi}_h, \delta\chi_h\} \in W_{0h}$ *and* $\forall\ \{\boldsymbol{\eta}_h, \omega_h\} \in W_{0h}$, *we have*

$$\alpha \int_\Omega \delta\boldsymbol{\xi}_h \cdot \boldsymbol{\eta}_h \, dx + \nu \int_\Omega \nabla\delta\boldsymbol{\xi}_h \cdot \nabla\boldsymbol{\eta}_h \, dx = \alpha \int_\Omega \delta\mathbf{v}_h \cdot \boldsymbol{\eta}_h \, dx + \nu \int_\Omega \nabla\delta\mathbf{v}_h \cdot \nabla\boldsymbol{\eta}_h \, dx$$
$$+ \int_\Omega (\delta\mathbf{v}_h \cdot \nabla)\mathbf{v}_h \cdot (\boldsymbol{\eta}_h + \nabla\omega_h) \, dx + \int_\Omega (\mathbf{v}_h \cdot \nabla)\delta\mathbf{v}_h \cdot (\boldsymbol{\eta}_h + \nabla\omega_h) \, dx. \tag{5.90}$$

Since $\{\boldsymbol{\xi}_h, \chi_h\} \in W_{0h}$, we have, from (5.90) (taking $\{\boldsymbol{\eta}_h, \omega_h\} = \{\boldsymbol{\xi}_h, \chi_h\}$),

$$\alpha \int_\Omega \boldsymbol{\xi}_h \cdot \delta\boldsymbol{\xi}_h \, dx + \nu \int_\Omega \nabla\boldsymbol{\xi}_h \cdot \nabla\delta\boldsymbol{\xi}_h \, dx = \alpha \int_\Omega \boldsymbol{\xi}_h \cdot \delta\mathbf{v}_h \, dx + \nu \int_\Omega \nabla\boldsymbol{\xi}_h \cdot \nabla\delta\mathbf{v}_h \, dx$$
$$+ \int_\Omega (\delta\mathbf{v}_h \cdot \nabla)\mathbf{v}_h \cdot (\boldsymbol{\xi}_h + \nabla\chi_h) \, dx + \int_\Omega (\mathbf{v}_h \cdot \nabla)\delta\mathbf{v}_h \cdot (\boldsymbol{\xi}_h + \nabla\chi_h) \, dx. \tag{5.91}$$

From (5.91) it then follows that $J'_h(\mathbf{v}_h, \phi_h)$ may be identified with the linear mapping from W_{0h} to $\mathbb{R}$ defined by

$$\langle J'_h(\mathbf{v}_h, \phi_h), \{\boldsymbol{\eta}_h, \omega_h\}\rangle = \alpha \int_\Omega \boldsymbol{\xi}_h \cdot \boldsymbol{\eta}_h \, dx + \nu \int_\Omega \nabla\boldsymbol{\xi}_h \cdot \nabla\boldsymbol{\eta}_h \, dx$$
$$+ \int_\Omega (\boldsymbol{\eta}_h \cdot \nabla)\mathbf{v}_h \cdot (\boldsymbol{\xi}_h + \nabla\chi_h) \, dx + \int_\Omega (\mathbf{v}_h \cdot \nabla)\boldsymbol{\eta}_h \cdot (\boldsymbol{\xi}_h + \nabla\chi_h) \, dx,$$
$$\forall\ \{\boldsymbol{\eta}_h, \omega_h\} \in W_{0h}. \tag{5.92}$$

5.6.4. *Conjugate gradient solution of the least-squares problem* (5.86)–(5.88)

In this section we shall describe a conjugate gradient solution of the least-squares problem (5.86)–(5.88). The algorithm to be considered is in fact a discrete analogue of algorithm (5.71)–(5.78) in Sec. 5.5.3, given by the following.

Step 0: Initialization

$$\{\mathbf{u}_h^0, \psi_h^0\} \in W_{gh} \text{ given:} \tag{5.93}$$

then compute $\{\mathbf{z}_h^0, \theta_h^0\}$ *as the solution of the variational problem*

$$\alpha \int_\Omega \mathbf{z}_h^0 \cdot \boldsymbol{\eta}_h \, dx + \nu \int_\Omega \nabla \mathbf{z}_h^0 \cdot \nabla \boldsymbol{\eta}_h \, dx = \langle J_h'(\mathbf{u}_h^0, \psi_h^0), \{\boldsymbol{\eta}_h, \omega_h\}\rangle,$$

$$\forall \{\boldsymbol{\eta}_h, \omega_h\} \in W_{0h}, \{\mathbf{z}_h^0, \theta_h^0\} \in W_{0h}, \tag{5.94}$$

where J_h' *is given by* (5.92), *and set*

$$\{\mathbf{w}_h^0, \tau_h^0\} = \{\mathbf{z}_h^0, \theta_h^0\}. \tag{5.95}$$

For $m \geq 0$, *assuming* $\{\mathbf{u}_h^m, \psi_h^m\} \in W_{gh}$, $\{\mathbf{z}_h^m, \theta_h^m\}$, $\{\mathbf{w}_h^m, \tau_h^m\} \in W_{0h}$ *known, compute* $\{\mathbf{u}_h^{m+1}, \psi_h^{m+1}\}$, $\{\mathbf{z}_h^{m+1}, \theta_h^{m+1}\}$, $\{\mathbf{w}_h^{m+1}, \tau_h^{m+1}\}$ *by the following.*

Step 1: Descent

$$\lambda^m = \operatorname*{Arg\,min}_{\lambda \in \mathbb{R}} J_h(\mathbf{u}_h^m - \lambda \mathbf{w}_h^m, \psi_h^m - \lambda \tau_h^m), \tag{5.96}$$

$$\mathbf{u}_h^{m+1} = \mathbf{u}_h^m - \lambda^m \mathbf{w}_h^m, \tag{5.97}$$

$$\psi_h^{m+1} = \psi_h^m - \lambda^m \tau_h^m. \tag{5.98}$$

Step 2: Construction of the new descent direction. Define $\{\mathbf{z}_h^{m+1}, \theta_h^{m+1}\}$ *as the solution of the variational problem*

$$\alpha \int_\Omega \mathbf{z}_h^{m+1} \cdot \boldsymbol{\eta}_h \, dx + \nu \int_\Omega \nabla \mathbf{z}_h^{m+1} \cdot \nabla \boldsymbol{\eta}_h \, dx = \langle J_h'(\mathbf{u}_h^{m+1}, \psi_h^{m+1}), \{\boldsymbol{\eta}_h, \omega_h\}\rangle,$$

$$\forall \{\boldsymbol{\eta}_h, \omega_h\} \in W_{0h}, \quad \{\mathbf{z}_h^{m+1}, \theta_h^{m+1}\} \in W_{0h}, \tag{5.99}$$

where $J_h'(u_h^{m+1}, \psi_h^{m+1})$ *is obtained from* (5.92), *and then*

$$\gamma_{m+1} = \frac{\alpha \int_\Omega \mathbf{z}_h^{m+1} \cdot (\mathbf{z}_h^{m+1} - \mathbf{z}_h^m) \, dx + \nu \int_\Omega \nabla \mathbf{z}_h^{m+1} \cdot \nabla(\mathbf{z}_h^{m+1} - \mathbf{z}_h^m) \, dx}{\alpha \int_\Omega |\mathbf{z}_h^m|^2 \, dx + \nu \int_\Omega |\nabla \mathbf{z}_h^m|^2 \, dx}, \tag{5.100}$$

$$\mathbf{w}_h^{m+1} = \mathbf{z}_h^{m+1} + \gamma_{m+1} \mathbf{w}_h^m, \tag{5.101}$$

$$\tau_h^{m+1} = \theta_h^{m+1} + \gamma_{m+1} \tau_h^m, \tag{5.102}$$

$m = m + 1$ *go to* (5.96).

It is clear that each iteration of algorithm (5.93)–(5.102) requires the solution of several discrete Stokes problems:

(i) one to obtain $\{\boldsymbol{\xi}_h^{m+1}, \chi_h^{m+1}\}$ from $\{\mathbf{u}_h^{m+1}, \psi_h^{m+1}\}$; in fact this is (5.88) with $\mathbf{v}_h = \mathbf{u}_h^{m+1}$, $\phi_h = \psi_h^{m+1}$;
(ii) problem (5.99) to obtain $\{\mathbf{z}_h^{m+1}, \theta_h^{m+1}\}$ from $\{\mathbf{u}_h^{m+1}, \psi_h^{m+1}\}$, $\{\boldsymbol{\xi}_h^{m+1}, \chi_h^{m+1}\}$;
(iii) the solution of the one-dimensional problem (5.96) requires several calculations of J_h, i.e., several solutions of the state equation (5.88) (which is a discrete Stokes problem).

The solution of these discrete Stokes problems will be discussed in Sec. 5.7.

5.7. On efficient solvers for the Stokes problem

In this section we shall discuss the solution of the various Stokes problems encountered in the previous sections, by either direct or iterative methods.

5.7.1. *The continuous case*

We follow B.G. 4P. [1].

5.7.1.1. *Synopsis.* From Secs. 5.4 and 5.5 it follows that efficient solvers for the Stokes problem

$$\alpha \mathbf{u} - \nu \Delta \mathbf{u} + \nabla p = \mathbf{f} \text{ in } \Omega, \qquad \nabla \cdot \mathbf{u} = 0 \text{ in } \Omega, \qquad \mathbf{u} = \mathbf{g} \text{ on } \Gamma \quad (5.103)$$

will be advantageous for the numerical solution of the steady and unsteady Navier–Stokes equations. In (5.103), $\alpha = 0$ for the steady case and $\alpha > 0$ for the unsteady case. We shall describe a method for solving (5.103). This method reduces the solution of the Stokes problem (5.103) to the solution of a finite number of Dirichlet problems and to the solution of a boundary integral equation. The finite element implementation of this method will be discussed in Sec. 5.7.2 (see also Glowinski and Pironneau [3], Glowinski, Periaux, and Pironneau [1], and B.G. 4P. [1, Sec. 8]). The reader mainly interested in applications (or frightened off by too much functional analysis!) can skip the rest of Sec. 5.7.1 and move on to Sec. 5.7.2.

5.7.1.2. *Principle of the method.* Let us define

$$\mathscr{H}^{1/2}(\Gamma) = \left\{\mu \in H^{1/2}(\Gamma), \int_\Gamma \mu \, d\Gamma = 0\right\}.$$

For the definition and properties of the Sobolev spaces $H^s(\Gamma)$, $s \in \mathbb{R}$, see Lions and Magenes [1] and Necas [1].

The decomposition properties of the Stokes problem follow directly from:

Theorem 5.6. *Let $\lambda \in H^{-1/2}(\Gamma)$ and let the operator $A: H^{-1/2}(\Gamma) \to H^{1/2}(\Gamma)$ be defined implicitly by the following cascade of Dirichlet problems:*

$$\Delta p_\lambda = 0 \text{ in } \Omega, \qquad p_\lambda = \lambda \text{ on } \Gamma, \qquad (5.104)$$

$$\alpha \mathbf{u}_\lambda - \nu \Delta \mathbf{u}_\lambda = -\nabla p_\lambda \text{ in } \Omega, \qquad \mathbf{u}_\lambda = \mathbf{0} \text{ on } \Gamma, \qquad (5.105)$$

$$-\Delta \psi_\lambda = \nabla \cdot \mathbf{u}_\lambda \text{ in } \Omega, \qquad \psi_\lambda = 0 \text{ on } \Gamma, \qquad (5.106)$$

and then

$$A\lambda = -\left.\frac{\partial \psi_\lambda}{\partial n}\right|_\Gamma. \qquad (5.107)$$

Then A is an isomorphism from $H^{-1/2}(\Gamma)/\mathbb{R}$ onto $\mathscr{H}^{1/2}(\Gamma)$. Moreover, the bilinear form $a(\cdot, \cdot)$ defined by

$$a(\lambda, \mu) = \langle A\lambda, \mu \rangle, \qquad \forall\, \mu \in H^{-1/2}(\Gamma) \qquad (5.108)$$

(where $\langle \cdot, \cdot \rangle$ *denotes the duality pairing between* $H^{1/2}(\Gamma)$ *and* $H^{-1/2}(\Gamma)$*) is continuous, symmetric, and* $H^{-1/2}(\Gamma)/\mathbb{R}$*-elliptic.*

We recall that a bilinear form defined over $V \times V$, where V is a Hilbert space, is V-elliptic if there exists $\beta > 0$ such that $a(v, v) \geq \beta \|v\|^2, \forall\, v \in V$. For the proof of Theorem 5.6, see Glowinski, Periaux, and Pironneau [1] (a variant of it concerning the biharmonic problem is avaiable in Glowinski and Pironneau [1]).

EXERCISE 5.4. Prove Theorem 5.6.

Application of Theorem 5.6 *to the solution of the Stokes problem* (5.103). Let us define $p_0, \mathbf{u}_0, \psi_0$ as the solutions of, respectively,

$$\Delta p_0 = \nabla \cdot \mathbf{f} \text{ in } \Omega, \qquad p_0 = 0 \text{ on } \Gamma, \tag{5.109}$$

$$\alpha \mathbf{u}_0 - \nu \Delta \mathbf{u}_0 = \mathbf{f} - \nabla p_0 \text{ in } \Omega, \qquad \mathbf{u}_0 = \mathbf{g} \text{ on } \Gamma, \tag{5.110}$$

$$-\Delta \psi_0 = \nabla \cdot \mathbf{u}_0 \text{ in } \Omega, \qquad \psi_0 = 0 \text{ on } \Gamma. \tag{5.111}$$

The following result is established in Glowinski, Periaux, and Pironneau [1]:

Theorem 5.7. *Let* $\{\mathbf{u}, p\}$ *be a solution of the Stokes problem* (5.103). *The trace* $\lambda = p|_\Gamma$ *is the unique solution of the linear variational equation*

$$\text{(E)} \qquad \lambda \in H^{-1/2}(\Gamma)/\mathbb{R}, \quad \langle A\lambda, \mu \rangle = \left\langle \frac{\partial \psi_0}{\partial n}, \mu \right\rangle, \quad \forall\, \mu \in H^{-1/2}(\Gamma)/\mathbb{R}. \tag{5.112}$$

EXERCISE 5.5. Prove Theorem 5.7.

From the above theorem it follows that the solution of the Stokes problem (5.103) has been reduced to $2N + 3$ Dirichlet problems[21] ($N + 2$ to obtain ψ_0, $N + 1$ to obtain $\{\mathbf{u}, p\}$ once λ is known) and to the solution of (E) which is a kind of boundary integral equation. The main difficulty in this approach is the fact that the (pseudo-differential) operator A is not known explicitely; actually the mixed finite element approximation of Sec. 5.3 has been precisely introduced (in Glowinski and Pironneau [4], [5]) to overcome this difficulty, extending to the Stokes problem an idea discussed in Glowinski and Pironneau [1] for the biharmonic problem.

5.7.2. *The discrete case*

5.7.2.1. *Introduction: formulation of the basic problem.* From Secs. 5.4 and 5.6 it follows that efficient solvers for the discrete Stokes problem:

---

[21] $N = 2$ if $\Omega \subset \mathbb{R}^2$; $N = 3$ if $\Omega \subset \mathbb{R}^3$.

Find $\{\mathbf{u}_h, \psi_h\} \in W_{gh}$ *such that*, $\forall\ \{\mathbf{v}_h, \phi_h\} \in W_{0h}$, *we have*

$$\alpha \int_\Omega \mathbf{u}_h \cdot \mathbf{v}_h\, dx + \nu \int_\Omega \nabla \mathbf{u}_h \cdot \nabla \mathbf{v}_h\, dx = \int_\Omega \mathbf{f}_{0h} \cdot \mathbf{v}_h\, dx + \int_\Omega \mathbf{f}_{1h} \cdot (\mathbf{v}_h + \nabla \phi_h)\, dx \tag{5.113}$$

will be very useful tools for solving the discrete steady and unsteady Navier–Stokes problems by the methods discussed in the previous sections. The content of this section is in fact the discrete analogue of Sec. 5.7.1 and very closely follows Glowinski and Pironneau [3], [5], Glowinski, Periaux, and Pironneau [1], B.G. 4P. [1, Sec. 8.6.2], Bristeau, Glowinski, Mantel, Periaux, Perrier, and Pironneau [1, Sec. 5.2], and Periaux [2].

5.7.2.2. *The space* M_h. Using the notation of Sec. 5.3, let us introduce M_h as a *complementary subspace* of H_{0h}^1 in H_h^1, i.e.,

$$H_h^1 = H_{0h}^1 \oplus M_h;$$

we set $N_h = \dim(M_h)$.

Various M_h may be used, but in the following we shall only consider M_h defined by

$$H_h^1 = H_{0h}^1 \oplus M_h;\ \textit{then},\ \forall\ \mu_h \in M_h,\ \textit{we have}\ \mu_{h|T} = 0,\ \forall\ T \in \mathscr{T}_h \ \textit{such that}\ \partial T \cap \partial\Omega = \varnothing. \tag{5.114}$$

The space M_h is uniquely defined by (5.114), and we observe that if $\mu_h \in M_h$, then μ_h vanishes outside a neighborhood of $\partial\Omega$ whose measure $\to 0$ with h. Therefore M_h can "almost" be considered to be a boundary space, and in fact it will play the role played by $H^{-1/2}(\Gamma)$ in Sec. 5.7.1. We also observe that $\mu_h \in M_h$ is completely determined by its trace on $\partial\Omega$, and in fact by the values it takes at the boundary nodes.

5.7.2.3. *Equivalence of* (5.113) *with a variational problem in* M_h. Let $\lambda_h \in M_h$; from λ_h we define $p_h, \mathbf{u}_h, \psi_h$ as the solutions of the cascade of discrete variational Dirichlet problems:

Find $p_h \in H_h^1$ *such that* $p_h - \lambda_h \in H_{0h}^1$ *and*

$$\int_\Omega \nabla p_h \cdot \nabla q_h\, dx = 0, \qquad \forall\ q_h \in H_{0h}^1. \tag{5.115}$$

Find $\mathbf{u}_h \in V_{0h}$ *such that*

$$\alpha \int_\Omega \mathbf{u}_h \cdot \mathbf{v}_h\, dx + \nu \int_\Omega \nabla \mathbf{u}_h \cdot \nabla \mathbf{v}_h\, dx = -\int_\Omega \nabla p_h \cdot \mathbf{v}_h\, dx, \qquad \forall\ \mathbf{v}_h \in V_{0h}. \tag{5.116}$$

Find $\psi_h \in H_{0h}^1$ such that

$$\int_\Omega \nabla \psi_h \cdot \nabla \phi_h\, dx = \int_\Omega \nabla \cdot \mathbf{u}_h \phi_h\, dx, \qquad \forall\ \phi_h \in H_{0h}^1. \tag{5.117}$$

Then from (5.115)–(5.117) we define $a_h: M_h \times M_h \to \mathbb{R}$ by

$$a_h(\lambda_h, \mu_h) = -\int_\Omega (\nabla\psi_h + \mathbf{u}_h)\cdot\nabla\mu_h\,dx. \tag{5.118}$$

The following analogue of Theorem 5.6 of Sec. 5.7.1.2 is proved in Glowinski, Periaux, and Pironneau [1]:

Theorem 5.8. *Assume that*

$$\forall\, T \in \mathscr{T}_h,\ T \text{ has at most one edge supported by } \partial\Omega; \tag{5.119}$$

then $a_h(\cdot,\cdot)$ is bilinear, symmetric, and positive definite over $(M_h/\mathbb{R}_h) \times (M_h/\mathbb{R}_h)$, where

$$\mathbb{R}_h = \{\mu_h \in M_h,\ \mu_h = \text{constant on } \partial\Omega\}.$$

EXERCISE 5.6. Prove Theorem 5.8.

Referring back to Sec. 5.7.2.1, it is easily proved (see Glowinski, Periaux, and Pironneau [1]) that (5.113) has a unique solution which is also the unique solution of the minimization problem

$$\operatorname*{Min}_{\{\mathbf{v}_h,\phi_h\}\in W_{gh}} \left\{\frac{\alpha}{2}\int_\Omega |\mathbf{v}_h|^2\,dx + \frac{\nu}{2}\int_\Omega |\nabla\mathbf{v}_h|^2\,dx - \int_\Omega \mathbf{f}_{0h}\cdot\mathbf{v}_h\,dx - \int_\Omega \mathbf{f}_{1h}\cdot(\mathbf{v}_h + \nabla\phi_h)\,dx\right\}. \tag{5.120}$$

Since

$$W_{gh} = \left\{\{\mathbf{v}_h,\phi_h\} \in V_{gh} \times H^1_{0h},\ \int_\Omega \nabla\phi_h\cdot\nabla q_h\,dx = \int_\Omega \nabla\cdot\mathbf{v}_h q_h\,dx\ \forall\, q_h \in H^1_h\right\},$$

(5.120) is a linearly constrained minimization problem (the number of linearly independent constraints is $\dim(H^1_h)$). We can therefore associate with (5.120) a Lagrangian functional (see, e.g., G.L.T. [1, Chapter 2], [3, Chapter 2])

$$\mathscr{L}_h: V_h \times H^1_h \times H^1_h \to \mathbb{R}$$

defined by

$$\mathscr{L}_h(\mathbf{v}_h,\phi_h,q_h) = j_h(\mathbf{v}_h,\phi_h) + \int_\Omega \nabla\phi_h\cdot\nabla q_h\,dx - \int_\Omega \nabla\cdot\mathbf{v}_h q_h\,dx, \tag{5.121}$$

where

$$j_h(\mathbf{v}_h,\phi_h) = \frac{\alpha}{2}\int_\Omega |\mathbf{v}_h|^2\,dx + \frac{\nu}{2}\int_\Omega |\nabla\mathbf{v}_h|^2\,dx - \int_\Omega \mathbf{f}_{0h}\cdot\mathbf{v}_h\,dx - \int_\Omega \mathbf{f}_{1h}\cdot(\mathbf{v}_h + \nabla\phi_h)\,dx. \tag{5.122}$$

Since (5.120) is a finite-dimensional linearly constrained minimization problem whose solution exists, there is a Lagrange multiplier related to (5.120)–(5.122) (cf., e.g., Rockafellar [1] and also Chapter I, Sec. 7.4.2), i.e., there exists $p_h \in H_h^1$ such that $\{\mathbf{u}_h, \psi_h, p_h\}$ is a saddle point of $\mathscr{L}_h$ over $V_{gh} \times H_{0h}^1 \times H_h^1$ (where $\{\mathbf{u}_h, \psi_h\}$ is the solution of (5.113), (5.120)).

Since $\mathscr{L}_h$ is extremal at $\{\mathbf{u}_h, \psi_h, p_h\}$, we obtain

$$\int_\Omega \nabla p_h \cdot \nabla \phi_h \, dx = \int_\Omega \mathbf{f}_{1h} \cdot \nabla \phi_h \, dx, \qquad \forall\, \phi_h \in H_{0h}^1, \quad p_h \in H_h^1, \tag{5.123}$$

$$\alpha \int_\Omega \mathbf{u}_h \cdot \mathbf{v}_h \, dx + \nu \int_\Omega \nabla \mathbf{u}_h \cdot \nabla \mathbf{v}_h \, dx + \int_\Omega \nabla p_h \cdot \mathbf{v}_h \, dx$$
$$= \int_\Omega (\mathbf{f}_{0h} + \mathbf{f}_{1h}) \cdot \mathbf{v}_h \, dx, \ \forall\, \mathbf{v}_h \in V_{0h}, \quad \mathbf{u}_h \in V_{gh}, \tag{5.124}$$

$$\int_\Omega \nabla \psi_h \cdot \nabla q_h \, dx = \int_\Omega \nabla \cdot \mathbf{u}_h q_h \, dx, \qquad \forall\, q_h \in H_h^1. \tag{5.125}$$

Relations (5.123)–(5.125) characterize $\{\mathbf{u}_h, \psi_h\}$ as the solution of (5.113), (5.120). An important conclusion is that the multiplier p_h appears from (5.124) as the discrete pressure. In fact it is proved in Glowinski and Pironneau [2] that if (5.119) holds, then p_h is uniquely defined in $H_h^1/\mathbb{R}$.

The key result concerning Sec. 5.7.2 is the following theorem which links the data $\mathbf{g}_h\ (=\mathbf{u}_{h|\Gamma})$, $\mathbf{f}_{0h}$, $\mathbf{f}_{1h}$ and the trace of the discrete pressure p_h:

Theorem 5.9. *Let p_h be the above discrete pressure and λ_h the component of p_h in M_h. Then if* (5.119) *holds, λ_h is the unique solution in $M_h/\mathbb{R}_h$ of the linear variational problem* (*discrete analogue of* (E) *in Sec.* 5.7.1.2):

$$(\mathrm{E}_h)\ \lambda_h \in M_h/\mathbb{R}_h,\ a_h(\lambda_h, \mu_h) = \int_\Omega (\nabla \psi_{0h} + \mathbf{u}_{0h}) \cdot \nabla \mu_h \, dx, \qquad \forall\, \mu_h \in M_h/\mathbb{R}_h, \tag{5.126}$$

where p_{0h}, $\mathbf{u}_{0h}$, ψ_{0h} are, respectively, solutions of

$$\int_\Omega \nabla p_{0h} \cdot \nabla q_h \, dx = \int_\Omega \mathbf{f}_{1h} \cdot \nabla q_h \, dx, \qquad \forall\, q_h \in H_{0h}^1, \quad p_{0h} \in H_{0h}^1, \tag{5.127}$$

$$\alpha \int_\Omega \mathbf{u}_{0h} \cdot \mathbf{v}_h \, dx + \nu \int_\Omega \nabla \mathbf{u}_{0h} \cdot \nabla \mathbf{v}_h \, dx$$
$$= \int_\Omega (\mathbf{f}_{0h} + \mathbf{f}_{1h} - \nabla p_{0h}) \cdot \mathbf{v}_h \, dx, \ \forall\, \mathbf{v}_h \in V_{0h}, \quad \mathbf{u}_{0h} \in V_{gh}, \tag{5.128}$$

$$\int_\Omega \nabla \psi_{0h} \cdot \nabla \phi_h \, dx = \int_\Omega \nabla \cdot \mathbf{u}_{0h} \phi_h \, dx, \qquad \forall\, \phi_h \in H_{0h}^1, \quad \psi_{0h} \in H_{0h}^1. \tag{5.129}$$

Theorem 5.9 (whose proof is given in Glowinski, Periaux, and Pironneau [1]) is in fact a variant of Theorem 3.2 in Glowinski and Pironneau [1] concerning approximate solution of the biharmonic problem.

The solution of (E_h) by direct and iterative methods is discussed in Secs. 5.7.2.4 and 5.7.2.5, respectively.

5.7.2.4. *Solution of* (E_h) *by a direct method.* This section follows B.G. 4P. [1], Glowinski and Pironneau [3] ,and Glowinski, Periaux, and Pironneau [1].

A. *Construction of a linear system equivalent to* (E_h)

General: As before, M_h is defined by (5.114); let

$$B_h = \{w_i\}_{i=1}^{N_h}$$

be a basis of M_h. Then, $\forall\ \mu_h \in M_h$,

$$\mu_h = \sum_{i=1}^{N_h} \mu_i w_i, \tag{5.130}$$

and from now on we shall write

$$\mathbf{r}_h \mu_h = \{\mu_1, \ldots, \mu_{N_h}\} \in \mathbb{R}^{N_h}. \tag{5.131}$$

In practice B_h is defined by

$$B_h = \{w_i\}_{i=1}^{N_h} \tag{5.132}$$

and (see Fig. 5.2)

$$\forall\ i = 1, \ldots, N_h, \quad w_i \in M_h, \quad w_i(P_i) = 1, \quad w_i(Q) = 0 \qquad \forall\ Q \text{ vertex of } \mathscr{T}_h, \quad Q \neq P_i, \tag{5.133}$$

where we have implicitly assumed (though in practice this is not strictly necessary) that the boundary nodes are numbered first. With this choice for B_h, we have $\mu_i = \mu_h(P_i)$ in (5.130), (5.131). Then (E_h) is equivalent to the linear system

$$\sum_{j=1}^{N_h} a_h(w_j, w_i)\lambda_j = \int_\Omega (\nabla\psi_{0h} + \mathbf{u}_{0h}) \cdot \nabla w_i\, dx, \qquad 1 \le i \le N_h. \tag{5.134}$$

Let $a_{ij} = a_h(w_j, w_i)$, $\mathbf{A}_h = (a_{ij})_{1 \le i, j \le N_h}$, $b_i = \int_\Omega (\nabla\psi_{0h} + \mathbf{u}_{0h}) \cdot \nabla w_i\, dx$, $\mathbf{b}_h = \{b_i\}_{i=1}^{N_h}$. The matrix $\mathbf{A}_h$ is full and symmetric, positive semidefinite. If (5.119) holds, then 0 is a single eigenvalue of $\mathbf{A}_h$; furthermore, if B_h is defined by (5.132), (5.133), then

$$\operatorname{Ker}(\mathbf{A}_h) = \{\mathbf{y} \,|\, \mathbf{y} \in \mathbb{R}^{N_h}, y_1 = y_2 = \cdots = y_{N_h}\}. \tag{5.135}$$

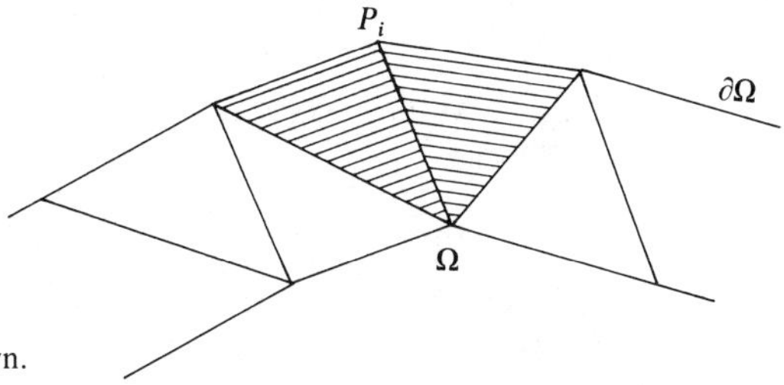

Figure 5.2. The support of w_i is shown.

As to the conditioning of $\mathbf{A}_h$ restricted to $R(\mathbf{A}_h)$ ($=\mathbb{R}^{N_h} - \mathrm{Ker}(\mathbf{A}_h)$), it can be shown that the *condition number* $\nu(\mathbf{A}_h)$ is of order h^{-2} if (5.119) holds. In fact, by analogy with Glowinski and Pironneau [1, Sec. 4] it is reasonable to conjecture that $\nu(\mathbf{A}_h) = O(h^{-1})$, but we have not been able to obtain this estimate.

Construction of $\mathbf{b}_h$. To compute the right-hand side of (5.134), it is necessary to solve the four (five if $\Omega \subset \mathbb{R}^3$) approximate Dirichlet problems (5.127)–(5.129). Due to the choice (5.114) for M_h, the integrals in the definition of the right-hand side of (5.134) involve functions whose supports are in the neighborhood of $\partial\Omega$ only (see Fig. 5.2).

Construction of $\mathbf{A}_h$. $\mathbf{A}_h$ is constructed column by column according to the relation $a_{ij} = a_h(w_j, w_i)$. To compute the jth column of $\mathbf{A}_h$, we solve (5.115)–(5.117) with $\lambda_h = w_j$ and compute a_{ij} from (5.118). Thus four discrete Dirichlet problems must be solved for each column (five in $\mathbb{R}^3$). Since the matrix $\mathbf{A}_h$ is symmetric, we may restrict i to be greater or equal to j. Incidentally, the above observation regarding the choice of M_h still holds, so that the integrals occurring in the a_{ij} are not very costly to compute.

B. *Solution of* (E_h) *by the Cholesky method*. Assume that (5.119), (5.133) hold. Then it may be shown from (5.135) that the submatrix $\overline{\mathbf{A}}_h = (a_{ij})_{1 \le i, j \le N_h - 1}$ is symmetric and positive definite. Therefore we may proceed as follows:

Take $\lambda_{N_h} = 0$ and solve

$$\overline{\mathbf{A}}_h \bar{\mathbf{r}}_h \lambda_h = \bar{\mathbf{b}}_h \tag{5.136}$$

(where $\bar{\mathbf{r}}_h \lambda_h = \{\lambda_1, \ldots, \lambda_{N_h - 1}\}$, $\bar{\mathbf{b}}_h = \{b_1, \ldots, b_{N_h - 1}\}$) by the Cholesky method via a factorization

$$\overline{\mathbf{A}}_h = \overline{\mathbf{L}}_h \overline{\mathbf{L}}_h^t (\text{or } \overline{\mathbf{A}}_h = \overline{\mathbf{L}}_h \overline{\mathbf{D}}_h \overline{\mathbf{L}}_h^t), \tag{5.137}$$

where $\overline{\mathbf{L}}_h$ is lower triangular and nonsingular (and $\overline{\mathbf{D}}_h$ is diagonal).

Let us review the subproblems which arise in the computation of $\{\mathbf{u}_h, p_h\}$ via (E_h) if the Cholesky method is used:

(i) the four (five if $\Omega \subset \mathbb{R}^3$) approximate Dirichlet problems (5.127)–(5.129):

(ii) $4(N_h - 1)$ approximate Dirichlet problems to construct $\overline{\mathbf{A}}_h$ ($5(N_h - 1)$ if $\Omega \subset \mathbb{R}^3$);

(iii) two triangular systems to compute λ_h, namely:

$$\overline{\mathbf{L}}_h \bar{\mathbf{y}}_h = \bar{\mathbf{b}}_h, \qquad \overline{\mathbf{L}}_h^t \bar{\mathbf{r}}_h \lambda_h = \bar{\mathbf{y}}_h,$$

(iv) three approximate Dirichlet problems to obtain p_h and $\mathbf{u}_h$ from λ_h (four if $\Omega \subset \mathbb{R}^3$).

In practice the matrices of the approximate Dirichlet problems should be factorized once and for all. There are two symmetric positive-definite matrices, one to approximate $-\Delta$ using *affine* elements, one to approximate $\alpha I - \nu\Delta$

using *quadratic* elements (or affine on $\tilde{\mathscr{T}}_h$ if (5.21) is used); an alternative is to use very efficient iterative solvers to solve these approximate Dirichlet problems, like those described in Glowinski, Periaux, and Pironneau [2], Glowinski, Mantel, Periaux, Pironneau, and Poirier [1], and Glowinski, Mantel, Periaux, Perrier, and Pironneau [1], for example.

Remark 5.9. We complete Remark 5.3 of Sec. 5.4.4.3. If one takes $\theta = \frac{1}{2}$ (resp., $\theta = \frac{1}{3}$) in the alternating-direction method (5.53)–(5.55) (resp., (5.56)–(5.59)), it follows from the decomposition properties of the Stokes problem discussed in Secs. 5.7.1 and 5.7.2 that the Dirichlet operators[22] $2I/k - (\nu/2)\Delta$ (resp., $2[2I/k - (\nu/3)\Delta]$) are involved in the Stokes step(s) (5.54) (resp., (5.57), (5.59)).

Since the nonlinear step (5.55) (resp., (5.58)) involves the operator[22] $2I/k - (\nu/2)\Delta$(resp., $2I/k - (\nu/3)\Delta$) (if, for example, a least-squares conjugate gradient method[23] is used), we see that *with the above values of* θ, *the same solvers can be used for the Stokes steps and the nonlinear steps of methods* (5.53)–(5.55) and (5.56)–(5.59); with other values of θ, this would no longer be true and about twice as much computer storage would be required.

5.7.2.5 *Solution of* (E_h) *by a conjugate gradient method*: *further remarks.* We follow Glowinski and Pironneau [3], [5] and Glowinski, Periaux, and Pironneau [1]. In Sec. 5.7.2.4 we have described a direct method for solving (E_h) (in fact the equivalent linear system (5.134)); actually a possible alternative is to solve (E_h) (and therefore (5.113)) by a conjugate gradient algorithm; this possibility follows from the fact that the bilinear form $a_h(\cdot, \cdot)$ in (E_h) (cf. (5.126)) is symmetric and positive semidefinite (from Theorem 5.8) (or equivalently, the matrix $\mathbf{A}_h$ in the linear system (5.134) is symmetric and positive semidefinite).

Using the notation of Sec. 5.7.2.4, a possible conjugate gradient algorithm is defined by the following.

Step 0: *Initialization*

$$\lambda_h^0 \in M_h \text{ arbitrarily given}; \tag{5.138}$$

$$\mathbf{g}_h^0 = \mathbf{A}_h \mathbf{r}_h \lambda_h^0 - \mathbf{b}_h, \tag{5.139}$$

$$\mathbf{z}_h^0 = \mathbf{g}_h^0. \tag{5.140}$$

For $n \geq 0$, *assuming that* λ_h^n, $\mathbf{g}_h^n$, $\mathbf{z}_h^n$ *are known, we compute* λ_h^{n+1}, $\mathbf{g}_h^{n+1}$, $\mathbf{z}_h^{n+1}$ *by the following.*

[22] In fact their discrete analogous.

[23] Like those described in Secs. 3.3, 3.4.

Step 1: Descent

$$\rho_n = \frac{(\mathbf{z}_h^n, \mathbf{g}_h^n)_h}{(\mathbf{A}_h \mathbf{z}_h^n, \mathbf{z}_h^n)_h} \quad \left(= \frac{\|\mathbf{g}_h^n\|_h^2}{(\mathbf{A}_h \mathbf{z}_h^n, \mathbf{z}_h^n)_h}\right), \tag{5.141}$$

$$\mathbf{r}_h \lambda_h^{n+1} = \mathbf{r}_h \lambda_h^n - \rho_n \mathbf{z}_h^n, \tag{5.142}$$

$$\mathbf{g}_h^{n+1} = \mathbf{g}_h^n - \rho_n \mathbf{A}_h \mathbf{z}_h^n. \tag{5.143}$$

Step 2: Construction of the new descent direction

$$\gamma_{n+1} = \frac{\|\mathbf{g}_h^{n+1}\|_h^2}{\|\mathbf{g}_h^n\|_h^2} \quad \left(= \frac{(\mathbf{g}_h^{n+1}, \mathbf{g}_h^{n+1} - \mathbf{g}_h^n)_h}{\|\mathbf{g}_h^n\|_h^2}\right), \tag{5.144}$$

$$\mathbf{z}_h^{n+1} = \mathbf{g}_h^{n+1} + \gamma_{n+1} \mathbf{z}_h^n, \tag{5.145}$$

$n = n + 1$ *go to* (5.141).

In (5.138)–(5.145), $(\cdot, \cdot)_h$ and $\|\cdot\|_h$ stand for the standard Euclidian scalar product of $\mathbb{R}^{N_h}$ and the corresponding norm, respectively (but one could use other scalar products and norms).

With the matrix $\mathbf{A}_h$ symmetric and positive semidefinite, one can show that the sequence $\{\lambda_h^n\}_{n \geq 0}$ converges[24] to λ_h, solution of (E_h); the component of λ_h in $\mathbb{R}_h$ is that of λ_h^0. Implementing (5.138)–(5.145) requires the solution of *four* Dirichlet problems at *each* iteration (*five* if $\Omega \subset \mathbb{R}^3$) to compute $\mathbf{A}_h \mathbf{z}_h^n$ using the relation

$$(\mathbf{A}_h \mathbf{r}_h \mu_{1h}, \mathbf{r}_h \mu_{2h})_h = a_h(\mu_{1h}, \mu_{2h}), \qquad \forall\, \mu_{1h}, \mu_{2h} \in M_h$$

and also the definition of $a_h(\cdot, \cdot)$ (given by (5.115)–(5.118)); let us show the details the computation of $\mathbf{A}_h \mathbf{z}_h^n$.

With $\mathbf{z}_h^n = \{z_i^n\}_{i=1}^{N_h}$ known (from (5.140) or (5.145)), we first define $\tilde{z}_h^n \in M_h$ by

$$\tilde{z}_h^n = \mathbf{r}_h^{-1} \mathbf{z}_h^n \quad \left(= \sum_{i=1}^{N_h} z_i^n w_i\right), \tag{5.146}$$

and then from $\tilde{z}_h^n$ we compute $\pi_h^n, \boldsymbol{\omega}_h^n, \chi_h^n$ by

$$\pi_h^n \in H_h^1, \quad \pi_h^n - \tilde{z}_h^n \in H_{0h}^1, \quad \int_\Omega \nabla \pi_h^n \cdot \nabla q_h \, dx = 0, \qquad \forall\, q_h \in H_{0h}^1, \tag{5.147}$$

$$\boldsymbol{\omega}_h^n \in V_{0h}, \quad \alpha \int_\Omega \boldsymbol{\omega}_h^n \cdot \mathbf{v}_h \, dx + \nu \int_\Omega \nabla \boldsymbol{\omega}_h^n \cdot \nabla \mathbf{v}_h \, dx = -\int_\Omega \nabla \pi_h^n \cdot \mathbf{v}_h \, dx \quad \forall\, \mathbf{v}_h \in V_{0h}, \tag{5.148}$$

$$\chi_h^n \in H_{0h}^1, \qquad \int_\Omega \nabla \chi_h^n \cdot \nabla \phi_h \, dx = \int_\Omega \nabla \cdot \boldsymbol{\omega}_h^n \phi_h \, dx, \qquad \forall\, \phi_h \in H_{0h}^1. \tag{5.149}$$

---

[24] In at most $N_h - 1$ iterations (assuming that roundoff errors are not taken into account).

Then from (5.118) we have

$$(\mathbf{A}_h \mathbf{z}_h^n, \mathbf{r}_h \mu_h)_h = a_h(\tilde{z}_h^n, \mu_h) = -\int_\Omega (\nabla \chi_h^n + \boldsymbol{\omega}_h^n) \cdot \nabla \mu_h \, dx \quad \left(= \sum_{i=1}^{N_h} (\mathbf{A}_h \mathbf{z}_h^n)_i \mu_i\right), \tag{5.150}$$

which implies

$$(\mathbf{A}_h \mathbf{z}_h^n, \mathbf{z}_h^n)_h = a_h(\tilde{z}_h^n, \tilde{z}_h^n) = -\int_\Omega (\nabla \chi_h^n + \boldsymbol{\omega}_h^n) \cdot \nabla \tilde{z}_h^n \, dx. \tag{5.151}$$

Here, also, one should factorize the matrices of the approximate Dirichlet problems.

Remark 5.10. The conjugate gradient algorithm (5.138)–(5.145) used to solve the linear problem (E_h) is the classical one (see Hestenes [2] for a detailed analysis of conjugate gradient methods applied to the solution of linear problems); the scalar ρ_n occurring in (5.141) is the solution of the single-variable minimization problem

$$\rho_n \in \mathbb{R}, \qquad I_h(\mathbf{r}_h \lambda_h^n - \rho_n \mathbf{z}_h^n) \le I_h(\mathbf{r}_h \lambda_h^n - \rho \mathbf{z}_h^n), \qquad \forall\, \rho \in \mathbb{R},$$

where $\forall\ \mathbf{y}_h \in \mathbb{R}^{N_h}$,

$$I_h(\mathbf{y}_h) = \tfrac{1}{2}(\mathbf{A}_h \mathbf{y}_h, \mathbf{y}_h)_h - (\mathbf{b}_h, \mathbf{y}_h)_h.$$

We recall that

$$\begin{aligned}
(\mathbf{A}_h \mathbf{z}_h^i, \mathbf{z}_h^j)_h &= 0, \qquad \forall\, i, j, \quad i \ne j,\\
(\mathbf{g}_h^i, \mathbf{g}_h^j)_h &= 0, \qquad \forall\, i, j, \quad i \ne j,\\
(\mathbf{g}_h^i, \mathbf{z}_h^j)_h &= 0, \qquad \forall\, i, j, \quad i > j;
\end{aligned} \tag{5.152}$$

the various formulas in (5.141), (5.144) then follow from (5.152).

Remark 5.11. Our numerical experiments show that the conjugate gradient algorithm (5.138)–(5.145) converges quickly, even on domains with corners. Typically, convergence is obtained in $O(\sqrt{N_h})$ iterations.

Remark 5.12. The direct and iterative methods described in Secs. 5.7.2.4 and 5.7.2.5, respectively, have been applied to the solution of two- and three-dimensional problems. We recommend the direct method if the Stokes problem has to be solved many times on a given domain. On the other hand, if the Stokes problem is to be solved a few times only or if N_h (the number of boundary nodes) is very large and requires a complicated *out of core* procedure, we recommend the conjugate gradient method (5.138)–(5.145).

Remark 5.13. The direct method of Sec. 5.7.2.4 for solving the Stokes problem (5.103) appears as a *coupling* between *ordinary finite element methods* and those *panel methods* (discussed in, e.g., Hunt [1]) for solving *boundary integral equations*.

5.8. Another mixed finite element method for the Stokes and Navier–Stokes problems: applications

5.8.1. *Introduction*

The finite element method introduced in Sec. 5.3 seems unnatural at first glance (despite Remark 5.1), but it has very interesting decomposition properties resulting in efficient direct or iterative Stokes solvers. Actually the above finite element approximation is a variant of a more classical one (advocated by Hood and Taylor [1]) which will be discussed below; we shall also discuss several methods for solving the corresponding approximate Stokes problems.

5.8.2. *Formulation of the basic problem: principle of the method*

Let us consider the following Navier–Stokes problem (with $\alpha \geq 0$):

$$\begin{aligned} &\alpha\mathbf{u} - \nu\Delta\mathbf{u} + (\mathbf{u}\cdot\nabla)\mathbf{u} + \nabla p = \mathbf{f} \text{ in } \Omega, \\ &\nabla\cdot\mathbf{u} = 0 \text{ in } \Omega, \\ &\mathbf{u} = \mathbf{g} \text{ on } \Gamma \left(\text{with} \int_\Gamma \mathbf{g}\cdot\mathbf{n}\, d\Gamma = 0\right). \end{aligned} \tag{5.153}$$

The incompressibility condition $\nabla\cdot\mathbf{u} = 0$ is equivalent to

$$\int_\Omega \nabla\cdot\mathbf{u}q\, dx = 0, \qquad \forall\, q \in H^1(\Omega). \tag{5.154}$$

From (5.154) it then follows that a possible variational formulation of the Navier–Stokes problem (5.153) is

$$\begin{aligned} &\{\mathbf{u}, p\} \in (H^1(\Omega))^2 \times L^2(\Omega), \qquad \mathbf{u} = \mathbf{g} \text{ on } \Gamma, \\ &\alpha\int_\Omega \mathbf{u}\cdot\mathbf{v}\, dx + \nu\int_\Omega \nabla\mathbf{u}\cdot\nabla\mathbf{v}\, dx + \int_\Omega (\mathbf{u}\cdot\nabla)\mathbf{u}\cdot\mathbf{v}\, dx + \int_\Omega \nabla p\cdot\mathbf{v}\, dx = \int_\Omega \mathbf{f}\cdot\mathbf{v}\, dx, \\ &\qquad \forall\, \mathbf{v} \in (H_0^1(\Omega))^2, \\ &\int_\Omega \nabla\cdot\mathbf{u}q\, dx = 0, \qquad \forall\, q \in H^1(\Omega). \end{aligned} \tag{5.155}$$

The mixed finite element approximation of (5.153) which follows is derived directly from (5.155).

5.8.3. *A mixed finite element approximation of the Navier–Stokes problem* (5.153)

We use the notation of Sec. 5.3.3.1; formulation (5.155) clearly suggests the following (well-known) mixed approximation of the Navier–Stokes problem (5.153):

Find $\{\mathbf{u}_h, p_h\} \in V_{gh} \times H_h^1$ *such that*

$$\alpha \int_\Omega \mathbf{u}_h \cdot \mathbf{v}_h \, dx + \nu \int_\Omega \nabla \mathbf{u}_h \cdot \nabla \mathbf{v}_h \, dx + \int_\Omega (\mathbf{u}_h \cdot \nabla)\mathbf{u}_h \cdot \mathbf{v}_h \, dx + \int_\Omega \nabla p_h \cdot \mathbf{v}_h \, dx$$

$$= \int_\Omega \mathbf{f}_h \cdot \mathbf{v}_h \, dx, \qquad \forall\, \mathbf{v}_h \in V_{0h}, \tag{5.156}$$

$$\int_\Omega \nabla \cdot \mathbf{u}_h q_h \, dx = 0, \qquad \forall\, q_h \in H_h^1.$$

5.8.4. *Convergence results*

5.8.4.1. *Case of the Stokes problem. Uniqueness of the discrete pressure.* The discrete Stokes problem associated with (5.156) is defined by:

Find $\{\mathbf{u}_h, p_h\} \in V_{gh} \times H_h^1$ *such that*

$$\alpha \int_\Omega \mathbf{u}_h \cdot \mathbf{v}_h \, dx + \nu \int_\Omega \nabla \mathbf{u}_h \cdot \nabla \mathbf{v}_h \, dx + \int_\Omega \nabla p_h \cdot \mathbf{v}_h \, dx = \int_\Omega \mathbf{f}_h \cdot \mathbf{v}_h \, dx, \qquad \forall\, \mathbf{v}_h \in V_{0h},$$

$$\int_\Omega \nabla \cdot \mathbf{u}_h q_h \, dx = 0, \qquad \forall\, q_h \in H_h^1. \tag{5.157}$$

From Chapter I, Sec. 7.4.2 it follows that (5.157) has at least one solution but with $\mathbf{u}_h$ uniquely defined. Actually, if condition (5.119) holds, then $\{\mathbf{u}_h, p_h\}$ is uniquely defined in $V_{gh} \times (H_h^1/\mathbb{R})$.

Exercise 5.7. Prove that the results of Chapter I, Sec. 7.4.2 can be applied to problem (5.157).

With regard to the convergence of the approximate solutions, it is proved in Bercovier and Pironneau [1] that if the hypotheses on $(\mathcal{T}_h)_h$ are those of Lemma 5.1 and Theorems 5.4 and 5.5 of Sec. 5.3.3.3, then the estimates (5.26), (5.27), (5.29), (5.31), (5.32),[25] (5.33), and (5.34)[25] hold.

5.8.4.2. *Case of the Navier–Stokes problem.* It has been proved by Le Tallec [3] that the convergence results mentioned in Sec. 5.8.4.1 also hold for the Navier–Stokes problems (5.153), (5.156).

---

[25] With $\psi_h = 0$.

5.8.5. *Application to the solution of the unsteady Navier–Stokes problem*

5.8.5.1. *A semidiscrete approximation of the unsteady Navier–Stokes equations.* We suppose that the time-dependent problem under consideration is still defined by (5.39)–(5.42); we can then reduce (5.39)–(5.42) to a system of ordinary differential equations which is defined by:

Find $\{\mathbf{u}_h(t), p_h(t)\} \in V_{gh}(t) \times (H_h^1/\mathbb{R})$ *such that*

$$\int_\Omega \frac{\partial \mathbf{u}_h}{\partial t} \cdot \mathbf{v}_h \, dx + \nu \int_\Omega \nabla \mathbf{u}_h \cdot \nabla \mathbf{v}_h \, dx + \int_\Omega (\mathbf{u}_h \cdot \nabla)\mathbf{u}_h \cdot \mathbf{v}_h \, dx + \int_\Omega \nabla p_h \cdot \mathbf{v}_h \, dx$$

$$= \int_\Omega \mathbf{f}_h \cdot \mathbf{v}_h \, dx, \qquad \forall\, \mathbf{v}_h \in V_{0h}, \tag{5.158}$$

$$\int_\Omega \nabla \cdot \mathbf{u}_h q_h \, dx = 0, \qquad \forall\, q_h \in H_h^1,$$

$\mathbf{u}_h(0) = \mathbf{u}_{0h} \in V_h$ *given* ($\mathbf{u}_{0h}$: *approximation of* $\mathbf{u}_0$).

In (5.158) the space $V_{gh}(t)$ is defined by (5.45). The approximate problem (5.158) is not directly suitable for computation, and for this purpose a convenient time discretization is required in order to obtain a fully discrete approximate problem; several such time discretization schemes are given in the following sections (actually these schemes are variants of the schemes described in Sec. 5.4 whose notations are retained).

5.8.5.2. *An ordinary implicit scheme.* This scheme is a variant of scheme (5.46), (5.47) (see Sec. 5.4.1) and is defined as follows ($\{\mathbf{u}_h^n, p_h^n\}$ approximates $\{\mathbf{u}(nk), p(nk)\}$, where $k = \Delta t$):

$$\mathbf{u}_h^0 = \mathbf{u}_{0h} \quad (\textit{see } (5.158)); \tag{5.159}$$

then for $n \geq 0$, *we obtain* $\{\mathbf{u}_h^{n+1}, p_h^{n+1}\}$ *from* $\mathbf{u}_h^n$ *by solving*

$$\int_\Omega \frac{\mathbf{u}_h^{n+1} - \mathbf{u}_h^n}{k} \cdot \mathbf{v}_h \, dx + \nu \int_\Omega \nabla \mathbf{u}_h^{n+1} \cdot \nabla \mathbf{v}_h \, dx + \int_\Omega (\mathbf{u}_h^{n+1} \cdot \nabla)\mathbf{u}_h^{n+1} \cdot \mathbf{v}_h \, dx$$

$$+ \int_\Omega \nabla p_h^{n+1} \cdot \mathbf{v}_h \, dx = \int_\Omega \mathbf{f}_h^{n+1} \cdot \mathbf{v}_h \, dx, \qquad \forall\, \mathbf{v}_h \in V_{0h},$$

$$\int_\Omega \nabla \cdot \mathbf{u}_h^{n+1} q_h \, dx = 0, \qquad \forall\, q_h \in H_h^1, \quad \{\mathbf{u}_h^{n+1}, p_h^{n+1}\} \in V_{gh}^{n+1} \times H_h^1/\mathbb{R}, \tag{5.160}$$

where $V_{gh}^m = V_{gh}(mk)$, $\mathbf{f}_h^m = \mathbf{f}_h(mk)$.

To obtain $\{\mathbf{u}_h^{n+1}, p_h^{n+1}\}$ from $\mathbf{u}_h^n$, we have to solve a finite-dimensional nonlinear problem which is a particular case of (5.156) (with $\alpha = 1/k$). The above scheme has a time truncation error in $O(k)$ and appears to be unconditionally stable.

5.8.5.3. *A Crank–Nicholson implicit scheme.* This scheme (which is a variant of (5.48)–(5.50)) is defined as follows:

$$\mathbf{u}_h^0 = \mathbf{u}_{0h}; \tag{5.161}$$

then for $n \geq 0$, *we obtain* $\mathbf{u}_h^{n+1}$ *from* $\mathbf{u}_h^n$ *by solving*

$$\int_\Omega \frac{\mathbf{u}_h^{n+1} - \mathbf{u}_h^n}{k} \cdot \mathbf{v}_h \, dx + \nu \int_\Omega \nabla \mathbf{u}_h^{n+1/2} \cdot \nabla \mathbf{v}_h \, dx + \int_\Omega (\mathbf{u}_h^{n+1/2} \cdot \nabla)\mathbf{u}_h^{n+1/2} \cdot \mathbf{v}_h \, dx$$
$$+ \int_\Omega \nabla p_h^{n+1/2} \cdot \mathbf{v}_h \, dx = \int_\Omega \mathbf{f}_h^{n+1/2} \cdot \mathbf{v}_h \, dx, \qquad \forall \mathbf{v}_h \in V_{0h},$$
$$\int_\Omega \nabla \cdot \mathbf{u}_h^{n+1/2} q_h \, dx = 0, \qquad \forall\, q_h \in H_h^1, \quad \{\mathbf{u}_h^{n+1/2}, p_h^{n+1/2}\} \in V_{gh}^{n+1/2} \times H_h^1/\mathbb{R} \tag{5.162}$$

where

$$\mathbf{u}_h^{n+1/2} = \tfrac{1}{2}(\mathbf{u}_h^n + \mathbf{u}_h^{n+1})$$

and $V_{gh}^{n+1/2} = V_{gh}((n + \frac{1}{2})k)$.

Using the relation $\mathbf{u}_h^{n+1} = \mathbf{u}_h^n + 2(\mathbf{u}_h^{n+1/2} - \mathbf{u}_h^n)$, we can eliminate $\mathbf{u}_h^{n+1}$ in (5.162) and therefore reduce this problem to a particular case of (5.156) (with $\alpha = 2/k$). The above scheme has a time truncation error in $O(k^2)$ and appears to be unconditionally stable.

5.8.5.4. *A two-step implicit scheme.* This scheme is defined by:

$$\mathbf{u}_h^0 = \mathbf{u}_{0h}, \mathbf{u}_h^1 \textit{ given}; \tag{5.163}$$

then for $n \geq 1$, *we obtain* $\{\mathbf{u}_h^{n+1}, p_h^{n+1}\}$ *from* $\mathbf{u}_h^n, \mathbf{u}_h^{n-1}$ *by solving*

$$\int_\Omega \frac{3\mathbf{u}_h^{n+1} - 4\mathbf{u}_h^n + \mathbf{u}_h^{n-1}}{2k} \cdot \mathbf{v}_h \, dx + \nu \int_\Omega \nabla \mathbf{u}_h^{n+1} \cdot \nabla \mathbf{v}_h \, dx$$
$$+ \int_\Omega (\mathbf{u}_h^{n+1} \cdot \nabla)\mathbf{u}_h^{n+1} \cdot \mathbf{v}_h \, dx + \int_\Omega \nabla p_h^{n+1} \cdot \mathbf{v}_h \, dx = \int_\Omega \mathbf{f}_h^{n+1} \cdot \mathbf{v}_h \, dx, \qquad \forall\, \mathbf{v}_h \in V_{0h},$$
$$\int_\Omega \nabla \cdot \mathbf{u}_h^{n+1} q_h \, dx = 0, \qquad \forall\, q_h \in H_h^1, \quad \{\mathbf{u}_h^{n+1}, p_h^{n+1}\} \in V_{gh}^{n+1} \times H_h^1/\mathbb{R}. \tag{5.164}$$

To obtain $\mathbf{u}_h^1$ from $\mathbf{u}_h^0$, we may use one of the two schemes discussed in Secs. 5.8.5.2 and 5.8.5.3. From (5.164) it follows that $\mathbf{u}_h^{n+1}$ is the solution of a particular problem (5.156) (with $\alpha = 3/2k$). The above scheme has a time truncation error in $O(k^2)$ and appears to be unconditionally stable.

5.8.5.5. *Alternating-direction methods.* In this section we consider variants of those alternating-direction methods discussed in Sec. 5.4.4. Actually all the comments and remarks made in Sec. 5.4.4.3 also hold for the schemes which follow.

A. *A Peaceman–Rachford alternating-direction method for solving the unsteady Navier–Stokes equations.* This variant of (5.53)–(5.55) (see Sec. 5.4.4.1) is defined as follows (with $0 < \theta < 1$):

$$\mathbf{u}_h^0 = \mathbf{u}_{0h}; \tag{5.165}$$

then for $n \geq 0$, *we obtain* $\{\mathbf{u}_h^{n+1/2}, p_h^{n+1/2}\}, \mathbf{u}_h^{n+1}$ *from* $\mathbf{u}_h^n$ *by solving*

$$\begin{gathered}
\int_\Omega \frac{\mathbf{u}_h^{n+1/2} - \mathbf{u}_h^n}{k/2} \cdot \mathbf{v}_h \, dx + \theta\nu \int_\Omega \nabla \mathbf{u}_h^{n+1/2} \cdot \nabla \mathbf{v}_h \, dx + \int_\Omega \nabla p_h^{n+1/2} \cdot \mathbf{v}_h \, dx \\
+ (1-\theta)\nu \int_\Omega \nabla \mathbf{u}_h^n \cdot \nabla \mathbf{v}_h \, dx + \int_\Omega (\mathbf{u}_h^n \cdot \nabla)\mathbf{u}_h^n \cdot \mathbf{v}_h \, dx = \int_\Omega \mathbf{f}_h^{n+1/2} \cdot \mathbf{v}_h \, dx, \qquad \forall\, \mathbf{v}_h \in V_{0h}, \\
\int_\Omega \nabla \cdot \mathbf{u}_h^{n+1/2} q_h \, dx = 0, \qquad \forall\, q_h \in H_h^1, \quad \{\mathbf{u}_h^{n+1}, p_h^{n+1/2}\} \in V_{gh}^{n+1/2} \times H_h^1/\mathbb{R},
\end{gathered} \tag{5.166}$$

and then

$$\begin{gathered}
\int_\Omega \frac{\mathbf{u}_h^{n+1} - \mathbf{u}_h^{n+1/2}}{k/2} \cdot \mathbf{v}_h + (1-\theta)\nu \int_\Omega \nabla \mathbf{u}_h^{n+1} \cdot \nabla \mathbf{v}_h \, dx \\
+ \int_\Omega (\mathbf{u}_h^{n+1} \cdot \nabla)\mathbf{u}_h^{n+1} \cdot \mathbf{v}_h \, dx + \theta\nu \int_\Omega \nabla \mathbf{u}_h^{n+1/2} \cdot \nabla \mathbf{v}_h \, dx + \int_\Omega \nabla p_h^{n+1/2} \cdot \mathbf{v}_h \, dx \\
= \int_\Omega \mathbf{f}_h^{n+1} \cdot \mathbf{v}_h \, dx, \qquad \forall\, \mathbf{v}_h \in V_{0h}, \quad \mathbf{u}_h^{n+1} \in V_{gh}^{n+1}.
\end{gathered} \tag{5.167}$$

B. *A second alternating-direction method.* This is the variant of (5.166), (5.167) defined as follows (again with $0 < \theta < 1$):

$$\mathbf{u}_h^0 = \mathbf{u}_{0h}; \tag{5.168}$$

then for $n \geq 0$, *we obtain* $\{\mathbf{u}_h^{n+1/4}, p_h^{n+1/4}\}, \mathbf{u}_h^{n+3/4}, \{\mathbf{u}_h^{n+1}, p_h^{n+1}\}$ *from* $\mathbf{u}_h^n$ *by solving*:

$$\begin{gathered}
\int_\Omega \frac{\mathbf{u}_h^{n+1/4} - \mathbf{u}_h^n}{k/4} \cdot \mathbf{v}_h \, dx + (1-\theta)\nu \int_\Omega \nabla \mathbf{u}_h^{n+1/4} \cdot \nabla \mathbf{v}_h \, dx + \int_\Omega \nabla p_h^{n+1/4} \cdot \mathbf{v}_h \, dx \\
+ \theta\nu \int_\Omega \nabla \mathbf{u}_h^n \cdot \nabla \mathbf{v}_h \, dx + \int_\Omega (\mathbf{u}_h^n \cdot \nabla)\mathbf{u}_h^n \cdot \mathbf{v}_h \, dx = \int_\Omega \mathbf{f}_h^{n+1/4} \cdot \mathbf{v}_h \, dx, \qquad \forall\, \mathbf{v}_h \in V_{0h}, \\
\int_\Omega \nabla \cdot \mathbf{u}_h^{n+1/4} q_h \, dx = 0, \qquad \forall\, q_h \in H_h^1, \quad \{\mathbf{u}_h^{n+1/4}, p_h^{n+1/4}\} \in V_{gh}^{n+1/4} \times H_h^1/\mathbb{R};
\end{gathered} \tag{5.169}$$

then

$$\int_\Omega \frac{\mathbf{u}_h^{n+3/4} - \mathbf{u}_h^{n+1/4}}{k/2} \cdot \mathbf{v}_h \, dx + \theta\nu \int_\Omega \nabla \mathbf{u}_h^{n+3/4} \cdot \nabla \mathbf{v}_h \, dx$$

$$+ \int_\Omega (\mathbf{u}_h^{n+3/4} \cdot \nabla)\mathbf{u}_h^{n+3/4} \cdot \mathbf{v}_h \, dx + (1-\theta)\nu \int_\Omega \nabla \mathbf{u}_h^{n+1/4} \cdot \nabla \mathbf{v}_h \, dx$$

$$+ \int_\Omega \nabla p_h^{n+1/4} \cdot \mathbf{v}_h \, dx = \int_\Omega \mathbf{f}_h^{n+3/4} \cdot \mathbf{v}_h \, dx, \qquad \forall\, \mathbf{v}_h \in V_{0h}, \quad \mathbf{u}_h^{n+3/4} \in V_{gh}^{n+3/4}; \tag{5.170}$$

and finally

$$\int_\Omega \frac{\mathbf{u}_h^{n+1} - \mathbf{u}_h^{n+3/4}}{k/4} \cdot \mathbf{v}_h \, dx + (1-\theta)\nu \int_\Omega \nabla \mathbf{u}_h^{n+1} \cdot \nabla \mathbf{v}_h \, dx + \int_\Omega \nabla p_h^{n+1} \cdot \mathbf{v}_h \, dx$$

$$+ \theta\nu \int_\Omega \nabla \mathbf{u}_h^{n+3/4} \cdot \nabla \mathbf{v}_h \, dx + \int_\Omega (\mathbf{u}_h^{n+3/4} \cdot \nabla)\mathbf{u}_h^{n+3/4} \cdot \mathbf{v}_h \, dx$$

$$= \int_\Omega \mathbf{f}_h^{n+1} \cdot \mathbf{v}_h \, dx, \qquad \forall \mathbf{v}_h \in V_{0h}, \quad \int_\Omega \nabla \cdot \mathbf{u}_h^{n+1} q_h \, dx = 0, \qquad \forall\, q_h \in H_h^1,$$

$$\{\mathbf{u}_h^{n+1}, p_h^{n+1}\} \in V_{gh}^{n+1} \times H_h^1/\mathbb{R}. \tag{5.171}$$

5.8.6. *Least-squares conjugate gradient solution of the discrete Navier–Stokes problem* (5.156)

5.8.6.1. *Generalities.* The various discrete Navier–Stokes problems that we have encountered in this section are all particular cases of problem (5.156); for this reason we shall study in some detail the least-squares formulation and the conjugate gradient solution of (5.156). Actually the following considerations very closely follow Secs. 5.5 and 5.6.

5.8.6.2. *A least-squares formulation of* (5.156). A natural least-squares formulation of (5.156) is as follows:

Find $\mathbf{u}_h \in \tilde{V}_{gh}$ *such that*

$$J_h(\mathbf{u}_h) \le J_h(\mathbf{v}_h), \qquad \forall\, \mathbf{v}_h \in \tilde{V}_{gh}, \tag{5.172}$$

with

$$\tilde{V}_{gh} = \left\{ \mathbf{v}_h \in V_{gh}, \int_\Omega \nabla \cdot \mathbf{v}_h q_h \, dx = 0, \quad \forall\, q_h \in H_h^1 \right\}, \tag{5.173}$$

$$J_h(\mathbf{v}_h) = \frac{\alpha}{2} \int_\Omega |\xi_h(\mathbf{v}_h)|^2 \, dx + \frac{\nu}{2} \int_\Omega |\nabla \xi_h(\mathbf{v}_h)|^2 \, dx, \tag{5.174}$$

where, in (5.174), $\xi_h(\mathbf{v}_h)$ $(=\xi_h)$ is a function of $\mathbf{v}_h$ through the state equation

$$\alpha \int_\Omega \xi_h \cdot \boldsymbol{\eta}_h \, dx + \nu \int_\Omega \nabla \xi_h \cdot \nabla \boldsymbol{\eta}_h \, dx + \int_\Omega \nabla \pi_h \cdot \boldsymbol{\eta}_h \, dx = \alpha \int_\Omega \mathbf{v}_h \cdot \boldsymbol{\eta}_h \, dx$$

$$+ \nu \int_\Omega \nabla \mathbf{v}_h \cdot \nabla \boldsymbol{\eta}_h \, dx + \int_\Omega (\mathbf{v}_h \cdot \nabla)\mathbf{v}_h \cdot \boldsymbol{\eta}_h \, dx - \int_\Omega \mathbf{f}_h \cdot \boldsymbol{\eta}_h \, dx, \qquad \forall\, \boldsymbol{\eta}_h \in V_{0h},$$

$$\int_\Omega \nabla \cdot \xi_h q_h \, dx = 0, \qquad \forall\, q_h \in H_h^1, \quad \xi_h \in V_{0h}, \quad \pi_h \in H_h^1/\mathbb{R}. \tag{5.175}$$

We observe that ξ_h is obtained from $\mathbf{v}_h$ through the solution of a discrete Stokes problem. The above least-squares formulation is justified by the following:

Proposition 5.3. *Suppose that* $\{\mathbf{u}_h, p_h\} \in V_{gh} \times (H_h^1/\mathbb{R})$ *is a solution of* (5.156); *then* $\mathbf{u}_h$ *is also a solution of the least-squares problem* (5.175) *and, if* (5.119) *holds, for the corresponding pair* $\{\xi_h, \pi_h\}$ *we have*

$$\xi_h = \mathbf{0}, \qquad \pi_h = -p_h; \tag{5.176}$$

we also have $J_h(\mathbf{u}_h) = 0$.

5.8.6.3. *Conjugate gradient solution of the least-squares problem* (5.172).

A. *Description of the conjugate gradient algorithm.* A conjugate gradient algorithm for solving the least-squares problem (5.172) is defined as follows.

Step 0: Initialization

$$\mathbf{u}_h^0 \in \tilde{V}_{gh} \text{ given}; \tag{5.177}$$

then compute $\mathbf{z}_h^0$ *as the solution of the linear variational equation*

$$\alpha \int_\Omega \mathbf{z}_h^0 \cdot \boldsymbol{\eta}_h \, dx + \nu \int_\Omega \nabla \mathbf{z}_h^0 \cdot \nabla \boldsymbol{\eta}_h \, dx = \langle J_h'(\mathbf{u}_h^0), \boldsymbol{\eta}_h \rangle, \qquad \forall\, \boldsymbol{\eta}_h \in \tilde{V}_{0h}, \quad \mathbf{z}_h^0 \in \tilde{V}_{0h} \tag{5.178}$$

and set

$$\mathbf{w}_h^0 = \mathbf{z}_h^0. \tag{5.179}$$

For $m \geq 0$, *assuming that* $\mathbf{u}_h^m, \mathbf{z}_h^m, \mathbf{w}_h^m$ *are known, compute* $\mathbf{u}_h^{m+1}, \mathbf{z}_h^{m+1}, \mathbf{w}_h^{m+1}$ *by the following.*

Step 1: Descent

$$\text{Compute } \lambda_m = \operatorname*{Arg\,min}_{\lambda \in \mathbb{R}} J_h(\mathbf{u}_h^m - \lambda \mathbf{w}_h^m), \tag{5.180}$$

$$\mathbf{u}_h^{m+1} = \mathbf{u}_h^m - \lambda_m \mathbf{w}_h^m. \tag{5.181}$$

Step 2: Construction of the new descent direction. Define $\mathbf{z}_h^{m+1}$ *as the solution of*

$$\mathbf{z}_h^{m+1} \in \tilde{V}_{0h},$$

$$\alpha \int_\Omega \mathbf{z}_h^{m+1} \cdot \boldsymbol{\eta}_h \, dx + \nu \int_\Omega \nabla \mathbf{z}_h^{m+1} \cdot \nabla \boldsymbol{\eta}_h \, dx = \langle J_h'(\mathbf{u}_h^{m+1}), \boldsymbol{\eta}_h \rangle, \qquad \forall \boldsymbol{\eta}_h \in \tilde{V}_{0h}; \tag{5.182}$$

then (Polak–Ribière strategy)

$$\gamma_{m+1} = \frac{\alpha \int_\Omega \mathbf{z}_h^{m+1} \cdot (\mathbf{z}_h^{m+1} - \mathbf{z}_h^m) \, dx + \nu \int_\Omega \nabla \mathbf{z}_h^{m+1} \cdot \nabla(\mathbf{z}_h^{m+1} - \mathbf{z}_h^m) \, dx}{\alpha \int_\Omega |\mathbf{z}_h^m|^2 \, dx + \nu \int_\Omega |\nabla \mathbf{z}_h^m|^2 \, dx}, \tag{5.183}$$

and finally

$$\mathbf{w}_h^{m+1} = \mathbf{z}_h^{m+1} + \gamma_{m+1} \mathbf{w}_h^m, \tag{5.184}$$

$m = m + 1$ *go to* (5.180).

B. *Calculation of* $J_h'(\mathbf{u}_h^{m+1})$ *and* $\mathbf{z}_h^{m+1}$*: further comments.* A most important step in order to use algorithm (5.177)–(5.184) is the calculation of $J_h'(\mathbf{u}_h^{m+1})$; another important step is the calculation of $\mathbf{z}_h^{m+1}$. Due to the importance of these steps, we shall give the corresponding calculations in detail. We have

$$\delta J = \langle J_h'(\mathbf{v}_h), \delta \mathbf{v}_h \rangle = \alpha \int_\Omega \boldsymbol{\xi}_h \cdot \delta \boldsymbol{\xi}_h \, dx + \nu \int_\Omega \nabla \boldsymbol{\xi}_h \cdot \nabla \delta \boldsymbol{\xi}_h \, dx; \tag{5.185}$$

to express $\delta \boldsymbol{\xi}_h$ as a function of $\delta \mathbf{v}_h$, we observe that (5.175) has the following variational formulation:

$\boldsymbol{\xi}_h \in \tilde{V}_{0h}$ *and* $\forall\, \boldsymbol{\eta}_h \in \tilde{V}_{0h}$, *we have*

$$\alpha \int_\Omega \boldsymbol{\xi}_h \cdot \boldsymbol{\eta}_h \, dx + \nu \int_\Omega \nabla \boldsymbol{\xi}_h \cdot \nabla \boldsymbol{\eta}_h \, dx = \alpha \int_\Omega \mathbf{v}_h \cdot \boldsymbol{\eta}_h \, dx + \nu \int_\Omega \nabla \mathbf{v}_h \cdot \nabla \boldsymbol{\eta}_h \, dx$$
$$+ \int_\Omega (\mathbf{v}_h \cdot \nabla) \mathbf{v}_h \cdot \boldsymbol{\eta}_h \, dx - \int_\Omega \mathbf{f}_h \cdot \boldsymbol{\eta}_h \, dx. \tag{5.186}$$

EXERCISE 5.8. Prove the equivalence between (5.175) and (5.186).

Relation (5.186), in turn, implies

$\delta \boldsymbol{\xi}_h \in \tilde{V}_{0h}$ *and* $\forall\, \boldsymbol{\eta}_h \in \tilde{V}_{0h}$, *we have*

$$\alpha \int_\Omega \delta \boldsymbol{\xi}_h \cdot \boldsymbol{\eta}_h \, dx + \nu \int_\Omega \nabla \delta \boldsymbol{\xi}_h \cdot \nabla \boldsymbol{\eta}_h \, dx = \alpha \int_\Omega \delta \mathbf{v}_h \cdot \boldsymbol{\eta}_h \, dx + \nu \int_\Omega \nabla \delta \mathbf{v}_h \cdot \nabla \boldsymbol{\eta}_h \, dx$$
$$+ \int_\Omega (\delta \mathbf{v}_h \cdot \nabla) \mathbf{v}_h \cdot \boldsymbol{\eta}_h \, dx + \int_\Omega (\mathbf{v}_h \cdot \nabla) \delta \mathbf{v}_h \cdot \boldsymbol{\eta}_h \, dx. \tag{5.187}$$

Since $\xi_h \in \tilde{V}_{0h}$, we can take $\boldsymbol{\eta}_h = \xi_h$ in (5.187); from (5.185) it then follows that

$$\delta J_h = \langle J_h'(\mathbf{v}_h), \delta \mathbf{v}_h \rangle = \alpha \int_\Omega \xi_h \cdot \delta \mathbf{v}_h \, dx + \nu \int_\Omega \nabla \xi_h \cdot \nabla \delta \mathbf{v}_h \, dx$$
$$+ \int_\Omega (\delta \mathbf{v}_h \cdot \nabla) \mathbf{v}_h \cdot \xi_h \, dx + \int_\Omega (\mathbf{v}_h \cdot \nabla) \delta \mathbf{v}_h \cdot \xi_h \, dx. \tag{5.188}$$

Finally from (5.188) it follows that $J_h'(\mathbf{v}_h)$ can be identified with the linear functional on $\tilde{V}_{0h}$ defined by

$$\boldsymbol{\eta}_h \to \alpha \int_\Omega \xi_h \cdot \boldsymbol{\eta}_h \, dx + \nu \int_\Omega \nabla \xi_h \cdot \nabla \boldsymbol{\eta}_h \, dx + \int_\Omega (\boldsymbol{\eta}_h \cdot \nabla) \mathbf{v}_h \cdot \xi_h \, dx$$
$$+ \int_\Omega (\mathbf{v}_h \cdot \nabla) \boldsymbol{\eta}_h \cdot \xi_h \, dx. \tag{5.189}$$

From (5.189) it then follows that to compute $\mathbf{z}_h^{m+1}$ from $\mathbf{u}_h^{m+1}$, we first have to solve the discrete Stokes problem (5.175), (5.186) (with $\mathbf{v}_h = \mathbf{u}_h^{m+1}$) to obtain ξ_h^{m+1}; then from (5.182), (5.189) we obtain $\mathbf{z}_h^{m+1}$ via the solution of

$$\mathbf{z}_h^{m+1} \in \tilde{V}_{0h} \text{ and } \forall\, \boldsymbol{\eta}_h \in \tilde{V}_{0h},$$
$$\alpha \int_\Omega \mathbf{z}_h^{m+1} \cdot \boldsymbol{\eta}_h \, dx + \nu \int_\Omega \nabla \mathbf{z}_h^{m+1} \cdot \nabla \boldsymbol{\eta}_h \, dx = \alpha \int_\Omega \xi_h^{m+1} \cdot \boldsymbol{\eta}_h \, dx$$
$$+ \nu \int_\Omega \nabla \xi_h^{m+1} \cdot \nabla \boldsymbol{\eta}_h \, dx + \int_\Omega (\boldsymbol{\eta}_h \cdot \nabla) \mathbf{u}_h^{m+1} \cdot \xi_h^{m+1} \, dx$$
$$+ \int_\Omega (\mathbf{u}_h^{m+1} \cdot \nabla) \boldsymbol{\eta}_h \cdot \xi_h^{m+1} \, dx. \tag{5.190}$$

Actually (cf. Exercise 5.8) problem (5.190) is equivalent to the discrete Stokes problem

$$\alpha \int_\Omega \mathbf{z}_h^{m+1} \cdot \boldsymbol{\eta}_h \, dx + \nu \int_\Omega \nabla \mathbf{z}_h^{m+1} \cdot \nabla \boldsymbol{\eta}_h \, dx + \int_\Omega \nabla \omega_h^{m+1} \cdot \boldsymbol{\eta}_h \, dx$$
$$= \alpha \int_\Omega \xi_h^{m+1} \cdot \boldsymbol{\eta}_h \, dx + \nu \int_\Omega \nabla \xi_h^{m+1} \cdot \nabla \boldsymbol{\eta}_h \, dx + \int_\Omega (\boldsymbol{\eta}_h \cdot \nabla) \mathbf{u}_h^{m+1} \cdot \xi_h^{m+1} \, dx$$
$$+ \int_\Omega (\mathbf{u}_h^{m+1} \cdot \nabla) \boldsymbol{\eta}_h \cdot \xi_h^{m+1} \, dx, \qquad \forall\, \boldsymbol{\eta}_h \in V_{0h},$$
$$\int_\Omega \nabla \cdot \mathbf{z}_h^{m+1} q_h \, dx = 0, \qquad \forall\, q_h \in H_h^1, \quad \{\mathbf{z}_h^{m+1}, \omega_h^{m+1}\} \in V_{0h} \times H_h^1/\mathbb{R}; \tag{5.191}$$

ω_h^{m+1} is clearly a (discrete) pressure. To summarize, at each iteration of the conjugate gradient algorithm (5.177)–(5.184), we have to solve several discrete Stokes problems, namely:

(i) the discrete Stokes problem (5.175), (5.186) with $\mathbf{v}_h = \mathbf{u}_h^{m+1}$ to obtain ξ_h^{m+1} from $\mathbf{u}_h^{m+1}$;
(ii) the discrete Stokes problem (5.182), (5.190), (5.191) to obtain $\mathbf{z}_h^{m+1}$ from ξ_h^{m+1};
(iii) the solution of the one-dimensional problem (5.180) requires several calculation of J_h, i.e., several solutions of the state equation (5.175), (5.186).

Thus, efficient solvers for the discrete Stokes problem (5.157) will be important tools for solving the Navier–Stokes equations by algorithm (5.177)–(5.184) via the least-squares formulation (5.172); actually these solvers will also be useful for solving the Stokes parts of the alternating-direction algorithms (5.165)–(5.167) and (5.168)–(5.171) of Sec. 5.8.5.5.

5.8.7. *On numerical solvers for the discrete Stokes problem* (5.157)

5.8.7.1. *Synopsis.* From Secs. 5.8.5 and 5.8.6 it follows that efficient solvers for the discrete Stokes problem (5.157), i.e.:

Find $\{\mathbf{u}_h, p_h\} \in V_{gh} \times H_h^1$ *such that*

$$\alpha \int_\Omega \mathbf{u}_h \cdot \mathbf{v}_h \, dx + \nu \int_\Omega \nabla \mathbf{u}_h \cdot \nabla \mathbf{v}_h \, dx + \int_\Omega \nabla p_h \cdot \mathbf{v}_h \, dx$$

$$= \int_\Omega \mathbf{f}_h \cdot \mathbf{v}_h \, dx, \qquad \forall\, \mathbf{v}_h \in V_{0h}, \int_\Omega \nabla \cdot \mathbf{u}_h q_h \, dx = 0, \qquad \forall\, q_h \in H_h^1,$$

will be advantageous if one wishes to solve the steady and unsteady Navier–Stokes equations through the finite element approximation of Sec. 5.8.3.

In the following sections we shall discuss the solution of (5.157) by either direct or iterative methods.

5.8.7.2. *Direct solution of* (5.157). We are going to give a linear system equivalent to (5.157); we recall that

$$H_h^1 = \{q_h | q_h \in C^0(\bar{\Omega}),\, q_{h|T} \in P_1, \qquad \forall\, T \in \mathcal{T}_h\}.$$

We now define $\tilde{H}_h^1$ by either

$$\tilde{H}_h^1 = \{z_h | z_h \in C^0(\bar{\Omega}), \quad z_{h|T} \in P_2, \qquad \forall\, T \in \mathcal{T}_h\} \text{ if } V_h \text{ is defined by (5.18),} \tag{5.192$_1$}$$

or

$$\tilde{H}_h^1 = \{z_h | z_h \in C^0(\bar{\Omega}), \quad z_{h|T} \in P_1, \qquad \forall\, T \in \tilde{\mathcal{T}}_h\} \text{ if } V_h \text{ is defined by (5.21).} \tag{5.192$_2$}$$

We clearly have $V_h = \tilde{H}_h^1 \times \tilde{H}_h^1$; we also have $V_{0h} = \tilde{H}_{0h}^1 \times \tilde{H}_{0h}^1$, where

$$\tilde{H}_{0h}^1 = \tilde{H}_h^1 \cap H_0^1(\Omega) = \{z_h | z_h \in \tilde{H}_h^1,\, z_{h|\Gamma} = 0\}. \tag{5.193}$$

We now define $\Sigma_h = \{P \in \bar{\Omega}, P \text{ vertex of } T \in \mathscr{T}_h\}$; we have $\dim(H_h^1) = \text{Card}(\Sigma_h)$, and we define M_h by $M_h = \dim(H_h^1)$.

Suppose that $\Sigma_h = \{P_i\}_{i=1}^{M_h}$; we associate to Σ_h the vector basis of H_h^1 defined by

$$\mathscr{B}_h = \{w_i\}_{i=1}^{M_h}. \tag{5.194}$$

For $i = 1, \ldots, M_h$,

$$w_i \in H_h^1,\ w_i(P_i) = 1, \quad w_i(P_j) = 0, \qquad \forall\, j \neq i, \quad 1 \leq j \leq M_h \tag{5.195}$$

We then have

$$q_h = \sum_{i=1}^{M_h} q_h(P_i) w_i, \qquad \forall\, q_h \in H_h^1. \tag{5.196}$$

We similarly define

$$\begin{aligned}\tilde{\Sigma}_h &= \{Q \in \bar{\Omega}, Q \text{ vertex of } T \in \mathscr{T}_h, \text{ or } Q \text{ midpoint of an edge of } T \in \mathscr{T}_h\}\\ &= \{Q \in \bar{\Omega}, Q \text{ vertex of } T \in \tilde{\mathscr{T}}_h\},\end{aligned} \tag{5.197}$$

$$\tilde{\Sigma}_{0h} = \{P \in \tilde{\Sigma}_h, P \notin \Gamma\}, \tag{5.198}$$

$$N_h = \text{Card}(\tilde{\Sigma}_h) = \dim(\tilde{H}_h^1), \tag{5.199}$$

$$N_{0h} = \text{Card}(\tilde{\Sigma}_{0h}) = \dim(\tilde{H}_{0h}^1). \tag{5.200}$$

We suppose that

$$\tilde{\Sigma}_{0h} = \{Q_i\}_{i=1}^{N_{0h}}, \tag{5.201}$$

$$\tilde{\Sigma}_h = \tilde{\Sigma}_{0h} \cup \{Q_i\}_{i=N_{0h}+1}^{N_h}, \tag{5.202}$$

and we associate to $\tilde{\Sigma}_{0h}$, $\tilde{\Sigma}_h$ the following vector bases of $\tilde{H}_{0h}^1$ and $\tilde{H}_h^1$, respectively:

$$\tilde{\mathscr{B}}_{0h} = \{\tilde{w}_j\}_{j=1}^{N_{0h}}, \tag{5.203}$$

$$\tilde{\mathscr{B}}_h = \{\tilde{w}_j\}_{j=1}^{N_h} \tag{5.204}$$

with $\tilde{w}_j$ defined by:

$$\forall\, j = 1, \ldots, N_h, \text{ we have } \tilde{w}_j \in \tilde{H}_h^1,\ \tilde{w}_j(Q_j) = 1, \quad \tilde{w}_j(Q_k) = 0, \qquad \forall\, k \neq j. \tag{5.205}$$

We then have $\forall\, \mathbf{v}_h \in V_h$ (resp., $\forall\, \mathbf{v}_h \in V_{0h}$),

$$\mathbf{v}_h = \sum_{j=1}^{N_h} \mathbf{v}_h(Q_j) \tilde{w}_j \tag{5.206}$$

(resp.,

$$\mathbf{v}_h = \sum_{j=1}^{N_{0h}} \mathbf{v}_h(Q_j) \tilde{w}_j). \tag{5.207}$$

Let $\{\mathbf{u}_h, p_h\}$ be the solution of (5.157). We have

$$\mathbf{u}_h = \sum_{j=1}^{N_{0h}} \mathbf{u}_h(Q_j)\tilde{w}_j + \sum_{j=N_{0h}+1}^{N_h} \mathbf{g}_h(Q_j)\tilde{w}_j; \tag{5.208}$$

if

$$\mathbf{u}_h = \{u_{1h}, u_{2h}\} \quad (\text{with } u_{1h}, u_{2h} \in \tilde{H}_h^1), \tag{5.209}$$

$$\mathbf{g}_h = \{g_{1h}, g_{2h}\}, \tag{5.210}$$

it follows from (5.208) that we have

$$\forall\, r = 1, 2 \qquad u_{rh} = \sum_{j=1}^{N_{0h}} u_{rh}(Q_j)\tilde{w}_j + \sum_{j=N_{0h}+1}^{N_h} g_{rh}(Q_j)\tilde{w}_j. \tag{5.211}$$

Similarly we have

$$p_h = \sum_{k=1}^{M_h} p_h(P_k) w_k. \tag{5.212}$$

Using the above relations, the discrete Stokes problem (5.157) is equivalent to the following linear system of $2N_{0h} + M_h$ equations:

$$\begin{aligned}
&\sum_{j=1}^{N_{0h}} \left(\alpha \int_\Omega \tilde{w}_j \tilde{w}_i\, dx + \nu \int_\Omega \nabla\tilde{w}_j \cdot \nabla\tilde{w}_i\, dx\right) u_{1h}(Q_j) \\
&\qquad + \sum_{k=1}^{M_h} \left(\int_\Omega \frac{\partial w_k}{\partial x_1}\, \tilde{w}_i\, dx\right) p_h(P_k) \\
&\quad = \int_\Omega f_{1h}\tilde{w}_i\, dx - \sum_{j=N_{0h}+1}^{N_h} \left(\alpha \int_\Omega \tilde{w}_j\tilde{w}_i\, dx + \nu \int_\Omega \nabla\tilde{w}_j \cdot \nabla\tilde{w}_i\, dx\right) g_{1h}(Q_j), \\
&\quad \forall\, i, \quad 1 \le i \le N_{0h},
\end{aligned} \tag{5.213$_1$}$$

$$\begin{aligned}
&\sum_{j=1}^{N_{0h}} \left(\alpha \int_\Omega \tilde{w}_j \tilde{w}_i\, dx + \nu \int_\Omega \nabla\tilde{w}_j \cdot \nabla\tilde{w}_i\, dx\right) u_{2h}(Q_j) \\
&\qquad + \sum_{k=1}^{M_h} \left(\int_\Omega \frac{\partial w_k}{\partial x_2}\, \tilde{w}_i\, dx\right) p_h(P_k) \\
&\quad = \int_\Omega f_{2h}\tilde{w}_i\, dx - \sum_{j=N_{0h}+1}^{N_h} \left(\alpha \int_\Omega \tilde{w}_j\tilde{w}_i\, dx + \nu \int_\Omega \nabla\tilde{w}_j \cdot \nabla\tilde{w}_i\, dx\right) g_{2h}(Q_j), \\
&\quad \forall\, i, \quad 1 \le i \le N_{0h},
\end{aligned} \tag{5.213$_2$}$$

$$\begin{aligned}
&\sum_{j=1}^{N_{0h}} \left[\left(\int_\Omega \frac{\partial \tilde{w}_j}{\partial x_1} w_l\, dx\right) u_{1h}(Q_j) + \left(\int_\Omega \frac{\partial \tilde{w}_j}{\partial x_2} w_l\, dx\right) u_{2h}(Q_j)\right] \\
&\quad = - \sum_{j=N_{0h}+1}^{N_h} \left[\left(\int_\Omega \frac{\partial \tilde{w}_j}{\partial x_1} w_l\, dx\right) g_{1h}(Q_j) + \left(\int_\Omega \frac{\partial \tilde{w}_j}{\partial x_2} w_l\, dx\right) g_{2h}(Q_j)\right] \\
&\quad \forall\, l, \quad 1 \le l \le M_h,
\end{aligned} \tag{5.214}$$

whose unknowns are $\{\{u_{1h}(Q_j), u_{2h}(Q_j)\}\}_{j=1}^{N_{0h}}$, $\{p_h(P_k)\}_{k=1}^{M_h}$ (and where $\mathbf{f}_h = \{f_{1h}, f_{2h}\}$).

Let us define $\mathbf{U}_h \in \mathbb{R}^{2N_{0h}}$ and $\mathbf{P}_h \in \mathbb{R}^{M_h}$ by

$$\mathbf{U}_h = \{\{u_{1h}(Q_j), u_{2h}(Q_j)\}\}_{j=1}^{N_{0h}}, \quad \mathbf{P}_h = \{p_h(P_k)\}_{k=1}^{M_h}; \tag{5.215}$$

since $\tilde{w}_j \in H_0^1(\Omega)$, $\forall\, j = 1, \ldots, N_{0h}$, we have (Green's formula)

$$\int_\Omega \frac{\partial \tilde{w}_j}{\partial x_r} w_l \, dx = -\int_\Omega \tilde{w}_j \frac{\partial w_l}{\partial x_r} \, dx,$$
$$\forall\, r = 1, 2, \quad \forall\, j = 1, \ldots, N_{0h}, \quad \forall\, l = 1, \ldots, M_h. \tag{5.216}$$

Multiplying the equation (5.214) by -1, it follows from (5.215), (5.216) that the linear system (5.213), (5.214) has the following block representation:

$$\mathbf{A}_h \mathbf{U}_h + \mathbf{B}_h^t \mathbf{P}_h = \mathbf{b}_h, \ \mathbf{B}_h \mathbf{U}_h = \mathbf{c}_h, \tag{5.217}$$

where $\mathbf{A}_h$ is a $2N_{0h} \times 2N_{0h}$ symmetric positive-definite sparse matrix and where we have (from (5.214)) $\mathbf{c}_h \in R(\mathbf{B}_h)$; such linear systems have been considered in Chapter I, Sec. 7.4.2.

The matrix associated with (5.217), i.e.,

$$\mathscr{A}_h = \begin{pmatrix} \mathbf{A}_h & \mathbf{B}_h^t \\ \mathbf{B}_h & \mathbf{0} \end{pmatrix}$$

is symmetric, indefinite, and sparse; we actually have

$$\operatorname{Ker}(\mathscr{A}_h) = \{\mathbf{0}\} \times \operatorname{Ker}(\mathbf{B}_h^t),$$

and if (5.119) holds, we have

$$\operatorname{Ker}(\mathbf{B}_h^t) = \{\mathbf{Q}_h \mid \mathbf{Q}_h = \{c\}_{k=1}^{M_h}, c \in \mathbb{R}\}, \tag{5.218}$$

i.e.,

$$\dim \operatorname{Ker}(\mathbf{B}_h^t) = 1 \Leftrightarrow \operatorname{Rank}(\mathscr{A}_h) = 2N_{0h} + M_h - 1.$$

From (5.218) it then follows that we can trivially reduce (5.217) to a system whose matrix is nonsingular (and still symmetric, sparse, and indefinite) by taking (for example) the M_hth component of $\mathbf{P}_h$ equal to zero.

The direct solution of systems like (5.217) (by sophisticated variants of the Gauss elimination method) has inspired several authors; let us mention among them Parlett and Reid [1]. Bunch and Parlett [1], Aasen [1], Bunch and Kaufman [1], Barwell and George [1], Duff, Munksgaard, Nielsen, and Reid [1], and Duff [1]; see also the references therein. The methods originating from the above references are quite storage demanding, which can be a major drawback when trying to solve very complicated flow problems (actually these methods can also be used—theoretically at least—for solving the discrete Stokes problem (5.113) of Sec. 5.7.2.1).

The solution of (5.157) by iterative methods (less storage demanding than the above direct methods) will be discussed in the following sections.

5.8.7.3. *Iterative methods for solving the discrete Stokes problem* (5.157)

5.8.7.3.1. *Generalities: synopsis.* In Sec. 5.8.7.2 we have shown that the discrete Stokes problem (5.157) is in fact equivalent to a linear system (namely (5.217)) having the following very special structure (we have slightly modified the notation of Sec. 5.8.7.2 and also of Chapter I, Sec. 7.4.2):

$$\mathbf{A}\mathbf{x} + \mathbf{B}^t\boldsymbol{\lambda} = \mathbf{b}, \quad \mathbf{B}\mathbf{x} = \mathbf{c}, \tag{5.219}$$

where $\mathbf{A}$ is an $N \times N$ symmetric positive-definite matrix, $\mathbf{B}$ is an $M \times N$ matrix, $\mathbf{b} \in \mathbb{R}^N$, $\mathbf{c} \in R(\mathbf{B})$.

From Chapter I, Sec. 7.4.2 it follows that (5.219) has a unique solution $\{\mathbf{x}, \hat{\boldsymbol{\lambda}}\}$ in $\mathbb{R}^N \times R(\mathbf{B})$; it also follows from Chapter I, Sec. 7.4.2 that

$$\{\mathbf{x}', \boldsymbol{\lambda}'\} \in \mathbb{R}^N \times \mathbb{R}^M$$

is a solution of (5.219) if and only if

$$\mathbf{x}' = \mathbf{x}, \quad \boldsymbol{\lambda}' = \hat{\boldsymbol{\lambda}} + \boldsymbol{\mu}, \boldsymbol{\mu} \in \operatorname{Ker}(\mathbf{B}^t). \tag{5.220}$$

The linear system (5.219) can also be written

$$\begin{pmatrix} \mathbf{A} & \mathbf{B}^t \\ \mathbf{B} & \mathbf{0} \end{pmatrix} \begin{pmatrix} \mathbf{x} \\ \boldsymbol{\lambda} \end{pmatrix} = \begin{pmatrix} \mathbf{b} \\ \mathbf{c} \end{pmatrix}; \tag{5.221}$$

since the matrix in (5.221) is symmetric and indefinite, this linear system can be solved iteratively by those *Lanczos methods* (generalizing the standard conjugate gradient methods) introduced in Lanczos [1] and some improved versions of which are discussed in Paige and Saunders [1] and Parlett [1]. Actually, from our numerical experiments[26] these Lanczos methods seem quite delicate to use and in the following sections we shall concentrate on iterative methods taking into account the fact that $\mathbf{A}$ is a symmetric and positive-definite matrix. More precisely, in Sec. 5.8.7.3.2 (resp., 5.8.7.3.3) we shall consider the solution of (5.219) by an Uzawa conjugate gradient algorithm (resp., by an augmented Lagrangian method); then in Secs. 5.8.7.3.4 and 5.8.7.3.5 we shall apply these methods to the solution of the discrete Stokes problem (5.157).

Finally, in Sec. 5.8.7.4 we shall discuss the solution of the continuous Stokes problem

$$\alpha\mathbf{u} - \nu\Delta\mathbf{u} + \nabla p = \mathbf{f} \text{ in } \Omega, \quad \nabla \cdot \mathbf{u} = 0 \text{ in } \Omega, \quad \mathbf{u} = \mathbf{g} \text{ on } \Gamma \tag{5.222}$$

by the continuous analogues of the algorithms of Secs. 5.8.7.3.2 and 5.8.7.3.3.

5.8.7.3.2. *Iterative solution of* (5.219) *by an Uzawa conjugate gradient algorithm.* From (5.219) it follows that $\boldsymbol{\lambda}$ is a solution of the linear system, in $\mathbb{R}^M$,

$$\mathbf{B}\mathbf{A}^{-1}\mathbf{B}^t\boldsymbol{\lambda} = \mathbf{B}\mathbf{A}^{-1}\mathbf{b} - \mathbf{c}. \tag{5.223}$$

[26] On discrete Stokes problems precisely.

We clearly have

$$\operatorname{Ker}(\mathbf{B}\mathbf{A}^{-1}\mathbf{B}^t) = \operatorname{Ker}(\mathbf{B}^t), \tag{5.224}$$

which, since the matrix $\mathbf{B}\mathbf{A}^{-1}\mathbf{B}^t$ is symmetric, implies that

$$R(\mathbf{B}\mathbf{A}^{-1}\mathbf{B}^t) = R(\mathbf{B}). \tag{5.225}$$

Since $\mathbf{c}$ and $\mathbf{B}\mathbf{A}^{-1}\mathbf{b}$ belong to $R(\mathbf{B})$, it follows from (5.225) that (5.223) has a solution; in fact it has a unique solution $\hat{\boldsymbol{\lambda}}$ in $R(\mathbf{B})$ since $\mathbf{B}\mathbf{A}^{-1}\mathbf{B}^t$ is an isomorphism from $R(\mathbf{B})$ onto $R(\mathbf{B})$.

Let us define $\mathscr{B}$ and $\boldsymbol{\beta}$ by

$$\mathscr{B} = \mathbf{B}\mathbf{A}^{-1}\mathbf{B}^t, \qquad \boldsymbol{\beta} = \mathbf{B}\mathbf{A}^{-1}\mathbf{b} - \mathbf{c};$$

we then have

$$\mathscr{B}\boldsymbol{\lambda} = \boldsymbol{\beta}. \tag{5.226}$$

Since $\mathscr{B}$ is symmetric and positive semidefinite and since $\boldsymbol{\beta} \in R(\mathbf{B})$, we can solve (5.226) (i.e., (5.219)) by the following conjugate gradient algorithm (where $(\cdot, \cdot)$ denotes the usual scalar product of $\mathbb{R}^M$ and $|\cdot|$ denotes the corresponding norm).

Step 0: Initialization

$$\boldsymbol{\lambda}^0 \in \mathbb{R}^M \textit{ arbitrarily given}; \tag{5.227}$$

$$\mathbf{g}^0 = \mathscr{B}\boldsymbol{\lambda}^0 - \boldsymbol{\beta}, \tag{5.228}$$

$$\mathbf{w}^0 = \mathbf{g}^0. \tag{5.229}$$

Then for $n \geq 0$, we obtain $\boldsymbol{\lambda}^{n+1}$, $\mathbf{g}^{n+1}$, $\mathbf{w}^{n+1}$ from $\boldsymbol{\lambda}^n$, $\mathbf{g}^n$, $\mathbf{w}^n$ by the following.

Step 1: Descent

$$\rho_n = \frac{(\mathbf{g}^n, \mathbf{w}^n)}{(\mathscr{B}\mathbf{w}^n, \mathbf{w}^n)} = \frac{|\mathbf{g}^n|^2}{(\mathscr{B}\mathbf{w}^n, \mathbf{w}^n)}, \tag{5.230}$$

$$\boldsymbol{\lambda}^{n+1} = \boldsymbol{\lambda}^n - \rho_n \mathbf{w}^n. \tag{5.231}$$

Step 2: New descent direction

$$\mathbf{g}^{n+1} = \mathscr{B}\boldsymbol{\lambda}^{n+1} - \boldsymbol{\beta}, \tag{5.232}$$

$$\gamma_n = \frac{|\mathbf{g}^{n+1}|^2}{|\mathbf{g}^n|^2}, \tag{5.233}$$

$$\mathbf{w}^{n+1} = \mathbf{g}^{n+1} + \gamma_n \mathbf{w}^n, \tag{5.234}$$

$n = n + 1$ go to (5.230).

From (5.231), (5.232) it follows that we also have

$$\mathbf{g}^{n+1} = \mathbf{g}^n - \rho_n \mathscr{B}\mathbf{w}^n, \tag{5.235}$$

which we should use instead of (5.232) in practice.

Concerning the convergence of algorithm (5.227)–(5.234) we have:

Proposition 5.4. *Let* M_0 *be the dimension of* $R(\mathscr{B})$[27] $(=R(\mathbf{B}))$; *then there exists* $n_0 \leq M_0$ *such that* $\boldsymbol{\lambda}^{n_0}$ *is a solution of* (5.223), (5.226); *more precisely we have*

$$\boldsymbol{\lambda}^{n_0} = \hat{\boldsymbol{\lambda}} + \boldsymbol{\lambda}_2^0, \tag{5.236}$$

where $\boldsymbol{\lambda}_2^0$ *is the component of* $\boldsymbol{\lambda}^0$ *belonging to* $\text{Ker}(\mathbf{B}^t)$ *in the decomposition*

$$\mathbb{R}^M = R(\mathbf{B}) \oplus \text{Ker}(\mathbf{B}^t).$$

Exercise 5.9. Prove Proposition 5.4.

Hint: First prove that

$$(\mathbf{g}^i, \mathbf{g}^j) = 0, \qquad \forall\, i, j, \quad i \neq j,$$

$$(\mathbf{g}^i, \mathbf{w}^j) = 0, \qquad \forall\, i, j, \quad i > j,$$

$$(\mathscr{B}\mathbf{w}^i, \mathbf{w}^j) = 0, \qquad \forall\, i, j, \quad i \neq j,$$

and use relation (5.235).

Remark 5.14. If $\boldsymbol{\lambda}^0 \in R(\mathbf{B})$ ($\boldsymbol{\lambda}^0 = \mathbf{0}$ for example), it follows from Proposition 5.4 that algorithm (5.227)–(5.234) converges to $\hat{\boldsymbol{\lambda}}$ in a finite number $(\leq M_0)$ of iterations.

Remark 5.15. The above finite termination result holds if roundoff errors are not considered. However, even with roundoff errors, a practical convergence can be obtained in much less than M_0 iterations, due to the superlinear convergence of conjugate gradient algorithms (see Polak [1] and Hestenes [2] for more details on the convergence of conjugate gradient methods).

For M_0 sufficient large, numerical experiments show a practical convergence in $O(\sqrt{M_0})$ for most applications (this is the case, in particular, for algorithm (5.138)–(5.145) in Sec. 5.7.2.5.).

Practical implementation of algorithm (5.227)–(5.234). In practice we should use (5.227)–(5.234) as follows.

Step 0: *Initialization*

$$\boldsymbol{\lambda}^0 \in \mathbb{R}^M, \tag{5.237}$$

$$\mathbf{x}^0 = \mathbf{A}^{-1}(\mathbf{b} - \mathbf{B}^t\boldsymbol{\lambda}^0), \tag{5.238}$$

$$\mathbf{g}^0 = \mathbf{c} - \mathbf{B}\mathbf{x}^0, \quad \mathbf{w}^0 = \mathbf{g}^0. \tag{5.239}$$

Then for $n \geq 0$ *we define* $\boldsymbol{\lambda}^{n+1}, \mathbf{g}^{n+1}, \mathbf{w}^{n+1}$ *from* $\boldsymbol{\lambda}^n, \mathbf{g}^n, \mathbf{w}^n$ *as follows.*

---

[27] $M_0 =$ rank of $\mathbf{B}$

Step 1: Descent

$$\boldsymbol{\xi}^n = \mathbf{A}^{-1}\mathbf{B}^t\mathbf{w}^n, \tag{5.240}$$

$$\boldsymbol{\eta}^n = \mathbf{B}\boldsymbol{\xi}^n \ (= \mathscr{B}\mathbf{w}^n), \tag{5.241}$$

$$\rho_n = \frac{|\mathbf{g}^n|^2}{(\boldsymbol{\eta}^n, \mathbf{w}^n)}, \tag{5.242}$$

$$\boldsymbol{\lambda}^{n+1} = \boldsymbol{\lambda}^n - \rho_n \mathbf{w}^n. \tag{5.243}$$

Step 2: New descent direction

$$\mathbf{g}^{n+1} = \mathbf{g}^n - \rho_n \boldsymbol{\eta}^n, \tag{5.244}$$

$$\gamma_n = \frac{|\mathbf{g}^{n+1}|^2}{|\mathbf{g}^n|^2}, \tag{5.245}$$

$$\mathbf{w}^{n+1} = \mathbf{g}^{n+1} + \gamma_n \mathbf{w}^n, \tag{5.246}$$

$n = n + 1$ *go to* (5.240).

Once $\{\boldsymbol{\lambda}^n\}_{n\geq 0}$ has converged to the $\boldsymbol{\lambda}$ solution of (5.226) (and (5.223)), we obtain the corresponding $\mathbf{x}$ in (5.219) via the relation

$$\mathbf{x} = \mathbf{A}^{-1}(\mathbf{b} - \mathbf{B}^t\boldsymbol{\lambda}). \tag{5.247}$$

From (5.237)–(5.246) it follows that each iteration of algorithm (5.227)–(5.234) will require the solution of a linear system whose matrix is $\mathbf{A}$; since $\mathbf{A}$ is symmetric and positive definite, one may use a Cholesky factorization $\mathbf{A} = \mathbf{L}\mathbf{L}^t$ of $\mathbf{A}$ (with $\mathbf{L}$ lower triangular) done once and for all; each iteration of (5.237)–(5.246) will therefore require the solution of two triangular linear systems, which is not a very costly operation.

Relation with Uzawa algorithms. We would now like to justify the title of this section (Sec. 5.8.7.3.2); to achieve this goal, let us consider the solution of (5.223), (5.226) by a standard gradient method with a constant step ρ (>0); such an algorithm is written as follows:

$$\boldsymbol{\lambda}^0 \in \mathbb{R}^M; \tag{5.248}$$

then for $n \geq 0$ *we define* $\boldsymbol{\lambda}^{n+1}$ *from* $\boldsymbol{\lambda}^n$ *by*

$$\boldsymbol{\lambda}^{n+1} = \boldsymbol{\lambda}^n - \rho(\mathscr{B}\boldsymbol{\lambda}^n - \boldsymbol{\beta}). \tag{5.249}$$

Since $\mathscr{B} = \mathbf{B}\mathbf{A}^{-1}\mathbf{B}^t$, (5.249) is in fact equivalent to

$$\mathbf{x}^n = \mathbf{A}^{-1}(\mathbf{b} - \mathbf{B}^t\boldsymbol{\lambda}^n), \tag{5.250}$$

$$\boldsymbol{\lambda}^{n+1} = \boldsymbol{\lambda}^n + \rho(\mathbf{B}\mathbf{x}^n - \mathbf{c}). \tag{5.251}$$

In (5.248), (5.250), (5.251) we recognize an Uzawa algorithm (see, e.g., G.L.T. [3, Chapter 2]) applied to the solution of the linear system (5.219); algorithm (5.227)–(5.234) (in its equivalent formulation (5.237)–(5.246)) is clearly a sophisticated variant of the Uzawa algorithm (5.248), (5.250), (5.251).

Algorithm (5.248), (5.250), (5.251) is a particular case of a general family of algorithms for computing the saddle points of Lagrangian functionals; in the present problem the Lagrangian functional is defined by

$$\mathscr{L}(\mathbf{y}, \boldsymbol{\mu}) = \tfrac{1}{2}((\mathbf{A}\mathbf{y}, \mathbf{y})) - ((\mathbf{b}, \mathbf{y})) + (\boldsymbol{\mu}, \mathbf{B}\mathbf{y} - \mathbf{c}) \tag{5.252}$$

(with $((\cdot, \cdot))$ the usual scalar product of $\mathbb{R}^N$), and the associated saddle-point problem is as follows:

Find $\{\mathbf{x}, \boldsymbol{\lambda}\} \in \mathbb{R}^N \times \mathbb{R}^M$ *such that*

$$\mathscr{L}(\mathbf{x}, \boldsymbol{\mu}) \leq \mathscr{L}(\mathbf{x}, \boldsymbol{\lambda}) \leq \mathscr{L}(\mathbf{y}, \boldsymbol{\lambda}), \qquad \forall \{\mathbf{y}, \boldsymbol{\mu}\} \in \mathbb{R}^N \times \mathbb{R}^M. \tag{5.253}$$

As usual (see G.L.T. [1], [3], Cea [1], [2], Rockafellar [1], and Ekeland and Temam [1] for more details) the *primal problem* associated with (5.253) is defined by

$$\operatorname*{Inf}_{\mathbf{y} \in \mathbb{R}^N} \operatorname*{Sup}_{\boldsymbol{\mu} \in \mathbb{R}^M} \mathscr{L}(\mathbf{y}, \boldsymbol{\mu}) \tag{5.254}$$

and the *dual problem* by

$$\operatorname*{Sup}_{\boldsymbol{\mu} \in \mathbb{R}^M} \operatorname*{Inf}_{\mathbf{y} \in \mathbb{R}^N} \mathscr{L}(\mathbf{y}, \boldsymbol{\mu}). \tag{5.255}$$

We then have:

Proposition 5.5. *The primal problem reduces to*

$$\mathbf{x} \in H, \quad J(\mathbf{x}) \leq J(\mathbf{y}), \qquad \forall\, \mathbf{y} \in H, \tag{5.256}$$

where

$$H = \{\mathbf{y} \mid \mathbf{y} \in \mathbb{R}^N, \mathbf{B}\mathbf{y} - \mathbf{c} = \mathbf{0}\}, \tag{5.257}$$

$$J(\mathbf{y}) = \tfrac{1}{2}((\mathbf{A}\mathbf{y}, \mathbf{y})) - ((\mathbf{b}, \mathbf{y})), \qquad \forall\, \mathbf{y} \in \mathbb{R}^N; \tag{5.258}$$

the dual problem is defined by

$$\boldsymbol{\lambda} \in \mathbb{R}^M, \quad J^*(\boldsymbol{\lambda}) \leq J^*(\boldsymbol{\mu}), \qquad \forall\, \boldsymbol{\mu} \in \mathbb{R}^M, \tag{5.259}$$

where

$$J^*(\boldsymbol{\mu}) = \tfrac{1}{2}(\mathbf{B}\mathbf{A}^{-1}\mathbf{B}^t\boldsymbol{\mu}, \boldsymbol{\mu}) - (\mathbf{B}\mathbf{A}^{-1}\mathbf{b} - \mathbf{c}, \boldsymbol{\mu}) \tag{5.260}$$

(i.e., the dual problem is the linear system (5.223), (5.226)).

PROOF. From (5.252), (5.258) we have

$$\operatorname*{Sup}_{\boldsymbol{\mu} \in \mathbb{R}^M} \mathscr{L}(\mathbf{y}, \boldsymbol{\mu}) = J(\mathbf{y}) + \operatorname*{Sup}_{\boldsymbol{\mu} \in \mathbb{R}^M} (\boldsymbol{\mu}, \mathbf{B}\mathbf{y} - \mathbf{c}); \tag{5.261}$$

we clearly have

$$\operatorname*{Sup}_{\boldsymbol{\mu} \in \mathbb{R}^M} (\boldsymbol{\mu}, \mathbf{B}\mathbf{y} - \mathbf{c}) = \begin{cases} 0 & \text{if } \mathbf{y} \in H, \\ +\infty & \text{if } \mathbf{y} \notin H. \end{cases} \tag{5.262}$$

From (5.262) it follows that the functional

$$\mathbf{y} \to \operatorname*{Sup}_{\boldsymbol{\mu} \in \mathbb{R}^M} (\boldsymbol{\mu}, \mathbf{B}\mathbf{y} - \mathbf{c})$$

is in fact the indicator functional I_H of the affine subspace H; since $\mathbf{c} \in R(\mathbf{B})$ implies that $H \neq \emptyset$, we have I_H convex, proper, and l.s.c.. Hence the primal problem reduces to

$$\operatorname*{Inf}_{\mathbf{y} \in \mathbb{R}^N} \{J(\mathbf{y}) + I_H(\mathbf{y})\}, \tag{5.263}$$

which is equivalent to (5.256) (from Chapter I, Sec. 2).

Now consider the minimization problem

$$\operatorname*{Inf}_{\mathbf{y} \in \mathbb{R}^N} \mathscr{L}(\mathbf{y}, \boldsymbol{\mu}); \tag{5.264}$$

it has a unique solution $\mathbf{x}(\boldsymbol{\mu})$ which is also the solution of the linear system

$$\mathbf{A}\mathbf{x}(\boldsymbol{\mu}) = \mathbf{b} - \mathbf{B}^t\boldsymbol{\mu}. \tag{5.265}$$

From (5.252), (5.265) we then have

$$\operatorname*{Inf}_{\mathbf{y} \in \mathbb{R}^N} \mathscr{L}(\mathbf{y}, \boldsymbol{\mu}) = -\tfrac{1}{2}(\mathbf{B}\mathbf{A}^{-1}\mathbf{B}^t\boldsymbol{\mu}, \boldsymbol{\mu}) + (\boldsymbol{\mu}, \mathbf{B}\mathbf{A}^{-1}\mathbf{b} - \mathbf{c}) - \tfrac{1}{2}((\mathbf{b}, \mathbf{A}^{-1}\mathbf{b})), \tag{5.266}$$

which implies that the dual problem (5.255) reduces, in fact, to (5.259). □

5.8.7.3.3. *Iterative solution of* (5.219) *by an augmented Lagrangian algorithm.* This section closely follows Fortin and Glowinski [3].

Principle and description of the method. We complete Remark 7.5 of Chapter I, Sec. 7.4.2.

Let r be a positive parameter; we observe that the linear system (5.219) is in fact equivalent to

$$\mathbf{A}\mathbf{x} + r\mathbf{B}^t(\mathbf{B}\mathbf{x} - \mathbf{c}) + \mathbf{B}^t\boldsymbol{\lambda} = \mathbf{b} \qquad \mathbf{B}\mathbf{x} = \mathbf{c}, \tag{5.267}$$

and that (5.267) characterizes $\{\mathbf{x}, \boldsymbol{\lambda}\}$ as a saddle point over $\mathbb{R}^N \times \mathbb{R}^M$ of the (augmented) Lagrangian functional $\mathscr{L}_r\colon \mathbb{R}^N \times \mathbb{R}^M \to \mathbb{R}$ defined by

$$\mathscr{L}_r(\mathbf{y}, \boldsymbol{\mu}) = \mathscr{L}(\mathbf{y}, \boldsymbol{\mu}) + \frac{r}{2}|\mathbf{B}\mathbf{y} - \mathbf{c}|^2, \tag{5.268}$$

with $\mathscr{L}(\cdot, \cdot)$ defined as in (5.252).

Applying the standard Uzawa algorithm to the saddle-point problem:

Find $\{\mathbf{x}, \boldsymbol{\lambda}\} \in \mathbb{R}^N \times \mathbb{R}^M$ *such that*

$$\mathscr{L}_r(\mathbf{x}, \boldsymbol{\mu}) \leq \mathscr{L}_r(\mathbf{x}, \boldsymbol{\lambda}) \leq \mathscr{L}_r(\mathbf{y}, \boldsymbol{\lambda}), \qquad \forall \{\mathbf{y}, \boldsymbol{\mu}\} \in \mathbb{R}^N \times \mathbb{R}^M, \tag{5.269}$$

we obtain:

$$\boldsymbol{\lambda}^0 \in \mathbb{R}^M \textit{ given}; \tag{5.270}$$

then for $n \geq 0$ *we define* $\mathbf{x}^n, \boldsymbol{\lambda}^{n+1}$ *from* $\boldsymbol{\lambda}^n$ *by:*

Find $\mathbf{x}^n \in \mathbb{R}^N$ *such that*

$$\mathscr{L}_r(\mathbf{x}^n, \boldsymbol{\lambda}^n) \le \mathscr{L}_r(\mathbf{y}, \boldsymbol{\lambda}^n), \qquad \forall\, \mathbf{y} \in \mathbb{R}^N, \tag{5.271}$$

$$\boldsymbol{\lambda}^{n+1} = \boldsymbol{\lambda}^n + \rho(\mathbf{B}\mathbf{x}^n - \mathbf{c}); \tag{5.272}$$

actually (5.271) is equivalent to the linear system

$$(\mathbf{A} + r\mathbf{B}^t\mathbf{B})\mathbf{x}^n = \mathbf{b} + r\mathbf{B}^t\mathbf{c} - \mathbf{B}^t\boldsymbol{\lambda}^n. \tag{5.271$'$}$$

Convergence of algorithm (5.270)–(5.272). Concerning the convergence of algorithm (5.270)–(5.272), we have:

Proposition 5.6. *We have convergence of algorithm* (5.270)–(5.272), $\forall\, \boldsymbol{\lambda}^0 \in \mathbb{R}^M$, *if and only if*

$$0 < \rho < 2\left(r + \frac{1}{\Lambda_{M_0}}\right), \tag{5.273}$$

where Λ_{M_0} *is the largest eigenvalue of* $\mathbf{A}^{-1}\mathbf{B}^t\mathbf{B}$. *If* (5.273) *holds, then*

$$\lim_{n \to +\infty} \{\mathbf{x}^n, \boldsymbol{\lambda}^n\} = \{\mathbf{x}, \hat{\boldsymbol{\lambda}} + \boldsymbol{\lambda}_2^0\}, \tag{5.274}$$

where $\{\mathbf{x}, \hat{\boldsymbol{\lambda}})$ *is the unique solution of* (5.219) *in* $\mathbb{R}^N \times R(\mathbf{B})$ *and* $\boldsymbol{\lambda}_2^0$ *is the same as in the statement of Proposition* 5.4.

PROOF. Let us define $\bar{\mathbf{x}}^n, \bar{\boldsymbol{\lambda}}^n$ by

$$\bar{\mathbf{x}}^n = \mathbf{x}^n - \mathbf{x}, \qquad \bar{\boldsymbol{\lambda}}^n = \boldsymbol{\lambda}^n - (\hat{\boldsymbol{\lambda}} + \boldsymbol{\lambda}_2^0). \tag{5.275}$$

We have $\mathbb{R}^M = R(\mathbf{B}) \oplus \mathrm{Ker}(\mathbf{B}^t)$, and

$$\boldsymbol{\lambda}^n = \boldsymbol{\lambda}_1^n + \boldsymbol{\lambda}_2^n,$$

where $\lambda_1^n \in R(\mathbf{B})$, $\lambda_2^n \in \mathrm{Ker}(\mathbf{B}^t)$ (the above decomposition is unique).

From (5.272) it follows that $\boldsymbol{\lambda}^{n+1} - \boldsymbol{\lambda}^n \in R(\mathbf{B})$, which shows that

$$\boldsymbol{\lambda}_2^n = \boldsymbol{\lambda}_2^0, \quad \forall\, n, \qquad \bar{\boldsymbol{\lambda}}^n \in R(\mathbf{B}), \qquad \forall\, n. \tag{5.276}$$

By subtraction we obtain, from (5.267), (5.271)$'$, (5.272),

$$(\mathbf{A} + r\mathbf{B}^t\mathbf{B})\bar{\mathbf{x}}^n + \mathbf{B}^t\bar{\boldsymbol{\lambda}}^n = \mathbf{0}, \tag{5.277}$$

$$\bar{\boldsymbol{\lambda}}^{n+1} = \bar{\boldsymbol{\lambda}}^n + \rho\mathbf{B}\bar{\mathbf{x}}^n. \tag{5.278}$$

Combining (5.277) and (5.278), by elimination of $\bar{\mathbf{x}}^n$ we obtain

$$\mathbf{A}^{-1}\mathbf{B}^t\bar{\boldsymbol{\lambda}}^{n+1} = \mathbf{A}^{-1}\mathbf{B}^t\bar{\boldsymbol{\lambda}}^n - \rho\mathbf{A}^{-1}\mathbf{B}^t\mathbf{B}(\mathbf{I} + r\mathbf{A}^{-1}\mathbf{B}^t\mathbf{B})^{-1}\mathbf{A}^{-1}\mathbf{B}^t\bar{\boldsymbol{\lambda}}^n. \tag{5.279}$$

Let us define $\{\mathbf{z}^n\}_{n \ge 0}$ by $\mathbf{z}^n = \mathbf{A}^{-1}\mathbf{B}^t\bar{\boldsymbol{\lambda}}^n$; it follows from (5.276) that

$$\mathbf{z}^n \in R(\mathbf{A}^{-1}\mathbf{B}^t\mathbf{B}), \qquad \forall\, n, \tag{5.280}$$

and from (5.279) that

$$\mathbf{z}^{n+1} = (\mathbf{I} - \rho\mathbf{A}^{-1}\mathbf{B}^t\mathbf{B}(\mathbf{I} + r\mathbf{A}^{-1}\mathbf{B}^t\mathbf{B})^{-1})\mathbf{z}^n. \tag{5.281}$$

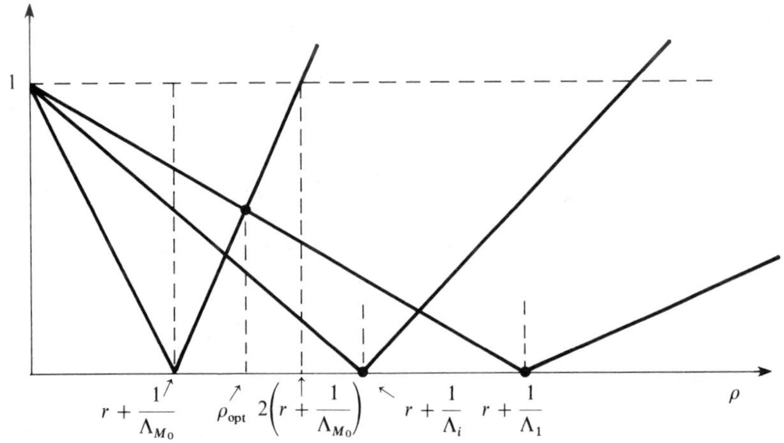

Figure 5.3

Since $R(\mathbf{B})$ and $R(\mathbf{A}^{-1}\mathbf{B}^t\mathbf{B})$ are isomorphic (via $\mathbf{A}^{-1}\mathbf{B}^t$), the convergence of $\mathbf{z}^n$ to $\mathbf{0}$ will imply (5.274) (with the same speed of convergence). The eigenvalues of $\mathbf{A}^{-1}\mathbf{B}^t\mathbf{B}$ are non-negative; let $\{\Lambda_i\}_{i=1}^{M_0}$ be the strictly positive eigenvalues of $\mathbf{A}^{-1}\mathbf{B}^t\mathbf{B}$, ordered such that

$$0 < \Lambda_1 \leq \Lambda_2 \leq \cdots \leq \Lambda_{M_0}.$$

From the properties of $\mathbf{A}$ and $\mathbf{B}^t\mathbf{B}$, there exists, a vector basis $\{\mathbf{w}_i\}_{i=1}^{M_0}$ of $R(\mathbf{A}^{-1}\mathbf{B}^t\mathbf{B})$ such that

$$\mathbf{w}_i \text{ is an eigenvector of } \mathbf{A}^{-1}\mathbf{B}^t\mathbf{B} \text{ associated with } \Lambda_i, \ \forall\, i = 1, \ldots, M_0,$$

$$((\mathbf{A}\mathbf{w}_i, \mathbf{w}_j)) = 0, \qquad \forall\, i, j, \quad 1 \leq i, j \leq M_0, \quad i \neq j. \tag{5.282}$$

Since $\mathbf{z}^n \in R(\mathbf{A}^{-1}\mathbf{B}^t\mathbf{B})$, we have $\mathbf{z}^n = \sum_{i=1}^{M_0} z_i^n \mathbf{w}_i$, and from (5.281)

$$\forall\, n \text{ and } \forall\, i = 1, \ldots, M_0, \text{ we have}$$

$$z_i^{n+1} = \frac{1 + (r - \rho)\Lambda_i}{1 + r\Lambda_i}\, z_i^n. \tag{5.283}$$

Relation (5.283) implies that we have convergence of $\mathbf{z}^n$ to $\mathbf{0}$, $\forall\, \boldsymbol{\lambda}^0 \in \mathbb{R}^M$, if and only if

$$\left|\frac{1 + (r - \rho)\Lambda_i}{1 + r\Lambda_i}\right| < 1, \qquad \forall\, i = 1, \ldots, M_0. \tag{5.284}$$

We have (5.284) if and only if ρ obeys (5.273); this clearly appears from Fig. 5.3, where we have shown the graphs of the functions

$$\rho \to \left|\frac{1 + (r - \rho)\Lambda_i}{1 + r\Lambda_i}\right|$$

for several Λ_i. □

As a corollary of Proposition 5.6 we have:

Proposition 5.7. *Suppose that $\rho = r$; then for r sufficiently large the convergence ratio of algorithm* (5.270)–(5.272) *is of order* $1/r$.

PROOF. The above result follows directly from (5.283) which shows that if $\rho = r$, then

$\forall\, n$ and $\forall\, i = 1, \ldots, M_0$, we have

$$|z_i^{n+1}| \leq \frac{1}{1 + r\Lambda_1} |z_i^n|. \qquad \square$$

Remark 5.16. From (5.283) (and Fig. 5.3) the optimal value of ρ is given by

$$\rho_{\text{opt}} = r + \frac{r(\Lambda_1 + \Lambda_{M_0}) + 2}{2r\Lambda_1\Lambda_{M_0} + (\Lambda_1 + \Lambda_{M_0})};$$

in fact if r is sufficiently large, our numerical experiments show that there is no practical gain in using $\rho = \rho_{\text{opt}}$ instead of $\rho = r$.

Remark 5.17. In Fortin and Glowinski [3] one will find a conjugate gradient variant of algorithm (5.270)–(5.272) (reducing to algorithm (5.237)–(5.246) if $r = 0$); actually if $\rho = r$ and r is sufficiently large, algorithm (5.270)–(5.272) has a very fast convergence and does not require a conjugate gradient acceleration.

Remark 5.18. Theoretically, the optimal strategy on r is to take this parameter as large as possible; in practice there is an upper limit due to roundoff errors (see Chapter VI, Sec. 4.3 for further comments). Actually some authors advocate the use of algorithm (5.270)–(5.272) with a sequence of parameters $\{r_n\}_{n \geq 0}$, obeying $\lim_{n \to +\infty} r_n = +\infty$, instead of a constant parameter r; in the particular case of algorithm (5.270)–(5.272), this strategy could be extremely costly since each iteration would require the solution of a linear system whose matrix depends upon n.

5.8.7.3.4. *Application of the conjugate gradient algorithm* (5.227)–(5.234) *to the solution of the discrete Stokes problem* (5.157). Instead of (5.227)–(5.234) we shall use its equivalent formulation (5.237)–(5.246). The courageous (or skeptical) reader will verify that we obtain an algorithm whose variational description is given by the following.

Step 0: Initialization

$$p_h^0 \in H_h^1 \text{ arbitrarily given}; \tag{5.285}$$

compute $\mathbf{u}_h^0 \in V_{gh}$ *such that*

$$\alpha \int_\Omega \mathbf{u}_h^0 \cdot \mathbf{v}_h \, dx + \nu \int_\Omega \nabla \mathbf{u}_h^0 \cdot \nabla \mathbf{v}_h \, dx = \int_\Omega \mathbf{f}_h \cdot \mathbf{v}_h \, dx - \int_\Omega \nabla \mathbf{p}_h^0 \cdot \mathbf{v}_h \, dx, \quad \forall\, \mathbf{v}_h \in V_{0h}; \tag{5.286}$$

compute $z_h^0 \in H_h^1$ *such that*

$$\int_\Omega z_h^0 q_h \, dx = \int_\Omega \nabla \cdot \mathbf{u}_h^0 q_h \, dx, \qquad \forall \, q_h \in H_h^1, \tag{5.287}$$

$$w_h^0 = z_h^0. \tag{5.288}$$

Then for $n \geq 0$ *compute* p_h^{n+1}, z_h^{n+1}, w_h^{n+1} *from* p_h^n, z_h^n, w_h^n *as follows*

Step 1: Descent

Compute $\boldsymbol{\chi}_h^n \in V_{0h}$ such that

$$\alpha \int_\Omega \boldsymbol{\chi}_h^n \cdot \mathbf{v}_h \, dx + \nu \int_\Omega \nabla \boldsymbol{\chi}_h^n \cdot \nabla \mathbf{v}_h \, dx = - \int_\Omega \nabla w_h^n \cdot \mathbf{v}_h \, dx, \qquad \forall \, \mathbf{v}_h \in V_{0h}; \tag{5.289}$$

then

$$\rho_n = \frac{\int_\Omega z_h^n w_h^n \, dx}{\int_\Omega \nabla \cdot \boldsymbol{\chi}_h^n w_h^n \, dx} = \frac{\|z_h^n\|_{L^2(\Omega)}^2}{\int_\Omega \nabla \cdot \boldsymbol{\chi}_h^n w_h^n \, dx}, \tag{5.290}$$

$$p_h^{n+1} = p_h^n - \rho_n w_h^n. \tag{5.291}$$

Step 2: New descent direction

Compute $z_h^{n+1} \in H_h^1$ *such that*

$$\int_\Omega z_h^{n+1} q_h \, dx = \int_\Omega z_h^n q_h \, dx - \rho_n \int_\Omega \nabla \cdot \boldsymbol{\chi}_h^n q_h \, dx, \qquad \forall \, q_h \in H_h^1, \tag{5.292}$$

$$\gamma_n = \frac{\|z_h^{n+1}\|_{L^2(\Omega)}^2}{\|z_h^n\|_{L^2(\Omega)}^2}, \tag{5.293}$$

$$w_h^{n+1} = z_h^{n+1} + \gamma_n w_h^n \tag{5.294}$$

$n = n + 1$ *go to* (5.289).

Thus each iteration of (5.285)–(5.294) costs:

(i) two uncoupled discrete Dirichlet problems[28] (associated with $\alpha I - \nu\Delta$) to obtain $\boldsymbol{\chi}_h^n$ from w_h^n via (5.289);
(ii) the linear problem (5.292) to obtain z_h^{n+1} from z_h^n and $\boldsymbol{\chi}_h^n$.

Remark 5.19. Algorithm (5.285)–(5.294) is a discrete variant of algorithm (13.17)–(13.27) in Glowinski and Pironneau [3] (see also Sec. 5.8.7.4, where the solution of the continuous Stokes problem (5.222) by a continuous analogue of (5.285)–(5.294) will be discussed).

[28] Three in dimension 3.

Remark 5.20. We can avoid the solution, at each iteration, of the linear problem (5.292) by using, over H_h^1, instead of the $L^2(\Omega)$-scalar product, the following scalar product[29]

$$(y_h, q_h)_h = \sum_{i=1}^{M_h} y_h(P_i)q_h(P_i), \qquad \forall\, y_h, q_h \in H_h^1, \tag{5.295}$$

where (cf. Sec. 5.8.7.2) $\{P_i\}_{i=1}^{M_h}$ is the set of the vertices of $\mathscr{T}_h$. If one uses the scalar product (5.295), one would have to replace (5.287), (5.290), (5.292), (5.293) by

$$z_h^0 \in H_h^1, \quad (z_h^0, q_h)_h = \int_\Omega \nabla \cdot \mathbf{u}_h^0 q_h\, dx, \qquad \forall\, q_h \in H_h^1, \tag{5.287$'$}$$

$$\rho_n = \frac{(z_h^n, w_h^n)_h}{\int_\Omega \nabla \cdot \boldsymbol{\chi}_h^n w_h^n\, dx} = \frac{(z_h^n, z_h^n)_h}{\int_\Omega \nabla \cdot \boldsymbol{\chi}_h^n w_h^n\, dx}, \tag{5.290$'$}$$

$$z_h^{n+1} \in H_h^1, \quad (z_h^{n+1}, q_h)_h = (z_h^n, q_h)_h - \rho_n \int_\Omega \nabla \cdot \boldsymbol{\chi}_h^n q_h\, dx, \qquad \forall\, q_h \in H_h^1, \tag{5.292$'$}$$

$$\gamma_n = \frac{(z_h^{n+1}, z_h^{n+1})_h}{(z_h^n, z_h^n)_h}, \tag{5.293$'$}$$

respectively. To be more precise, from (5.292)′, (5.295) we have

$$z_h^{n+1}(P_i) = z_h^n(P_i) - \rho_n \int_\Omega \nabla \cdot \boldsymbol{\chi}_h^n w_i\, dx, \qquad \forall\, i = 1, \ldots, M_h, \tag{5.296}$$

where w_i is the ith element of the vector basis $\mathscr{B}_h$ of H_h^1 defined by (5.194), (5.195).

5.8.7.3.5. *Application of the augmented Lagrangian algorithm* (5.270)–(5.272) *to the solution of the discrete Stokes problem* (5.157). There are in fact several possibilities for implementing algorithm (5.270)–(5.272), depending upon the scalar product equipping H_h^1; a most natural choice is the following variant of (5.295):

$$(y_h, q_h)_h = \sum_{i=1}^{M_h} \omega_i y_h(P_i)q_h(P_i), \qquad \forall\, y_h, q_h \in H_h^1, \tag{5.297}$$

where $\{P_i\}_{i=1}^{M_h}$ is the set of the vertices of $\mathscr{T}_h$ and

$$\omega_i = \int_\Omega w_i\, dx$$

(with w_i as in (5.296)); we have, in fact,

$$\omega_i = \tfrac{1}{3} \text{ meas } (\bar{\Omega}_i),$$

[29] We can also use scalar products other than (5.295) (like (5.297) in Sec. 5.8.7.3.5, for example).

where $\overline{\Omega}_i$ is the union of those triangles of $\mathscr{T}_h$ having P_i as a common vertex (we may consider that $(y_h, q_h)_h$ is obtained from $\int_\Omega y_h q_h \, dx$ through an approximate numerical integration procedure, namely the two-dimensional trapezoidal rule).

Now, if $\mathbf{v}_h \in V_h$, we define the approximate divergence operator $\operatorname{div}_h: V_h \to H_h^1$ by

$$\operatorname{div}_h \mathbf{v}_h = \sum_{i=1}^{M_h} \frac{1}{\omega_i} \left(\int_\Omega \nabla \cdot \mathbf{v}_h w_i \, dx \right) w_i; \tag{5.298}$$

the approximate incompressibility condition

$$\int_\Omega \nabla \cdot \mathbf{v}_h q_h \, dx = 0, \qquad \forall \, q_h \in H_h^1$$

is then equivalent to

$$\operatorname{div}_h \mathbf{v}_h = 0. \tag{5.299}$$

The augmented Lagrangian algorithm (5.270)–(5.272) applied to the solution of the discrete Stokes problem (5.157) then takes the following form (with $r > 0$, $\rho > 0$):

$$p_h^0 \in H_h^1 \textit{ arbitrarily given}; \tag{5.300}$$

then for $n \geq 0$, *we obtain* $\mathbf{u}_h^n$ *and* p_h^{n+1} *from* p_h^n *by the following*:

Compute $\mathbf{u}_h^n \in V_{gh}$ *such that*

$$\alpha \int_\Omega \mathbf{u}_h^n \cdot \mathbf{v}_h \, dx + \nu \int_\Omega \nabla \mathbf{u}_h^n \cdot \nabla \mathbf{v}_h \, dx + r(\operatorname{div}_h \mathbf{u}_h^n, \operatorname{div}_h \mathbf{v}_h)_h$$

$$= \int_\Omega \mathbf{f}_h \cdot \mathbf{v}_h \, dx - \int_\Omega \nabla p_h^n \cdot \mathbf{v}_h \, dx, \qquad \forall \, \mathbf{v}_h \in V_{0h}; \tag{5.301}$$

then

$$p_h^{n+1} = p_h^n - \rho \operatorname{div}_h \mathbf{u}_h^n. \tag{5.302}$$

A good choice for ρ is (still) $\rho = r$.

Thus each iteration of algorithm (5.300)–(5.302) costs the solution of the linear problem (5.301) which is clearly related to the continuous operator[30]

$$\mathbf{v} \to \alpha \mathbf{v} - \nu \Delta \mathbf{v} - r \nabla (\nabla \cdot \mathbf{v}), \tag{5.303}$$

which is a variant of the linear elasticity operator. We have to observe that (unlike the operator $\mathbf{v} \to \alpha \mathbf{v} - \nu \Delta \mathbf{v}$) the operator (5.303) couples the components of $\mathbf{v}$ (through the term $-r\nabla(\nabla \cdot \mathbf{v})$); a similar property holds for the

[30] Actually this operator also plays an important role in the compressible Navier–Stokes equations.

discrete variant of the operator (5.303) occurring in (5.301). Such coupling makes the solution of (5.301) much more costly than, for example, the solution of (5.289) in algorithm (5.285)–(5.294) of Sec. 5.8.7.3.4.

Remarks 5.16, 5.17, and 5.18 of Sec. 5.8.7.3.3 still hold for algorithm (5.300)–(5.302).

A continuous analogue of algorithm (5.300)–(5.302) will be discussed in Sec. 5.8.7.4.

Some variants of the above methods for solving the Stokes and Navier–Stokes problems can be found in Fortin and Thomasset [2], Thomasset [1], and Segal [1].

5.8.7.4. *Some remarks on the iterative solution of the continuous Stokes problem* (5.222)

5.8.7.4.1. *Motivation.* Suppose that an infinite-dimensional problem (P) has been approximated by a finite-dimensional problem (P_h); also suppose that a given algorithm for solving (P_h) is in fact the discrete analogue of an algorithm for solving (P). In many cases the convergence of the infinite-dimensional algorithm will imply the convergence of the discrete one in a number of iterations independent (or "almost" independent) of the discretization parameter h; it is this observation which has motivated our study of the continuous analogues of the algorithms of Secs. 5.8.7.3.2 and 5.8.7.3.3.

The following analysis closely follows Glowinski and Pironneau [3, Secs. 13.3.2, 13.3.3].

5.8.7.4.2. *A fundamental theoretical result.* From now on, Ω is a bounded domain of $\mathbb{R}^N$ ($N = 2, 3$ in practice) and Γ $(= \partial\Omega)$ is regular (say Lipschitz continuous). We define $H \subset L^2(\Omega)$ by

$$H = \left\{ q \,|\, q \in L^2(\Omega),\ \int_\Omega q(x)\, dx = 0 \right\}.$$

The convergence of the iterative methods, to be described in the sequel, will follow from:

Theorem 5.10. *Let* $\mathscr{A}: L^2(\Omega) \to L^2(\Omega)$ *be defined by*

$$q \in L^2(\Omega), \tag{5.304}$$

$$\alpha\mathbf{v} - \nu\Delta\mathbf{v} = -\nabla q \text{ in } \Omega, \quad \mathbf{v} \in (H_0^1(\Omega))^N \quad (\text{which implies } \mathbf{v}|_\Gamma = \mathbf{0}), \tag{5.305}$$

$$\mathscr{A}q = \nabla \cdot \mathbf{v}. \tag{5.306}$$

Then $\mathscr{A}$ *is H-elliptic, self-adjoint, and automorphic from H onto H* (*i.e.,* $\exists\, \beta > 0$ *such that* $(\mathscr{A}q, q)_{L^2} \geq \beta\|q\|_{L^2}^2,\ \forall\, q \in H$).

The proof of Theorem 5.10 can be found in Crouzeix [2].

Remark 5.21. The discrete forms of $\mathscr{A}$ are, in general, *full* matrices.

5.8.7.4.3. *Gradient methods and variants.* Let us define $\mathbf{H}_g^1 \subset (H^1(\Omega))^N$ by[31]

$$\mathbf{H}_g^1 = \{\mathbf{v} \mid \mathbf{v} \in (H^1(\Omega))^N, \mathbf{v} = \mathbf{g} \text{ on } \Gamma\}. \tag{5.307}$$

Now let $\{\mathbf{u}, p\} \in \mathbf{H}_g^1 \times L^2(\Omega)/\mathbb{R}$ be the solution of the Stokes problem

$$\alpha\mathbf{u} - \nu\Delta\mathbf{u} + \nabla p = \mathbf{f} \text{ in } \Omega, \quad \nabla \cdot \mathbf{u} = 0 \text{ in } \Omega, \quad \mathbf{u} = \mathbf{g} \text{ on } \Gamma, \tag{5.308}$$

and $\mathbf{u}_0$ be the solution of the (vector) Dirichlet problem

$$\alpha\mathbf{u}_0 - \nu\Delta_0 = \mathbf{f} \text{ in } \Omega, \quad \mathbf{u}_0 = \mathbf{g} \text{ on } \Gamma. \tag{5.309}$$

Problem (5.309) has a unique solution, and by subtracting (5.309) and (5.308), we have

$$\alpha(\mathbf{u} - \mathbf{u}_0) - \nu\Delta(\mathbf{u} - \mathbf{u}_0) = -\nabla p \text{ in } \Omega, \quad \mathbf{u} - \mathbf{u}_0 \in (H_0^1(\Omega))^N.$$

Hence (from Theorem 5.10), $\mathscr{A}p = \nabla \cdot (\mathbf{u} - \mathbf{u}_0) = -\nabla \cdot \mathbf{u}_0$. In other words, the pressure is the unique solution, in $L^2(\Omega)/\mathbb{R}$, of

$$\mathscr{A}p = -\nabla \cdot \mathbf{u}_0. \tag{5.310}$$

Due to the properties of $\mathscr{A}$ (see Theorem 5.10), it is natural to solve (5.310) (and therefore the continuous Stokes problem (5.308)) by iterative methods such as the method of steepest descent (or even better, by a conjugate gradient method).

Gradient method with fixed step size. For a given $\rho > 0$, consider the following algorithm:

$$p^0 \in L^2(\Omega) \text{ given arbitrarily;} \tag{5.311}$$

then for $n \geq 0$, p^n *given, compute*

$$p^{n+1} = p^n - \rho(\mathscr{A}p^n + \nabla \cdot \mathbf{u}_0). \tag{5.312}$$

In practice one has to replace (5.312) by

$$\alpha\mathbf{u}^n - \nu\Delta\mathbf{u}^n = \mathbf{f} - \nabla p^n \text{ in } \Omega, \quad \mathbf{u}^n = \mathbf{g} \text{ on } \Gamma, \tag{5.312a}$$

$$p^{n+1} = p^n - \rho\nabla \cdot \mathbf{u}^n. \tag{5.312b}$$

Remark 5.22. To solve (5.312a), we have to solve N independent Dirichlet problem for $\alpha I - \nu\Delta$ (in practice, $N = 2, 3$).

Remark 5.23. The previous method is close to the artificial compressibility method of Chorin and Yanenko.

We now discuss the convergence of algorithm (5.311), (5.312); we have, indeed, the following result:

[31] We suppose that $\mathbf{g} \in (H^{1/2}(\Omega))^N$ with $\int_\Omega \mathbf{g} \cdot \mathbf{n}\, d\Gamma = 0$.

Theorem 5.11. *If in* (5.311), (5.312) *we have*

$$0 < \rho < 2\frac{\nu}{N}, \tag{5.313}$$

then, $\forall\, p^0 \in L^2(\Omega)$, *we have*

$$\lim_{n \to +\infty} \{\mathbf{u}^n, p^n\} = \{\mathbf{u}, p\} \quad \text{in } (H^1(\Omega))^N \times L^2(\Omega), \text{ strongly}, \tag{5.314}$$

where $\{\mathbf{u}, p\}$ *is the solution of the Stokes problem* (5.308) *with* $\int_\Omega p\, dx = \int_\Omega p^0\, dx$. *Moreover, the convergence is linear.*[32]

PROOF. It suffices (from (5.312a)) to prove the above convergence results for the sequence $\{p^n\}_{n\geq 0}$; from (5.312b) we have

$$\int_\Omega p^n\, dx - \int_\Omega p^{n+1}\, dx = \rho \int_\Omega \nabla \cdot \mathbf{u}^n\, dx = \rho \int_\Gamma \mathbf{u}^n \cdot \mathbf{n}\, d\Gamma = \int_\Gamma \mathbf{g} \cdot \mathbf{n}\, d\Gamma = 0$$

which implies

$$\int_\Omega p^n\, dx = \int_\Omega p^0\, dx, \qquad \forall\, n \geq 1. \tag{5.315}$$

Now consider that solution $\{\mathbf{u}, p\}$ of the Stokes problem (5.308) such that

$$\int_\Omega p\, dx = \int_\Omega p^0\, dx, \tag{5.316}$$

and define $\bar{\mathbf{u}}^n$ $(\in (H_0^1(\Omega)^N)$ and $\bar{p}^n$ by

$$\bar{\mathbf{u}}^n = \mathbf{u}^n - \mathbf{u}, \qquad \bar{p}^n = p^n - p;$$

then from (5.315), (5.316) we have

$$\bar{p}^n \in H, \qquad \forall\, n. \tag{5.317}$$

We also have (from (5.310))

$$p = p - \rho(\mathscr{A}p + \nabla \cdot \mathbf{u}_0). \tag{5.318}$$

Subtracting (5.318) from (5.312) (resp., (5.308) from (5.312a)), we obtain

$$\bar{p}^{n+1} = \bar{p}^n - \rho \mathscr{A} \bar{p}^n \tag{5.319}$$

(resp.,

$$\bar{\mathbf{u}}^n \in (H_0^1(\Omega))^N, \quad \alpha\bar{\mathbf{u}}^n - \nu\Delta\bar{\mathbf{u}}^n = -\nabla\bar{p}^n \text{ in } \Omega). \tag{5.320}$$

Let us denote $\|\cdot\|_{L^2(\Omega)}$ by $|\cdot|$; from (5.319) we have

$$|\bar{p}^{n+1}|^2 = |\bar{p}^n|^2 - 2\rho(\mathscr{A}\bar{p}^n, \bar{p}^n)_{L^2(\Omega)} + \rho^2 |\mathscr{A}\bar{p}^n|^2. \tag{5.321}$$

We actually have (from (5.320))

$$\mathscr{A}\bar{p}^n = \nabla \cdot \bar{\mathbf{u}}^n \tag{5.322}$$

from which it follows that

$$(\mathscr{A}\bar{p}^n, \bar{p}^n)_{L^2(\Omega)} = \int_\Omega \nabla \cdot \bar{\mathbf{u}}^n \bar{p}^n\, dx = -\langle \bar{\mathbf{u}}^n, \nabla \bar{p}^n \rangle \tag{5.323}$$

---

[32] i.e., at least as fast as the convergence of a *geometric* sequence.

(where $\langle \cdot, \cdot \rangle$ denotes the duality between $(H_0^1(\Omega))^N$ and $(H^{-1}(\Omega))^N$); it then follows from (5.320), (5.323) that

$$(\mathscr{A}\bar{p}^n, \bar{p}^n)_{L^2(\Omega)} = \alpha \int_\Omega |\bar{\mathbf{u}}^n|^2 \, dx + \nu \int_\Omega |\nabla \bar{\mathbf{u}}^n|^2 \, dx$$

which, combined with (5.321), implies that

$$|\bar{p}^{n+1}|^2 = |\bar{p}^n|^2 - 2\rho\left(\alpha \int_\Omega |\bar{\mathbf{u}}^n|^2 \, dx + \nu \int_\Omega |\nabla \bar{\mathbf{u}}^n|^2 \, dx\right) + \rho^2 \int_\Omega |\nabla \cdot \bar{\mathbf{u}}^n|^2 \, dx. \quad (5.324)$$

Let $\mathbf{v} \in (H^1(\Omega))^N$; we have (if $\mathbf{v} = \{v_i\}_{i=1}^N$)

$$\begin{aligned}
\int_\Omega |\nabla \cdot \mathbf{v}|^2 \, dx &= \sum_{i=1}^N \int_\Omega \left|\frac{\partial v_i}{\partial x_i}\right|^2 dx + 2 \sum_{\substack{1 \le i, j \le N \\ i \ne j}} \int_\Omega \frac{\partial v_i}{\partial x_i} \frac{\partial v_j}{\partial x_j} \, dx \\
&\le \sum_{i=1}^N \int_\Omega \left|\frac{\partial v_i}{\partial x_i}\right|^2 dx + \sum_{\substack{1 \le i, j \le N \\ i \ne j}} \int_\Omega \left(\left|\frac{\partial v_i}{\partial x_i}\right|^2 + \left|\frac{\partial v_j}{\partial x_j}\right|^2\right) dx \\
&= N \sum_{i=1}^N \int_\Omega \left|\frac{\partial v_i}{\partial x_i}\right|^2 dx \le N \int_\Omega |\nabla \mathbf{v}|^2 \, dx \\
&\le \frac{N}{\nu}\left(\alpha \int_\Omega |\mathbf{v}|^2 \, dx + \nu \int_\Omega |\nabla \mathbf{v}|^2 \, dx\right);
\end{aligned}$$

we finally obtain

$$\alpha \int_\Omega |\mathbf{v}|^2 \, dx + \nu \int_\Omega |\nabla \mathbf{v}|^2 \, dx \ge \frac{\nu}{N} \int_\Omega |\nabla \cdot \mathbf{v}|^2 \, dx, \qquad \forall\, \mathbf{v} \in (H^1(\Omega))^N. \quad (5.325)$$

From (5.324), (5.325) it then follows that

$$|\bar{p}^{n+1}|^2 \le |\bar{p}^n|^2 - \rho\left(\frac{2\nu}{N} - \rho\right)\int_\Omega |\nabla \cdot \bar{\mathbf{u}}^n|^2 \, dx,$$

i.e.,

$$|\bar{p}^{n+1}|^2 \le |\bar{p}^n|^2 - \rho\left(\frac{2\nu}{N} - \rho\right)|\mathscr{A}\bar{p}^n|^2. \quad (5.326)$$

Since $\mathscr{A}$ is an automorphism from H onto H (cf. Theorem 5.10), we have

$$|\bar{p}^n| \le \|\mathscr{A}^{-1}\| \, |\mathscr{A}\bar{p}^n|. \quad (5.327)$$

If condition (5.313) holds, from (5.326), (5.327) we have

$$|\bar{p}^{n+1}| \le \left(1 - \rho\left(\frac{2\nu}{N} - \rho\right)\|\mathscr{A}^{-1}\|^{-2}\right)^{1/2} |\bar{p}^n| \quad (5.328)$$

with

$$0 \le 1 - \rho\left(\frac{2\nu}{N} - \rho\right)\|\mathscr{A}^{-1}\|^{-2} < 1. \quad (5.329)$$

Relations (5.328), (5.329) prove the convergence properties stated above (and give moreover an estimate of the speed of convergence).

Variants of (5.311), (5.312). In Crouzeix [2] one can find variants of (5.311), (5.312) where a sequence of parameters $\{\rho_n\}_{n\geq 0}$ (cyclic, in particular) is used instead of a fixed ρ. Accelerating method of Tchebycheff type can also be found in the above reference for (5.311), (5.312).

Steepest descent and minimal residual procedures for (5.311), (5.312) can be found in Fortin and Glowinski [3] and Fortin and Thomasset [2]. Each of these methods requires N uncoupled Dirichlet problems for $\alpha I - \nu\Delta$ to be solved at each iteration. However, these variants of (5.311), (5.312) seem less efficient (at least their discrete analogues) than the conjugate gradient method of Sec. 5.8.7.4.4 which, by the way, is only slightly costlier to implement.

5.8.7.4.4. *A conjugate gradient method*. From Daniel [1] it follows that one may solve the Stokes problem (5.308) via (5.310) by a conjugate gradient method. Such a conjugate gradient algorithm is defined as follows.

Step 0: Initialization

$$p^0 \in L^2(\Omega) \text{ arbitrarily given;} \tag{5.330}$$

then solve

$$\alpha\mathbf{u}^0 - \nu\Delta\mathbf{u}^0 = \mathbf{f} - \nabla p^0 \text{ in } \Omega, \quad \mathbf{u}^0 = \mathbf{g} \text{ on } \Gamma, \tag{5.331}$$

and set

$$z^0 = \nabla\cdot\mathbf{u}^0 \tag{5.332}$$

$$w^0 = z^0. \tag{5.333}$$

Then for $n \geq 0$, *compute* $p^{n+1}, z^{n+1}, w^{n+1}$ *from* p^n, z^n, w^n *as follows.*

Step 1: Descent

$$\rho_n = \frac{(z^n, w^n)_{L^2(\Omega)}}{(\mathscr{A}w^n, w^n)_{L^2(\Omega}} = \frac{\|z^n\|^2_{L^2(\Omega)}}{(\mathscr{A}w^n, w^n)_{L^2(\Omega)}}, \tag{5.334}$$

$$p^{n+1} = p^n - \rho_n w^n \tag{5.335}$$

Step 2: New descent direction

$$z^{n+1} = z^n - \rho_n \mathscr{A}w^n, \tag{5.336}$$

$$\gamma_n = \frac{\|z^{n+1}\|^2_{L^2(\Omega)}}{\|z^n\|^2_{L^2(\Omega)}}, \tag{5.337}$$

$$w^{n+1} = z^{n+1} + \gamma_n w^n, \tag{5.338}$$

$n = n + 1$ *go to* (5.334).

To implement (5.330)–(5.338), it is necessary to know $\mathscr{A}w^n$. From Theorem 5.10 (see Sec. 5.8.7.4.2), $\mathscr{A}w^n$ can be obtained by

$$\alpha\chi^n - \nu\Delta\chi^n = -\nabla w^n \quad \text{in } \Omega, \qquad \chi^n \in (H_0^1(\Omega))^N \quad (\Rightarrow \chi^n = \mathbf{0} \quad \text{on } \Gamma), \tag{5.339}$$

$$\mathscr{A}w^n = \nabla \cdot \chi^n. \tag{5.340}$$

Thus each iteration costs N uncoupled problems for $\alpha I - \nu\Delta$. The strong convergence of $\{\mathbf{u}^n, p^n\}$ to $\{\mathbf{u}, p\}$ can be shown as in Theorem 5.11.

Remark 5.24. Due to the H-ellipticity of $\mathscr{A}$, *it is not necessary to precondition* (i.e., *to scale*) the conjugate gradient algorithm above.

Remark 5.25. Algorithm (5.330)–(5.338) is clearly a continuous analogue of algorithm (5.285)–(5.294) in Sec. 5.8.7.3.4.

5.8.7.4.5. *Augmented Lagrangian methods.* We now consider the continuous analogues of the methods described in Sec. 5.8.7.3.5. Therefore, let $r > 0$; we note that the Stokes problem (5.222) (and (5.308)) is equivalent to

$$\alpha\mathbf{u} - \nu\Delta\mathbf{u} - r\nabla(\nabla \cdot \mathbf{u}) + \nabla p = \mathbf{f} \text{ in } \Omega, \quad \nabla \cdot \mathbf{u} = 0 \text{ in } \Omega, \quad \mathbf{u} = \mathbf{g} \text{ on } \Gamma. \tag{5.341}$$

It is then natural to generalize algorithm (5.311), (5.312) by:

$$p^0 \in L^2(\Omega) \textit{ arbitrarily given}; \tag{5.342}$$

then for $n \geq 0$, p^n *being known, we compute* $\mathbf{u}^n$, p^{n+1} *by*

$$\alpha\mathbf{u}^n - \nu\Delta\mathbf{u}^n - r\nabla(\nabla \cdot \mathbf{u}^n) = \mathbf{f} - \nabla p^n \text{ in } \Omega, \quad \mathbf{u}^n = \mathbf{g} \text{ on } \Gamma, \tag{5.343}$$

$$p^{n+1} = p^n - \rho\nabla \cdot \mathbf{u}^n, \qquad \rho > 0. \tag{5.344}$$

For the convergence of (5.342)–(5.344), we can prove—by a variant of the proof of Theorem 5.11—the following:

Theorem 5.12. *If in* (5.342)–(5.344), ρ *satisfies*

$$0 < \rho < 2\left(r + \frac{\nu}{N}\right), \tag{5.345}$$

then, $\forall\, p^0 \in L^2(\Omega)$, *one has*

$$\lim_{n \to +\infty} \{\mathbf{u}^n, p^n\} = \{\mathbf{u}, p\} \quad \textit{in } (H^1(\Omega))^N \times L^2(\Omega) \textit{ strongly},$$

where $\{\mathbf{u}, p\}$ *is the solution of the Stokes problem* (5.222), (5.308) *with* $\int_\Omega p\, dx = \int_\Omega p^0\, dx$. *Moreover, the convergence is linear.*

Remark 5.26. The above result can be made more precise by observing that

$$p^{n+1} - p = (I - \rho(rI + \mathscr{A}^{-1})^{-1})(p^n - p). \tag{5.346}$$

Let us prove (5.346); using the notation in the proof of Theorem 5.11, we have

$$\bar{p}^{n+1} = \bar{p}^n - \rho\nabla \cdot \bar{u}^n, \tag{5.347}$$

$$\alpha\bar{\mathbf{u}}^n - \nu\Delta\bar{\mathbf{u}}^n - r\nabla(\nabla \cdot \bar{\mathbf{u}}^n) = -\nabla\bar{p}^n \text{ in } \Omega, \quad \bar{\mathbf{u}}^n \in (H_0^1(\Omega))^N. \tag{5.348}$$

In fact (5.348) can also be written

$$\alpha\bar{\mathbf{u}}^n - \nu\Delta\bar{\mathbf{u}}^n = -\nabla(\bar{p}^n - r\nabla \cdot \bar{\mathbf{u}}^n) \text{ in } \Omega, \quad \bar{\mathbf{u}}^n \in (H_0^1(\Omega))^N; \tag{5.349}$$

since $\nabla \cdot \bar{\mathbf{u}}^n \in H$, from (5.349) (and from the definition of $\mathscr{A}$) we have

$$\nabla \cdot \bar{\mathbf{u}}^n = \mathscr{A}(\bar{p}^n - r\nabla \cdot \bar{\mathbf{u}}^n)$$

i.e.,

$$\nabla \cdot \bar{\mathbf{u}}^n = (I + r\mathscr{A})^{-1}\mathscr{A}\bar{p}^n. \tag{5.350}$$

Combining (5.350) with (5.347), we obtain

$$\bar{p}^{n+1} = (I - \rho(I + r\mathscr{A})^{-1}\mathscr{A})\bar{p}^n. \tag{5.351}$$

We have

$$I - \rho(I + r\mathscr{A})^{-1}\mathscr{A} = (rI + \mathscr{A}^{-1})^{-1}((r - \rho)I + \mathscr{A}^{-1}), \tag{5.352}$$

and (5.352) yields

$$\|I - \rho(I + r\mathscr{A})^{-1}\mathscr{A}\| \leq \frac{1}{r}(|r - \rho| + \|\mathscr{A}^{-1}\|). \tag{5.353}$$

From (5.351), (5.353) it follows that for the classical choice $\rho = r$, we have

$$\|p^{n+1} - p\|_{L^2(\Omega)} \leq \frac{\|\mathscr{A}^{-1}\|}{r}\|p^n - p\|_{L^2(\Omega)}. \tag{5.354}$$

Therefore, if r is large enough and if $\rho = r$, the convergence ratio of algorithm (5.342)–(5.344) is of order $1/r$.

Remark 5.27. The system (5.343) is closely related to the linear elasticity system. Once it is discretized by finite differences or finite elements (as in Sec. 5.8.7.3.5), it can be solved using a Cholesky factorization $\mathbf{LL}^t$ or $\mathbf{LDL}^t$, done once and for all.

Remark 5.28. Algorithm (5.342)–(5.344) has the drawback of requiring the solution of a system of N partial differential equations coupled (if $r > 0$) by $r\nabla(\nabla\cdot)$, while this is not so for algorithms of Secs. 5.8.7.4.3 and 5.8.7.4.4. Hence, much more computer storage is required.

Remark 5.29. By inspecting (5.354), it seems that one should take $\rho = r$ and r as large as possible. However, (5.343) and its discrete forms would be ill conditioned if r is too large. In practice, if (5.343) is solved by a direct method (Gauss, Cholesky), one should take r in the range of $10^2\nu$ to $10^5\nu$. In such cases, and if $\rho = r$, the convergence of (5.342)–(5.344) is extremely fast (about three iterations). Under such conditions, it is not necessary to use a conjugate gradient accelerating scheme.

5.9. Numerical experiments

In this section we shall present the results of some numerical experiments obtained using the methods of the above sections. Further numerical results obtained using the same methods may be found in B.G.4P. [1], Le Tallec [2], Periaux [2], Glowinski, Mantel, Periaux, and Pironneau [1], Glowinski, Mantel, Periaux, Perrier, and Pironneau [1]; here we follow Bristeau, Glowinski, Mantel, Periaux, Perrier, and Pironneau [1, Sec. 6].

In Sec. 5.9.1 we shall describe the results related to flows in a channel whose section presents a sudden enlargement due to a step; in Sec. 5.9.2 the numerical results will be related to a flow at Re = 250, around and inside an idealized nozzle at high incidence.

5.9.1. *Flows in a channel with a step*

We consider the solution of the Navier–Stokes equations for the flows of incompressible viscous fluids in the channel with a step of Fig. 5.4.

In order to compare our results with those of Hutton [1], we have considered flows at Re = 100 and 191; the computational domain and the boundary condition are also those of Hutton, *loc. cit.*, i.e., $\mathbf{u} = \mathbf{0}$ on the channel walls and Poiseuille flows upstream and downstream. We have used the space discretization of Sec. 5.3 with V_h defined by (5.21) in Sec. 5.3.3.1, i.e., $\mathbf{u}_h$ (resp., p_h) piecewise linear on $\mathscr{T}_h$ (resp., $\tilde{\mathscr{T}}_h$); both triangulations $\mathscr{T}_h$ and $\tilde{\mathscr{T}}_h$ are shown on Fig. 5.4 on which we have also indicated the number of nodes, finite elements, and nonzero elements in the Cholesky factors of the discrete analogue of $-\Delta$ (resp., $I/\Delta t - \nu\Delta$) associated with $\mathscr{T}_h$ (resp., $\tilde{\mathscr{T}}_h$). As we can see from these numbers, we are really dealing with fairly large matrices; the second of these matrices would have been even larger if we had used (on $\mathscr{T}_h$) a piecewise quadratic approximation for the velocity. Figure 5.4 also shows the refinement of both triangulations close to the step corner and also behind the step.

The steady-state solutions have been obtained via the time integration of the fully discrete Navier–Stokes equations, using those schemes described in Sec. 5.4. We have used, most particularly, the backward implicit scheme of Sec. 5.4.1 and also the alternating-direction schemes of Secs. 5.4.4.1 and 5.4.4.2. The numerical tests have been performed with $k = \Delta t = 0.4$. On Fig. 5.5 we have indicated the stream lines of the computed solutions showing very clearly a *recirculation* zone whose size increases with Re. If H is the height of the step, we observe that the length of the recirculation zone is approximately $6H$ at Re = 100 and $8H$ at Re = 191, in good agreement with the

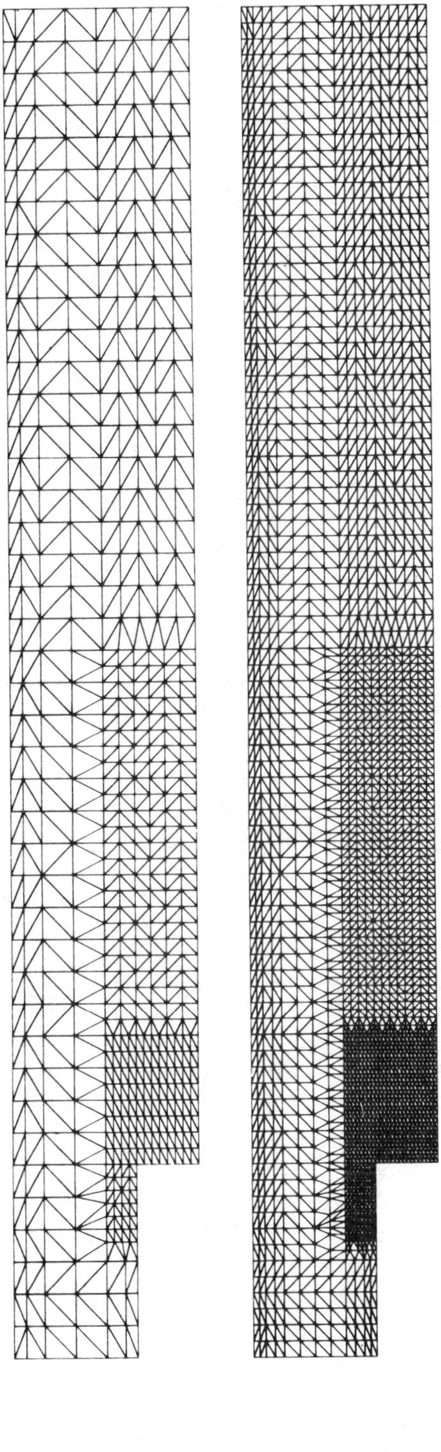

	$\mathcal{T}_h$	$\tilde{\mathcal{T}}_h$
Nodes	619	2346
Triangles	1109	4436
Cholesky's coefficients	21654	154971

Figure 5.4

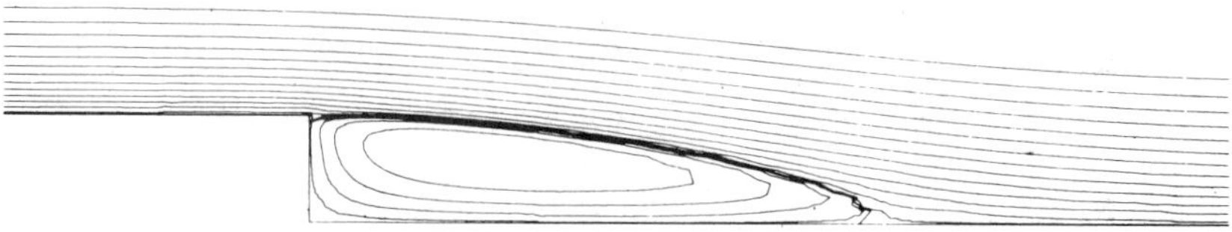

Re = 100

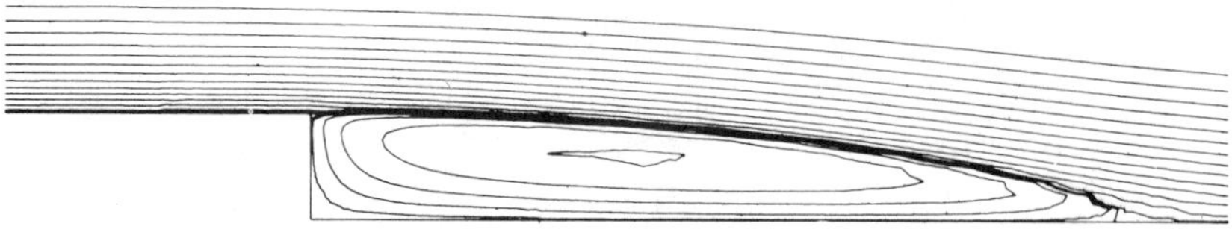

Re = 191

Figure 5.5. Stream lines.

results obtained by other authors, and, particularly, Hutton [1]. Finally, in Fig. 5.6, we have shown the pressure distributions corresponding to the two above Reynolds numbers.

From these numerical experiments it appears that the alternating-direction schemes are half as costly (in terms of computational time) as the nonsplit schemes; moreover, the incompressibility condition (measured by $\|\nabla \cdot \mathbf{u}_h\|_{L^2(\Omega)}$) is also better satisfied with these alternating-direction schemes; as a last remark we would like to mention that the least-squares solution of the nonlinear step in the above alternating-direction methods requires about three conjugate gradient iterations[33] (despite this small number of iterations the cost of the nonlinear step is about 10 times larger than the cost of the Stokes step(s)).

In conclusion, on the basis of the above examples, we advocate the alternating-direction schemes even for solving the steady-state problems.

5.9.2. *Unsteady flow around and inside an idealized nozzle at high incidence*

We again follow Periaux [2] and Bristeau, Glowinski, Mantel, Periaux, Perrier, and Pironneau [1] in this subsection. The problem under consideration is the unsteady flow of an incompressible viscous fluid at Re = 250 (see also Glowinski, Mantel, Periaux, Perrier, and Pironneau [1] and Glowinski, Mantel, Periaux, and Pironneau [1] for computations at Re = 750) around and inside an idealized nozzle at high incidence (Re has been computed using the distance between the two walls of the nozzle as the characteristic length). Figures 5.7 and 5.8 show an enlarged view of the nozzle close to the air intake. The computational domain is clearly bounded, and appropriate boundary conditions have been prescribed in the far field. The velocity vector at ∞ (in the actual problem) is horizontal, and the angle of attack of the nozzle is 40° (which is quite large at that Reynold's number); another feature of our

[33] At least for this class of problems.

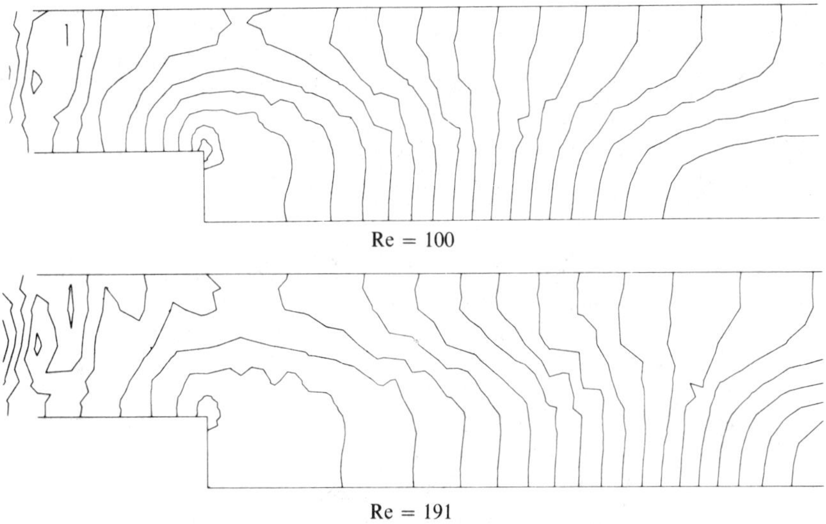

Figure 5.6. Isopressure lines.

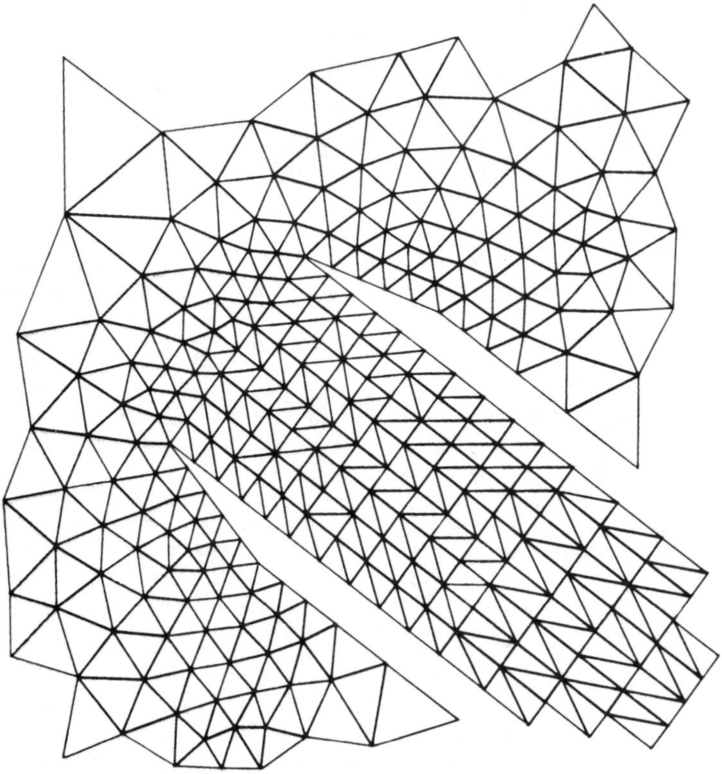

Figure 5.7. Details of $\mathcal{T}_h$ close to the air intake.

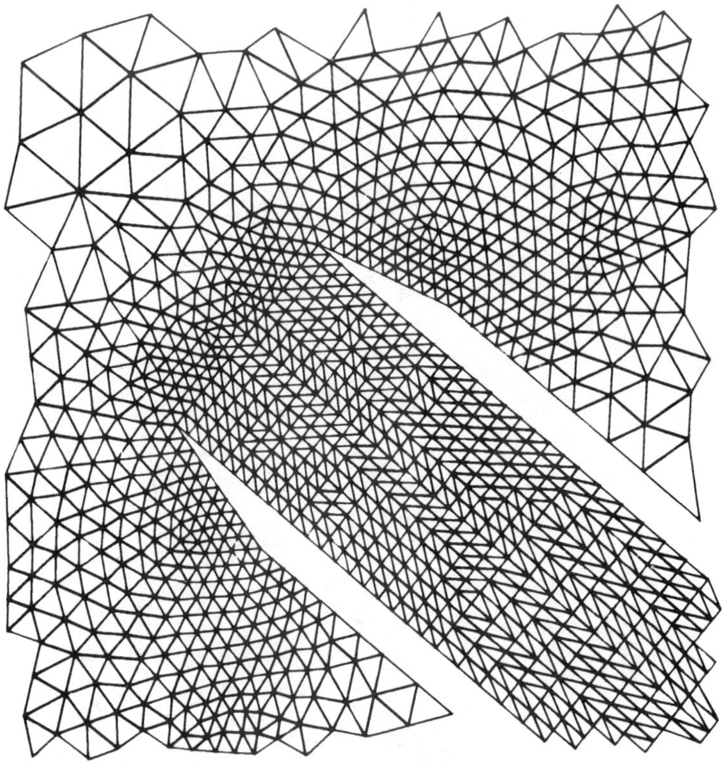

Figure 5.8. Details of $\tilde{\mathscr{T}}_h$ close to the air intake.

problem is that a given flow has been prescribed (via a velocity distribution) on a cross section of the nozzle in order to simulate a suction phenomenon (due to an engine). For the space discretization we have used the same finite element method as in Sec. 5.9.1, and for the time discretization, the Crank–Nicholson scheme of Sec. 5.4.2, with $k = \Delta t = 0.1$; we have used a triangulation $\mathscr{T}_h$ (resp., $\tilde{\mathscr{T}}_h$) consisting of 1458 (resp., 5832) triangles and 795 (resp., 3049) nodes. Details of $\mathscr{T}_h$ and $\tilde{\mathscr{T}}_h$ are shown on Figs. 5.7 and 5.8. Let us mention that the number of nonzero elements in the Cholesky factors, corresponding to $\mathscr{T}_h$ and $-\Delta$ (resp., $\tilde{\mathscr{T}}_h$ and $2I/k - \nu\Delta$) is 101 370 (resp., 314 685). Further computational details about the present problem are given in Periaux [2]. As mentioned before, a Crank–Nicholson scheme has been used, taking as the initial value the solution of the corresponding (discrete) Stokes problem (creeping flow).

In Fig. 5.9–5.12 we have indicated the distributions at various time steps of the velocity, stream function, pressure, and vorticity, respectively. It is interesting to observe the formation of eddies at the leading edges and their propagation inside and outside the nozzle.

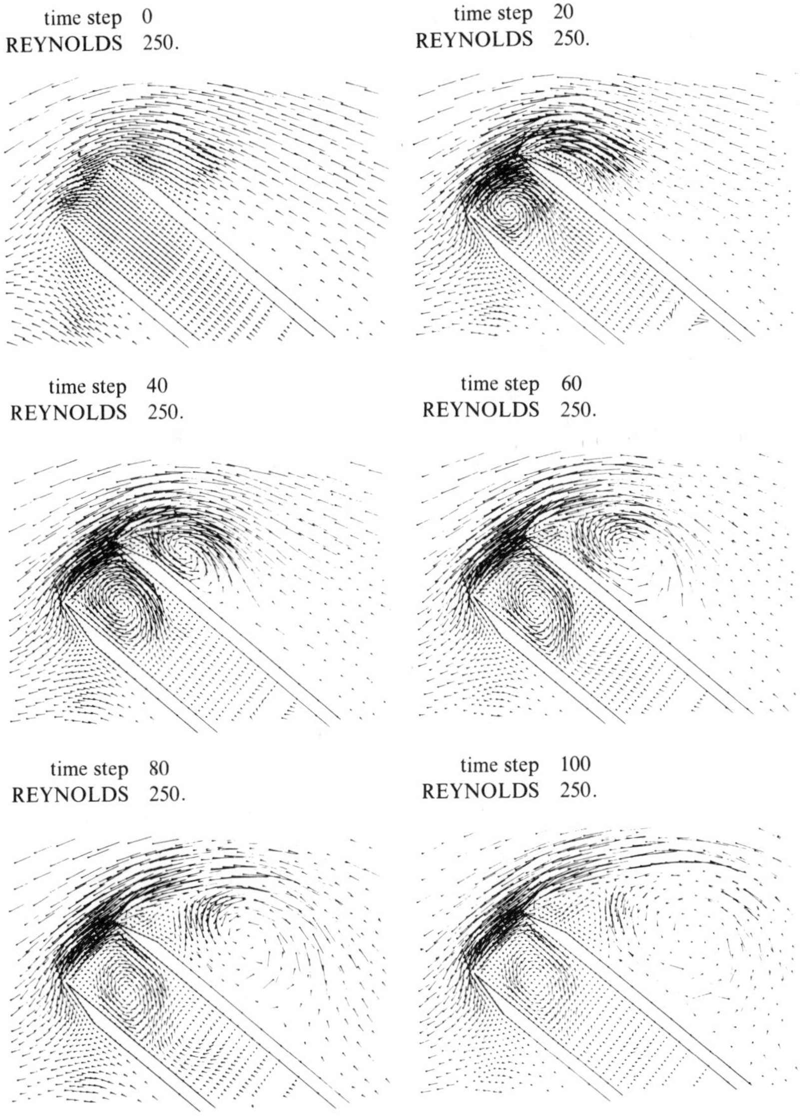

Figure 5.9. Unsteady flow for an air intake at high incidence (velocity distribution).

As may be guessed, the computer time required for solving such a large time-dependent nonlinear problem (about 7000 unknowns) on such complicated geometry is important; it actually required several hours of IBM 370–168 to run 100 time steps; however, we have the feeling (on the basis of the comparisons made for the problems of Sec. 5.9.1) that an alternating-direction scheme would substantially reduce the computer time and could

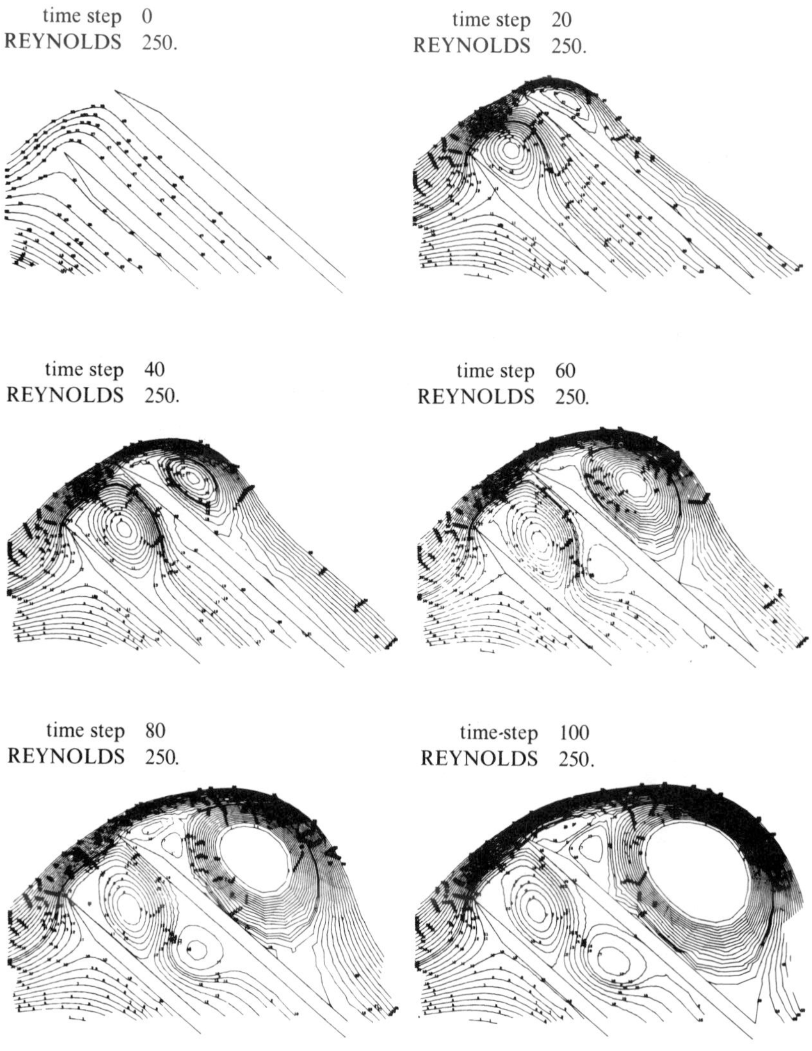

Figure 5.10. Unsteady flow for an air intake at high incidence (stream lines).

provide results of better quality (actually this prediction has been supported by recent numerical experiments at Re = 750 and 1500).

As mentioned above, the results of computations at Re = 750 for the same geometry are given in Glowinski, Mantel, Periaux, Perrier, and Pironneau [1] and Glowinski, Mantel, Periaux, and Pironneau [1] (refined triangulations are necessary). Let us also mention that the convergence at each time step of our conjugate gradient algorithm is usually obtained in three to five iterations. To conclude, let us mention that a very important saving in compu-

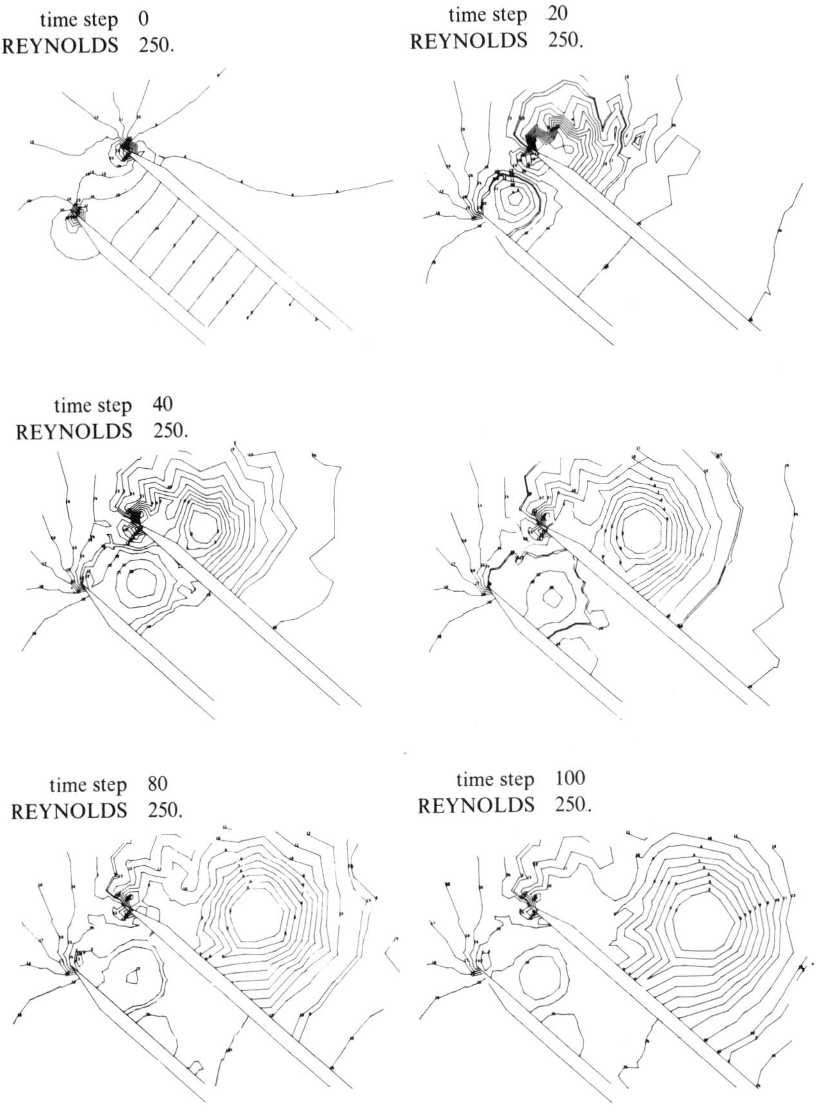

Figure 5.11. Unsteady flow for an air intake at high incidence (isopressure lines).

tational time and storage can be obtained from the solution of our very large linear systems by sophisticated iterative methods like those based on incomplete factorizations and, again, conjugate gradient schemes (see, e.g., Manteuffel [1], [2], Glowinski, Mantel, Periaux, Pironneau, and Poirier [1], Glowinski, Mantel, Periaux, Perrier, and Pironneau [1] and the references therein).

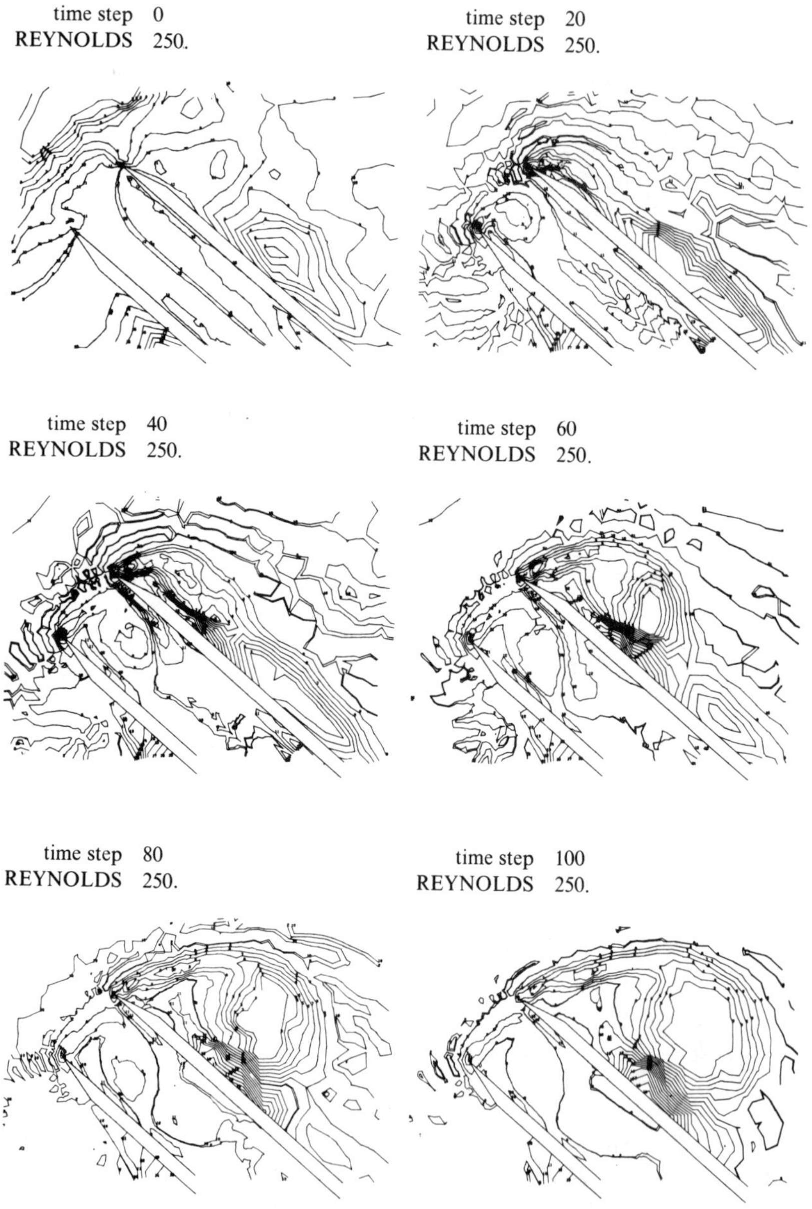

Figure 5.12. Unsteady flow for an air intake at high incidence (isovorticity lines).

5.10. Further comments on Sec. 5 and conclusion

In Sec. 5 we have presented several numerical methods for solving the steady and unsteady Navier–Stokes equations for incompressible viscous flows in the pressure–velocity formulation; the machinery behind these methods may seem a bit complicated, but in fact the resulting methodology is extremely robust for the following reasons:

(i) The finite element formulation allows us to handle, in a stable fashion (stable because a variational formulation is used), the Navier–Stokes equations on complicated geometries; moreover, we have used finite element methods leading to error estimates of optimal order for both pressure and velocity.
(ii) Fully implicit schemes are used for the time discretization since semi-implicit and explicit schemes resulted in very time-consuming methods.
(iii) The resulting nonlinear finite-dimensional problems are solved, via a nonlinear least-squares formulation, by a conjugate gradient method with scaling (i.e., with preconditioning), the scaling operator being a discrete Poisson operator or a discrete Stokes operator, depending on whether or not alternate-direction methods are used. Once again, stability is reinforced, since least-squares formulations act as convexifiers of the original problem (locally at least). Moreover, conjugate gradient algorithms usually have a superlinear convergence (practically quadratic in many cases).
(iv) The only linear systems to be solved are related to discrete Poisson and Stokes problems; using the finite element methods of Secs. 5.3 and 5.8, we have been able to reduce the solution of the discrete Stokes problems to the solution of linear subproblems of smaller size associated with symmetric positive-definite matrices, which is also a guaranty of numerical stability (these matrices are obviously factorized once and for all if direct methods of solutions are chosen).

From the above properties, the resulting algorithms are very reliable and well suited for simulating separated flows with recirculation in complicated geometries. There is no theoretical difficulties in applying the above techniques to the solution of three-dimensional problems; in this latter case, or in very complicated two-dimensional problems, for solving the very large linear systems involved in our methodology we definitely used sophisticated iterative techniques resulting in a very important saving of computational time and storage.

In the present methodology, upwinding has been avoided, since we feared that the corresponding artificial viscosity may "kill" some of the small-scale phenomenon possibly associated with flows at high Reynold's numbers. However, in Appendix 2 one may find several references concerning upwinding and also the description (with some numerical tests) of a new upwinded finite element method of high accuracy which may be useful in computational fluid dynamics.

6. Further Comments on Chapter VII and Conclusion

6.1. Comments

In Secs. 1–5 we have considered the solution of nonlinear problems by least-squares methods; actually linear programming techniques have also been used for solving linear and nonlinear boundary-value problems for partial differential equations. In this direction, we should mention Rosen [1], [2] and the survey paper by Cheung [1] (see also the references in these papers). In our opinion least-squares methods are more natural than linear programming methods since many boundary-value problems for partial differential equations have a natural formulation in Hilbert functional spaces; therefore a least-squares formulation using the corresponding metric of these spaces (or of their dual space) is a most natural approach.

Closely related to least-squares method is the *method of weighted residuals*; a basic reference on this subject is Finlayson [1] (see also Zienkiewicz [1]).

In Sec. 3.5 we have described a continuation technique for solving nonlinear problems; this technique has been used in Sec. 3.5.4 for solving the test problem—with turning point—(3.70). For a general theory on the approximation of nonlinear problems with turning points and/or bifurcation points, we refer to Brezzi, Rappaz, and Raviart [1]–[3].

In Sec. 4 we have discussed the least-squares solution of a particular family of compressible flow problems; we refer to Chattot [1] where closely related least-squares methods have also been used for solving nonlinear compressible flow problems.

In Sec. 5 we have discussed the numerical solution of the Navier–Stokes equations for incompressible viscous fluids by finite element methods. We have chosen to work with the primitive variables (i.e., pressure and velocity) and with continuous approximations of these variables; we refer to Argyris and Dunne, [1] for a discussion of the finite element solution of the Navier–Stokes equations using a stream function formulation (see also Reinhart [1] and Bristeau, Glowinski, Periaux, Perrier, and Pironneau [1, Sec. 4.4], and for a survey of finite difference methods, see Roache [1]). For finite element approximation of the Navier–Stokes equations using discontinuous finite element approximations of the velocity and/or the pressure, see Girault and Raviart [1] and the survey paper by Raviart [3].

Concerning alternating-direction and fractional step methods, we would like to mention that the symmetrization process applied in (5.56)–(5.59) and (5.168)–(5.171) is due to Strang [2]; splitting methods for the incompressible Navier–Stokes equations are mathematically analyzed in Temam [1] and Beale and Majda [1]. Basic references concerning alternating-direction methods for solving initial-value problems are[34] Beam and Warming [1], [2]

[34] See also the references therein and Godlewsky [1] and Hayes [1], [2].

and Warming and Beam [1]–[3]; these references are in fact strongly motivated by the numerical solution of the compressible Navier–Stokes equations.

Returning to the direct solution (discussed in Sec. 5.8.7.2) of the discrete Stokes problem (5.157), i.e.:

Find $\{\mathbf{u}_h, p_h\} \in V_{gh} \times H_h^1$ *such that*

$$\alpha \int_\Omega \mathbf{u}_h \cdot \mathbf{v}_h \, dx + \nu \int_\Omega \nabla \mathbf{u}_h \cdot \nabla \mathbf{v}_h \, dx + \int_\Omega \nabla p_h \cdot \mathbf{v}_h \, dx = \int_\Omega \mathbf{f}_h \cdot \mathbf{v}_h \, dx, \qquad \forall\, \mathbf{v}_h \in V_{0h},$$

$$\int_\Omega \nabla \cdot \mathbf{u}_h q_h \, dx = 0, \qquad \forall\, q_h \in H_h^1,$$

we have seen (in Sec. 5.8.6.2) that (5.157) is in fact equivalent to:

Find $\mathbf{u}_h \in \tilde{V}_{gh}$ *such that*

$$\alpha \int_\Omega \mathbf{u}_h \cdot \mathbf{v}_h \, dx + \nu \int_\Omega \nabla \mathbf{u}_h \cdot \nabla \mathbf{v}_h \, dx = \int_\Omega \mathbf{f}_h \cdot \mathbf{v}_h \, dx, \qquad \forall\, \mathbf{v}_h \in \tilde{V}_{0h}, \tag{6.1}$$

where

$$\tilde{V}_{0h} = \left\{ \mathbf{v}_h \in V_{0h}, \int_\Omega \nabla \cdot \mathbf{v}_h q_h \, dx = 0, \quad \forall\, q_h \in H_h^1 \right\},$$

$$\tilde{V}_{gh} = \left\{ \mathbf{v}_h \in V_{gh}, \int_\Omega \nabla \cdot \mathbf{v}_h q_h \, dx = 0, \quad \forall\, q_h \in H_h^1 \right\}$$

(H_h^1, V_{0h} and V_{gh} have been defined in Sec. 5.3.3.1). Local vector bases of $\tilde{V}_{0h}$ and $\tilde{V}_{gh}$ have been found by F. Hecht [1]; they reduce the solution of (5.157) to that of a linear system whose matrix is symmetric, positive definite, and sparse. The solution of this system directly gives $\mathbf{u}_h$.

As a last comment concerning the Navier–Stokes equations, we would like to mention that the exponential stretching method of Sec. 4.8 can be applied (and has been used, in fact, by several authors) to the solution of the Navier–Stokes equations on very large space domains (of $\mathbb{R}^2$, at least).

6.2. Conclusion

In this chapter we have discussed the solution of several nonlinear problems in fluid dynamics by least-squares and conjugate gradient methods. Actually the methods described in this chapter can also be (and have been) applied to the solution of other nonlinear problems. A very important ingredient in the methodology that we have discussed is the choice of a proper norm to use in the least-squares formulation. In fact, after discretization, the corresponding conjugate gradient methods can be classified as conjugate gradient algorithms with scaling (or preconditioning). Related techniques have been used (in a non-least-squares context) to solve linear and nonlinear boundary-value problems (much simpler than those in this chapter) by, e.g., Bartels and Daniel [1], Axelsson [1], Douglas and Dupont [2], and Concus, Golub, and O'Leary [1].

The problems in this chapter are nonlinear, and several steps in the method we have described lead to mathematically open questions.

The conjugate gradient method has been chosen, since in some circumstances (see Polak [1] and Daniel [1] for more details) it may achieve a superlinear convergence, like Newton's method, in fact. Compared to Newton's method, the conjugate gradient method with scaling has the advantage of requiring at each iteration the solution of linear systems related to a matrix which is always the same during the iterative process; hence this matrix can be prefactorized (by Cholesky's method, for example).

APPENDIX I

A Brief Introduction to Linear Variational Problems

1. Introduction

At various points in the previous chapters we have tried to reduce the solution of nonlinear variational problems to that of sequences of linear variational problems, clearly showing thus the importance of the latter subject. Despite the fact that the theoretical and numerical solutions of linear partial differential equations via linear variational formulations have motivated several excellent books and monographs (let us mention, among others, Mikhlin [1], [2], Lions [2], Necas [1], Aubin [2], Ciarlet [2], Oden and Reddy [1] and Rektoris [1]; see also the references therein[1]) we felt obliged (for completeness) to give a brief introductory account of this very important subject. For brevity we shall concentrate on elliptic problems (for linear parabolic and hyperbolic problems see, e.g., Lions [3]).

In Sec. 2 we shall give the formulation of a standard family of abstract linear variational problems and shall prove existence and uniqueness results.

In Sec. 3, which is concerned with the internal approximation of the variational problems of Sec. 2, we shall prove some general approximation results.

Finally, in Sec. 4 we shall discuss the variational formulation of some boundary-value problems for linear elliptic partial differential operators (including problems in linear elasticity); some brief comments on the numerical approximation of these problems by finite element methods will complete the discussion.

2. A Family of Linear Variational Problems

2.1. Functional context

We consider:

(i) a real Hilbert space V with scalar product $(\cdot, \cdot)$ and associated norm $\|\cdot\|$;

(ii) V^*, the topological dual space of V;

---

[1] See also Schultz [1] and Prenter [1].

(iii) a bilinear form $a: V \times V \to \mathbb{R}$, continuous and V-elliptic (i.e., $\exists\, \alpha > 0$ such that $a(v, v) \geq \alpha\|v\|^2$, $\forall\, v \in V$); $a(\cdot, \cdot)$ is possibly nonsymmetric (i.e., we do not assume that $a(u, v) = a(v, u)$, $\forall\, u, v \in V$);
(iv) a linear continuous functional $L: V \to \mathbb{R}$.

2.2. Formulation of the problem

The fundamental linear variational problem under consideration is formulated as follows:

Find $u \in V$ such that

$$a(u, v) = L(v), \qquad \forall\, v \in V. \tag{2.1) \quad (P}$$

2.3. Existence and uniqueness results for (P)

The following theorem is known as the *Lax–Milgram Theorem*.

Theorem 2.1. *Under the above hypotheses on V, $a(\cdot, \cdot)$, and $L(\cdot)$, problem* (P) *has a unique solution.*

PROOF. We first prove the uniqueness and then the existence.

(1) *Uniqueness.* Let u_1 and u_2 be solutions of (P). We then have

$$a(u_1, v) = L(v), \qquad \forall\, v \in V, \quad u_1 \in V, \tag{2.2}$$

$$a(u_2, v) = L(v), \qquad \forall\, v \in V, \quad u_2 \in V. \tag{2.3}$$

By substracting, from (2.2), (2.3) we obtain

$$a(u_2 - u_1, v) = 0, \qquad \forall\, v \in V; \tag{2.4}$$

taking $v = u_2 - u_1$ in (2.4) and using the V-ellipticity of $a(\cdot, \cdot)$, we obtain

$$\alpha\|u_2 - u_1\|^2 \leq a(u_2 - u_1, u_2 - u_1) = 0,$$

which proves that $u_1 = u_2$, since $\alpha > 0$.

(2) *Existence.* From the Riesz representation theorem for Hilbert spaces,[2] there exists a unique $l \in V$ such that

$$L(v) = (l, v), \qquad \forall\, v \in V. \tag{2.5}$$

---

[2] The Riesz representation theorem states that, $\forall\, F \in V^*$ (i.e., F is linear and continuous from V to $\mathbb{R}$), there exists a unique $f \in V$ such that

$$F(v) = (f, v), \qquad \forall\, v \in V.$$

For a proof of this fundamental theorem, see, e.g., Yosida [1], Oden [1], or any textbook on functional analysis or Hilbert spaces.

Now suppose that v has been fixed in $a(v, w)$; since the mapping from V to $\mathbb{R}$ defined by

$$w \to a(v, w)$$

is linear and continuous, from the above Riesz theorem there exists a unique element of V, denoted by $A(v)$, such that

$$a(v, w) = (A(v), w), \qquad \forall\, v, w \in V. \tag{2.6}$$

Let us prove the linearity of this operator A; on the one hand we have

$$a(\lambda_1 v_1 + \lambda_2 v_2, w) = (A(\lambda_1 v_1 + \lambda_2 v_2), w), \qquad \forall\, \lambda_1, \lambda_2 \in \mathbb{R}, \quad \forall\, v_1, v_2, w \in V; \tag{2.7}$$

on the other hand, from the bilinearity of $a(\cdot, \cdot)$, we have

$$\begin{aligned} a(\lambda_1 v_1 + \lambda_2 v_2, w) &= \lambda_1 a(v_1, w) + \lambda_2 a(v_2, w) \\ &= \lambda_1(A(v_1), w) + \lambda_2(A(v_2), w) \\ &= (\lambda_1 A(v_1) + \lambda_2 A(v_2), w). \end{aligned} \tag{2.8}$$

From (2.7), (2.8) it follows that

$$A(\lambda_1 v_1 + \lambda_2 v_2) = \lambda_1 A(v_1) + \lambda_2 A(v_2),$$

which shows the linearity of A;[3] from this linearity property we shall use the notation $A(v) = Av$.

Let us show the continuity of A; from the continuity of $a(\cdot, \cdot)$ there exists M such that

$$|a(v, w)| \leq M\|v\| \, \|w\|, \qquad \forall\, v, w \in V,$$

i.e. (from (2.6)),

$$|(Av, w)| \leq M\|v\| \, \|w\|, \qquad \forall\, v, w \in V. \tag{2.9}$$

Taking $w = Av$ in (2.9), we have

$$\|Av\| \leq M\|v\|,$$

which shows the continuity of A (and also that $\|A\| \leq M$).[4] It follows from (2.5), (2.6) that (P) is equivalent to the linear problem in V,

$$Au = l, \tag{2.10}$$

which is, in turn, equivalent to finding u such that

$$u = u - \rho(Au - l) \quad \text{for some } \rho > 0; \tag{2.11}$$

to solve the fixed-point problem (2.11), consider the mapping $W_\rho: V \to V$ defined by

$$W_\rho(v) = v - \rho(Av - l); \tag{2.12}$$

if $v_1, v_2 \in V$, we have

$$\|W_\rho(v_2) - W_\rho(v_1)\|^2 = \|v_2 - v_1\|^2 - 2\rho a(v_2 - v_1, v_2 - v_1) + \rho^2\|A(v_2 - v_1)\|^2.$$

---

[3] One can easily prove that $a(\cdot, \cdot)$ symmetric (i.e., $a(v, w) = a(w, v)$, $\forall\, v, w \in V$) implies the symmetry of A (i.e., $(Av, w) = (Aw, v)$, $\forall\, v, w \in V$).

[4] We also have $\alpha \leq \|A\|$, since $\forall\, v \in V$, we have $\alpha\|v\|^2 \leq a(v, v) = (Av, v) \leq \|A\| \, \|v\|^2$.

Hence we have

$$\|W_\rho(v_2) - W_\rho(v_1)\|^2 \le (1 - 2\rho\alpha + \rho^2\|A\|^2)\|v_2 - v_1\|^2;$$

then W_ρ is strictly and uniformly contracting if $0 < \rho < 2\alpha/\|A\|^2$. By taking ρ in this range, we have a unique solution of the fixed-point problem (2.11) which implies the existence of a solution for (P) and completes the proof of the theorem. □

2.4. Remarks

Remark 2.1. The Lax–Milgram Theorem (Theorem 2.1) is a particular case of the *Lions–Stampacchia* Theorems (Theorems 3.1 and 4.1 of Chapter I, Secs. 3 and 4) (obtained by taking $K = V$ and $j \equiv 0$, respectively).

Remark 2.2. Suppose that $a(\cdot, \cdot)$ is a symmetric bilinear form over $V \times V$; we then have:

Proposition 2.1. *There is equivalence between problem* (P) *and the following problem*:

Find $u \in V$ *such that*

$$J(u) \le J(v), \qquad \forall\, v \in V, \tag{π}$$

where

$$J(v) = \tfrac{1}{2}a(v, v) - L(v). \tag{2.13}$$

PROOF. (1) (P) *implies* (π). Let u be the solution of (P) and let $v \in V$; we have

$$\begin{aligned} J(v) = J(u + v - u) &= \tfrac{1}{2}a(u + v - u, u + v - u) - L(u + v - u) \\ &= \tfrac{1}{2}a(u, u) - L(u) + a(u, v - u) - L(v - u) + \tfrac{1}{2}a(v - u, v - u) \\ &= J(u) + a(u, v - u) - L(v - u) + \tfrac{1}{2}a(v - u, v - u). \end{aligned} \tag{2.14}$$

Since u is solution of (P) and $a(\cdot, \cdot)$ is V-elliptic, we have

$$a(u, v - u) - L(v - u) = 0, \qquad \forall\, v \in V, \tag{2.15}$$

$$a(v - u, v - u) \ge 0, \qquad \forall\, v \in V; \tag{2.16}$$

from (2.14)–(2.16) it then follows that

$$J(v) - J(u) \ge 0, \qquad \forall\, v \in V,$$

i.e., u is a solution of (π).

(2) (π) *implies* (P). Let u be a solution of (π) and $v \in V$; since u is a solution of (π), we clearly have

$$\frac{J(u + tv) - J(u)}{t} \ge 0, \qquad \forall\, v \in V, \ \forall\, t > 0. \tag{2.17}$$

Since

$$\lim_{\substack{t\to 0\\ t>0}} \frac{J(u+tv)-J(u)}{t} = a(u, v) - L(v),$$

it follows from (2.17) that

$$a(u, v) - L(v) \geq 0, \qquad \forall\, v \in V. \tag{2.18}$$

Taking $-v$ instead of v in (2.18), we see that

$$a(u, v) - L(v) = 0, \qquad \forall\, v \in V,$$

i.e., u is a solution of (P). □

Remark 2.3. Suppose that V is a complex Hilbert space; then Theorem 2.1 still holds if we suppose that $a(\cdot, \cdot)$ is continuous, sesquilinear (i.e.,

$$a(\lambda_1 v_1 + \lambda_2 v_2, w) = \lambda_1 a(v_1, w) + \lambda_2 a(v_2, w), \qquad \forall\, v_1 v_2, w \in V, \quad \forall\, \lambda_1, \lambda_2 \in \mathbb{C},$$

$$a(v, \lambda_1 w_1 + \lambda_2 w_2) = \bar{\lambda}_2 a(v, w_1) + \bar{\lambda}_2 a(v, w_2), \qquad \forall\, v, w_1, w_2 \in V, \quad \forall\, \lambda_1, \lambda_2 \in \mathbb{C}),$$

and V-elliptic in the following sense[5]:

$$\mathscr{R}a(v, v) \geq \alpha \|v\|^2, \qquad \forall\, v \in V$$

with $\alpha > 0$.

Remark 2.4. The hypotheses on V, $a(\cdot, \cdot)$, and $L(\cdot)$ being those of Secs. 2.1, 2.2, and 2.3, we consider another real Hilbert space H, such that

$$V \subset H, \quad \text{the injection } \tau^6 \text{ from } V \text{ to } H \text{ is compact}^7; \tag{2.19}$$

we suppose that H is equipped with the scalar product $[\cdot, \cdot]$.

Using the *Riesz–Schauder theory* (see, e.g., Riesz and Nagy [1] and Yosida [1]), we can prove the following generalization of the Lax–Milgram Theorem (Theorem 2.1.).

Theorem 2.2. *Let $\lambda \in \mathbb{R}$ and let the hypotheses on V, $a(\cdot, \cdot)$, and $L(\cdot)$ be those of Secs. 2.1, 2.2, and 2.3. Then the linear variational problem:*

---

[5] $\mathscr{R}z$ = real part of $z \in \mathbb{C}$ ($\mathscr{R}z = x$ if $z = x + iy$ ($x, y \in \mathbb{R}$)).

[6] The injection operator $\tau: V \to H$ is defined by

$$\tau v = v, \qquad \forall\, v \in V.$$

[7] That is:

(i) τ is continuous from V to H.

(ii) Let X be a bounded set of V; then the closure $\bar{X}^H$ of X in H is a compact set of H.

This implies that if $\lim_{n\to+\infty} v_n = v$ weakly in V, then $\lim_{n\to+\infty} v_n = v$ strongly in H. See, e.g., Yosida [1] for a general definition of compact operators.

Find $u \in V$ such that

$$a(u, v) + \lambda[u, v] = L(v), \qquad \forall\, v \in V, \tag{2.20}$$

has a unique solution if we suppose that the following (eigenvalue) problem:
Find $w \in V$, $w \neq 0$ such that

$$a(w, v) + \lambda[w, v] = 0, \qquad \forall\, v \in V,$$

has no solution.

3. Internal Approximation of Problem (P)

In this section we shall study the approximation of (P) from an abstract axiomatic point of view.

3.1. Approximation of V

We suppose that we are given a parameter h converging to zero and a family $\{V_h\}_h$ of closed subspaces of V (in practice, the V_h are finite dimensional and h varies over a sequence). We suppose that $\{V_h\}_h$ satisfies the following condition:

There exist $\mathscr{V} \subset V$ such that $\overline{\mathscr{V}} = V$, and

$$r_h\colon \mathscr{V} \to V_h \text{ such that } \lim_{h \to 0} \|r_h v - v\| = 0, \qquad \forall\, v \in \mathscr{V}. \tag{3.1}$$

Remark 3.1. The terminology "internal approximation" is justified by the fact that $V_h \subset V$, $\forall\, h$.

3.2. Approximation of (P)

We approximate problem (P) by the following:
Find $u_h \in V_h$ such that

$$a(u_h, v_h) = L(v_h), \qquad \forall\, v_h \in V_h. \tag{P_h}$$

From Theorem 2.1 it then follows that:

Theorem 3.1. *Problem* (P_h) *has a unique solution.*

Remark 3.2. In most applications it will be necessary to replace $a(\cdot, \cdot)$ and L by $a_h(\cdot, \cdot)$ and L_n (usually defined, in practical cases, from $a(\cdot, \cdot)$ and L by a numerical integration procedure); on this important matter we refer to Ciarlet [1], [2], [3].

Remark 3.3. Suppose that $a(\cdot, \cdot)$ is *symmetric*; then we can easily prove that the approximate problem (P_h) is equivalent to the minimization problem:

Find $u_h \in V_h$ *such that*

$$J(u_h) \leq J(v_h), \qquad \forall\, v_h \in V_h, \tag{π_h}$$

where the functional J is still defined by (2.13).

3.3. Convergence results

The convergence results to be proved in a moment follow directly from:

Lemma 3.1 (Cea [3]). *Let* u *(resp.,* u_h*) be the solution of* (P) *(resp.,* (P_h)*); we then have*

$$\|u_h - u\| \leq \frac{\|A\|}{\alpha} \operatorname*{Inf}_{v_h \in V_h} \|v_h - u\|. \tag{3.2}$$

If $a(\cdot, \cdot)$ *is symmetric, then instead of* (3.2), *we have*

$$\|u_h - u\| \leq \left(\frac{\|A\|}{\alpha}\right)^{1/2} \operatorname*{Inf}_{v_h \in V_h} \|v_h - u\|. \tag{3.3}$$

PROOF. (i) *Proof of* (3.2). Let $v_h \in V_h$; since u (resp., u_h) is a solution of (P) (resp., (P_h)), and since $V_h \subset V$, we have

$$a(u, v_h - u_h) = L(v_h - u_h),$$

$$a(u_h, v_h - u_h) = L(v_h - u_h),$$

which, by substraction, imply

$$a(u_h - u, u_h - u) = a(u_h - u, v_h - u), \qquad \forall\, v_h \in V_h. \tag{3.4}$$

From (3.4) (and from the properties of $a(\cdot, \cdot)$ discussed in Sec. 2.3) it then follows that

$$\alpha\|u_h - u\|^2 \leq a(u_h - u, u_h - u) \leq \|A\|\,\|u_h - u\|\,\|v_h - u\|, \qquad \forall\, v_h \in V_h,$$

i.e.,

$$\alpha\|u_h - u\| \leq \|A\|\,\|v_h - u\|, \qquad \forall\, v_h \in V_h, \tag{3.5}$$

which, in turn, clearly implies (3.2).

(ii) *Proof of* (3.3). Now suppose that $a(\cdot, \cdot)$ is symmetric; let $v \in V$ and u be the solution of (P). From the definition of J (see (2.13)) and from the fact that u solves (P), we then have

$$\begin{aligned} J(v) &= J(u + v - u) = J(u) + a(u, v - u) - L(v - u) + \tfrac{1}{2}a(v - u, v - u) \\ &= J(u) + \tfrac{1}{2}a(v - u, v - u), \qquad \forall\, v \in V. \end{aligned} \tag{3.6}$$

Let u_h be the solution of (P_h); from Remark 3.3 we have

$$J(v_h) \geq J(u_h), \qquad \forall\, v_h \in V_h,$$

i.e.,

$$J(v_h) - J(u) \geq J(u_h) - J(u), \qquad \forall\, v_h \in V_h,$$

which, combined with (3.6), implies that

$$a(u_h - u, u_h - u) \leq a(v_h - u, v_h - u), \qquad \forall\, v_h \in V_h. \tag{3.7}$$

From (3.7) and from the properties of $a(\cdot, \cdot)$ we have

$$\alpha\|u_h - u\|^2 \leq \|A\| \, \|v_h - u\|^2, \qquad \forall \, v_h \in V_h; \tag{3.8}$$

(3.3) follows clearly from (3.8). □

Lemma 3.1 is quite useful (as shown in, e.g., Ciarlet [1], [2], [3]) for obtaining error estimates; actually it also implies the following:

Theorem 3.2. *The hypotheses on* V, a, *and* L *being those of Secs.* 2.1, 2.2, *and* 2.3, *and if we suppose that* $\{V_h\}_h$ *obeys* (3.1), *we have*

$$\lim_{h \to 0} \|u_h - u\| = 0, \tag{3.9}$$

where u *(resp.,* u_h*) is the solution of* (P) *(resp.,* (P$_h$)*).*

Proof. Let us consider $\varepsilon > 0$; since $\overline{\mathscr{V}} = V$, there exists $u_\varepsilon \in \mathscr{V}$ such that

$$\|u_\varepsilon - u\| \leq \frac{\alpha}{\|A\|}\varepsilon. \tag{3.10}$$

From (3.10) and Lemma 3.1 it follows that

$$\|u_h - u\| \leq \frac{\|A\|}{\alpha}\|r_h u_\varepsilon - u\| \leq \frac{\|A\|}{\alpha}(\|r_h u_\varepsilon - u_\varepsilon\| + \|u_\varepsilon - u\|)$$

$$\leq \frac{\|A\|}{\alpha}\|r_h u_\varepsilon - u_\varepsilon\| + \varepsilon, \qquad \forall \, \varepsilon > 0. \tag{3.11}$$

Combined with (3.1), relation (3.11) implies

$$0 \leq \limsup_{h \to 0} \|u_h - u\| \leq \varepsilon, \qquad \forall \, \varepsilon > 0. \tag{3.12}$$

The strong convergence result (3.9) easily follows from (3.12). □

3.4. A particular case of internal approximation: the method of Galerkin

A popular example of internal approximation is the *method of Galerkin* described below.

In this section we suppose that V is a separable real Hilbert space in the following sense:

$$\text{There exists a countable subset } \mathscr{B} = \{w_j\}_{j=1}^{+\infty} \text{ of } V \text{ such that the subspace } \mathscr{V} \text{ of } V \text{ generated by } \mathscr{B} \text{ is dense in } V^{8} \tag{3.13}$$

(we can always suppose that the w_j are linearly independent).

[8] $\mathscr{V} = \{v \in V, v = \sum_{k=1}^{\mu} \lambda_{j_k} w_{j_k} \text{ with } \mu \geq 1\}$.

For any integer $m \geq 1$ we define $\mathscr{B}_m$ by

$$\mathscr{B}_m = \{w_j\}_{j=1}^m$$

and V_m as the subspace of V generated by $\mathscr{B}_m$; if we denote by π_m the projection operator from V to V_m, it follows from the density property stated in (3.13) that

$$\lim_{m \to +\infty} \|\pi_m v - v\| = 0, \qquad \forall\, v \in V. \tag{3.14}$$

The Galerkin approximation (P_m) of problem (P) is then defined as follows:

Find $u_m \in V_m$ such that

$$a(u_m, v) = L(v), \qquad \forall\, v \in V_m. \tag{P_m}$$

Concerning the convergence of $\{u_m\}_m$ to the solution u of (P), we have the following:

Theorem 3.3. *We suppose that the hypotheses on V, a, and L are still those of Secs.* 2.1, 2.2, *and* 2.3; *we also suppose that* (3.13) *holds. We then have*

$$\lim_{m \to +\infty} \|u_m - u\| = 0, \tag{3.15}$$

where u (resp., u_m) is the solution of (P) *(resp.,* (P_m)).

PROOF. We can give a direct proof of Theorem 3.3, but in fact it suffices to apply Theorem 3.2 with $\mathscr{V} = V$, $h = 1/m$, $V_h = V_m$, and $r_h = \pi_m$. □

3.5. On the practical solution of the approximate problem (P_h)

A most important step toward the actual solution of problem (P) is the practical solution of (P_h).

We suppose that V_h is finite dimensional with

$$N_h = \dim V_h. \tag{3.16}$$

Let $\mathscr{B}_h = \{w_i\}_{i=1}^{N_h}$ be a vector basis of V_h; problem (P_h) is clearly equivalent to:

Find $u_h \in V_h$ such that

$$a(u_h, w_i) = L(w_i), \qquad \forall\, i = 1, \ldots, N_h. \tag{3.17}$$

Since $u_h \in V_h$, there exists a unique vector $\mathbf{\Lambda}_h = \{\lambda_j\}_{j=1}^{N_h} \in \mathbb{R}^{N_h}$ such that

$$u_h = \sum_{j=1}^{N_h} \lambda_j w_j. \tag{3.18}$$

Combining (3.17), (3.18), we find that u_h is obtained through the solution of the linear system

$$\sum_{j=1}^{N_h} a(w_j, w_i)\lambda_j = L(w_i) \qquad \text{for } i = 1, \ldots, N_h \tag{3.19}$$

whose unknowns are the $\lambda_j, j = 1, \ldots, N_h$.

The linear system (3.19) can also be written as follows:

$$\mathbf{A}_h \mathbf{\Lambda}_h = \mathbf{F}_h, \tag{3.20}$$

where $\mathbf{F}_h = \{L(w_i)\}_{i=1}^{N_h}$ and where the matrix $\mathbf{A}_h$ is defined by

$$\mathbf{A}_h = (a(w_j, w_i))_{1 \le i, j \le N_h}. \tag{3.21}$$

It is quite easy to show that the V-ellipticity of $a(\cdot, \cdot)$ implies that $\mathbf{A}_h$ is positive definite; moreover, the symmetry of $a(\cdot, \cdot)$ implies (from (3.21)) that $\mathbf{A}_h$ is symmetric.

Solving the linear system (3.20) can be achieved by various direct or iterative methods; in Sec. 4.5 of this appendix we shall give several references concerning the solution of the large sparse linear systems obtained from the approximation of partial differential equations.

4. Application to the Solution of Elliptic Problems for Partial Differential Operators

4.1. A trivial example in $L^2(\Omega)$

4.1.1. *Formulation of the problem: existence and uniqueness results*

In order to illustrate the generalities of Secs. 2 and 3, we have chosen, to begin with, a trivial example.

Let Ω be a domain[9] of $\mathbb{R}^N$ (possibly unbounded); we consider the problem (P) associated with the triple $\{V, a, L\}$ defined as follows:

(i) We take $V = L^2(\Omega)$; it is a classical result that $L^2(\Omega)$ equipped with the scalar product[10]

$$(v, w) = \int_\Omega v(x)w(x)\, dx, \qquad \forall\, v, w \in L^2(\Omega), \tag{4.1}$$

and the corresponding norm

$$\|v\| = \left(\int_\Omega |v(x)|^2\, dx\right)^{1/2}, \qquad \forall\, v \in L^2(\Omega), \tag{4.2}$$

[9] i.e., an open connected subset of $\mathbb{R}^N$.

[10] We consider real-valued functions only.

is a Hilbert space (in (4.1), (4.2) we have used the notation $dx = dx_1 \cdots dx_N$).

(ii) We define $a: L^2(\Omega) \times L^2(\Omega) \to \mathbb{R}$ by

$$a(v, w) = \int_\Omega a_0(x)v(x)w(x)\,dx, \qquad \forall\, v, w \in L^2(\Omega), \tag{4.3}$$

with

$$a_0 \in L^\infty(\Omega), \qquad a_0(x) \geq \alpha > 0 \text{ a.e. on } \Omega; \tag{4.4}$$

$a(\cdot, \cdot)$ is clearly bilinear. From (4.3), (4.4) we have

$$|a(v, w)| \leq \|a_0\|_{L^\infty(\Omega)}\|v\|_{L^2(\Omega)}\|w\|_{L^2(\Omega)}, \qquad \forall\, v, w \in L^2(\Omega), \tag{4.5}$$

and

$$a(v, v) \geq \alpha\|v\|^2_{L^2(\Omega)}, \qquad \forall\, v \in L^2(\Omega), \tag{4.6}$$

which imply that $a(\cdot, \cdot)$ is continuous and $L^2(\Omega)$-elliptic, respectively,

(iii) Let $f \in L^2(\Omega)$; we finally define $L: L^2(\Omega) \to \mathbb{R}$ by

$$L(v) = \int_\Omega fv\,dx, \qquad \forall\, v \in L^2(\Omega); \tag{4.7}$$

the linear form L is continuous over $L^2(\Omega)$ (actually, from the Riesz representation theorem, any linear continuous functional from $L^2(\Omega)$ to $\mathbb{R}$ has a unique representation of type (4.7)).

From Theorem 2.1 (see Sec. 2.3) and from the above properties of $L^2(\Omega)$, $a(\cdot, \cdot)$, and $L(\cdot)$, it follows that the corresponding problem (P), i.e.:

Find $u \in L^2(\Omega)$ such that

$$\int_\Omega a_0(x)u(x)v(x)\,dx = \int_\Omega f(x)v(x)\,dx, \qquad \forall\, v \in L^2(\Omega), \tag{4.8}$$

has a unique solution. Actually we do not need Theorem 2.1 to see that problem (4.8) has a unique solution which is given by

$$u = \frac{f}{a_0}. \tag{4.9}$$

Remark 4.1. In the particular case of the bilinear form $a(\cdot, \cdot)$ given by (4.3), (4.4), the operator A introduced in Sec. 2.3 is explicitly given by

$$Av = a_0 v, \qquad \forall\, v \in L^2(\Omega)$$

(and A^{-1} by

$$A^{-1}v = \frac{v}{a_0}, \qquad \forall\, v \in L^2(\Omega));$$

furthermore, we have

$$\|A\| = \|a_0\|_{L^\infty(\Omega)}$$

(and

$$\|A^{-1}\| = \left\|\frac{1}{a_0}\right\|_{L^\infty(\Omega)}\Bigg).$$

4.1.2. *Approximation of problem* (4.8)

From (4.9) we know the exact solution of problem (4.8); however, for pedagogical purposes, we shall study the approximation of (4.8) by a method of finite element type, giving us an opportunity to apply those general principles—discussed in Sec. 3—concerning the approximation of linear variational problems of type (P).

For simplicity we suppose that Ω is a bounded domain of $\mathbb{R}^N$.

To approximate problem (4.8) we first introduce a finite family $\{\Omega_i\}_{i=1}^{N_h}$ such that

$$\Omega_i \subset \Omega, \qquad \forall\, i = 1, \ldots, N_h, \tag{4.10$_1$}$$

$\forall\, i = 1, \ldots, N_h$, Ω_i is a measurable connected subset of Ω such

$$\text{that } \int_{\Omega_i} dx > 0, \tag{4.10$_2$}$$

$$\Omega_i \cap \Omega_j = \emptyset, \qquad \forall\, 1 \le i, j \le N_h, \quad i \ne j, \tag{4.10$_3$}$$

$$\bigcup_{i=1}^{N_h} \bar{\Omega}_i = \bar{\Omega}; \tag{4.10$_4$}$$

we define h by

$$h = \sup_i \delta_i, \tag{4.11}$$

where δ_i = diameter of Ω_i (we recall that

$$\text{diameter } \Omega_i = \sup_{\{x, y\} \in \Omega_i \times \Omega_i} \text{distance}(x, y));$$

Figure 4.1 illustrates a particular decomposition obeying (4.10) of a domain $\Omega \subset \mathbb{R}^2$.

We now define

$$\begin{aligned} &\mathscr{B}_h = \{w_i\}_{i=1}^{N_h}, \text{ where } \forall\, i = 1, \ldots, N_h, \\ &w_i \text{ is the characteristic function of } \Omega_i \end{aligned} \tag{4.12}$$

Figure 4.1

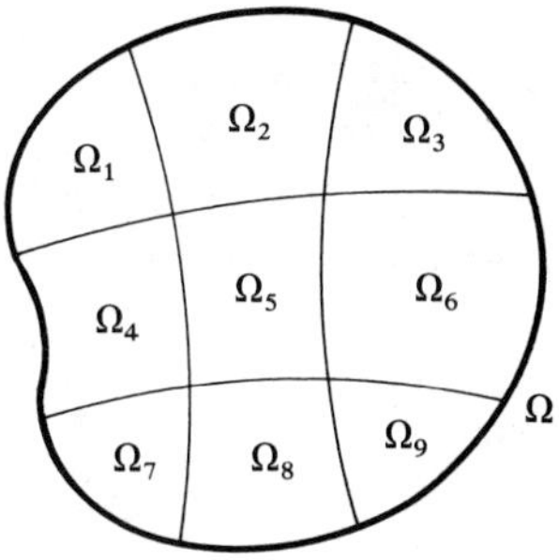

(we then have

$$w_i(x) = \begin{cases} 1 & \text{if } x \in \Omega_i, \\ 0 & \text{if } x \notin \Omega_i, \end{cases} \tag{4.13}$$

and $w_i \in L^p(\Omega)$, $\forall\, i = 1, \ldots, N_h$, $\forall\, p$ such that $1 \leq p \leq +\infty$).

We now define V_h as the subspace of $L^2(\Omega)$ (and in fact of $L^p(\Omega)$, $\forall\, p$, $1 \leq p \leq +\infty$) generated by $\mathscr{B}_h$; we then have

$$V_h = \left\{ v_h \mid v_h = \sum_{i=1}^{N_h} v_i w_i,\ v_i \in \mathbb{R}, \qquad \forall\, i = 1, \ldots, N_h \right\}, \tag{4.14}$$

and also $\dim V_h = N_h$. We observe that if $v_h \in V_h$, then v_h is piecewise constant.

The problem (P_h) corresponding to (4.8) and V_h is defined by:

Find $u_h \in V_h$ such that

$$\int_\Omega a_0 u_h v_h \, dx = \int_\Omega f v_h \, dx, \qquad \forall\, v_h \in V_h; \tag{4.15}$$

problem (4.15) *has a unique solution.*

Concerning the convergence of $\{u_h\}_h$ to the solution u of (4.8), we have:

Proposition 4.1. *We suppose that $f \in L^2(\Omega)$ and that a_0 satisfies* (4.4); *we then have*

$$\lim_{h \to 0} \|u_h - u\|_{L^2(\Omega)} = 0, \tag{4.16}$$

where u_h (resp., u) is the solution of (4.15) *(resp.,* (4.8)).

Proof. To apply Theorem 3.2 of Sec. 3.3, it suffices to find $\mathscr{V}$ and r_h obeying (3.1). Since

$$\overline{C^0(\overline{\Omega})}^{L^2(\Omega)} = L^2(\Omega),$$

we can take $\mathscr{V} = C^0(\overline{\Omega})$. We define then r_h as follows:

$$r_h v \in V_h, \qquad \forall\, v \in C^0(\overline{\Omega}), \tag{4.17$_1$}$$

$$r_h v = \sum_{i=1}^{N_h} v(P_i) w_i, \tag{4.17$_2$}$$

where, $\forall\, i = 1, \ldots, N_h$, $P_i \in \Omega_i$; using the uniform continuity of v on $\overline{\Omega}$, we can easily prove that[11]

$$\lim_{h \to 0} \|r_h v - v\|_{L^2(\Omega)} = 0, \qquad \forall\, v \in C^0(\overline{\Omega}), \tag{4.18}$$

which completes the proof of the proposition. □

Suppose that

$$u_h = \sum_{i=1}^{N_h} u_i w_i; \tag{4.19}$$

the approximate problem (4.15) is then equivalent to the linear system

$$\sum_{j=1}^{N_h} u_j \int_\Omega a_0 w_j w_i \, dx = \int_\Omega f w_i \, dx, \qquad 1 \le i \le N_h. \tag{4.20}$$

From (4.10), (4.13) it follows that (4.20) reduces to

$$u_i \int_{\Omega_i} a_0 \, dx = \int_{\Omega_i} f \, dx, \qquad \forall\, i = 1, \ldots, N_h,$$

i.e.,

$$u_i = \frac{\int_{\Omega_i} f \, dx}{\int_{\Omega_i} a_0 \, dx}, \qquad \forall\, i = 1, \ldots, N_h. \tag{4.21}$$

Remark 4.2. In practice the computation of u_i via relation (4.21) would require the computation of the $2N_h$ integrals below:

$$\int_{\Omega_i} f \, dx, \quad \int_{\Omega_i} a_0 \, dx, \qquad 1 \le i \le N_h. \tag{4.22}$$

As mentioned in Sec. 3.2, Remark 3.2, the approximate calculation of the integrals (4.22) would introduce an extra error; also, as mentioned in the above remark, the effect of numerical integration on the solution of elliptic variational problems for partial differential operators is discussed in Ciarlet [1], [2], [3] (see also Strang and Fix [1] and Oden and Reddy [1]).

Remark 4.3. If $a_0 \equiv 1$, the solution of (4.8) is clearly $u = f$ and that of the corresponding approximate problem (4.15) is given by:

$$u_h = \sum_{i=1}^{N_h} \frac{\int_{\Omega_i} u \, dx}{\int_{\Omega_i} dx} w_i; \tag{4.23}$$

[11] We have, in fact, a stronger convergence result since

$$\lim_{h \to 0} \|r_h v - v\|_{L^\infty(\Omega)} = 0.$$

it is well known from integration theory that if u_h is obtained from u by (4.23), we have

$$\lim_{h\to 0} \|u_h - u\|_{L^p(\Omega)} = 0, \qquad \forall\, u \in L^p(\Omega), \quad 1 \le p < +\infty, \tag{4.24}$$

$$\lim_{h\to 0} \|u_h - u\|_{L^\infty(\Omega)} = 0, \qquad \forall\, u \in C^0(\overline{\Omega}) \tag{4.25}$$

(if $u \in L^\infty(\Omega)$ we still have (4.24) for all p such that $1 \le p < +\infty$; we also have[12]

$$\lim_{h\to 0} u_h = u \text{ weakly} * \text{ in } L^\infty(\Omega)). \tag{4.26}$$

Returning to the case $a_0 \not\equiv 1$, but still obeying (4.4), we can directly prove (using integration theory) that if $u_h \in V_h$ is given by

$$u_h = \sum_{i=1}^{N_h} \left(\frac{\int_{\Omega_i} f\, dx}{\int_{\Omega_i} a_0\, dx} \right) w_i, \tag{4.27}$$

then

$$\lim_{h\to 0} \left\| u_h - \frac{f}{a_0} \right\|_{L^p(\Omega)} = 0, \qquad \forall\, f \in L^p(\Omega), \quad 1 \le p < +\infty; \tag{4.28}$$

we can also prove that if $f \in L^\infty(\Omega)$, then (4.28) still holds for all $p \in [1, +\infty[$ and that

$$\lim_{h\to 0} u_h = \frac{f}{a_0} \text{ weakly} * \text{ in } L^\infty(\Omega)$$

(if $a_0, f \in C^0(\overline{\Omega})$, we have

$$\left.\lim_{h\to 0} \left\| u_h - \frac{f}{a_0} \right\|_{L^\infty(\Omega)} = 0 \right).$$

4.2. Solution of Neumann problems for second-order elliptic partial differential operators

In this section we shall discuss the formulation, and the solution via variational methods, of Neumann problems for second-order elliptic partial differential operators; such problems play a very important role in various branches of physics and mechanics.

---

[12] Relation (4.26) means that

$$\lim_{h\to 0} \int_\Omega u_h v\, dx = \int_\Omega uv\, dx, \qquad \forall\, v \in L^1(\Omega);$$

see, e.g., Yosida [1] for the properties of the weak$*$ topologies.

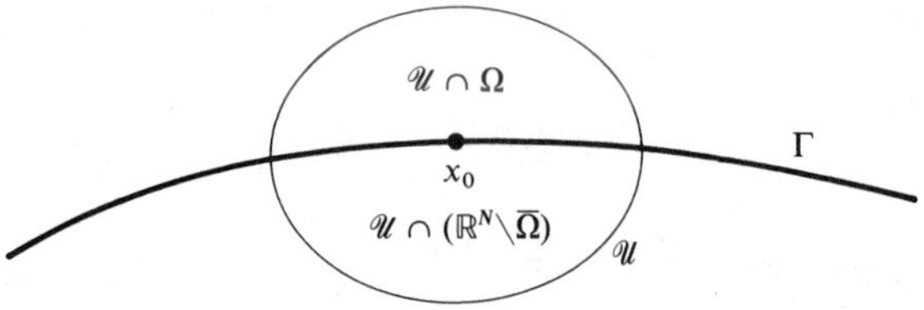

Figure 4.2

The finite element approximation of these Neumann's problems will be discussed in Sec. 4.5.

4.2.1. *The classical formulation*

Let Ω be a (possibly unbounded) domain of $\mathbb{R}^N$ (we denote by x $(=\{x_i\}_{i=1}^N)$ the generic point of $\mathbb{R}^N$); we denote by Γ the boundary $\partial\Omega$ of Ω, and we suppose that Γ is a reasonably smooth manifold with Ω locally on one side of Γ (the last property means that if $x_0 \in \Gamma$ and if we consider a sufficiently small neighborhood $\mathscr{U}$ of x_0 in $\mathbb{R}^N$, we have the situation of Fig. 4.2).

Let $\bar{\bar{\mathbf{A}}}(x)$ be a linear operator from $\mathbb{R}^N$ to $\mathbb{R}^N$ depending upon x over Ω, and let us denote by $\mathbf{\nabla}$ the vector $\{\partial/\partial x_i\}_{i=1}^N$; we consider the following Neumann problem:

Find u such that

$$-\mathbf{\nabla} \cdot (\bar{\bar{\mathbf{A}}}\mathbf{\nabla} u) + a_0 u = f \text{ in } \Omega, \qquad (\bar{\bar{\mathbf{A}}}\mathbf{\nabla} u) \cdot \mathbf{n} = g \text{ on } \Gamma, \tag{4.29}$$

where a_0, f (resp., g) are given functions defined over Ω (resp., Γ), where $\mathbf{n}$ is the outward unit vector normal at Γ, and where $\mathbf{a} \cdot \mathbf{b}$ denotes the usual scalar product of $\mathbb{R}^N$, i.e.,

$$\mathbf{a} \cdot \mathbf{b} = \sum_{i=1}^N a_i b_i, \qquad \forall\, \mathbf{a}, \mathbf{b} \in \mathbb{R}^N, \quad \mathbf{a} = \{a_i\}_{i=1}^N, \quad \mathbf{b} = \{b_i\}_{i=1}^N.$$

Remark 4.4. The Neumann problem (4.29) is written using the so-called *divergence formulation*; if $\bar{\bar{\mathbf{A}}}$ is a differentiable function of x such that

$$\bar{\bar{\mathbf{A}}}(x) = (a_{ij}(x))_{1 \le i, j \le N},$$

from

$$\mathbf{\nabla} \cdot (\bar{\bar{\mathbf{A}}}\mathbf{\nabla} v) = \sum_{i=1}^N \frac{\partial}{\partial x_i} \sum_{j=1}^N a_{ij} \frac{\partial v}{\partial x_j},$$

we have

$$\mathbf{\nabla} \cdot (\bar{\bar{\mathbf{A}}}\mathbf{\nabla} v) = \sum_{1 \le i, j \le N} a_{ij} \frac{\partial^2 v}{\partial x_i\, \partial x_j} + \sum_{1 \le i, j \le N} \frac{\partial a_{ij}}{\partial x_i} \frac{\partial v}{\partial x_j}, \tag{4.30}$$

i.e., a quite explicit formulation of the operator

$$v \to \mathbf{\nabla} \cdot (\bar{\bar{\mathbf{A}}}\mathbf{\nabla} v).$$

Too many engineers and physicists have a tendency to approximate (4.29) (mostly by finite difference methods) using (4.30) to evaluate $\nabla \cdot (\bar{\bar{\mathbf{A}}}\nabla u)$; actually this approach can be very dangerous since some stability properties of the original problem can be lost during the approximation process, leading to a non-well-posed approximate problem (even if the original problem is well posed); moreover, there is, in general, a conservation principle behind a divergence formulation, and it is of fundamental importance to preserve this conservation property. Another reason for using the divergence formulation is that if $\bar{\bar{\mathbf{A}}}$ is not differentiable, the divergence formulation may still be meaningful, unlike the product form in (4.30). Actually the variational formulation of problem (4.29)—given in Sec. 4.2.2—provides a quite elegant way to overcome those difficulties related to the possible nondifferentiability of $\bar{\bar{\mathbf{A}}}$; also, in some sense, it preserves the above conservation principles.

Remark 4.5. If $\bar{\bar{\mathbf{A}}} = \mathbf{I}$, i.e., if

$$a_{ij}(x) = \begin{cases} 1 & \text{if } i = j, \\ 0 & \text{if } i \neq j, \end{cases}$$

the Neumann problem (4.29) reduces to

$$-\Delta u + a_0 u = f \quad \text{in } \Omega, \qquad \frac{\partial u}{\partial n} = g \quad \text{on } \Gamma, \tag{4.31}$$

where the Laplace operator Δ is defined by

$$\Delta = \sum_{i=1}^{N} \frac{\partial^2}{\partial x_i^2},$$

and where $\partial/\partial n$ denotes the normal derivative on Γ (defined by $\partial v/\partial n = \nabla v \cdot \mathbf{n}$).

4.2.2. *A variational formulation of the Neumann problem* (4.29)

Let v be a smooth function defined over $\bar{\Omega}$; multiplying the first relation (4.29) by v and integrating over Ω, we obtain

$$-\int_\Omega \nabla \cdot (\bar{\bar{\mathbf{A}}}\nabla u) v \, dx + \int_\Omega a_0 uv \, dx = \int_\Omega fv \, dx. \tag{4.32}$$

Let us now consider a vector function $x \to \mathbf{V}(x)$, defined over $\bar{\Omega}$ and taking its values in $\mathbb{R}^N$; we recall (Green–Ostrogradsky formula) that

$$\int_\Omega \mathbf{V} \cdot \nabla v \, dx + \int_\Omega v \nabla \cdot \mathbf{V} \, dx = \int_\Omega \nabla \cdot (v\mathbf{V}) \, dx = \int_\Gamma v \mathbf{V} \cdot \mathbf{n} \, d\Gamma, \tag{4.33}$$

where $d\Gamma$ denotes the superficial measure along Γ. Taking $\mathbf{V} = \bar{\bar{\mathbf{A}}}\nabla u$ in (4.33), we obtain

$$-\int_\Omega \nabla \cdot (\bar{\bar{\mathbf{A}}}\nabla u) v \, dx = \int_\Omega (\bar{\bar{\mathbf{A}}}\nabla u) \cdot \nabla v \, dx - \int_\Gamma (\bar{\bar{\mathbf{A}}}\nabla u) \cdot \mathbf{n} \, v d\Gamma,$$

which, combined with (4.32) and with the second relation (4.29), implies

$$\int_\Omega (\bar{\mathbf{A}}\nabla u)\cdot\nabla v\,dx + \int_\Omega a_0 uv\,dx = \int_\Omega fv\,dx + \int_\Gamma gv\,d\Gamma; \tag{4.34}$$

the function v in (4.32), (4.34) is usually called a *test function*. We observe that (4.34)—unlike formulation (4.29)—involves only first-order derivatives of the unknown function u.

Conversely, it is proved in, e.g., Lions [2], [6] and Necas [1], that if (4.34) holds, $\forall\, v \in \mathscr{V}$, where[13]

$$\mathscr{V} = \{v \mid v \in C^1(\bar{\Omega}),\ v \text{ has a compact support in } \bar{\Omega}\}, \tag{4.35}$$

then u is in some sense a solution of the Neumann problem (4.29).

At this point, if one wishes to go further, it is necessary to introduce the Sobolev space $H^1(\Omega)$ defined by

$$H^1(\Omega) = \left\{v \mid v \in L^2(\Omega),\ \frac{\partial v}{\partial x_i} \in L^2(\Omega), \quad \forall\, i = 1, \ldots, N\right\}^{14}; \tag{4.36}$$

the derivatives in (4.36) are taken in the distribution sense, i.e.,

$$\int_\Omega \frac{\partial v}{\partial x_i}\phi\,dx = -\int_\Omega v\frac{\partial \phi}{\partial x_i}\,dx, \qquad \forall\, \phi \in \mathscr{D}(\Omega), \tag{4.37}$$

where

$$\mathscr{D}(\Omega) = \{\phi \mid \phi \in C^\infty(\Omega),\ \phi \text{ has a compact support in } \Omega\}. \tag{4.38}$$

For more details and further properties of the Sobolev spaces, see, e.g., Lions and Magenes [1], Necas [1], Adams [1], and Oden and Reddy [1].

Equipped with the scalar product

$$(v, w)_{H^1(\Omega)} = \int_\Omega (\nabla v\cdot\nabla w + vw)\,dx \tag{4.39}$$

and the corresponding norm

$$\|v\|_{H^1(\Omega)} = \left(\int_\Omega (|\nabla v|^2 + |v|^2)\,dx\right)^{1/2}, \tag{4.40}$$

the space $H^1(\Omega)$ is an Hilbert space.

In the sequel frequent use will be made of the following results (proved in the above references on Sobolev spaces):

---

[13] If Ω is bounded, then $\mathscr{V} = C^1(\bar{\Omega})$.

[14] We observe that if Ω is bounded (resp., unbounded), $H^1(\Omega)$ contains (resp., does not contain) the functions constant over Ω.

Proposition 4.2. *We suppose that Γ is sufficiently smooth (Lipschitz continuous, for example) with Ω locally on one side of Γ. We then have*

$$\overline{\mathscr{D}(\bar{\Omega})}^{H^1(\Omega)} = H^1(\Omega), \tag{4.41}$$

where

$$\mathscr{D}(\bar{\Omega}) = \{v \mid v \in C^\infty(\bar{\Omega}),\ v \text{ has a compact support in } \bar{\Omega}\}.[15]$$

Let $v \in \mathscr{D}(\bar{\Omega})$; we define the *trace operator* γ_0 as the linear mapping defined by

$$\gamma_0 v = v|_\Gamma; \tag{4.42}$$

the following proposition can be proved:

Proposition 4.3. *There exists a constant $c(\Omega)$ such that*

$$\forall\, v \in \mathscr{D}(\bar{\Omega}), \qquad \|\gamma_0 v\|_{L^2(\Gamma)} \le c(\Omega)\|v\|_{H^1(\Omega)}. \tag{4.43}$$

As a corollary to Propositions 4.2 and 4.3, there exists a linear continuous operator from $H^1(\Omega)$ to $L^2(\Gamma)$ (the trace operator) whose restriction to $\mathscr{D}(\bar{\Omega})$ coincides with γ_0; we still use the notation γ_0 for that trace operator and we have

$$\forall\, v \in H^1(\Omega), \qquad \|\gamma_0 v\|_{L^2(\Gamma)} \le c(\Omega)\|v\|_{H^1(\Omega)}, \tag{4.44}$$

with $c(\Omega)$ as in (4.43). Actually, when no confusion is likely to arise, we shall simply write $\gamma_0 v = v$.

Returning to the Neumann problem (4.29) and to (4.34), we suppose that the following hypotheses concerning $\bar{\bar{\mathbf{A}}}$, a_0, f, g hold:

$$f \in L^2(\Omega), \qquad g \in L^2(\Gamma), \tag{4.45}$$

$$a_0 \in L^\infty(\Omega), \qquad a_0(x) \ge \alpha_0 > 0 \text{ a.e. on } \Omega, \tag{4.46}$$

$a_{ij} \in L^\infty(\Omega)$, $\forall\, 1 \le i, j \le N$; there exists $\alpha > 0$ such that

$$\bar{\bar{\mathbf{A}}}(x)\boldsymbol{\xi} \cdot \boldsymbol{\xi} \ge \alpha|\boldsymbol{\xi}|^2 \text{ a.e. on } \Omega, \qquad \forall\, \boldsymbol{\xi} = \{\xi_i\}_{i=1}^N \in \mathbb{R}^N$$
$$\left(\text{with } |\boldsymbol{\xi}| = \left(\sum_{i=1}^N \xi_i^2\right)^{1/2}\right). \tag{4.47}$$

We now define $a: H^1(\Omega) \times H^1(\Omega) \to \mathbb{R}$ and $L: H^1(\Omega) \to \mathbb{R}$ by

$$a(v, w) = \int_\Omega (\bar{\bar{\mathbf{A}}}\nabla v) \cdot \nabla w\, dx + \int_\Omega a_0 v w\, dx, \qquad \forall\, v, w \in H^1(\Omega), \tag{4.48}$$

$$L(v) = \int_\Omega f v\, dx + \int_\Gamma g \gamma_0 v\, d\Gamma, \qquad \forall\, v \in H^1(\Omega), \tag{4.49}$$

respectively; $a(\cdot, \cdot)$ (resp., L) is clearly bilinear (resp., linear).

[15] If Ω is bounded, we have $\mathscr{D}(\bar{\Omega}) = C^\infty(\bar{\Omega})$.

From (4.40), (4.44), (4.45), (4.49), and from the Schwarz inequality in $L^2(\Omega)$ and $L^2(\Gamma)$, we have

$$\begin{aligned}|L(v)| &\leq \|f\|_{L^2(\Omega)}\|v\|_{L^2(\Omega)} + \|g\|_{L^2(\Gamma)}\|\gamma_0 v\|_{L^2(\Gamma)} \\ &\leq (\|f\|_{L^2(\Omega)} + c(\Omega)\|g\|_{L^2(\Gamma)})\|v\|_{H^1(\Omega)}, \quad \forall\, v \in H^1(\Omega), \end{aligned} \tag{4.50}$$

which shows the continuity of L.

We now define

$$|\bar{\bar{\mathbf{A}}}(x)| = \operatorname*{Sup}_{\xi \in \mathbb{R}^N - \{\mathbf{0}\}} \frac{|\bar{\bar{\mathbf{A}}}(x)\xi|}{|\xi|};$$

from (4.47) we clearly find that

$$\text{the function } x \to |\bar{\bar{\mathbf{A}}}(x)| \text{ belongs to } L^\infty(\Omega); \tag{4.51}$$

we denote by $\|\bar{\bar{\mathbf{A}}}\|_{L^\infty(\Omega)}$ the L^∞-norm of the above function; we then have, from (4.48) and from the Schwarz inequality,

$$\begin{aligned}|a(v, w)| &\leq \|\bar{\bar{\mathbf{A}}}\|_{L^\infty(\Omega)}\left(\int_\Omega |\nabla v|^2\, dx\right)^{1/2}\left(\int_\Omega |\nabla w|^2\, dx\right)^{1/2} \\ &\quad + \|a_0\|_{L^\infty(\Omega)}\|v\|_{L^2(\Omega)}\|w\|_{L^2(\Omega)} \\ &\leq \operatorname{Max}(\|\bar{\bar{\mathbf{A}}}\|_{L^\infty(\Omega)}, \|a_0\|_{L^\infty(\Omega)})\|v\|_{H^1(\Omega)}\|w\|_{H^1(\Omega)}, \quad \forall\, v, w \in H^1(\Omega). \end{aligned} \tag{4.52}$$

Relation (4.52) implies the continuity of $a(\cdot, \cdot)$.

Finally, from (4.46), (4.47) we have

$$a(v, v) \geq \operatorname{Min}(\alpha, \alpha_0)\|v\|^2_{H^1(\Omega)}, \quad \forall\, v \in H^1(\Omega), \tag{4.53}$$

which shows the $H^1(\Omega)$-ellipticity of $a(\cdot, \cdot)$.

From the above properties of $H^1(\Omega)$, $a(\cdot, \cdot)$, and $L(\cdot)$ we can apply Theorem 2.1 of Sec. 2.3 to prove:

Proposition 4.4. *The linear variational problem:*

Find $u \in H^1(\Omega)$ such that

$$a(u, v) = L(v), \forall\, v \in H^1(\Omega), \tag{4.54}$$

has a unique solution.

Remark 4.6. Suppose that $\bar{\bar{\mathbf{A}}}$ is symmetric (i.e., $a_{ij}(x) = a_{ji}(x)$, $\forall\, 1 \leq i, j \leq N$, a.e. on Ω); this, in turn, implies the symmetry of the bilinear form $a(\cdot, \cdot)$. From Proposition 2.1 of Sec. 2.4 it then follows that there is equivalence between (4.54) and the minimization problem:

Find $u \in H^1(\Omega)$ such that

$$J(u) \leq J(v), \qquad \forall\, v \in H^1(\Omega), \tag{4.55}$$

where

$$J(v) = \frac{1}{2}\int_\Omega (\bar{\bar{\mathbf{A}}}\nabla v)\cdot\nabla v\,dx + \frac{1}{2}\int_\Omega a_0 v^2\,dx - \int_\Omega fv\,dx - \int_\Gamma gv\,d\Gamma. \tag{4.56}$$

Problem (4.55), (4.56) is a typical problem of the classical calculus of variations.

At this step, the natural question that arises is the following: *Is the solution of the linear variational problem* (4.54) *a solution of the Neumann problem* (4.29)?

To discuss this property, first consider $v \in \mathscr{D}(\Omega)$; since $\mathscr{D}(\Omega) \subset H^1(\Omega)$, we have

$$a(u, v) = L(v), \qquad \forall\, v \in \mathscr{D}(\Omega),$$

or more explicitly, taking into account the fact that $\gamma_0 v = 0$, $\forall\, v \in \mathscr{D}(\Omega)$,

$$\int_\Omega (\bar{\bar{\mathbf{A}}}\nabla u)\cdot\nabla v\,dx + \int_\Omega a_0 uv\,dx = \int_\Omega fv\,dx, \qquad \forall\, v \in \mathscr{D}(\Omega), \tag{4.57}$$

i.e., in a distribution sense (see L. Schwartz [1], Necas [1], and Lions [1], [6])

$$\langle \bar{\bar{\mathbf{A}}}\nabla u, \nabla v\rangle + \langle a_0 u, v\rangle = \langle f, v\rangle, \qquad \forall\, v \in \mathscr{D}(\Omega), \tag{4.58}$$

where $\langle\cdot,\cdot\rangle$ denotes the duality pairing between $\mathscr{D}'(\Omega)$ and $\mathscr{D}(\Omega)$. From the properties of the differentiation in $\mathscr{D}'(\Omega)$ (see Schwartz, Lions, and Necas, *loc. cit.*), we have

$$-\langle \nabla\cdot(\bar{\bar{\mathbf{A}}}\nabla u), v\rangle + \langle a_0 u, v\rangle = \langle f, v\rangle, \qquad \forall\, v \in \mathscr{D}(\Omega),$$

i.e., in a distribution sense,

$$-\nabla\cdot(\bar{\bar{\mathbf{A}}}\nabla u) + a_0 u = f. \tag{4.59}$$

A rigorous proof of the fact that the solution u of (4.54) satisfies the boundary condition in (4.29) is definitely beyond the scope of this appendix[16]; therefore we shall give only a formal proof of this property. Taking v in $\mathscr{D}(\bar{\Omega})$, from (4.59) we have

$$-\int_\Omega \nabla\cdot(\bar{\bar{\mathbf{A}}}\nabla u)v\,dx + \int_\Omega a_0 uv\,dx = \int_\Omega fv\,dx, \qquad \forall\, v \in \mathscr{D}(\bar{\Omega}); \tag{4.60}$$

from the Green–Ostrogradsky formula[17] (4.33) it then follows that

$$\int_\Omega (\bar{\bar{\mathbf{A}}}\nabla u)\cdot\nabla v\,dx + \int_\Omega a_0 uv\,dx = \int_\Omega fv\,dx + \int_\Gamma (\bar{\bar{\mathbf{A}}}\nabla u)\cdot\mathbf{n}v\,d\Gamma, \qquad \forall\, v \in \mathscr{D}(\bar{\Omega}),$$

[16] For this proof, see Lions and Necas, *loc. cit.*

[17] This is precisely the most crucial step, since here we do not justify the fact that (4.33) can be applied in this context.

i.e.,

$$a(u, v) = \int_\Omega fv\, dx + \int_\Gamma (\overline{\overline{\mathbf{A}}}\nabla u) \cdot \mathbf{n}\, v d\Gamma, \qquad \forall\, v \in \mathscr{D}(\overline{\Omega}). \tag{4.61}$$

Comparing (4.61) to (4.54), we obtain

$$\int_\Gamma gv\, d\Gamma = \int_\Gamma (\overline{\overline{\mathbf{A}}}\nabla u) \cdot \mathbf{n} v\, d\Gamma, \qquad \forall\, v \in \mathscr{D}(\overline{\Omega}); \tag{4.62}$$

actually it is proved in the above references that (4.62) implies

$$(\overline{\overline{\mathbf{A}}}\nabla u) \cdot \mathbf{n} = g \quad \text{on } \Gamma,$$

which is precisely the boundary condition in (4.29). Thus the solution u of the variational problem (4.54) is also a solution of the Neumann problem (4.29); actually we have even more, since the following can be proved:

Proposition 4.5. *If we suppose that f, g, a_0, and $\overline{\overline{\mathbf{A}}}$ satisfy* (4.45)–(4.47), *then the Neumann problem* (4.29) *has a unique solution belonging to $H^1(\Omega)$, which is also the unique solution of the linear variational problem* (4.54).

Proposition 4.5 is quite important since it suggests the use of (4.54) to solve the Neumann problem (4.29) (for the numerical solution in particular).

4.2.3. *Remarks and further comments*

4.2.3.1. *Some remarks.*

Remark 4.7. Let us consider $N + 1$ functions f_i, $i = 0, \ldots, N$, such that $f_i \in L^2(\Omega)$, $\forall\, i = 0, \ldots, N$; with $\{f_i\}_{i=0}^N$ we associate the linear functional $L: H^1(\Omega) \to \mathbb{R}$ defined by

$$\begin{aligned} L(v) &= \int_\Omega f_0 v\, dx + \sum_{i=1}^N \int_\Omega f_i \frac{\partial v}{\partial x_i}\, dx \\ &= \int_\Omega f_0 v\, dx + \int_\Omega \mathbf{f} \cdot \nabla v\, dx, \qquad \forall\, v \in H^1(\Omega), \end{aligned} \tag{4.63}$$

where $\mathbf{f} = \{f_i\}_{i=1}^N$. The linear functional (4.63) is clearly continuous over $H^1(\Omega)$; actually, using the Schwarz inequality, we have

$$\begin{aligned} |L(v)| &\le \|f_0\|_{L^2(\Omega)} \|v\|_{L^2(\Omega)} + \sum_{N=1}^N \|f_i\|_{L^2(\Omega)} \left\| \frac{\partial v}{\partial x_i} \right\|_{L^2(\Omega)} \\ &\le \left(\sum_{i=0}^N \|f_i\|^2_{L^2(\Omega)} \right)^{1/2} \left(\|v\|^2_{L^2(\Omega)} + \sum_{i=1}^N \left\| \frac{\partial v}{\partial x_i} \right\|^2_{L^2(\Omega)} \right)^{1/2} \\ &= \left(\sum_{i=0}^N \|f_i\|^2_{L^2(\Omega)} \right)^{1/2} \|v\|_{H^1(\Omega)}, \qquad \forall\, v \in H^1(\Omega). \end{aligned} \tag{4.64}$$

In fact it is shown in, e.g., Lions [2] that any continuous linear functional over $H^1(\Omega)$ has a (nonunique) representation of type (4.63).

We now consider the linear variational problem:

Find $u \in H^1(\Omega)$ such that

$$a(u, v) = L(v), \qquad \forall\, v \in H^1(\Omega), \tag{4.65}$$

where $a(\cdot, \cdot)$ (resp., $L(\cdot)$) is defined by (4.48) (resp., (4.63)); if (4.46), (4.47) still hold, problem (4.65) has a unique solution. Taking $v \in \mathscr{D}(\Omega)$ in (4.65), we should prove, as in Sec. 4.2.2, that u is a solution (in the sense of distributions) of the linear partial differential equation (of elliptic type)

$$-\nabla \cdot (\bar{\bar{\mathbf{A}}} \nabla u) + a_0 u = f_0 - \nabla \cdot \mathbf{f}; \tag{4.66}$$

concerning the boundary conditions, we should prove (at least in a formal way) that u satisfies

$$(\bar{\bar{\mathbf{A}}} \nabla u) \cdot \mathbf{n} = \mathbf{f} \cdot \mathbf{n} \quad \text{on } \Gamma. \tag{4.67}$$

We observe that if $\mathbf{f}$ is the more general function of $(L^2(\Omega))^N$, then $\mathbf{f} \cdot \mathbf{n}$ is meaningless on Γ; however, it can be proved that if

$$\mathbf{f} \in (L^2(\Omega))^N, \qquad \nabla \cdot \mathbf{f} \in L^2(\Omega),$$

then a mathematical sense can be given to $\mathbf{f} \cdot \mathbf{n}$ on Γ and to the boundary condition (4.67).

Remark 4.8. Some application lead us to consider linear functionals $L(\cdot)$ like

$$L(v) = \int_\Omega fv\, dx + \int_\Gamma gv\, d\Gamma + \int_\gamma hv\, d\gamma, \tag{4.68}$$

where γ is a measurable part of a $(N - 1)$ manifold contained in Ω. We suppose that f, g, and h belong to $L^2(\Omega)$, $L^2(\Gamma)$, and $L^2(\gamma)$, respectively, which implies that $L(\cdot)$ is continuous over $H^1(\Omega)$; therefore if we suppose that $a(\cdot, \cdot)$ and $L(\cdot)$ are defined by (4.48) and (4.68), respectively, and that (4.46), (4.47) still hold, then the variational problem:

Find $u \in H^1(\Omega)$ such that

$$a(u, v) = L(v), \qquad \forall\, v \in H^1(\Omega), \tag{4.69}$$

has a unique solution; this solution is also the unique solution, in $H^1(\Omega)$, of

$$-\nabla \cdot (\bar{\bar{\mathbf{A}}} \nabla u) + a_0 u = f + T(h) \quad \text{(in the sense of distributions)},$$
$$(\bar{\bar{\mathbf{A}}} \nabla u) \cdot \mathbf{n} = g \quad \text{on } \Gamma, \tag{4.70}$$

where the distribution $T(h)$ is defined by

$$\langle T(h), v \rangle = \int_\gamma hv\, d\gamma, \qquad \forall\, v \in \mathscr{D}(\Omega).$$

We illustrate the above considerations by the particular case where $N = 1$, $\Omega =]0, 1[$ and where

$$L(v) = \int_0^1 fv\,dx + \lambda v(0) + \mu v(1) + \nu v(x_0), \qquad \forall\, v \in H^1(0, 1), \tag{4.71}$$

$$a(v, w) = \int_0^1 a_1(x)\frac{dv}{dx}\frac{dw}{dx}\,dx + \int_0^1 a_0 vw\,dx, \qquad \forall\, v, w \in H^1(0, 1), \tag{4.72}$$

with

$$f \in L^2(0, 1);\quad \lambda, \mu, \nu \in \mathbb{R};\quad x_0 \in]0, 1[;\quad a_1 \in L^\infty(0, 1),\, a_1(x) \geq \alpha > 0$$
$$\text{a.e. on }]0, 1[;\quad a_0 \in L^\infty(0, 1),\, a_0(x) \geq \alpha_0 > 0 \text{ a.e. on }]0, 1[. \tag{4.73}$$

Since $H^1(0, 1) \subset C^0[0, 1]$ and the injection from $H^1(0, 1)$ to $C^0[0, 1]$ is continuous[18] (with $C^0[0, 1]$ equipped with the $L^\infty(0, 1)$-norm), (4.71) makes sense (we have to keep in mind that if $\Omega \subset \mathbb{R}^N$ with $N \geq 2$, then $H^1(\Omega) \not\subset C^0(\bar{\Omega})$).

From the hypotheses (4.73), the linear functional (4.71) (resp., the bilinear functional (4.72)) is continuous (resp., continuous and $H^1(0, 1)$-elliptic); the variational problem:

Find $u \in H^1(0, 1)$ such that

$$a(u, v) = L(v), \qquad \forall\, v \in H^1(0, 1), \tag{4.74}$$

has a unique solution which is also the unique solution in $H^1(0, 1)$ of the one-dimensional Neumann problem

$$-\frac{d}{dx}\left(a_1\frac{du}{dx}\right) + a_0 u = f + \nu\delta_{x_0} \text{ in }]0, 1[,$$

$$\left(a_1\frac{du}{dx}\right)_{x=0} = -\lambda, \quad \left(a_1\frac{du}{dx}\right)_{x=1} = \mu, \tag{4.75}$$

where δ_{x_0} denotes the Dirac measure at x_0.

Remark 4.9. Actually the hypothesis

$$a_0(x) \geq \alpha_0 > 0 \text{ a.e. on } \Omega \text{ (cf. (4.46))}$$

is quite strong and can be replaced by the weaker one,

$$a_0(x) \geq 0 \text{ a.e. on } \Omega, \int_\Omega a_0(x)\,dx > 0; \tag{4.76}$$

---

[18] In fact the injection from $H^1(0, 1)$ to $C^0[0, 1]$ is compact. We can also prove that

$$\|v\|_{C^0[0,1]}\left(= \operatorname*{Max}_{x\in[0,1]} |v(x)|\right) \leq \sqrt{2}\,\|v\|_{H^1(0,1)}, \qquad \forall\, v \in H^1(0, 1).$$

on the other hand, in view of applications (in fluid mechanics, for example) it is quite important to discuss the solution of the Neumann problem (4.29) in the important case where $a_0 = 0$.

Before discussing (in Sec. 4.2.3.2) the solution of the Neumann problem (4.29) under the above hypotheses, we shall prove a quite useful (and classical) lemma:

Lemma 4.1. *Let Ω be a bounded domain of $\mathbb{R}^N$, and let us consider the bilinear form $a: H^1(\Omega) \times H^1(\Omega) \to \mathbb{R}$, where*

$$a(v, w) = a_1(v, w) + a_0(v, w), \qquad \forall\, v, w \in H^1(\Omega); \tag{4.77}$$

we suppose that

$$a_1(v, w) = \int_\Omega (\overline{\overline{\mathbf{A}}}\nabla v) \cdot \nabla w \, dx, \qquad \forall\, v, w \in H^1(\Omega), \tag{4.78}$$

where $\overline{\overline{\mathbf{A}}}$ satisfies (4.47), *and also*

$$\begin{aligned} &a_0(\cdot, \cdot) \text{ is bilinear, continuous, and positive semidefinite} \\ &(\text{i.e., } a_0(v, v) \geq 0, \quad \forall\, v \in H^1(\Omega)), \text{ and such that} \\ &a_0(v, v) = 0,\ v = \text{constant, implies } v = 0. \end{aligned} \tag{4.79}$$

We then have the following property:

$$\text{the bilinear form } a(\cdot, \cdot) \text{ is } H^1(\Omega)\text{-elliptic.} \tag{4.80}$$

PROOF. The proof of Lemma 4.1 is quite classical and follows directly from the compactness property below (known as Rellich's Theorem).

Since Ω is bounded, we have:

$$\text{the injection from } H^1(\Omega) \text{ to } L^2(\Omega) \text{ is compact,} \tag{4.81}$$

which implies, as a corollary,

$$\text{If } \lim_{n \to +\infty} v_n = v \text{ weakly in } H^1(\Omega), \quad \text{then } \lim_{n \to +\infty} \|v_n - v\|_{L^2(\Omega)} = 0. \tag{4.82}$$

The $H^1(\Omega)$-ellipticity of $a(\cdot, \cdot)$ indicates the existence of $\gamma > 0$ such that

$$a(v, v) \geq \gamma \|v\|^2_{H^1(\Omega)}, \qquad \forall\, v \in H^1(\Omega). \tag{4.83}$$

Suppose that (4.83) is not true; there is equivalence between the fact that (4.83) does not hold and the existence of a sequence $\{v_n\}_{n \geq 0}$ in $H^1(\Omega)$ such that

$$\|v_n\|_{H^1(\Omega)} = 1, \qquad \forall\, n, \qquad \lim_{n \to +\infty} a(v_n, v_n) = 0. \tag{4.84}$$

Since—from (4.84)—the sequence $\{v_n\}_{n \geq 0}$ is bounded in the Hilbert space $H^1(\Omega)$, we can extract from $\{v_n\}_{n \geq 0}$ a subsequence—still denoted by $\{v_n\}_{n \geq 0}$—such that

$$\lim_{n \to +\infty} v_n = v^* \text{ weakly in } H^1(\Omega).^{19} \tag{4.85}$$

[19] This follows from the fact, proved in, e.g., Yosida [1], that the closed, convex, bounded sets of the Hilbert spaces are weakly compact (here $\{v_n\}_{n \geq 0}$ is contained in the closed ball of $H^1(\Omega)$, whose center is 0 and radius 1).

We observe (see part (2) of the proof of Theorem 5.2 in Chapter I, Sec. 5.4) that (4.84), (4.85), combined with the continuity of $a(\cdot, \cdot)$ and its positive-semidefinite property, imply

$$0 \leq a(v^*, v^*) \leq \liminf_{n \to +\infty} a(v_n, v_n) = \lim_{n \to +\infty} a(v_n, v_n) = 0,$$

i.e.,

$$a(v^*, v^*) = 0. \tag{4.86}$$

If we make (4.86) explicit, from (4.77), (4.78) we have

$$\int_\Omega (\overline{\mathbf{A}}\nabla v^*) \cdot \nabla v^* \, dx + a_0(v^*, v^*) = 0. \tag{4.87}$$

Combining (4.47), (4.79), and (4.87), we obtain

$$\nabla v^* = \mathbf{0} \text{ a.e. on } \Omega \quad (\Leftrightarrow v^* = \text{constant on } \Omega), \tag{4.88}$$

$$a_0(v^*, v^*) = 0, \tag{4.89}$$

and (4.79), (4.88), (4.89), in turn, imply that

$$v^* = 0. \tag{4.90}$$

We observe that (4.47) and the second relation (4.84) (resp., (4.82), (4.85), (4.90)) imply

$$\lim_{n \to +\infty} \nabla v_n = \mathbf{0} \text{ strongly in } (L^2(\Omega))^N \tag{4.91}$$

(resp.,

$$\lim_{n \to +\infty} v_n = 0 \text{ stongly in } L^2(\Omega)). \tag{4.92}$$

From (4.91), (4.92) we have

$$\lim_{n \to +\infty} v_n = 0 \text{ strongly in } H^1(\Omega),$$

i.e.,

$$\lim_{n \to +\infty} \|v_n\|_{H^1(\Omega)} = 0. \tag{4.93}$$

Actually there is a contradiction between (4.93) and the first relation (4.84); therefore (4.84) does not hold, or equivalently, (4.83) holds. □

Applications of Lemma 4.1 are given in the following sections.

4.2.3.2. *Some applications of Lemma* 4.1. We now apply Lemma 4.1 to the solution of the Neumann problem (4.29) if either a_0 satisfies (4.76) or $a_0 = 0$ on Ω. More precisely, we have the following propositions:

Proposition 4.6. *We consider the Neumann problem* (4.29) *with* f, g, $\overline{\mathbf{A}}$ *still obeying* (4.45), (4.47), *respectively; if we suppose that* Ω *is bounded and that*

$$a_0 \in L^\infty(\Omega), \quad a_0(x) \geq 0 \text{ a.e. on } \Omega, \quad \int_\Omega a_0(x)\, dx > 0, \tag{4.94}$$

then (4.29) *has a unique solution in* $H^1(\Omega)$, *which is also the unique solution of the variational problem*:

Find $u \in H^1(\Omega)$ *such that*

$$\int_\Omega (\bar{\bar{\mathbf{A}}}\nabla u) \cdot \nabla v \, dx + \int_\Omega a_0 uv \, dx = \int_\Omega fv \, dx + \int_\Gamma gv \, d\Gamma, \qquad \forall\, v \in H^1(\Omega). \tag{4.95}$$

PROOF. It suffices to prove that the bilinear form occurring in (4.95) is $H^1(\Omega)$-elliptic. This follows directly from Lemma 4.1 and from the fact that if $v = \text{constant} = C$, then

$$\int_\Omega a_0(x) \, dx > 0$$

and

$$\int_\Omega a_0(x) v^2 \, dx = C^2 \int_\Omega a_0(x) \, dx = 0$$

imply $C = 0$, i.e., $v = 0$. □

Proposition 4.7. *We consider the Neumann problem* (4.29) *with* Ω *bounded and* **A** *still obeying* (4.47); *if we suppose that* $a_0 = 0$, *then* (4.29) *has a unique solution* u *in* $H^1(\Omega)/\mathbb{R}$[20] *if and only if*

$$\int_\Omega f \, dx + \int_\Gamma g \, d\Gamma = 0; \tag{4.96}$$

u *is also the unique solution in* $H^1(\Omega)/\mathbb{R}$ *of the variational problem*:

Find $u \in H^1(\Omega)$ *such that*

$$\int_\Omega (\bar{\bar{\mathbf{A}}}\nabla u) \cdot \nabla v \, dx = \int_\Omega fv \, dx + \int_\Gamma gv \, d\Gamma, \qquad \forall\, v \in H^1(\Omega). \tag{4.97}$$

PROOF. For clarity we divide the proof into several steps.

Step 1. Suppose that $a_0 = 0$; if u is a solution of (4.29) and if C is a constant, it is clear from

$$\nabla(u + C) = \nabla u$$

that $u + C$ is also a solution of (4.29).

If u is a solution of (4.29), we can show, as in Sec. 4.2.2, that (4.97) holds; taking $v = 1$ in (4.97), we obtain (4.96).

Step 2. Consider the bilinear form over $H^1(\Omega) \times H^1(\Omega)$ defined by

$$\tilde{a}(v, w) = \int_\Omega \nabla v \cdot \nabla w \, dx + \left(\int_\Omega v \, dx\right)\left(\int_\Omega w \, dx\right), \qquad \forall\, v, w \in H^1(\Omega); \tag{4.98}$$

---

[20] This means that u is determined in $H^1(\Omega)$ only to within an arbitrary constant.

the bilinear form $\tilde{a}(\cdot, \cdot)$ is clearly continuous, and from Lemma 4.1, it is $H^1(\Omega)$-elliptic (it suffices to observe that if $v = \text{constant} = C$, then

$$0 = \left(\int_\Omega v\,dx\right)^2 = C^2\,(\text{meas}(\Omega))^2 \Rightarrow C = v = 0).$$

From these properties, $v \to (\tilde{a}(v, v))^{1/2}$ defines, over $H^1(\Omega)$, a norm equivalent to the usual $H^1(\Omega)$-norm defined by (4.40).

Step 3. We now consider the space

$$V_1 = \left\{v \mid v \in H^1(\Omega), \int_\Omega v(x)\,dx = 0\right\};$$

V_1 being the kernel of the linear continuous functional

$$v \to \int_\Omega v(x)\,dx$$

is a closed subspace of $H^1(\Omega)$. If we suppose that $H^1(\Omega)$ has been equipped with the scalar product defined by $\tilde{a}(\cdot, \cdot)$ (see (4.98)), it follows from Step 2, and from the definition of V_1, that over V_1,

$$v \to \left(\int_\Omega |\nabla v|^2\,dx\right)^{1/2}$$

defines a norm equivalent to the $H^1(\Omega)$-norm (4.49); henceforth we shall endow V_1 with the following scalar product:

$$\{v, w\} \to \int_\Omega \nabla v \cdot \nabla w\,dx.$$

From these properties of V_1, and from (4.47), the variational problem:
Find $u \in V_1$ such that

$$\int_\Omega (\bar{\bar{\mathbf{A}}}\nabla u) \cdot \nabla v\,dx = \int_\Omega fv\,dx + \int_\Gamma gv\,d\Gamma, \qquad \forall\, v \in V_1, \tag{4.99}$$

has a unique solution.

Step 4. Returning to $H^1(\Omega)$ equipped with its usual product, let us introduce $V_0 \subset H^1(\Omega)$ defined by

$$V_0 = \{v \mid v = \text{constant over } \Omega\}; \tag{4.100}$$

if $v \in V_0^\perp$, we have

$$0 = \int_\Omega \nabla v \cdot \nabla c\,dx + \int_\Omega vc\,dx = c\int_\Omega v\,dx, \qquad \forall\, c \in \mathbb{R},$$

which shows that $V_0^\perp = V_1$. We then have

$$H^1(\Omega) = V_0 \oplus V_1,$$

and for any $v \in H^1(\Omega)$, we have a unique decomposition

$$v = v_0 + v_1, \quad v_i \in V_i, \qquad \forall\, i = 0, 1.$$

Step 5. From (4.96), (4.99) it follows that we also have

$$\int_\Omega (\bar{\bar{\mathbf{A}}}\nabla u) \cdot \nabla(v + c)\, dx = \int_\Omega f(v + c)\, dx + \int_\Gamma g(v + c)\, d\Gamma,$$
$$\forall\, v \in V_1, \quad \forall\, c \in \mathbb{R}, \quad u \in V_1. \quad (4.101)$$

From the results of Step 4, relation (4.101) implies that u is a solution of (4.97) (but the only one belonging to V_1); actually if we consider a second solution of (4.97), say u^*, we clearly have (from (4.47))

$$\alpha \int_\Omega |\nabla(u^* - u)|^2\, dx \leq \int_\Omega (\bar{\bar{\mathbf{A}}}\nabla(u^* - u)) \cdot \nabla(u^* - u)\, dx = 0. \quad (4.102)$$

From (4.102) it follows that $u^* - u = \text{const}$; this completes the proof of the proposition.

□

Remark 4.10. In many cases where $a_0 = 0$ in (4.29), one is more interested in ∇u than in u itself [this is the case, for example, in fluid mechanics (resp., electrostatics), where u would be a velocity potential (resp., an electrical potential) and ∇u (resp., $-\nabla u$) the corresponding velocity (resp., electrical field)]; in such cases, the fact that u is determined only to within an arbitrary constant does not matter, since $\nabla(u + c) = \nabla u, \forall\, c \in \mathbb{R}$.

4.3. Solution of Dirichlet problems for second-order elliptic partial differential operators

We shall now discuss the formulation and the solution via variational methods of Dirichlet problems for linear second-order elliptic partial differential operators. The finite element approximation of these problems will be discussed in Sec. 4.5.

4.3.1. *The classical formulation.*

With Ω, $\bar{\bar{\mathbf{A}}}$, a_0, f, g, and the notation as in Sec. 4.2.1, we consider the following Dirichlet problem:

$$-\nabla \cdot (\bar{\bar{\mathbf{A}}}\nabla u) + \nabla \cdot (\boldsymbol{\beta} u) + a_0 u = f \text{ in } \Omega, \qquad u = g \text{ on } \Gamma, \quad (4.103)$$

where $\boldsymbol{\beta}$ is a given vector function defined over Ω and taking its values in $\mathbb{R}^N$. Remark 4.4 of Sec. 4.2.1 still holds for the Dirichlet problem (4.103).

Remark 4.11. If $\bar{\bar{\mathbf{A}}} = \mathbf{I}$, $a_0 = 0$, $\boldsymbol{\beta} = \mathbf{0}$, the Dirichlet problem (4.103) reduces to

$$-\Delta u = f \textit{ in } \Omega, \qquad u = g \text{ on } \Gamma, \quad (4.104)$$

which is the classical Dirichlet problem for the Laplace operator Δ.

4.3.2. *A variational formulation of the Dirichlet problem* (4.103)

Let $v \in \mathscr{D}(\Omega)$ (where $\mathscr{D}(\Omega)$ is still defined by (4.38)); we then have

$$v = 0 \text{ on } \Gamma. \tag{4.105}$$

Multiplying the first relation (4.103) by v, we obtain (still using the Green–Ostrogradsky formula, and taking (4.105) into account)

$$\int_\Omega (\overline{\mathbf{A}}\nabla u) \cdot \nabla v \, dx - \int_\Omega u\boldsymbol{\beta} \cdot \nabla v \, dx + \int_\Omega a_0 uv \, dx = \int_\Omega fv \, dx, \qquad \forall\, v \in \mathscr{D}(\Omega). \tag{4.106}$$

Conversely it can be proved that if (4.106) holds, then u satisfies the second-order partial differential equation in (4.103) (at least in a distribution sense).

Let us now introduce the Sobolev space $H_0^1(\Omega)$ defined by

$$H_0^1(\Omega) = \overline{\mathscr{D}(\Omega)}^{H^1(\Omega)}; \tag{4.107}$$

if $\Gamma = \partial\Omega$ is sufficiently smooth, we also have

$$H_0^1(\Omega) = \{v \mid v \in H^1(\Omega),\ \gamma_0 v = 0\}, \tag{4.108}$$

where γ_0 is the trace operator introduced in Sec. 4.2.2. From (4.107), (4.108), $H_0^1(\Omega)$ is a closed subspace of $H^1(\Omega)$.

An important property of $H_0^1(\Omega)$ is the following:

Suppose that Ω is bounded in at least one direction of $\mathbb{R}^N$; then

$$v \to \left(\int_\Omega |\nabla v|^2 \, dx\right)^{1/2} \tag{4.109}$$

defines a norm over $H_0^1(\Omega)$ equivalent to the $H^1(\Omega)$-norm.

Property (4.109) holds, for example, for

$$\Omega =]\alpha, \beta[\times \mathbb{R}^{N-1} \quad \text{with } \alpha, \beta \in \mathbb{R}, \quad \alpha < \beta,$$

but does not hold for

$$\Omega = \{y \mid y = \{y_i\}_{i=1}^N, \quad y_i > 0, \qquad \forall\, i = 1, \ldots, N\}.$$

Returning to the Dirichlet problem (4.103), and to (4.106), we suppose that the following hypotheses on $\overline{\mathbf{A}}$, a_0, $\boldsymbol{\beta}$, f, and g hold:

$$f \in L^2(\Omega), \quad \exists\, \tilde{g} \in H^1(\Omega) \quad \text{such that } g = \gamma_0 \tilde{g}, \tag{4.110}$$

$$a_0 \in L^\infty(\Omega), \quad a_0(x) \geq \alpha_0 > 0 \text{ a.e. on } \Omega, \tag{4.111}$$

$$\overline{\mathbf{A}} \text{ satisfies (4.47)}, \tag{4.112}$$

$$\boldsymbol{\beta} \in (L^\infty(\Omega))^N, \quad \nabla \cdot \boldsymbol{\beta} = 0 \quad \text{(in the distribution sense)}. \tag{4.113}$$

We now define $a: H^1(\Omega) \times H^1(\Omega) \to \mathbb{R}$ and $L: H^1(\Omega) \to \mathbb{R}$ by

$$a(v, w) = \int_\Omega (\bar{\bar{\mathbf{A}}}\nabla v) \cdot \nabla w \, dx - \int_\Omega v\boldsymbol{\beta} \cdot \nabla w \, dx + \int_\Omega a_0 vw \, dx, \qquad \forall\, v, w \in H^1(\Omega), \tag{4.114}$$

$$L(v) = \int_\Omega fv \, dx, \qquad \forall\, v \in H^1(\Omega), \tag{4.115}$$

respectively; $a(\cdot, \cdot)$ (resp., L) is clearly bilinear continuous (resp., linear continuous).

Before discussing the variational formulation of the Dirichlet problem (4.103), we shall prove the following useful lemma:

Lemma 4.2. *Suppose that* $\boldsymbol{\beta}$ *satisfies* (4.113); *we then have*

$$\int_\Omega v\boldsymbol{\beta} \cdot \nabla w \, dx = -\int_\Omega w\boldsymbol{\beta} \cdot \nabla v \, dx, \qquad \forall\, v, w \in H_0^1(\Omega) \tag{4.116}$$

(i.e., the bilinear form

$$\{v, w\} \to \int_\Omega v\boldsymbol{\beta} \cdot \nabla w \, dx$$

is skew symmetric over $H_0^1(\Omega) \times H_0^1(\Omega)$).

PROOF. Let $v, w \in \mathscr{D}(\Omega)$; we have

$$\int_\Omega v\boldsymbol{\beta} \cdot \nabla w \, dx = \int_\Omega \boldsymbol{\beta} \cdot \nabla(vw) \, dx - \int_\Omega w\boldsymbol{\beta} \cdot \nabla v \, dx. \tag{4.117}$$

Since $vw \in \mathscr{D}(\Omega)$ and $\nabla \cdot \boldsymbol{\beta} = 0$, we also have

$$\int_\Omega \boldsymbol{\beta} \cdot \nabla(vw) \, dx = \langle \boldsymbol{\beta}, \nabla(vw) \rangle = -\langle \nabla \cdot \boldsymbol{\beta}, vw \rangle = 0 \tag{4.118}$$

(where $\langle \cdot, \cdot \rangle$ denotes the duality between $\mathscr{D}'(\Omega)$ and $\mathscr{D}(\Omega)$).

From (4.117), (4.118), we then have

$$\int_\Omega v\boldsymbol{\beta} \cdot \nabla w \, dx = -\int_\Omega w\boldsymbol{\beta} \cdot \nabla v \, dx, \qquad \forall\, v, w \in \mathscr{D}(\Omega). \tag{4.119}$$

From the *density* of $\mathscr{D}(\Omega)$ in $H_0^1(\Omega)$, (4.119) implies (4.116). □

Lemma 4.2 implies:

Proposition 4.8. *Suppose that* (4.111)–(4.113) *hold; then the bilinear form* $a(\cdot, \cdot)$ *defined by* (4.114) *is* $H_0^1(\Omega)$*-elliptic.*

PROOF. From (4.111), (4.112), (4.114) we have

$$a(v, v) \geq \operatorname{Min}(\alpha, \alpha_0)\|v\|^2_{H^1(\Omega)} - \int_\Omega v\boldsymbol{\beta} \cdot \nabla v \, dx, \qquad \forall\, v \in H_0^1(\Omega); \tag{4.120}$$

since $\boldsymbol{\beta}$ satisfies (4.113), from Lemma 4.2 we have

$$\int_\Omega v\boldsymbol{\beta} \cdot \nabla v \, dx = 0, \qquad \forall\, v \in H_0^1(\Omega). \tag{4.121}$$

Combining (4.120) and (4.121), we have

$$a(v, v) \geq \operatorname{Min}(\alpha, \alpha_0)\|v\|^2_{H^1(\Omega)}, \qquad \forall\, v \in H_0^1(\Omega),$$

i.e., the $H_0^1(\Omega)$-ellipticity of $a(\cdot, \cdot)$. □

Using the above results we now prove:

Proposition 4.9. *Suppose that* (4.110)–(4.115) *hold; then the variational problem*:

Find $u \in H^1(\Omega)$ such that $\gamma_0 u = g$ and

$$a(u, v) = L(v), \qquad \forall\, v \in H_0^1(\Omega) \tag{4.122}$$

has a unique solution. This solution is also the unique solution in $H^1(\Omega)$ of the Dirichlet problem (4.103).

PROOF. (1) *Uniqueness.* Suppose that (4.122) has two solutions u_1 and u_2; we then have

$$a(u_1, v) = L(v), \qquad \forall\, v \in H_0^1(\Omega),$$
$$a(u_2, v) = L(v), \qquad \forall\, v \in H_0^1(\Omega),$$

and by substraction,

$$a(u_2 - u_1, v) = 0, \qquad \forall\, v \in H_0^1(\Omega). \tag{4.123}$$

Since $u_1, u_2 \in H^1(\Omega)$ with $\gamma_0 u_1 = \gamma_0 u_2\ (=g)$, we have

$$u_2 - u_1 \in H^1(\Omega), \quad \gamma_0(u_2 - u_1) = 0,$$

i.e., $u_2 - u_1 \in H_0^1(\Omega)$. Taking $v = u_2 - u_1$ in (4.123) and using Proposition 4.8, we have

$$0 \leq \operatorname{Min}(\alpha, \alpha_0)\|u_2 - u_1\|^2_{H^1(\Omega)} \leq 0,$$

i.e., $u_2 = u_1$.

(2) *Existence.* We have (from (4.110)) $u = \gamma_0 \tilde{g}$, where $\tilde{g} \in H^1(\Omega)$; this leads to the introduction of $\bar{u} \in H_0^1(\Omega)$ such that

$$\bar{u} = u - \tilde{g} \quad (\Leftrightarrow u = \bar{u} + \tilde{g}). \tag{4.124}$$

There is clearly equivalence between (4.122) and the linear variational problem in $H_0^1(\Omega)$ below:

Find $\bar{u} \in H_0^1(\Omega)$ such that

$$a(\bar{u}, v) = L(v) - a(\tilde{g}, v), \qquad \forall\, v \in H_0^1(\Omega). \tag{4.125}$$

From Proposition 4.8, $a(\cdot, \cdot)$ is bilinear, continuous over $H_0^1(\Omega)$, and $H_0^1(\Omega)$-elliptic; moreover, the linear functional

$$v \to L(v) - a(\tilde{g}, v) \tag{4.126}$$

is clearly continuous over $H_0^1(\Omega)$. From the properties of $H_0^1(\Omega)$, $a(\cdot, \cdot)$, and of the linear functional (4.126), we can apply Theorem 2.1 of Sec. 2.3 to prove that (4.125) has a unique solution in $H_0^1(\Omega)$; this, in turn, implies (taking part (1) into account) that (4.122) has a unique solution as well.

(3) *The Solution u of* (4.122) *Satisfies* (4.103) *and Conversely*. Taking $v \in \mathscr{D}(\Omega)$ in (4.122), we find that u satisfies (4.103) in the sense of distributions. Conversely, if $u \in H^1(\Omega)$, with $\gamma_0 u = g$, satisfies (4.103), we can easily prove that

$$a(u, v) = L(v), \qquad \forall\, v \in \mathscr{D}(\Omega), \tag{4.127}$$

and using the density of $\mathscr{D}(\Omega)$ in $H_0^1(\Omega)$, we find that (4.127) also holds for all the v in $H_0^1(\Omega)$. □

Remark 4.12. Suppose that $\overline{\overline{\mathbf{A}}}$ is symmetric and that $\boldsymbol{\beta} = \mathbf{0}$; this implies the symmetry of the bilinear form $a(\cdot, \cdot)$. There is then equivalence between (4.122) and the minimization problem:

Find $u \in H_g$ such that

$$J(u) \le J(v), \qquad \forall\, v \in H_g, \tag{4.128}$$

where

$$J(v) = \frac{1}{2}\int_\Omega (\overline{\overline{\mathbf{A}}}\nabla v) \cdot \nabla v \, dx + \frac{1}{2}\int_\Omega a_0 v^2 \, dx - \int_\Omega fv \, dx \tag{4.129}$$

and

$$H_g = \{v \,|\, v \in H^1(\Omega), \gamma_0 v = g\}. \tag{4.130}$$

4.3.3. *Further remarks and comments*

Remarks 4.7 and 4.8 of Sec. 4.2.3.1 still hold for the Dirichlet problem (4.103) in that we can replace L defined by (4.115) by more complicated linear continuous functionals like the one in (4.63), or

$$L(v) = \int_\Omega fv \, dx + \int_\gamma hv \, d\gamma$$

with h and γ as in (4.68).

Concerning the case where $a_0 = 0$ in (4.103), (4.114), we have the following:

Proposition 4.10. *Suppose that Ω is bounded in at least one direction of $\mathbb{R}^N$; also suppose that $a_0 = 0$, the hypotheses on $\overline{\overline{\mathbf{A}}}$, $\boldsymbol{\beta}$, f, g remaining the same. Then the variational problem* (4.122) *still has a unique solution which is also the unique solution in $H^1(\Omega)$ of the Dirichlet problem* (4.103).

PROOF. It suffices to prove that the bilinear functional $a: H^1(\Omega) \times H^1(\Omega) \to \mathbb{R}$ defined by

$$a(v, w) = \int_\Omega (\overline{\overline{\mathbf{A}}}\nabla v) \cdot \nabla w \, dx - \int_\Omega v\boldsymbol{\beta} \cdot \nabla w \, dx$$

is $H_0^1(\Omega)$- elliptic. This follows from (4.112) (and (4.47)) and from Lemma 4.2 which implies that

$$a(v, v) \geq \alpha \int_\Omega |\nabla v|^2 \, dx, \qquad \forall\, v \in H_0^1(\Omega),$$

and from (4.109) which implies that $v \to (\int_\Omega |\nabla v|^2 \, dx)^{1/2}$ is a norm over $H_0^1(\Omega)$ equivalent to the $H^1(\Omega)$-norm. □

4.3.4. *Some comments on mixed boundary-value problems (Neumann–Dirichlet problems)*

We briefly discuss the solution (via variational methods) of elliptic problems for linear second-order partial differential operators combining the boundary conditions of Sec. 4.2 (Neumann's boundary conditions) and Secs. 4.3.1., 4.3.2, and 4.3.3 (Dirichlet's boundary conditions).

With Ω as in the above sections, we suppose that Γ $(=\partial\Omega)$ is the union of Γ_0, Γ_1 such that

$$\Gamma_0 \cup \Gamma_1 = \Gamma, \qquad \Gamma_0 \cap \Gamma_1 = \emptyset; \tag{4.131}$$

such a situation is described in Fig. 4.1 of Chapter II, Sec. 4.1.

We now consider the mixed boundary-value problem

$$-\nabla \cdot (\bar{\bar{\mathbf{A}}}\nabla u) + a_0 u = f \text{ in } \Omega, \qquad u = g_0 \quad \text{on } \Gamma_0, \qquad (\bar{\bar{\mathbf{A}}}\nabla u) \cdot \mathbf{n} = g_1 \text{ on } \Gamma_1, \tag{4.132}$$

where $\bar{\bar{\mathbf{A}}}$, a_0, f are as in Sec. 4.3.1 and where g_0, g_1 are given functions defined over Γ_0, Γ_1, respectively.

Taking a test function v "sufficiently smooth" and such that

$$v|_{\Gamma_0} = 0, \tag{4.133}$$

we obtain (still using the Green–Ostrogradsky formula (4.33)) that any solution of (4.132) satisfies

$$\int_\Omega (\bar{\bar{\mathbf{A}}}\nabla u) \cdot \nabla v \, dx + \int_\Omega a_0 uv \, dx = \int_\Omega fv \, dx + \int_{\Gamma_1} g_1 v \, d\Gamma. \tag{4.134}$$

To solve (4.132) by variational methods, we first introduce $a(\cdot, \cdot)$ and $L(\cdot)$ defined by

$$a(v, w) = \int_\Omega (\bar{\bar{\mathbf{A}}}\nabla v) \cdot \nabla w \, dx + \int_\Omega a_0 vw \, dx, \qquad \forall\, v, w \in H^1(\Omega), \tag{4.135}$$

$$L(v) = \int_\Omega fv \, dx + \int_{\Gamma_1} g_1 v \, d\Gamma, \qquad \forall\, v \in H^1(\Omega). \tag{4.136}$$

We suppose that a_0 and $\bar{\bar{\mathbf{A}}}$ still satisfy (4.46), (4.47) and that

$$f \in L^2(\Omega), \qquad g_1 \in L^2(\Gamma_1). \tag{4.137}$$

These hypotheses on $\overline{\overline{\mathbf{A}}}$, a_0, f, g_1 imply that $a(\cdot, \cdot)$ is bilinear continuous over $H^1(\Omega) \times H^1(\Omega)$ and $H^1(\Omega)$-elliptic and that L is a linear continuous functional over $H^1(\Omega)$.

We now introduce (motivated by (4.133)) the following subspace V_0 of $H^1(\Omega)$:

$$V_0 = \{v \mid v \in H^1(\Omega), \gamma_0 v = 0 \text{ a.e. on } \Gamma_0\}; \tag{4.138}$$

actually V_0 is a closed subspace of $H^1(\Omega)$.

Using a variant of the proof of Proposition 4.9 in Sec. 4.3.2, we should prove the following:

Proposition 4.11. *Suppose that there exists $\tilde{g}_0 \in H^1(\Omega)$ such that*

$$g_0 = \gamma_0 \tilde{g}_0 |_{\Gamma_0}; \tag{4.139}$$

also suppose that the above hypotheses on $\overline{\overline{\mathbf{A}}}$, a_0, f, g_1 hold. Then the variational problem:

Find $u \in H^1(\Omega)$ such that $\gamma_0 u = g_0$ on Γ_0 and

$$\int_\Omega (\overline{\overline{\mathbf{A}}}\nabla u) \cdot \nabla v \, dx + \int_\Omega a_0 uv \, dx = \int_\Omega fv \, dx + \int_{\Gamma_1} g_1 v \, d\Gamma, \qquad \forall\, v \in V_0 \tag{4.140}$$

has a unique solution, which is also the unique solution in $H^1(\Omega)$ of the mixed boundary-value problem (4.132).

For a proof see, e.g., Necas [1].

Most remarks of Secs. 4.2.3 and 4.3.3 still hold for (4.132), (4.140); in particular, if Ω is bounded and if $\int_{\Gamma_0} d\Gamma > 0$, then we can suppose that $a_0 = 0$ [this follows from the fact that if Ω is bounded, then

$$v \to \left(\int_\Omega |\nabla v|^2 \, dx\right)^{1/2}$$

defines a norm over V_0 which is equivalent to the $H^1(\Omega)$-norm (Lemma 4.1 can be used to prove this equivalence property)].

4.4. Solution of second-order elliptic problems with Fourier boundary conditions

4.4.1. *Synopsis*

In this section we shall discuss the solution—via variational methods—of the so-called *Fourier problem for linear second-order elliptic partial differential operators*; our interest in this problem is twofold:

(i) The Fourier problem occurs in the modelling of several *heat-transfer* phenomana.

(ii) The Neumann and Dirichlet problems discussed in Secs. 4.2 and 4.3, respectively, can be considered, in fact, to be particular cases[21] of the Fourier problem; this claim will be justified in the sequel.

4.4.2. *Classical formulations of the Fourier problem*

With Ω, $\bar{\bar{\mathbf{A}}}$, a_0, f, g as in Sec. 4.2, we consider the following boundary-value problem:

Find u such that

$$-\nabla \cdot (\bar{\bar{\mathbf{A}}}\nabla u) + a_0 u = f \text{ in } \Omega, \qquad (\bar{\bar{\mathbf{A}}}\nabla u)\cdot \mathbf{n} + ku = g \text{ on } \Gamma, \tag{4.141}$$

where k is a given function defined over Γ.

4.4.3. *A variational formulation of the Fourier problem* (4.141)

Let v be a smooth function defined over $\bar{\Omega}$; multiplying the first relation (4.141) by v and integrating over Ω, we obtain (using the Green–Ostrogradsky formula (4.33))

$$\int_\Omega (\bar{\bar{\mathbf{A}}}\nabla u)\cdot \nabla v\, dx + \int_\Omega a_0 uv\, dx = \int_\Omega fv\, dx + \int_\Gamma (\bar{\bar{\mathbf{A}}}\nabla u)\cdot \mathbf{n} v\, d\Gamma; \tag{4.142}$$

combining (4.142) and the boundary condition in (4.141), we finally obtain

$$\int_\Omega (\bar{\bar{\mathbf{A}}}\nabla u)\cdot \nabla v\, dx + \int_\Omega a_0 uv\, dx + \int_\Gamma kuv\, d\Gamma = \int_\Omega fv\, dx + \int_\Gamma gv\, d\Gamma. \tag{4.143}$$

Conversely, it can be proved that if (4.143) holds, $\forall\, v \in \mathscr{V}$, where $\mathscr{V}$ is still defined by (4.35), then u is in some sense a solution of the Fourier problem (4.141).

We now define $a\colon H^1(\Omega) \times H^1(\Omega) \to \mathbb{R}$ and $L\colon H^1(\Omega) \to \mathbb{R}$ by

$$a(v, w) = \int_\Omega [(\bar{\bar{\mathbf{A}}}\nabla v)\cdot \nabla w + a_0 vw]\, dx + \int_\Gamma kvw\, d\Gamma, \qquad \forall\, v, w \in H^1(\Omega), \tag{4.144}$$

$$L(v) = \int_\Omega fv\, dx + \int_\Gamma g\gamma_0 v\, d\Gamma, \qquad \forall\, v \in H^1(\Omega), \tag{4.145}$$

respectively. We suppose that the following hypotheses concerning $\bar{\bar{\mathbf{A}}}$, a_0, k, f, g hold:

$$f \in L^2(\Omega), \quad g \in L^2(\Gamma), \tag{4.146}$$

$$a_0 \in L^\infty(\Omega), \quad a_0(x) \geq \alpha_0 > 0 \text{ a.e. on } \Omega, \tag{4.147}$$

$$k \in L^\infty(\Gamma), \quad k(x) \geq 0 \text{ a.e. on } \Gamma, \tag{4.148}$$

$$\bar{\bar{\mathbf{A}}} \text{ satisfies (4.47).} \tag{4.149}$$

---

[21] Similarly for the mixed boundary-value problem of Sec. 4.3.4.

From the above hypotheses and from the continuity of the trace operator $\gamma_0: H^1(\Omega) \to L^2(\Gamma)$ (see Sec. 4.2.2), we find that $a(\cdot, \cdot)$ is bilinear continuous over $H^1(\Omega) \times H^1(\Omega)$ and $H^1(\Omega)$-elliptic, and that $L(\cdot)$ is linear continuous; we can therefore apply Theorem 2.1 of Sec. 2.3 to prove:

Proposition 4.12. *If the above hypotheses on* $\bar{\bar{\mathbf{A}}}$, a_0, k, f, g *hold, the linear variational problem:*

Find $u \in H^1(\Omega)$ *such that*

$$a(u, v) = L(v), \qquad \forall\, v \in H^1(\Omega) \tag{4.150}$$

has a unique solution; this solution is also the unique solution in $H^1(\Omega)$ *of the Fourier problem* (4.141).

Remark 4.13. Suppose that $\bar{\bar{\mathbf{A}}}$ is symmetric; this, in turn implies the symmetry of the bilinear form $a(\cdot, \cdot)$. From Proposition 2.1 of Sec. 2.4 it then follows that (4.150) is equivalent to the minimization problem:

Find $u \in H^1(\Omega)$ *such that*

$$J(u) \leq J(v), \qquad \forall\, v \in H^1(\Omega), \tag{4.151}$$

where

$$J(v) = \frac{1}{2}\int_\Omega [(\bar{\bar{\mathbf{A}}}\nabla v)\cdot\nabla v + a_0 v^2]\,dx + \frac{1}{2}\int_\Gamma kv^2\,d\Gamma - \int_\Omega fv\,dx - \int_\Gamma gv\,d\Gamma. \tag{4.152}$$

Remark 4.14. Using Lemma 4.1, we should prove that Proposition 4.6 of Sec. 4.2.3.2 still holds for problem (4.141), (4.150) in that if Ω is bounded and if

$$a_0 \in L^\infty(\Omega), \quad a_0(x) \geq 0 \text{ a.e. on } \Omega, \quad \int_\Omega a_0(x)\,dx > 0,$$

then problem (4.141) has a unique solution in $H^1(\Omega)$ which is also the unique solution of the variational problem (4.150).

Concerning the case $a_0 = 0$ on Ω, we shall prove the following:

Proposition 4.13. *We suppose that* Ω *is bounded and that* $a_0 = 0$ *on* Ω; *we also suppose that k satisfies*

$$k \in L^\infty(\Gamma), \quad k(x) \geq 0 \text{ a.e. on } \Gamma, \quad \int_\Gamma k(x)\,d\Gamma > 0. \tag{4.153}$$

Then, the hypotheses on f, g, $\bar{\bar{\mathbf{A}}}$ *being* (4.146), (4.149), *respectively, the variational problem:*

Find $u \in H^1(\Omega)$ *such that*

$$\int_\Omega (\bar{\bar{\mathbf{A}}}\nabla u)\cdot\nabla v\,dx + \int_\Gamma kuv\,d\Gamma = \int_\Omega fv\,dx + \int_\Gamma gv\,d\Gamma, \qquad \forall\, v \in H^1(\Omega), \tag{4.154}$$

has a unique solution which is also the unique solution in $H^1(\Omega)$ of the boundary-value problem

$$-\nabla\cdot(\bar{\bar{\mathbf{A}}}\nabla u) = f \text{ in } \Omega, \qquad (\bar{\bar{\mathbf{A}}}\nabla u)\cdot\mathbf{n} + ku = g \text{ on } \Gamma. \tag{4.155}$$

PROOF. It suffices to prove that the bilinear form occurring in (4.154) is $H^1(\Omega)$-elliptic. This follows directly from Lemma 4.1 (see Sec. 4.2.3.2) and from the fact that if $v =$ constant $= C$, then

$$\int_\Gamma k(x)\,d\Gamma > 0$$

and

$$\int_\Gamma k(x)v^2\,d\Gamma = C^2\int_\Gamma k(x)\,d\Gamma = 0$$

imply that $C = 0$, i.e., $v = 0$. □

4.4.4. *The Neumann and Dirichlet problems as particular cases of the Fourier problem*

We would like to justify the claims of Sec. 4.4.1 concerning the relationships between the Neumann and Dirichlet problems and the Fourier problem. Obtaining the Neumann problem (4.29) from the Fourier problem is quite trivial since it suffices to take $k = 0$ on Γ in (4.141) and (4.144); obtaining the Dirichlet problem from the Fourier problem is less simple.

First consider the variational formulation of the standard nonhomogeneous Dirichlet problem:

Find $u \in H^1(\Omega)$ such that $\gamma_0 u = g$, and

$$\int_\Omega [(\bar{\bar{\mathbf{A}}}\nabla u)\cdot\nabla v + a_0 uv]\,dx = L(v), \qquad \forall\, v \in H_0^1(\Omega), \tag{4.156}$$

where $g = \gamma_0\tilde{g}$, $\tilde{g} \in H^1(\Omega)$, and where L is linear continuous from $H^1(\Omega)$ to $\mathbb{R}$; taking L linear continuous from $H^1(\Omega)$ to $\mathbb{R}$ in (4.156) is not restrictive since any functional linear continuous over $H_0^1(\Omega)$ can be considered as the restriction over $H_0^1(\Omega)$ of a functional linear continuous over $H^1(\Omega)$.

With g and L as in (4.156) and $\varepsilon > 0$, we now consider the following particular Fourier problem:

Find $u_\varepsilon \in H^1(\Omega)$ such that

$$\int_\Omega [(\bar{\bar{\mathbf{A}}}\nabla u_\varepsilon)\cdot\nabla v + a_0 u_\varepsilon v]\,dx + \frac{1}{\varepsilon}\int_\Gamma u_\varepsilon v\,d\Gamma$$
$$= L(v) + \frac{1}{\varepsilon}\int_\Gamma gv\,d\Gamma, \qquad \forall\, v \in H^1(\Omega). \tag{4.157}$$

Remark 4.15. If $\bar{\bar{\mathbf{A}}}$ is symmetric, there is equivalence between (4.157) and the following minimization problem:

Find $u_\varepsilon \in H^1(\Omega)$ *such that*

$$J_\varepsilon(u_\varepsilon) \le J_\varepsilon(v), \qquad \forall\, v \in H^1(\Omega), \tag{4.158}$$

where

$$J_\varepsilon(v) = \frac{1}{2}\int_\Omega [(\overline{\overline{\mathbf{A}}}\nabla v)\cdot\nabla v + a_0 v^2]\,dx - L(v) + \frac{1}{2\varepsilon}\int_\Gamma (v-g)^2\,d\Gamma. \tag{4.159}$$

Thus (4.157), (4.158) is obtained from (4.156) by using a penalty procedure to handle the boundary condition $\gamma_0 u = g$ (see Chapter I, Sec. 7, for more details on penalty procedures).

The relationship between (4.156) and (4.157) follows from:

Proposition 4.14. *We suppose that* $\overline{\overline{\mathbf{A}}}$ *satisfies* (4.47) *and that* a_0 *satisfies*

$$a_0 \in L^\infty(\Omega), \quad a_0(x) \ge \alpha_0 > 0 \text{ a.e. on } \Omega \tag{4.160}$$

(*or possibly—if* Ω *is bounded—*

$$a_0 \in L^\infty(\Omega), \quad a_0(x) \ge 0 \text{ a.e. on } \Omega). \tag{4.161}$$

We then have

$$\lim_{\varepsilon\to 0} \|u_\varepsilon - u\|_{H^1(\Omega)} = 0, \tag{4.162}$$

where u (*resp.* u_ε) *is the solution of the Dirichlet problem* (4,156) (*resp., of the Fourier problem* (4.157)).

PROOF. (1) *A priori estimates for* $\{u_\varepsilon\}_\varepsilon$. Let us define $a(\cdot,\cdot)$ bilinear continuous from $H^1(\Omega)\times H^1(\Omega)$ into $\mathbb{R}$ by

$$a(v, w) = \int_\Omega [(\overline{\overline{\mathbf{A}}}\nabla v)\cdot\nabla w + a_0 vw]\,dx, \qquad \forall\, v, w \in H^1(\Omega); \tag{4.163}$$

taking $v = u_\varepsilon - u$ in (4.157), and using the notation of (4.163), we obtain

$$a(u_\varepsilon, u_\varepsilon - u) + \frac{1}{\varepsilon}\int_\Gamma (u_\varepsilon - g)^2\,d\Gamma = L(u_\varepsilon - u),$$

which, in turn, implies

$$a(u_\varepsilon - u, u_\varepsilon - u) + \frac{1}{\varepsilon}\|u_\varepsilon - g\|^2_{L^2(\Gamma)} = L(u_\varepsilon - u) - a(u, u_\varepsilon - u), \qquad \forall\, \varepsilon > 0. \tag{4.164}$$

Using (4.52) (which still holds), from (4.164) we obtain

$$a(u_\varepsilon - u, u_\varepsilon - u) + \frac{1}{\varepsilon}\|u_\varepsilon - g\|^2_{L^2(\Gamma)} \le C_1\|u_\varepsilon - u\|_{H^1(\Omega)}, \tag{4.165}$$

where

$$C_1 = \|L\| + \max(\|\overline{\overline{\mathbf{A}}}\|_{L^\infty(\Omega)}, \|a_0\|_{L^\infty(\Omega)})\|u\|_{H^1(\Omega)}.$$

If (4.160) holds, from (4.165) we obtain

$$\|u_\varepsilon - u\|_{H^1(\Omega)} \le \frac{C_1}{\min(\alpha, \alpha_0)}, \qquad \forall\, \varepsilon > 0. \tag{4.166}$$

If Ω is bounded and if (4.161) holds, we first observe that (4.165) can also be written as

$$a(u_\varepsilon - u, u_\varepsilon - u) + \|u_\varepsilon - u\|^2_{L^2(\Gamma)} + \frac{1-\varepsilon}{\varepsilon}\|u_\varepsilon - u\|^2_{L^2(\Gamma)} \le C_1\|u_\varepsilon - u\|_{H^1(\Omega)}; \tag{4.167}$$

then, since (from Lemma 4.1 and $\alpha > 0$)

$$v \to \left(\alpha \int_\Omega |\nabla v|^2\, dx + \|v\|^2_{L^2(\Gamma)}\right)^{1/2}$$

defines, over $H^1(\Omega)$, a norm equivalent to the usual $H^1(\Omega)$ norm (4.40), we have from (4.167),

$$\|u_\varepsilon - u\|_{H^1(\Omega)} \le C_2, \qquad \forall\, \varepsilon, \quad 0 < \varepsilon \le 1. \tag{4.168}$$

Both (4.166) and (4.168) imply the boundedness of $\{u_\varepsilon\}_\varepsilon$ in $H^1(\Omega)$. □

(2) *Weak convergence of* $\{u_\varepsilon\}_\varepsilon$. From the boundedness of $\{u_\varepsilon\}_\varepsilon$ in $H^1(\Omega)$, we have the existence of $u^* \in H^1(\Omega)$ and of an extracted subsequence—still denoted by $\{u_\varepsilon\}_\varepsilon$—such that

$$\lim_{\varepsilon \to 0} u_\varepsilon = u^* \text{ weakly in } H^1(\Omega); \tag{4.169}$$

from the continuity of γ_0 from $H^1(\Omega)$ into $L^2(\Gamma)$ and from (4.169), it then follows that

$$\lim_{\varepsilon \to 0} \gamma_0 u_\varepsilon = \gamma_0 u^* \text{ weakly in } L^2(\Gamma).^{22} \tag{4.170}$$

Since $\gamma_0 v = 0$, $\forall\, v \in H_0^1(\Omega)$, we have (from (4.157))

$$a(u_\varepsilon, v) = L(v), \qquad \forall\, v \in H_0^1(\Omega), \quad \forall\, \varepsilon > 0,$$

which, combined with (4.169), implies that at the limit,

$$a(u^*, v) = L(v), \qquad \forall\, v \in H_0^1(\Omega). \tag{4.171}$$

To show that $u^* = u$, it suffices to show (from (4.156), (4.171)) that $\gamma_0 u^* = g$; the boundedness of $\{u_\varepsilon\}_\varepsilon$ in $H^1(\Omega)$ and (4.164) imply

$$\lim_{\varepsilon \to 0} \|u_\varepsilon - g\|_{L^2(\Gamma)} = 0,$$

which combined with (4.170) implies $\gamma_0 u^* = g$. Thus we have proved that $u^* = u$ (since (4.156) has a unique solution, the whole $\{u_\varepsilon\}_\varepsilon$ converges to u).

(3) *Strong convergence of* $\{u_\varepsilon\}_\varepsilon$. From the weak convergence of $\{u_\varepsilon\}_\varepsilon$ to u in $H^1(\Omega)$, we observe that

$$\lim_{\varepsilon \to 0} \{L(u_\varepsilon - u) - a(u, u_\varepsilon - u)\} = 0. \tag{4.172}$$

[22] If Γ is bounded, we have, in fact, $\lim_{\varepsilon \to 0} \|\gamma_0(u_\varepsilon - u^*)\|_{L^2(\Gamma)} = 0$.

Since $a(v, v) \geq 0, \forall\, v \in H^1(\Omega)$, it follows from (4.164), (4.172) that

$$\lim_{\varepsilon \to 0} a(u_\varepsilon - u, u_\varepsilon - u) = 0, \tag{4.173}$$

$$\lim_{\varepsilon \to 0} \frac{1}{\varepsilon} \|u_\varepsilon - g\|^2_{L^2(\Gamma)} = 0. \tag{4.174}$$

If (4.160) (resp., (4.161)) holds, the strong convergence property (4.162) follows from (4.173) and from

$$a(u_\varepsilon - u, u_\varepsilon - u) \geq \operatorname{Min}(\alpha, \alpha_0)\|u_\varepsilon - u\|^2_{H^1(\Omega)}$$

(resp., from (4.173), (4.174), and from the fact that

$$v \to \left(\int_\Omega |\nabla v|^2\, dx + \int_\Gamma |v|^2\, d\Gamma\right)^{1/2}$$

defines a norm equivalent to the $H^1(\Omega)$-norm). □

Remark 4.16. It is possible to prove Proposition 4.14 using Theorem 7.1 of Chapter I, Sec. 7.3; we should take $V = H^1(\Omega)$, $a(\cdot, \cdot)$, and $L(\cdot)$ as above, and j and K defined by

$$j(v) = \frac{1}{2}\int_\Gamma |v - g|^2\, d\Gamma,$$

$$K = \{v \mid v \in H^1(\Omega), \gamma_0 v = g\},$$

respectively.

Remark 4.17. Let us consider the mixed boundary-value problem (4.132) (of Neumann–Dirichlet type), whose variational formulation is given by (4.140). In fact, the solution of this problem can also be obtained (if the usual hypotheses on $\overline{\mathbf{A}}$, a_0 hold) as the limit, as $\varepsilon \to 0$, of the solution u_ε of the following problem of Fourier type:

Find $u_\varepsilon \in H^1(\Omega)$ such that

$$\int_\Omega [(\overline{\mathbf{A}}\nabla u_\varepsilon)\cdot \nabla v + a_0 uv]\, dx + \frac{1}{\varepsilon}\int_{\Gamma_0} u_\varepsilon v\, d\Gamma = \int_\Omega fv\, dx + \int_{\Gamma_1} g_1 v\, d\Gamma$$
$$+ \frac{1}{\varepsilon}\int_{\Gamma_0} g_0 v\, d\Gamma, \qquad \forall\, v \in H^1(\Omega). \tag{4.175}$$

We observe that

$$\frac{1}{\varepsilon}\int_{\Gamma_0} vw\, d\Gamma = \int_\Gamma kvw\, d\Gamma, \qquad \forall\, v, w \in H^1(\Omega),$$

where k is defined by

$$k(x) = \frac{1}{\varepsilon} \quad \text{if } x \in \Gamma_0, 0 \quad \text{if } x \notin \Gamma_0.$$

The proof of the convergence of the solution u_ε of (4.175) to the solution u of (4.132), (4.140) is left to the reader as an exercise.

The above approximation properties are, in fact, of practical interest since many modern finite element codes for solving elliptic partial differential equation problems compute the solution of the (discrete) Dirichlet, Neumann, Neumann–Dirichlet, etc. problems via discrete Fourier formulations (derived from (4.157), (4.175)) with $\varepsilon > 0$ "very small." In the finite element code, MODULEF, for example, one uses $\varepsilon = 10^{-30}$ (see also Perronnet [1] for more details about MODULEF and further references).

From the practical interest of these approximations by penalization of the Dirichlet boundary condition $\gamma_0 u = g$ on Γ (or Γ_0), a quite natural problem which arises is to estimate the approximation error as a function of ε; the following two propositions give partial answers to this problem:

Proposition 4.15. *We suppose that the hypotheses on a_0, $\overline{\overline{\mathbf{A}}}$ are those of Proposition 4.14 and that*

$$L(v) = \int_\Omega fv\, dx, \quad \forall\, v \in H^1(\Omega), \tag{4.176}$$

where $f \in L^2(\Omega)$. We also suppose that the solution u of (4.156) satisfies

$$(\overline{\overline{\mathbf{A}}}\nabla u) \cdot \mathbf{n} \in L^2(\Gamma). \tag{4.177}$$

We then have

$$\|u_\varepsilon - u\|_{H^1(\Omega)} = O(\sqrt{\varepsilon}), \tag{4.178}$$

where u (resp., u_ε) is the solution of (4.156) (resp., (4.157)).

PROOF. Let us define $\lambda \in L^2(\Gamma)$ by

$$\lambda = (\overline{\overline{\mathbf{A}}}\nabla u) \cdot \mathbf{n};$$

since u is a solution of (4.156), we have (cf. Sec. 4.3)

$$-\nabla \cdot (\overline{\overline{\mathbf{A}}}\nabla u) + a_0 u = f \text{ in } \Omega. \tag{4.179}$$

Multiplying (4.179) by $v \in H^1(\Omega)$ and using the Green–Ostrogradsky formula (4.33), we obtain

$$a(u, v) = \int_\Omega fv\, dx + \int_\Gamma \lambda v\, d\Gamma, \qquad \forall\, v \in H^1(\Omega), \tag{4.180}$$

where $a(\cdot, \cdot)$ is still defined by (4.163). We also have

$$a(u_\varepsilon, v) + \frac{1}{\varepsilon}\int_\Gamma (u_\varepsilon - g)v\, d\Gamma = \int_\Omega fv\, dx, \qquad \forall\, v \in H^1(\Omega). \tag{4.181}$$

Taking $v = u_\varepsilon - u$ in (4.180), (4.181), by subtraction (also taking into account that $\gamma_0 u = g$), we obtain

$$a(u_\varepsilon - u, u_\varepsilon - u) + \frac{1}{\varepsilon}\int_\Gamma (u_\varepsilon - g)^2\, d\Gamma = -\int_\Gamma \lambda(u_\varepsilon - g)\, d\Gamma. \tag{4.182}$$

Since $a(v, v) \geq 0$, $\forall\, v \in H^1(\Omega)$, from (4.182) and from the Schwarz inequality in $L^2(\Gamma)$ we obtain

$$\|u_\varepsilon - g\|_{L^2(\Gamma)} \leq \varepsilon \|\lambda\|_{L^2(\Gamma)}. \tag{4.183}$$

Combining (4.182) and (4.183), we obtain, $\forall\, \varepsilon, 0 < \varepsilon \leq 1$,

$$a(u_\varepsilon - u, u_\varepsilon - u) + \int_\Gamma |u_\varepsilon - g|^2 \, d\Gamma \leq \varepsilon \|\lambda\|^2_{L^2(\Gamma)}$$

which clearly implies (4.178) (if (4.160) holds, we have

$$\|u_\varepsilon - u\|_{H^1(\Omega)} \leq \frac{\|\lambda\|_{L^2(\Gamma)}}{\sqrt{\operatorname{Min}(\alpha, \alpha_0)}} \varepsilon^{1/2}\Bigg).$$

Proposition 4.16. *The hypotheses on* a_0, $\bar{\bar{\mathbf{A}}}$, $L(\cdot)$ *being those of Proposition* 4.15, *we suppose that the solution u of* (4.156) *satisfies*[23]

$$(\bar{\bar{\mathbf{A}}}\nabla u) \cdot \mathbf{n} \in \gamma_0 H^1(\Omega). \tag{4.184}$$

We then have

$$\|u_\varepsilon - u\|_{H^1(\Omega)} = O(\varepsilon), \tag{4.185}$$

where u (resp., u_ε) is the solution of (4.156) (*resp.*, (4.157)).

PROOF. With $\lambda = (\bar{\bar{\mathbf{A}}}\nabla u) \cdot \mathbf{n}$, we define $u^1 \in H^1(\Omega)$ as the (unique) solution of the Dirichlet problem

$$\nabla \cdot (\bar{\bar{\mathbf{A}}}\nabla u^1) = 0 \text{ in } \Omega, \quad u^1 = -\lambda \text{ on } \Gamma; \tag{4.186}$$

we now define $w_\varepsilon \in H^1(\Omega)$ by

$$w_\varepsilon = u + \varepsilon u^1. \tag{4.187}$$

We clearly have

$$\|w_\varepsilon - u\|_{H^1(\Omega)} = O(\varepsilon). \tag{4.188}$$

From (4.188) then it follows that to prove (4.185) it suffices to prove that

$$\|w_\varepsilon - u_\varepsilon\|_{H^1(\Omega)} = O(\varepsilon).$$

Taking $v = w_\varepsilon - u_\varepsilon$ in (4.157), we have (with $a(\cdot, \cdot)$ still defined by (4.163))

$$a(u_\varepsilon, w_\varepsilon - u_\varepsilon) + \frac{1}{\varepsilon}\int_\Gamma (u_\varepsilon - g)(w_\varepsilon - u_\varepsilon)\, d\Gamma = \int_\Omega f(w_\varepsilon - u_\varepsilon)\, dx,$$

[23] We have $\gamma_0 H^1(\Omega) = \{\mu \mid \mu \in L^2(\Gamma), \exists\, \tilde{\mu} \in H^1(\Omega) \text{ such that } \mu = \gamma_0 \tilde{\mu}\}$; actually $\gamma_0 H^1(\Omega) = H^{1/2}(\Gamma)$ (for the definition and properties of the Sobolev spaces H^s with $s \in \mathbb{R}$, see, e.g., Adams [1], Lions and Magenes [1] and Necas [1]).

which we rewrite

$$a(w_\varepsilon - u_\varepsilon, w_\varepsilon - u_\varepsilon) + \frac{1}{\varepsilon}\int_\Gamma (w_\varepsilon - u_\varepsilon)^2\, d\Gamma$$
$$= a(w_\varepsilon, w_\varepsilon - u_\varepsilon) + \frac{1}{\varepsilon}\int_\Gamma (w_\varepsilon - g)(w_\varepsilon - u_\varepsilon)\, d\Gamma - \int_\Omega f(w_\varepsilon - u_\varepsilon)\, dx. \quad (4.189)$$

Combining (4.186), (4.187), and (4.189), we obtain

$$a(w_\varepsilon - u_\varepsilon, w_\varepsilon - u_\varepsilon) + \frac{1}{\varepsilon}\int_\Gamma (w_\varepsilon - u_\varepsilon)^2\, d\Gamma = \varepsilon a(u^1, w_\varepsilon - u_\varepsilon)$$
$$+ \left\{a(u, w_\varepsilon - u_\varepsilon) - \int_\Gamma \lambda(w_\varepsilon - u_\varepsilon)\, d\Gamma - \int_\Omega f(w_\varepsilon - u_\varepsilon)\, dx\right\}; \quad (4.190)$$

from (4.180), the second term in the right-hand side of (4.190) vanishes; we then have

$$a(w_\varepsilon - u_\varepsilon, w_\varepsilon - u_\varepsilon) + \frac{1}{\varepsilon}\int_\Gamma (w_\varepsilon - u_\varepsilon)^2\, d\Gamma = \varepsilon a(u^1, w_\varepsilon - u_\varepsilon)$$

which implies

$$a(w_\varepsilon - u_\varepsilon, w_\varepsilon - u_\varepsilon) + \frac{1}{\varepsilon}\int_\Gamma (w_\varepsilon - u_\varepsilon)^2\, d\Gamma$$
$$\leq \varepsilon \operatorname{Max}(\|\bar{\bar{\mathbf{A}}}\|_{L^\infty(\Omega)}, \|a_0\|_{L^\infty(\Omega)})\|u^1\|_{H^1(\Omega)}\|w_\varepsilon - u_\varepsilon\|_{H^1(\Omega)}. \quad (4.191)$$

By a now standard reasoning, (4.191) implies that

$$\|w_\varepsilon - u_\varepsilon\|_{H^1(\Omega)} = O(\varepsilon),$$

which completes the proof of the proposition. □

Remark 4.18. The hypotheses (4.177), (4.184) concerning the regularity of $(\bar{\bar{\mathbf{A}}}\nabla u) \cdot \mathbf{n}$ on Γ are often encountered in practice. If the a_{ij}, f, g are sufficiently smooth functions of their respective arguments, and if Γ is a sufficiently smooth curve or surface, properties (4.177), (4.184) will follow from regularity properties of the solutions of the Dirichlet problem (4.156) discussed in, e.g., Necas [1].

4.5. Finite element approximations of the Neumann, Dirichlet, and Fourier problems

4.5.1. *Synopsis*

We now consider the finite element approximation of the second-order elliptic problems discussed in Secs. 4.2, 4.3, and 4.4; actually the finite element techniques to be described in this section will also be used to approximate the second-order elliptic problem discussed in Sec. 4.6 (and have been used all along in this book to solve nonlinear problems much more complicated than those linear problems discussed in this appendix).

The following subsections cover only a small part of this very important subject; for more details (theoretical and practical) concerning finite element approximations, see Aubin [2], Ciarlet [1]–[3], Aziz and Babuska [1], Raviart and Thomas [1], Oden and Reddy [1] ,Mercier [1], Strang and Fix [1], Zienkiewicz [1], etc..

4.5.2. *Basic hypotheses: triangulations of* Ω *and fundamental discrete spaces*

For simplicity we suppose that Ω is a bounded polygonal domain of $\mathbb{R}^2$; we then define a family $\{\mathscr{T}_h\}_h$ of triangulations of Ω such that:

(i) $\forall\, h$, $\mathscr{T}_h$ is a finite collection of closed[24] triangles contained in $\overline{\Omega}$;
(ii) $\cup_{T \in \mathscr{T}_h} T = \overline{\Omega}$;
(iii) h is the maximal length of the edges of the $T \in \mathscr{T}_h$;
(iv) if $T, T' \in \mathscr{T}_h$, $T \neq T'$, we have

$$\mathring{T} \cap \mathring{T}' = \varnothing$$

and either $T \cap T' = \varnothing$ or T and T' have in common a whole edge or only one vertex.

Various triangulations for which conditions (i)–(iv) hold are shown in several chapters of this book (see also Appendix II).

From $\mathscr{T}_h$ we now define the two following finite-dimensional spaces:

$$H_h^1 = \{v_h \mid v_h \in C^0(\overline{\Omega}),\quad v_h|_T \in P_1, \qquad \forall\, T \in \mathscr{T}_h\}, \tag{4.192}$$

$$H_{0h}^1 = \{v_h \mid v_h \in H_h^1,\quad v_h = 0 \text{ on } \Gamma\}, \tag{4.193}$$

where P_1 is the space of the polynomials in two variables of degree less or equal to 1 (i.e., if $q \in P_1$, then $q(x_1, x_2) = \alpha x_1 + \beta x_2 + \gamma$, $\alpha, \beta, \gamma \in \mathbb{R}$).

Let ∇v_h be the gradient of v_h (in the sense of distributions); it can be shown that if $v_h \in H_h^1$, then

$$\nabla v_h = \sum_{T \in \mathscr{T}_h} \nabla(v_{h|T}) \chi_{\mathring{T}} \tag{4.194}$$

where $\chi_{\mathring{T}}$ is the characteristic function of the interior $\mathring{T}$ of T i.e.,

$$\chi_{\mathring{T}}(x) = \begin{cases} 1 & \text{if } x \in \mathring{T}, \\ 0 & \text{if } x \notin \mathring{T}. \end{cases}$$

From (4.192), (4.194) it then follows that ∇v_h is piecewise constant over Ω, and also that

$$H_{0h}^1 = H_h^1 \cap H_0^1(\Omega).$$

[24] i.e., if $T \in \mathscr{T}_h$, then $T = \mathring{T} \cup \partial T$ (with $\mathring{T}$ = interior of T, ∂T boundary of T).

Let Σ_h (resp., Σ_{0h}) be the (finite) set of the vertices of $\mathscr{T}_h$ (resp., of the vertices of $\mathscr{T}_h$ which do not belong to Γ); we denote by N_h (resp., N_{0h}) the number of elements of Σ_h (resp., Σ_{0h}) and we suppose that

$$\Sigma_{0h} = \{Q_i\}_{i=1}^{N_{0h}}, \qquad \Sigma_h = \Sigma_{0h} \cup \{Q_i\}_{i=N_{0h}+1}^{N_h} \tag{4.195}$$

(i.e., $Q_i \in \Gamma$ if $N_{0h+1} \leq i \leq N_h$).

We recall that any polynomial of P_1 is entirely defined by the values it takes at the three vertices of a triangle; furthermore, if $p, q \in P_1$ and satisfy $p(x) = q(x)$, $p(x') = q(x')$, where $x \neq x'$, then p and q coincide on the line xx'. From these properties it follows that Σ_h (resp., Σ_{0h}) is H_h^1 (resp., H_{0h}^1) unisolvent (in the sense of Ciarlet [1], [2], [3]), since

$$\forall\, \alpha = \{\alpha_i\}_i \in \mathbb{R}^{N_h} \text{ (resp., } \mathbb{R}^{N_{0h}}\text{), there exists a unique } v_h \in H_h^1$$
$$\text{(resp., } H_{0h}^1\text{) such that } v_h(Q_i) = \alpha_i, \qquad \forall\, i = 1, \ldots, N_h \text{ (resp., } \forall\, i = 1, \ldots, N_{0h}\text{)}. \tag{4.196}$$

To each $Q_i \in \Sigma_h$ we now associate the function w_i (uniquely) defined by

$$w_i \in H_h^1, \qquad w_i(Q_i) = 1, \quad w_i(Q_j) = 0 \quad \text{if } i \neq j. \tag{4.197}$$

It is then quite easy to show that $\mathscr{B}_h$ and $\mathscr{B}_{0h}$ defined by

$$\mathscr{B}_h = \{w_i\}_{i=1}^{N_h}, \tag{4.198}$$

$$\mathscr{B}_{0h} = \{w_i\}_{i=1}^{N_{0h}} \tag{4.199}$$

are bases of H_h^1 and H_{0h}^1, respectively; we indeed have

$$v_h = \sum_{i=1}^{N_h} v_h(Q_i) w_i, \qquad \forall\, v_h \in H_h^1, \tag{4.200}$$

and also

$$v_h = \sum_{i=1}^{N_{0h}} v_h(Q_i) w_i, \qquad \forall\, v_h \in H_{0h}^1. \tag{4.201}$$

We observe that w_i vanishes outside the union of those triangles of $\mathscr{T}_h$ having Q_i as a common vertex.

Remark 4.19. In this appendix we consider piecewise linear approximations on triangles only; for more sophisticated methods making use of higher-degree polynomials, quadrilateral elements, curved elements, and also for finite element methods for three-dimensional problems, see the references given in Sec. 4.5.1.

4.5.3. *Some fundamental results on finite element approximations*

4.5.3.1. *Generalites: synopsis.* In this section (Sec. 4.5.3, which closely follows Ciarlet [1], [2], [3] and Mercier [1]), we shall give some results about the approximation properties of the particular finite element method introduced in Sec. 4.5.2. We shall be particularly concerned with providing estimates for the finite element approximation errors; we shall use *Sobolev norms* to express these estimates.

4.5.3.2. *Definition and some basic properties of the Sobolev spaces* $W^{m,p}(\Omega)$. From the importance of Sobolev spaces for estimating finite element approximation errors, we shall complete Sec. 4.2.2 with the following definitions and properties.

Let m be a non-negative integer and p satisfies $1 \leq p \leq +\infty$; with Ω a domain of $\mathbb{R}^N$, we define the Sobolev space $W^{m,p}(\Omega)$ as

$$W^{m,p}(\Omega) = \{v \mid \partial^\alpha v \in L^p(\Omega), \qquad \forall\, \alpha, |\alpha| \leq m\}. \tag{4.202}$$

In (4.202), $\alpha = \{\alpha_1, \ldots, \alpha_N\}$ is a multiindex such that $\forall\, i = 1, \ldots, N$, α_i is a non-negative integer, $|\alpha| = \sum_{i=1}^N \alpha_i$, and $\partial^\alpha v$ is defined by

$$\partial^\alpha v = v \quad \text{if } |\alpha| = 0, \qquad \partial^\alpha v = \frac{\partial^{|\alpha|} v}{\partial x_1^{\alpha_1} \cdots \partial x_N^{\alpha_N}} \quad \text{if } |\alpha| \geq 1; \tag{4.203}$$

in (4.203), the derivatives are taken in a distribution sense. If $m = 0$, we clearly have, from (4.202), (4.203), $W^{0,p}(\Omega) = L^p(\Omega)$; if $p = 2$, one usually uses the notation

$$H^m(\Omega) = W^{m,2}(\Omega).$$

It can be proved that $W^{m,p}(\Omega)$ is a *Banach space* if equipped with the norm defined by

$$\|v\|_{W^{m,p}(\Omega)} = \left(\sum_{\substack{\alpha \\ 0 \leq |\alpha| \leq m}} \|\partial^\alpha v\|^p_{L^p(\Omega)} \right)^{1/p}. \tag{4.204}$$

For simplicity, one quite often uses the notation

$$\|v\|_{m,p,\Omega} = \|v\|_{W^{m,p}(\Omega)}, \tag{4.205$_1$}$$

and also

$$|v|_{m,p,\Omega} = \left(\sum_{\substack{\alpha \\ |\alpha| = m}} \|\partial^\alpha v\|^p_{L^p(\Omega)} \right)^{1/p}. \tag{4.205$_2$}$$

If $p = +\infty$, we should use the notation

$$\|v\|_{m,\infty,\Omega} = \|v\|_{W^{m,\infty}(\Omega)} = \max_{\substack{\alpha \\ 0 \leq |\alpha| \leq m}} \|\partial^\alpha v\|_{L^\infty(\Omega)},$$

$$|v|_{m,\infty,\Omega} = \max_{\substack{\alpha \\ |\alpha| = m}} \|\partial^\alpha v\|_{L^\infty(\Omega)}.$$

A very important property of the spaces $W^{m,p}(\Omega)$ is given by the following:

Proposition 4.17. *Let Ω be a bounded domain of $\mathbb{R}^N$; then if Γ $(=\partial\Omega)$ is sufficiently smooth,*[25] *and if*

$$m > \frac{N}{p}, \tag{4.206}$$

we have

$$W^{m,p}(\Omega) \subset C^0(\bar{\Omega}), \tag{4.207}$$

the injection from $W^{m,p}(\Omega)$ into $C^0(\bar{\Omega})$ being continuous.[26,27]

The above proposition is proved in, e.g., Adams [1]. We observe, most particularly, that if $N = 2$, then

$$W^{1,p}(\Omega) \subsetneq C^0(\bar{\Omega}) \quad \text{if } p > 2, \qquad W^{2,p}(\Omega) \subsetneq C^0(\bar{\Omega}) \quad \text{if } p > 1. \tag{4.208}$$

4.5.3.3. *Error estimates using Sobolev norms.* (I) *Local estimates.* Let T be a triangle; we define h_T and ρ_T by

$$h_T = \text{length of the largest edge of } T, \tag{4.209}$$

and

$$\rho_T = \text{supremum of the diameters of all the disks contained in } T, \tag{4.210}$$

respectively (we clearly have $h = \max_{T \in \mathcal{T}_h} h_T$).

Since T is a polygonal domain of $\mathbb{R}^2$, from Proposition 4.17 we have

$$W^{m,p}(\mathring{T}) \subsetneq C^0(T) \quad \text{if } m > \frac{2}{p}. \tag{4.211}$$

We denote by A_i, $i = 1, 2, 3$, the vertices of T, and we define $\pi_T: C^0(T) \to P_1$ by

$$\pi_T v \in P_1, \qquad \forall\, v \in C^0(T), \qquad (\pi_T v)(A_i) = v(A_i), \qquad \forall\, i = 1, 2, 3; \tag{4.212}$$

π_T is the linear interpolation operator associated with the three vertices of T.

Concerning the approximation properties of π_T, we have the following:

Proposition 4.18. *We consider $p, q \geq 1$, and m a non-negative integer such that*

$$W^{2,p}(\mathring{T}) \subsetneq C^0(T), \tag{4.213}$$

$$W^{2,p}(\mathring{T}) \subsetneq W^{m,q}(\mathring{T}). \tag{4.214}$$

Then there exists C independent of T such that

$$|v - \pi_T v|_{m,q,T} \leq C(\text{meas } T)^{1/q - 1/p} \frac{h_T^2}{\rho_T^m} |v|_{2,p,T}, \qquad \forall\, v \in W^{2,p}(\mathring{T}). \tag{4.215}$$

---

[25] Ω can be a polyhedral domain, for example.

[26] In fact, it is compact.

[27] In the sequel, we shall use the symbol $\subsetneq$ to denote the inclusion, with continuity of the injection.

Remark 4.20. In the particular case where $q = p$, (4.215) holds if $m = 0, 1, 2$, and then (4.215) reduces to

$$|v - \pi_T v|_{m,p,T} \leq C \frac{h_T^2}{\rho_T^m} |v|_{2,p,T}, \qquad \forall\, v \in W^{2,p}(\mathring{T}) \quad \text{if } m = 0, 1, 2. \tag{4.216}$$

4.5.3.4. *Error estimates using Sobolev norms.* (II) *Global estimates.* We use the notation of Sec. 4.5.2; in Sec. 4.5.2 we have introduced a family $\{\mathcal{T}_h\}_h$ of triangulations of the domain Ω. This family $\{\mathcal{T}_h\}_h$ is said to be regular if there exists a constant σ such that

$$\frac{h_T}{\rho_T} \leq \sigma, \qquad \forall\, T \in \mathcal{T}_h, \quad \forall\, h. \tag{4.217}$$

Remark 4.21. Condition (4.217) is equivalent to the following property

The angles of $\mathcal{T}_h$ are bounded from below by $\theta_0 > 0$, independent of h. (4.218)

We now define the interpolation operator $\pi_h\colon C^0(\bar{\Omega}) \to H_h^1$ by

$$\pi_h \in H_h^1, \qquad \forall\, v \in C^0(\bar{\Omega}), \qquad (\pi_h v)(Q_i) = v(Q_i), \qquad \forall\, Q_i \in \Sigma_h; \tag{4.219}$$

from (4.192), (4.193) we clearly have

$$\pi_h v \in H_{0h}^1, \qquad \forall\, v \in C^0(\bar{\Omega}), \quad v = 0 \text{ on } \Gamma. \tag{4.220}$$

We observe that

$$\pi_h v|_T = \pi_T(v|_T), \qquad \forall\, v \in C^0(\bar{\Omega}), \quad \forall\, T \in \mathcal{T}_h, \tag{4.221}$$

where π_T is defined by (4.212); we also observe (from (4.208)) that $\pi_h v$ can be defined if either $v \in W^{1,p}(\Omega)$ with $p > 2$ or $v \in W^{2,p}(\Omega)$ with $p > 1$.

Combining the above properties with the local estimates of Sec. 4.5.3.3 (most particularly (4.216)), we should prove Theorem 4.1, which provides global interpolation error estimates:

Theorem 4.1. *Suppose that $\{\mathcal{T}_h\}_h$ is a regular family of triangulation* (*i.e.*, (4.217), (4.218) *hold*). *We then have the following interpolation errors estimates*:

If $p > 2$ and $m = 0, 1$, then

$$\|v - \pi_h v\|_{m,p,\Omega} \leq C h^{1-m} |v|_{1,p,\Omega}, \qquad \forall\, v \in W^{1,p}(\Omega), \tag{4.222}$$

If $p > 1$ and $m = 0, 1$, then

$$\|v - \pi_h v\|_{m,p,\Omega} \leq C h^{2-m} |v|_{2,p,\Omega}, \qquad \forall\, v \in W^{2,p}(\Omega), \tag{4.223}$$

where C is a constant independent of v and h.

For a proof of Theorem 4.1, see Ciarlet [1], [2], [3].

EXERCISE 4.1. Suppose that in addition to (4.217), (4.218), the family $\{\mathcal{T}_h\}_h$ also satisfies the following property (which in fact implies (4.217), (4.218)):

$$\frac{h}{\min\limits_{T \in \mathcal{T}_h} h_T} \leq \beta, \qquad \forall\, h, \tag{4.224}$$

where β is a constant. Then prove that there exists a constant C independent of h such that

$$|v_h|_{1,p,\Omega} \leq \frac{C}{h} |v_h|_{0,p,\Omega}, \qquad \forall\, v_h \in H_h^1. \tag{4.225}$$

[Relation (4.225) plays a very important role in many theoretical questions concerning the finite element approximation of important problems (like the Stokes and Navier–Stokes problems discussed in Chapter VII, Sec. 5).]

4.5.3.5. *Some approximation properties of C^2 functions.* Let T be a triangle of $\mathbb{R}^2$ whose vertices are denoted by A_1, A_2, A_3 (with $A_i = \{a_{i1}, a_{i2}\}$). With π_T as in (4.212), we shall estimate, by elementary methods,

$$|\pi_T v - v|_{m,\infty,T} \quad \text{for } v \in C^2(T) \text{ and } m = 0, 1.$$

For more results in this direction, see e.g., Ciarlet and Wagschal [1].

Let $v \in C^0(T)$ and consider the linear interpolate $\pi_T v$; we have (with $v_i = v(A_i)$)

$$\pi_T v(x) = \sum_{i=1}^{3} v_i w_i(x), \qquad \forall\, x \in \mathbb{R}^2, \tag{4.226}$$

where the functions w_i are (uniquely) defined by

$$w_i \in P_1, \quad w_i(A_i) = 1, \quad w_i(A_j) = 0 \qquad \text{if } j \neq i, \quad i = 1, 2, 3. \tag{4.227}$$

Since the functions $x \to 1, x \to x_1, x \to x_2$ belong to P_1, we also have

$$\sum_{i=1}^{3} w_i(x) = 1, \qquad \forall\, x, \tag{4.228}$$

$$x_1 = \sum_{i=1}^{3} a_{i1} w_i(x), \tag{4.229}$$

$$x_2 = \sum_{i=1}^{3} a_{i2} w_i(x). \tag{4.230}$$

Properties (4.228)–(4.230) justify the fact that the $w_i(x)$ are very often called the *barycentric coordinates* of $x \in \mathbb{R}^2$, with respect to A_1, A_2, A_3.

Another property of the w_i is

$$w_i(x) \geq 0, \qquad \forall\, i = 1, 2, 3 \Leftrightarrow x \in T. \tag{4.231}$$

Suppose that $v \in C^2(T)$; we denote by D^2v the *Hessian matrix* of v, i.e.,

$$D^2 v = \begin{pmatrix} \dfrac{\partial^2 v}{\partial x_1^2} & \dfrac{\partial^2 v}{\partial x_1\,\partial x_2} \\ \dfrac{\partial^2 v}{\partial x_1\,\partial x_2} & \dfrac{\partial^2 v}{\partial x_2^2} \end{pmatrix}; \tag{4.232}$$

we then define $|D^2v(x)|$ by

$$|D^2 v(x)| = \underset{\boldsymbol{\xi} \in \mathbb{R}^2 - \{\mathbf{0}\}}{\text{Max}} \frac{|D^2 v(x)\boldsymbol{\xi}|}{|\boldsymbol{\xi}|} \tag{4.233}$$

(with $|\boldsymbol{\xi}| = \sqrt{\xi_1^2 + \xi_2^2}$ if $\boldsymbol{\xi} = \{\xi_1, \xi_2\}$); $|D^2v(x)|$ is clearly the *spectral radius* of $D^2v(x)$, and since $v \in C^2(T)$, we have

$$\|D^2 v\|_{0,\infty,T} = \underset{x \in T}{\text{Sup}}\, |D^2 v(x)| < +\infty. \tag{4.234}$$

We now prove a first approximation result:

Theorem 4.2. *Suppose that $v \in C^2(T)$; we then have*

$$|\pi_T v - v|_{0,\infty,T} = \underset{x \in T}{\text{Max}}\, |v(x) - \pi_T v(x)| \le \tfrac{1}{2} h_T^2 \|D^2 v\|_{0,\infty,T}. \tag{4.235}$$

PROOF. Using a second-order Taylor expansion, we have

$$v_i = v(A_i) = v(x) + \vec{\nabla} v(x) \cdot \overrightarrow{xA_i} + \tfrac{1}{2}\overrightarrow{xA_i} \cdot (D^2 v(\xi_i)\overrightarrow{xA_i}) \tag{4.236}$$

where $\xi_i \in [x, A_i]$ (therefore $\xi_i \in T$ if $x \in T$); combining (4.226) and (4.236), we obtain

$$\begin{aligned} \pi_T v(x) = \sum_{i=1}^{3} v_i w_i(x) &= v(x)\left(\sum_{i=1}^{3} w_i(x)\right) + \vec{\nabla} v(x) \cdot \left(\sum_{i=1}^{3} w_i(x)\overrightarrow{xA_i}\right) \\ &\quad + \frac{1}{2}\sum_{i=1}^{3} w_i(x)\overrightarrow{xA_i} \cdot (D^2 v(\xi_i)\overrightarrow{xA_i}), \end{aligned}$$

which, if one takes (4.228)–(4.230) into account, implies

$$\pi_T v(x) - v(x) = \frac{1}{2}\sum_{i=1}^{3} w_i(x)\overrightarrow{xA_i} \cdot (D^2 v(\xi_i)\overrightarrow{xA_i}). \tag{4.237}$$

If we suppose that $x \in T$, from (4.228), (4.231), (4.233), (4.234), we have

$$\begin{aligned} |\pi_T v(x) - v(x)| &\le \frac{1}{2}\left(\sum_{j=1}^{3} w_j(x)\right) \underset{i}{\text{Max}}\, |\overrightarrow{xA_i} \cdot (D^2 v(\xi_i)\overrightarrow{xA_i})| \\ &\le \frac{1}{2}\|D^2 v\|_{0,\infty,T}\, \underset{i}{\text{Max}}\, |\overrightarrow{xA_i}|^2. \end{aligned} \tag{4.238}$$

Since $\text{Max}_i\, |\overrightarrow{xA_i}| = h_T$, from (4.238) we obtain

$$|\pi_T v(x) - v(x)| \le \tfrac{1}{2}\|D^2 v\|_{0,\infty,T} h_T^2, \qquad \forall\, x \in T, \quad \forall\, v \in C^2(T),$$

which finally proves (4.235). □

EXERCISE 4.2. Prove that

$$\|\pi_T v - v\|_{0,\infty,T} \leq h_T |v|_{1,\infty,T}, \qquad \forall\, v \in C^1(T) \tag{4.239}$$

(we recall that $|v|_{1,\infty,T} = \operatorname{Max}_{x\in T} |\nabla v(x)|$).

We now consider the more complicated problem, which is obtaining estimates for $\|\nabla(\pi_T v - v)\|_{L^\infty(T)\times L^\infty(T)}$ $(= |\pi_T v - v|_{1,\infty,T})$; in this direction we have the following:

Theorem 4.3. *Suppose that $v \in C^2(T)$; we then have*

$$|\pi_T v - v|_{1,\infty,T} \leq 3 \frac{h_T}{\sin\theta_T} \|D^2 v\|_{0,\infty,T}, \tag{4.240}$$

where θ_T is the smallest angle of T.

PROOF. From $\pi_T v = \sum_{i=1}^3 v_i w_i$, we have

$$\nabla(\pi_T v) = \sum_{i=1}^3 v_i \nabla w_i,$$

which implies—combined with (4.236)—that

$$\nabla(\pi_T v(x)) = \nabla\left(\sum_{i=1}^3 w_i(x)\right) v(x) + \sum_{i=1}^3 \vec{\nabla} w_i(\vec{\nabla} v(x) \cdot \overrightarrow{xA_i})$$

$$+ \frac{1}{2} \sum_{i=1}^3 \nabla w_i (\overrightarrow{xA_i} \cdot (D^2 v(\xi_i)\overrightarrow{xA_i})). \tag{4.241}$$

From (4.228) we have that $\nabla(\sum_{i=1}^3 w_i) = \nabla 1 = \mathbf{0}$; now using elementary algebra we should prove that

$$\sum_{i=1}^3 \vec{\nabla} w_i(\vec{\nabla} v(x) \cdot \overrightarrow{xA_i}) = \vec{\nabla} v(x). \tag{4.242}$$

From these relations, (4.241) reduces to

$$\nabla(\pi_T v - v)(x) = \frac{1}{2} \sum_{i=1}^3 \nabla w_i(\overrightarrow{xA_i} \cdot (D^2 v(\xi_i)\overrightarrow{xA_i})), \tag{4.243}$$

which, in turn, implies

$$|\nabla(\pi_T v - v)(x)| \leq \frac{1}{2}\left(\sum_{i=1}^3 |\nabla w_i|\right) \operatorname*{Max}_i |\overrightarrow{xA_i} \cdot (D^2 v(\xi_i))\overrightarrow{xA_i}|$$

$$\leq \frac{1}{2}\left(\sum_{i=1}^3 |\nabla w_i|\right) \|D^2 v\|_{0,\infty,T} h_T^2, \qquad \forall\, x \in T. \tag{4.244}$$

To obtain (4.240), we must estimate $\sum_{i=1}^3 |\nabla w_i|$; actually we have (with m_T = measure of T)

$$|\nabla w_1| = \frac{|\overrightarrow{A_2A_3}|}{2m_T}, \quad |\nabla w_2| = \frac{|\overrightarrow{A_1A_3}|}{2m_T}, \quad |\nabla w_3| = \frac{|\overrightarrow{A_1A_2}|}{2m_T}. \tag{4.245}$$

From (4.245) it follows that we clearly have

$$|\nabla w_i| \leq \frac{h_T}{2m_T}, \qquad \forall\, i = 1, 2, 3,$$

which, combined with (4.244), implies

$$|\nabla(\pi_T v - v)(x)| \leq \frac{3}{4}\, \|D^2 v\|_{0,\infty,T} \frac{h_T^3}{m_T}, \qquad \forall\, x \in T. \tag{4.246}$$

But since the relation

$$4m_T \geq h_T^2 \sin\theta_T$$

always holds, from (4.246) we finally obtain

$$\operatorname*{Max}_{x\in T} |\nabla(\pi_T v - v)(x)| \leq 3\|D^2 v\|_{0,\infty,T} \frac{h_T}{\sin\theta_T},$$

i.e., relation (4.240). □

EXERCISE 4.3. Prove (4.242).

EXERCISE 4.4. Prove (4.245).

Theorems 4.2 and 4.3 give local approximation results; now suppose that Ω is a polygonal bounded domain of $\mathbb{R}^2$ and that $\{\mathcal{T}_h\}_h$ is a family of triangulations of Ω satisfying hypotheses (i)–(iv) of Sec. 4.5.2. With π_h the interpolation operator defined by (4.219), we would like to estimate $|\pi_h v - v|_{m,\infty,\Omega}$ for $m = 0, 1$ and v a sufficiently smooth function.

Using Theorems 4.2 and 4.3, and also the facts that

$$h = \operatorname*{Max}_{T\in\mathcal{T}_h} h_T,$$

$\theta_0 \leq \operatorname*{Min}_{T\in\mathcal{T}_h} \theta_T$ if $\{\mathcal{T}_h\}_h$ is a regular family of triangulations of Ω (i.e., if (4.217), (4.218) hold),

$$|v|_{m,\infty,\Omega} = \operatorname*{Max}_{T\in\mathcal{T}_h} |(v|_T)|_{m,\infty,T}, \qquad \forall\, v \in C^m(\overline{\Omega}),$$

$$\pi_h v|_T = \pi_T(v|_T), \qquad \forall\, v \in C^0(\overline{\Omega}),$$

we should easily prove the following:

Theorem 4.4. *Suppose that $\{\mathcal{T}_h\}_h$ is a regular family of triangulations of Ω and that the linear interpolation operator π_h is defined by* (4.219). *We then have*

$$\|\pi_h v - v\|_{0,\infty\Omega} \leq h|v|_{1,\infty,\Omega}, \qquad \forall\, v \in C^1(\overline{\Omega}), \tag{4.247}$$

$$\|\pi_h v - v\|_{0,\infty,\Omega} \leq \tfrac{1}{2}h^2\|D^2 v\|_{0,\infty,\Omega}, \qquad \forall\, v \in C^2(\overline{\Omega}), \tag{4.248}$$

$$|\pi_h v - v|_{1,\infty,\Omega} \leq 3\,\frac{h}{\sin\theta_0}\,\|D^2 v\|_{0,\infty,\Omega}, \qquad \forall\, v \in C^2(\overline{\Omega}) \tag{4.249}$$

($\|D^2 v\|_{0,\infty,\Omega}$ is defined from (4.234) by replacing T by Ω).

EXERCISE 4.5. Prove—if $\Omega \subset \mathbb{R}^N$—that

$$|v|_{2,\infty,\Omega} \leq \|D^2 v\|_{0,\infty,\Omega} \leq N|v|_{2,\infty,\Omega}, \qquad \forall\, v \in C^2(\bar{\Omega}). \tag{4.250}$$

4.5.4. *Application to the finite element solution of the Neumann and Fourier problems*

In this section we suppose that Ω is a bounded polygonal domain of $\mathbb{R}^2$.

4.5.4.1. *Synopsis.* The Neumann problem for a large class of second-order elliptic partial differential operators has been defined and discussed in Sec. 4.2. In Sec. 4.4 we have shown that the Neumann problem is in fact a particular case of the so-called Fourier problem defined in Sec. 4.4.2. For this reason, in this section (Sec. 4.5.4) we shall consider only the finite element approximation of the Fourier problem (4.141) and then the convergence properties of the approximate solutions.

The convergence of the approximate solutions follows from the general approximation results of Sec. 3, combined with the approximation properties stated in Sec. 4.5.3.

4.5.4.2. *Formulations of the continuous and approximate Fourier problems.* We recall (from Sec. 4.4) that the operator formulation of the Fourier problem is

$$-\nabla \cdot (\bar{\bar{\mathbf{A}}} u) + a_0 u = f \text{ in } \Omega, \qquad (\bar{\bar{\mathbf{A}}}\nabla u)\cdot \mathbf{n} + ku = g \text{ on } \Gamma, \tag{4.251}$$

the corresponding variational formulation being:

Find $u \in H^1(\Omega)$ such that

$$\int_\Omega (\bar{\bar{\mathbf{A}}}\nabla u)\cdot \nabla v\, dx + \int_\Omega a_0 uv\, dx + \int_\Gamma kuv\, d\Gamma$$
$$= \int_\Omega fv\, dx + \int_\Gamma gv\, d\Gamma, \qquad \forall\, v \in H^1(\Omega). \tag{4.252}$$

We suppose that the hypotheses on $\bar{\bar{\mathbf{A}}}$, a_0, k, f, g made in Sec. 4.4.3 still hold; then problem (4.251), (4.252) has a unique solution. To approximate (4.251), (4.252), we consider a family $\{\mathcal{T}_h\}_h$ of triangulations of Ω (as in Sec. 4.5.2) and the discrete space H_h^1 defined by (4.192). Then we approximate (4.251), (4.252) by the following finite-dimensional linear variational problem:

Find $u_h \in H_h^1$ such that

$$\int_\Omega (\bar{\bar{\mathbf{A}}}\nabla u_h)\cdot \nabla v_h\, dx + \int_\Omega a_0 u_h v_h\, dx + \int_\Gamma ku_h v_h\, d\Gamma$$
$$= \int_\Omega fv_h\, dx + \int_\Gamma gv_h\, d\Gamma, \qquad \forall\, v_h \in H_h^1; \tag{4.253}$$

problem (4.253) clearly has a unique solution.

4.5.4.3. *Convergence of the approximate solutions.* Concerning the convergence of $\{u_h\}_h$ to the solution of (4.252), we have:

Proposition 4.19. *Suppose that* $\overline{\overline{\mathbf{A}}}, a_0, k, f, g$ *satisfy* (4.146)–(4.149); *also suppose that* $\{\mathcal{T}_h\}_h$ *is a regular family of triangulations of* Ω (*in the sense of Sec.* 4.5.3.4). *We then have*

$$\lim_{h \to 0} \|u_h - u\|_{H^1(\Omega)} = 0, \tag{4.254}$$

where u_h (*resp.,* u) *is the solution of* (4.253) (*resp.,* (4.251), (4.252)).

PROOF. To apply Theorem 3.2 of Sec. 3.3 (with $V_h = H_h^1$), it suffices to find $\mathscr{V}$ and r_h obeying (3.1). Since (cf. Sec. 4.2.2, Proposition 4.2)

$$\overline{\mathscr{D}(\overline{\Omega})}^{H^1(\Omega)} = H^1(\Omega),$$

we can take $\mathscr{V} = \mathscr{D}(\overline{\Omega})$. We then define r_h by

$$r_h v \in H_h^1, \qquad \forall\, v \in C^0(\overline{\Omega}), \qquad (r_h v)(Q_i) = v(Q_i), \qquad \forall\, Q_i \in \Sigma_h, \tag{4.255}$$

where (cf. Sec. 4.5.2) Σ_h is the set of the vertices of $\mathcal{T}_h$; actually r_h coincides with the interpolation operator π_h defined by (4.219). Since $\{\mathcal{T}_h\}_h$ is a regular family of triangulations of Ω, it follows from Theorem 4.4 of Sec. 4.5.3.5 (we can also use Theorem 4.1 of Sec. 4.5.3.4) that

$$\lim_{h \to 0} \|r_h v - v\|_{H^1(\Omega)} = 0, \qquad \forall\, v \in \mathscr{D}(\overline{\Omega}),$$

which completes the proof of the proposition. □

The convergence result (4.254) still holds if

$$a_0 \in L^\infty(\Omega), \quad a_0 \geq 0, \quad \int_\Omega a_0(x)\, dx > 0$$

(resp., if

$$a_0 = 0, \quad k \in L^\infty(\Gamma), \quad k \geq 0, \quad \int_\Gamma k(x)\, d\Gamma > 0),$$

the other hypotheses on $\overline{\overline{\mathbf{A}}}, k, f, g$ (resp., $\overline{\overline{\mathbf{A}}}, f, g$) remaining the same.

4.5.4.4. *Formulation of the approximate Fourier problem* (4.253) *as a linear system. Comments on numerical integration.* We retain the notation of Sec. 4.5.2. Let $\mathscr{B}_h$ be the basis of H_h^1 defined by (4.197), (4.198); we then have

$$u_h = \sum_{j=1}^{N_h} u_h(Q_j) w_j,$$

where $\{Q_j\}_{j=1}^{N_h} = \Sigma_h$. Setting $u_j = u_h(Q_j)$, the approximate Fourier problem (4.253) is then equivalent to the following linear system:

$$\sum_{j=1}^{N_h} \left[\int_\Omega (\bar{\bar{\mathbf{A}}} \nabla w_j) \cdot \nabla w_i \, dx + \int_\Omega a_0 w_j w_i \, dx + \int_\Gamma k w_j w_i \, d\Gamma \right] u_j$$

$$= \int_\Omega f w_i \, dx + \int_\Gamma g w_i \, d\Gamma, \qquad 1 \leq i \leq N_h, \tag{4.256}$$

whose unknowns are $u_j, j = 1, \ldots, N_h$.

From the properties of $\bar{\bar{\mathbf{A}}}$, a_0, k, the matrix of the above system is positive definite; it is symmetric if $\bar{\bar{\mathbf{A}}}(x)$ is symmetric a.e. on Ω. From these properties, system (4.256) has a unique solution. The solution of linear systems like (4.256) will be considered in Sec. 4.5.7. What we would like to do in this section is to discuss the computational work required to obtain the above system (namely the associated matrix and the right-hand-side vector). For $1 \leq i, j \leq N_h$, define α_{ij} and β_i by

$$\alpha_{ij} = \int_\Omega [(\bar{\bar{\mathbf{A}}} \nabla w_j) \cdot \nabla w_i + a_0 w_j w_i] \, dx + \int_\Gamma k w_j w_i \, d\Gamma, \tag{4.257}$$

$$\beta_i = \int_\Omega f w_i \, dx + \int_\Gamma g w_i \, d\Gamma, \tag{4.258}$$

respectively. Then the linear system (4.256) can be written as follows:

$$\mathbf{A}_h \mathbf{U}_h = \boldsymbol{\beta}_h, \tag{4.259}$$

where $\mathbf{A}_h = (\alpha_{ij})_{1 \leq i, j \leq N_h}$, $\mathbf{U}_h = \{u_i\}_{i=1}^{N_h}$, $\boldsymbol{\beta}_h = \{\beta_i\}_{i=1}^{N_h}$. To obtain $\mathbf{A}_h$ and $\boldsymbol{\beta}_h$, we clearly have to compute a large number of integrals. Define Ω_i and Γ_i (see Fig. 4.3) by

$$\Omega_i = \text{support } (w_i), \qquad \Gamma_i = \partial\Omega_i;$$

the sets Ω_i and Γ_i are small if N_h is large (i.e., if h is small).

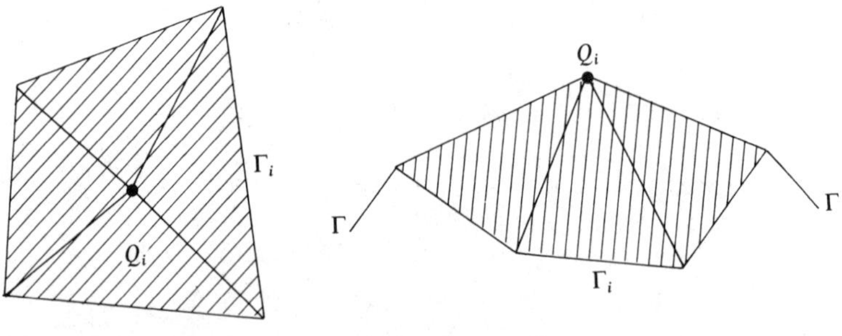

Figure 4.3. The sets Ω_i and Γ_i. (a) Q_i is not a boundary node, (b) Q_i is a boundary node.

An important simplification follows from the fact that in (4.257) (resp., (4.258)) we have

$$\int_{\Omega} = \int_{\Omega_i \cap \Omega_j}, \qquad \int_{\Gamma} = \int_{\Gamma \cap \Gamma_i \cap \Gamma_j}$$

(resp.

$$\int_{\Omega} = \int_{\Omega_i}, \qquad \int_{\Gamma} = \int_{\Gamma \cap \Gamma_i}\Bigg);$$

hence the above integrals have to be evaluated on a small number (which is zero most of the time) of triangles and edges of $\mathscr{T}_h$. The resulting matrix $\mathbf{A}_h$ is a sparse one. Since, in practice, the exact calculation of α_{ij} and β_i is generally impossible, numerical integration has to be used to approximate the above quantities. For simplicity we suppose that $\bar{\bar{\mathbf{A}}}$ and f (resp., k, g) are continuous over $\bar{\Omega}$ (resp., Γ) (the generalization to piecewise continuous functions is straightforward if we suppose that their discontinuities are supported by edges of $\mathscr{T}_h$ (resp., vertices of $\mathscr{T}_h$ belonging to Γ)).

Let T be a triangle of $\mathscr{T}_h$ contained in $\bar{\Omega}_i \cap \bar{\Omega}_j$ (resp., $\bar{\Omega}_i$); the calculation of α_{ij} and β_i requires the evaluation of

$$\int_T [(\bar{\bar{\mathbf{A}}}\nabla w_j) \cdot \nabla w_i + a_0 w_j w_i] \, dx \tag{4.260$_1$}$$

and

$$\int_T f w_i \, dx, \tag{4.260$_2$}$$

respectively. Denote the three vertices of T and its centroid by Q_{1T}, Q_{2T}, and Q_{3T}, and G_T, respectively. Remembering that ∇w_i, ∇w_j are constant over T and concentrating on simple numerical integration formulas, we may approximate the integral in $(4.260)_1$ by either

$$\text{meas}\,(T)\bar{\bar{\mathbf{A}}}(G_T)\nabla(w_j|_T) \cdot \nabla(w_i|_T) + \frac{1}{3}\,\text{meas}(T)\sum_{k=1}^{3} a_0(Q_{kT})w_j(Q_{kT})w_i(Q_{kT}), \tag{4.261$_1$}$$

or

$$\frac{1}{3}\,\text{meas}(T)\sum_{k=1}^{3}\{\bar{\bar{\mathbf{A}}}(Q_{kT})\nabla(w_j|_T) \cdot \nabla(w_i|_T) + a_0(Q_{kT})w_j(Q_{kT})w_i(Q_{kT})\}; \tag{4.261$_2$}$$

formula $(4.261)_2$ is obtained through the application of the two-dimensional trapezoidal rule to $(4.260)_1$. Similarly, we should approximate the integral in $(4.260)_2$ by

$$\frac{1}{3}\,\text{meas}(T)\sum_{k=1}^{3} f(Q_{kT})w_i(Q_{kT}); \tag{4.262}$$

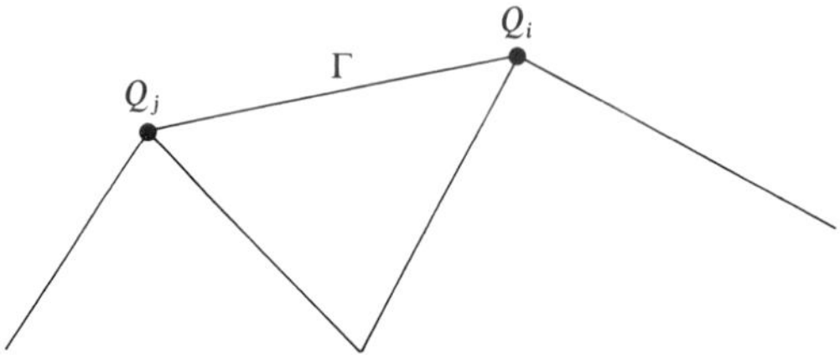

Figure 4.4

From (4.262) and from the properties of the w_i it then follows that $\int_\Omega f w_i \, dx$ will be approximated by

$$\frac{1}{3} \sum_{i=1}^{N_h} \omega_i f(Q_i),$$

where $\omega_i = \text{meas}(\Omega_i)$ and $\{Q_i\}_{i=1}^{N_h} = \Sigma_h$.

We now consider the calculation of the boundary integrals occurring in (4.257), (4.258); we have to evaluate integrals like (see Fig. 4.4)

$$\int_{Q_iQ_j} k w_i w_j \, d\Gamma, \quad \int_{Q_iQ_j} k w_i^2 \, d\Gamma, \quad \int_{Q_iQ_j} g w_i \, d\Gamma. \tag{4.263}$$

Again using the trapezoidal rule and remembering that $w_i(Q_i) = 1$, $w_i(Q_j) = 0$ if $i \neq j$, for the above integrals we obtain the following approximations:

$$0, \quad \tfrac{1}{2}|\overrightarrow{Q_iQ_j}|k(Q_i), \quad \tfrac{1}{2}|\overrightarrow{Q_iQ_j}|g(Q_i),$$

respectively. Using Simpson's rule, more elaborate approximations can be obtained for the integrals in (4.263).

EXERCISE 4.6. Prove that the discrete Fourier problem, obtained using numerical integration, is still well posed. Also prove the convergence of the approximate solutions.

For more details on numerical integration and its effect on the convergence of finite element methods, see Strang and Fix [1], Ciarlet [1]–[3], Oden and Reddy [1], and Bernadou and Boisserie [1].

Remark 4.22. Relations (2.261) require a knowledge of the first-order derivatives of the w_i. If T is a triangle whose (positively oriented) vertices are denoted by A, B, C, with coordinates $\{a_1, a_2\}$, $\{b_1, b_2\}$, $\{c_1, c_2\}$, respectively, and if v is a polynomial of P_1 such that

$$v(A) = v_A, \quad v(B) = v_B, \quad v(C) = v_C,$$

we indeed have

$$\frac{\partial v}{\partial x_1} = \frac{1}{2m(T)}\{v_A(b_2 - c_2) + v_B(c_2 - a_2) + v_C(a_2 - b_2)\},$$

$$\frac{\partial v}{\partial x_2} = -\frac{1}{2m(T)}\{v_A(b_1 - c_1) + v_B(c_1 - a_1) + v_C(a_1 - b_1)\}, \tag{4.264}$$

where

$$\begin{aligned} m(T) = \text{meas}(T) &= 0.5|\overrightarrow{CA} \times \overrightarrow{CB}| \\ &= 0.5|(a_1 - c_1)(b_2 - c_2) - (b_1 - c_1)(a_2 - c_2)|. \end{aligned} \tag{4.265}$$

EXERCISE 4.7. Prove (4.264), (4.265).

4.5.5. *Application to the finite element solution of the Dirichlet and mixed Neumann–Dirichlet problems*

With Ω as in Sec. 4.5.4, we now consider the finite element solution of Dirichlet and mixed Neumann–Dirichlet problems. Since (see Sec. 4.3.4 for more details) the Dirichlet problem appears as a particular case of the other problem, we shall consider the second one only (actually the extension to the mixed Fourier–Dirichlet problem is straightforward).

4.5.5.1. *Formulation of the continuous and discrete Neumann–Dirichlet problems.* We suppose that Γ $(=\partial\Omega)$ is the union of Γ_0, Γ_1 such that $\Gamma_0 \cup \Gamma_1 = \Gamma$, $\Gamma_0 \cap \Gamma_1 = \emptyset$; the continuous problem is then defined by

$$-\nabla \cdot (\overline{\overline{\mathbf{A}}}\nabla u) + a_0 u = f \text{ in } \Omega, \qquad u = g_0 \text{ on } \Gamma_0, \qquad (\overline{\overline{\mathbf{A}}}\nabla u) \cdot \mathbf{n} = g_1 \text{ on } \Gamma. \tag{4.266}$$

With V_0 defined by

$$V_0 = \{v \mid v \in H^1(\Omega), \quad v = 0 \text{ a.e. on } \Gamma_0\},$$

for (4.266) we have the following variational formulation:

Find $u \in H^1(\Omega)$ such that $u = g_0$ on Γ_0 and

$$\int_\Omega (\overline{\overline{\mathbf{A}}}\nabla u) \cdot \nabla v \, dx + \int_\Omega a_0 uv \, dx = \int_\Omega fv \, dx + \int_{\Gamma_1} g_1 v \, d\Gamma, \qquad \forall\, v \in V_0. \tag{4.267}$$

We suppose that the hypotheses on $\overline{\overline{\mathbf{A}}}$, a_0, f, g_0, g_1 are those made in Sec. 4.3.4; then problem (4.267) has a unique solution. To approximate (4.266), (4.267), we consider a family $\{\mathcal{T}_h\}_h$ of triangulations of Ω—as in Sec. 4.5.4.2—and we suppose that the following property holds:

$$\text{the points at the interface of } \Gamma_0 \text{ and } \Gamma_1 \text{ are vertices of } \mathcal{T}_h. \tag{4.268}$$

From the above property, the following subspace of H_h^1, defined by

$$V_{0h} = \{v_h \mid v_h \in H_h^1, \quad v_h = 0 \text{ on } \Gamma_0\}, \tag{4.269}$$

is also a subspace of V_0. Eventually we suppose that g_0 is continuous over $\bar{\Gamma}_0$ ($\bar{\Gamma}_0$: closure of Γ_0 in Γ). As the approximate problem we consider:

Find $u_h \in H_h^1$, $u_h(Q) = g_0(Q)$, $\forall\, Q \in \bar{\Gamma}_0 \cap \Sigma_h$, *such that*

$$\int_\Omega (\bar{\bar{\mathbf{A}}}\nabla u_h)\cdot\nabla v_h\,dx + \int_\Omega a_0 u_h v_h\,dx = \int_\Omega f v_h\,dx + \int_{\Gamma_1} g_1 v_h\,d\Gamma, \qquad \forall\, v_h \in V_{0h}. \tag{4.270}$$

If the above hypotheses hold, the approximate problem (4.270) has a unique solution.

4.5.5.2. *Convergence of the approximate solutions.* Concerning the convergence of $\{u_h\}_h$ to the solution of (4.270), we have:

Proposition 4.20. *Suppose that* $\bar{\bar{\mathbf{A}}}$, a_0, f, g_1 *satisfy the hypotheses made in* Sec. 4.3.4 *and that* $\{\mathcal{T}_h\}_h$ *is a regular family of triangulations of* Ω *(in the sense of* Sec. 4.5.3.4) *obeying* (4.268). *Also suppose (for simplicity) that* $g_0 = \tilde{g}_{0|\Gamma_0}$, *where* $\tilde{g}_0$ *is Lipschitz continuous over* $\bar{\Omega}$. *We then have*

$$\lim_{h\to 0} \|u_h - u\|_{H^1(\Omega)} = 0, \tag{4.271}$$

where u_h *(resp.,* u*) is the solution of* (4.270) *(resp.,* (4.266), (4.267)*).*

PROOF. We sketch the proof in the case where

$$g_0 = 0 \quad \text{on } \Gamma_0, \tag{4.272}$$

implying that one can take $\tilde{g}_0 = 0$ over $\bar{\Omega}$. If (4.272) holds, problems (4.267), (4.270) reduce to:

Find $u \in V_0$, *such that*

$$\int_\Omega (\bar{\bar{\mathbf{A}}}\nabla u)\cdot\nabla v\,dx + \int_\Omega a_0 uv\,dx = \int_\Omega fv\,dx + \int_{\Gamma_1} g_1 v\,d\Gamma, \qquad \forall\, v \in V_0; \tag{4.273}$$

Find $u_h \in V_{0h}$, *such that*

$$\int_\Omega (\bar{\bar{\mathbf{A}}}\nabla u_h)\cdot\nabla v_h\,dx + \int_\Omega a_0 u_h v_h\,dx = \int_\Omega f v_h\,dx + \int_{\Gamma_1} g_1 v_h\,d\Gamma, \qquad \forall\, v_h \in V_{0h}, \tag{4.274}$$

respectively.

To apply Theorem 3.2 of Sec. 3.3 (with V_h replaced by V_{0h}), it suffices to find $\mathcal{V}$ and r_h which obey (3.1). Define $\mathcal{V}$ by

$$\mathcal{V} = \{v \mid v \in C^\infty(\bar{\Omega}),\quad v = 0 \text{ in the neighborhood of } \Gamma_0\}.$$

Since Γ is polygonal and (from (4.268)) Γ_0 and Γ_1 are "nice" subsets of Γ, we have (from, e.g., Necas [1])

$$\overline{\mathcal{V}}^{H^1(\Omega)} = V_0.$$

If $v \in \mathscr{V}$ and if r_h is still defined by (4.255), we have—since (4.268) holds—$r_h v \in V_{0h}$; on the other hand, we still have (see the proof of Proposition 4.19 for more details)

$$\lim_{h \to 0} \|r_h v - v\|_{H^1(\Omega)} = 0, \qquad \forall\, v \in \mathscr{V},$$

which completes the proof of the proposition. □

EXERCISE 4.8. Prove Proposition 4.20 if $g_0 \neq 0$, with $g_0 = \tilde{g}_0|_{\Gamma_0}$, $\tilde{g}_0$ being Lipschitz continuous over $\overline{\Omega}$.

4.5.5.3. *Formulation of* (4.270) *as a linear system.* We retain the notation of Sec. 4.5.2.[28] Let $\mathscr{B}_h$ be the basis of H_h^1 defined by (4.197), (4.198); we then have

$$u_h = \sum_{j=1}^{N_{0h}} u_h(Q_j) w_j + \sum_{j=N_{0h}+1}^{N_h} g_0(Q_j) w_j, \tag{4.275}$$

where the vertices of $\mathscr{T}_h$ belonging to Γ_0 have been ordered from $N_{0h}+1$ to N_h. Therefore it is clear that

$$\mathscr{B}_{0h} = \{w_j\}_{j=1}^{N_{0h}} \text{ is a basis of } V_{0h}. \tag{4.276}$$

Again setting $u_h(Q_j) = u_j$, it follows from (4.275), (4.276) that the approximate problem (4.270) is equivalent to the linear system

$$\sum_{j=1}^{N_{0h}} \left[\int_\Omega \overline{\overline{\mathbf{A}}} \nabla w_j \cdot \nabla w_i \, dx + \int_\Omega a_0 w_j w_i \, dx \right] u_j$$
$$= \int_\Omega f w_i \, dx + \int_{\Gamma_1} g_1 w_i \, d\Gamma - \sum_{j=N_{0h}+1}^{N_h} \left[\int_\Omega \overline{\overline{\mathbf{A}}} \nabla w_j \cdot \nabla w_i \, dx + \int_\Omega a_0 w_j w_i \, dx \right]$$
$$\times\, g_0(Q_j), \qquad 1 \leq i \leq N_{0h}. \tag{4.277}$$

From the properties of $\overline{\overline{\mathbf{A}}}$, a_0, the matrix in (4.277) is positive definite (it is symmetric if $\overline{\overline{\mathbf{A}}}(x)$ is symmetric a.e. on Ω); from these properties, (4.277) therefore has a unique solution. All the comments and observations, made in Sec. 4.5.4.4 about the necessity of using numerical integration to obtain a computationable approximation of (4.277) still hold.

4.5.6. *Some comments on the relationships between finite element and finite difference methods*

It is well known (by the specialists, at least) that most finite difference methods can be viewed as particular case of finite element methods (and vice versa).

To justify this claim, we consider the following Dirichlet problem:

$$-\Delta u = f \text{ in } \Omega, \qquad u = g \text{ on } \Gamma, \tag{4.278}$$

---

[28] Except for N_{0h}, which is now defined by $N_{0h} = \dim V_{0h} =$ number of vertices of $\mathscr{T}_h$ which do not belong to $\overline{\Gamma}_0$.

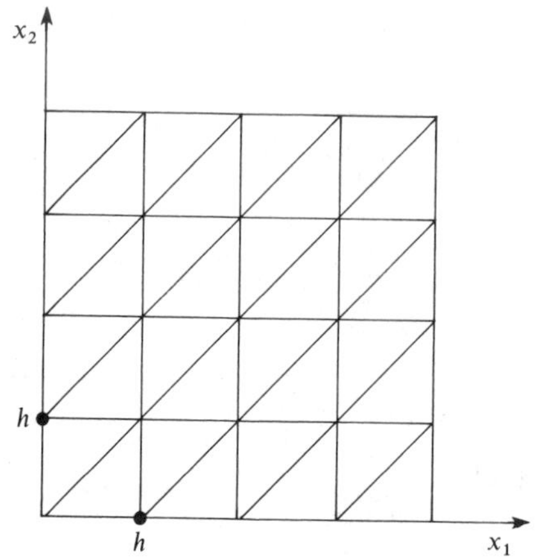

Figure 4.5

where $\Omega =]0, 1[\times]0, 1[$, $\Gamma = \partial\Omega$, where $f \in C^0(\bar{\Omega})$, and $g \in C^0(\Gamma)$ with $g = \tilde{g}|_\Gamma$, where $\tilde{g} \in H^1(\Omega)$. If the above hypotheses on f and g hold, then the Dirichlet problem (4.278) has a unique solution in $H^1(\Omega)$.

To approximate (4.278), we consider a positive integer N and define h by $h = 1/(N + 1)$. The triangulation $\mathscr{T}_h$ used to approximate (4.278) is as in Fig. 4.5 (where $N = 3$).

The vertices of $\mathscr{T}_h$ are the points Q_{ij} defined by

$$Q_{ij} = \{ih, jh\}, \qquad 0 \leq i, j \leq N + 1.$$

Applying the results of Sec. 4.5.5, we obtain (with the obvious notation) the following approximate problem:

$$\sum_{1 \leq k, l \leq N} \left(\int_\Omega \nabla w_{ij} \cdot \nabla w_{kl} \, dx \right) u_{kl} = \int_\Omega f w_{ij} \, dx$$

$$- \sum_{Q_{kl} \in \Gamma} \left(\int_\Omega \nabla w_{ij} \cdot \nabla w_{kl} \, dx \right) g(Q_{kl}), \qquad 1 \leq i, \quad j \leq N, \qquad (4.279)$$

where w_{kl} is the basis function associated with Q_{kl} by (4.197), (4.198).

Using the trapezoidal rule to evaluate $\int_\Omega f w_{ij} \, dx$ (see Sec. 4.5.4.4), we find (after simplification by h^2) that (4.279) reduces to (with $f_{ij} = f(Q_{ij})$)

$$-\frac{u_{i+1j} + u_{i-1j} + u_{ij+1} + u_{ij-1} - 4u_{ij}}{h^2} = f_{ij},$$

$$\text{for } 1 \leq i, j \leq N, \quad \text{with } u_{kl} = g(Q_{kl}) \quad \text{if } Q_{kl} \in \Gamma. \qquad (4.280)$$

Hence, the finite element approximate Dirichlet problem (4.279) reduces to the usual five-point finite difference scheme to approximate the Dirichlet problem (4.278). A similar conclusion holds for more complicated problems than (4.278) (a numerical integration procedure is required in most cases).

EXERCISE 4.9. Prove that (4.279) reduces to (4.280).

4.5.7. *Solution of the linear systems equivalent to the approximate problems*

4.5.7.1. *Synopsis.* In the previous sections we have discussed the finite element approximation of a fairly broad family of elliptic boundary-value problems.

It has been shown that the discrete problem is in fact equivalent to a linear system whose matrix is positive definite (possibly symmetric) and sparse. Actually the solution of the sparse linear systems, resulting from the finite element discretization of elliptic boundary-value problems by finite element methods, has inspired a large number of papers, and it is impossible to mention all of them. Some references have been given in Chapter VII, Sec. 5.8.7.2, and many other significant references can be found in Duff [2] and George [1] (see also Golub and Meurant [1], Mercier [1], Mercier and Pironneau [1], and Thomasset [1]).

In Secs. 4.5.7.2, 4.5.7.3, and 4.5.7.4 we shall consider the solution of the linear system

$$\mathbf{A}\mathbf{x} = \mathbf{b} \tag{4.281}$$

by Cholesky, over-relaxation, and conjugate gradient methods, respectively; we suppose that $\mathbf{A}$ is a symmetric and positive-definite real matrix.

4.5.7.2. *Solution of the linear system* (4.281) *by the method of Cholesky.* We present the standard Cholesky method only (for further comments, see the above references and also Ciarlet [4]).

Since $\mathbf{A} = (a_{ij})_{1 \le i, j \le N}$ is symmetric and positive definite, there exists a unique lower triangular matrix $\mathbf{L} = (l_{ij})_{1 \le i, j \le N}$ such that

$$l_{ii} > 0, \qquad \forall\, i = 1, \ldots, N, \qquad \mathbf{A} = \mathbf{L}\mathbf{L}^t. \tag{4.282}$$

The elements l_{ij} of the matrix $\mathbf{L}$ are given by

$$l_{ij} = 0 \qquad \text{if } j > i, \tag*{(4.283)$_1$}$$

$$l_{11} = \sqrt{a_{11}}, \quad l_{i1} = \frac{a_{i1}}{l_{11}} \qquad \text{if } i = 1, \ldots, N, \tag*{(4.283)$_2$}$$

if $j \ge 2$, then

$$l_{jj} = \sqrt{a_{jj} - \sum_{k=1}^{j-1} l_{jk}^2}, \quad l_{ij} = \frac{a_{ij} - \sum_{k=1}^{j-1} l_{ik} l_{jk}}{l_{jj}} \qquad \text{if } i > j. \tag*{(4.283)$_3$}$$

Once $\mathbf{L}$ has been obtained, solving (4.281) is trivial since it is equivalent to

$$\mathbf{L}\mathbf{y} = \mathbf{b}, \qquad \mathbf{L}^t\mathbf{x} = \mathbf{y}, \tag{4.284}$$

and both systems in (4.284) are easily solved since **L** and $\mathbf{L}^t$ are triangular matrices.

An important property of the above Cholesky factorization is that it preserves the band structure of **A**; more precisely, if **A** is a $2p + 1$-diagonal matrix (i.e., $a_{ij} = 0$ if $|i - j| > p$), then a similar property holds for **L**, since—from formulas (4.283)—we have $l_{ij} = 0$ if $i - j > p$. However, if **A** is a sparse matrix, **L** is usually less sparse than **A**.

EXERCISE 4.10. Prove the existence of **L** obeying (4.283).

4.5.7.3. *Solution of the linear system* (4.281) *by over-relaxation.* Since Chapter V is concerned with relaxation methods, in this section we shall only give some brief comments on their applications to the solution of linear systems like (4.281).

With ω a positive parameter, we consider the following iterative method (where $\mathbf{b} = \{b_i\}_{i=1}^N$, $\mathbf{x}^n = \{x_i^n\}_{i=1}^N$):

$$\mathbf{x}^0 = \{x_1^0, \ldots, x_N^0\} \text{ given arbitrarily;} \tag{4.285}$$

then for $n \geq 0$, $\mathbf{x}^n$ *being known, we compute* $\mathbf{x}^{n+1}$ *component by component, by*

$$x_i^{n+1} = (1 - \omega)x_i^n + \frac{\omega}{a_{ii}}\left(b_i - \sum_{j<i} a_{ij}x_j^{n+1} - \sum_{j>i} a_{ij}x_j^n\right), \qquad (4.286)_i$$

for $i = 1, \ldots, N$.

If $\omega > 1$ (resp., $\omega = 1$, $\omega < 1$), algorithm (4.285), (4.286) is an over-relaxation (resp., relaxation, under-relaxation) method.

From Chapter V, Sec. 5.4 it follows that algorithm (4.285), (4.286) converges, $\forall\, \mathbf{x}^0 \in \mathbb{R}^N$, to $\mathbf{x} = \mathbf{A}^{-1}\mathbf{b}$ if and only if $0 < \omega < 2$.

For an analysis of the speed of convergence see, e.g., Varga [1], Young [1], Isaacson and Keller [1], and Ciarlet [4].

The choice of ω in algorithm (4.285), (4.286) may be critical, and in general the optimal value of this parameter is unknown. However, there exists a method, due to D. Young, which allows the automatic adjustment of this parameter; theoretically this method, discussed in Varga [1] and Young [1], applies to a class of matrices less general than symmetric positive-definite matrices; however, we observed that it behaved quite well on some problems for which the hypotheses stated in the above references were not fullfilled.

4.5.7.4. *Solution of the linear system* (4.281) *by a conjugate gradient method.* The solution of linear and nonlinear problems by conjugate gradient methods has been discussed in several places in this book, and several references concerning these methods have been given. Concentrating on the solution of the linear system (4.281), a standard algorithm is as follows (with $\mathbf{S}$ a symmetric positive-definite matrix):

$$\mathbf{x}^0 \in \mathbb{R}^N \textit{ arbitrarily given,} \tag{4.287}$$

$$\mathbf{r}^0 = \mathbf{A}\mathbf{x}^0 - \mathbf{b}, \tag{4.288}$$

$$\mathbf{g}^0 = \mathbf{S}^{-1}\mathbf{r}^0, \tag{4.289}$$

$$\mathbf{w}^0 = \mathbf{g}^0. \tag{4.290}$$

Then, for $n \geq 0$, *assuming that* $\mathbf{x}^n, \mathbf{r}^n, \mathbf{g}^n, \mathbf{w}^n$ *are known and that* $\mathbf{r}^n \neq \mathbf{0}$, *we obtain* $\mathbf{x}^{n+1}, \mathbf{r}^{n+1}, \mathbf{g}^{n+1}, \mathbf{w}^{n+1}$ *by*

$$\rho_n = \frac{(\mathbf{r}^n, \mathbf{g}^n)}{(\mathbf{A}\mathbf{w}^n, \mathbf{w}^n)}, \tag{4.291}$$

$$\mathbf{x}^{n+1} = \mathbf{x}^n - \rho_n \mathbf{w}^n, \tag{4.292}$$

$$\mathbf{r}^{n+1} = \mathbf{r}^n - \rho_n \mathbf{A}\mathbf{w}^n. \tag{4.293}$$

If $\mathbf{r}^{n+1} = \mathbf{0}$, *then* $\mathbf{x}^{n+1} = \mathbf{x} = \mathbf{A}^{-1}\mathbf{b}$; *if* $\mathbf{r}^{n+1} \neq \mathbf{0}$, *then compute*

$$\mathbf{g}^{n+1} = \mathbf{S}^{-1}\mathbf{r}^{n+1}, \tag{4.294}$$

$$\gamma_{n+1} = \frac{(\mathbf{r}^{n+1}, \mathbf{g}^{n+1})}{(\mathbf{r}^n, \mathbf{g}^n)}, \tag{4.295}$$

$$\mathbf{w}^{n+1} = \mathbf{g}^{n+1} + \gamma_{n+1}\mathbf{w}^n. \tag{4.296}$$

Then $n = n + 1$ *and go to* (4.291).

In (4.287)–(4.296), $(\cdot, \cdot)$ denotes the usual scalar product of $\mathbb{R}^N$ (i.e., $(\mathbf{x}, \mathbf{y}) = \sum_{i=1}^N x_i y_i$ if $\mathbf{x} = \{x_i\}_{i=1}^N$, $\mathbf{y} = \{y_i\}_{i=1}^N$).

If roundoff errors are neglected, then, $\forall\ \mathbf{x}_0$, there exists $n_0 \leq N$ such that $\mathbf{x}^{n_0} = \mathbf{x} = \mathbf{A}^{-1}\mathbf{b}$.

The above matrix $\mathbf{S}$ is a scaling (or preconditioning) matrix; a proper choice of $\mathbf{S}$ can accelerate the convergence in a very substantial way (clearly, $\mathbf{S}$ has to be easier to handle than $\mathbf{A}$). It is clear that if one uses a Cholesky factorization of $\mathbf{S}$, it will be done once and for all, before running algorithm (4.287)–(5.296).

The solution of linear systems of equations, with A nonsymmetric, by methods of conjugate gradient type is discussed in, e.g., Concus and Golub [1] (see also Duff [2] and George [1]).

4.6. Application to an elliptic boundary-value problem arising from geophysics

4.6.1. *Synopsis: formulation of the problem*

In this section we shall discuss the solution, by the methods of the previous sections, of a linear elliptic boundary-value problem originating from geophysics. More precisely, this problem occurs in the modelling of the interactions between jets of electrons and protons emitted by the Sun and the Earth magnetosphere (see Blanc and Richmond [1] for more details). After an appropriate change or coordinates, the above problem takes the following formulation (where u is an *electrical potential* and $\Omega =]0, 1[\times]0, 1[$):

Find a function u satisfying

$$-\nabla \cdot (\bar{\bar{\mathbf{A}}}\nabla u) = f \text{ in } \Omega, \tag{4.297$_1$}$$

$$u(x_1, 1) = g_0(x_1) \quad \text{if } 0 < x_1 < 1, \tag{4.297$_2$}$$

$$u(0, x_2) = u(1, x_2) \quad \text{if } 0 < x_2 < 1, \tag{4.297$_3$}$$

$$(\bar{\bar{\mathbf{A}}}\nabla u \cdot \mathbf{n})(0, x_2) + (\bar{\bar{\mathbf{A}}}\nabla u \cdot \mathbf{n})(1, x_2) = 0 \quad \text{if } 0 < x_2 < 1, \tag{4.297$_4$}$$

$$(\bar{\bar{\mathbf{A}}}\nabla u \cdot \mathbf{n})(x_1, 0) - \frac{\partial}{\partial x_1}\left(k(x_1)\frac{\partial u}{\partial x_1}(x_1, 0)\right) = g_1(x_1) \quad \text{if } 0 < x_1 < 1, \tag{4.297$_5$}$$

$$u(0, 0) = u(1, 0); \tag{4.297$_6$}$$

in the above relations, $\bar{\bar{\mathbf{A}}}$, f and g_0, g_1, k are given functions, defined over Ω and Γ, respectively. Actually, if $\bar{\bar{\mathbf{A}}} = (a_{ij})_{1 \le i, j \le 2}$, we have, for $(4.297)_4$ and $(4.297)_5$, the following more explicit formulations:

$$\left(a_{11}\frac{\partial u}{\partial x_1} + a_{12}\frac{\partial u}{\partial x_2}\right)(0, x_2) = \left(a_{11}\frac{\partial u}{\partial x_1} + a_{12}\frac{\partial u}{\partial x_2}\right)(1, x_2) \quad \text{if } 0 < x_2 < 1, \tag{4.298}$$

$$-\left(a_{21}\frac{\partial u}{\partial x_1} + a_{22}\frac{\partial u}{\partial x_2}\right)(x_1, 0) - \frac{\partial}{\partial x_1}\left(k(x_1)\frac{\partial u}{\partial x_1}(x_1, 0)\right) = g_1(x_1) \quad \text{if } 0 < x_1 < 1. \tag{4.299}$$

The boundary conditions in problem (4.297) may seem rather complicated; actually, using an appropriate variational formulation, it will be seen in the following sections that problem (4.297) is almost as easy to solve as the Dirichlet, Neumann, and Fourier problems discussed earlier.

4.6.2. *Variational formulation of problem* (4.297)

Let v be a smooth function defined over $\bar{\Omega}$ (we may suppose that $v \in C^\infty(\bar{\Omega})$, for example). Multiplying $(4.297)_1$ by v and applying the Green–Ostrogradsky formula (4.33), we obtain

$$\int_\Omega (\bar{\bar{\mathbf{A}}}\nabla u) \cdot \nabla v \, dx = \int_\Omega fv \, dx + \int_\Gamma (\bar{\bar{\mathbf{A}}}\nabla u) \cdot \mathbf{n} v \, d\Gamma. \tag{4.300}$$

Now suppose that

$$v(x_1, 1) = 0 \quad \text{if } 0 < x_1 < 1, \qquad v(0, x_2) = v(1, x_2) \quad \text{if } 0 < x_2 < 1; \tag{4.301}$$

it follows from (4.301) and $(4.297)_4$ that the boundary integral in (4.300) reduces to

$$\int_{\Gamma_1} (\bar{\bar{\mathbf{A}}}\nabla u) \cdot \mathbf{n} v \, d\Gamma,$$

where $\Gamma_1 = \{x \mid x = \{x_1, x_2\},\ 0 < x_1 < 1,\ x_2 = 0\}$, implying, in turn, that (4.300) reduces to

$$\int_\Omega (\bar{\bar{\mathbf{A}}}\nabla u) \cdot \nabla v \, dx = \int_\Omega fv \, dx + \int_{\Gamma_1} (\bar{\bar{\mathbf{A}}}\nabla u) \cdot \mathbf{n} v \, d\Gamma. \tag{4.302}$$

Combining $(4.297)_5$ and (4.302), and using the second relation (4.301), we obtain (after integrating by parts over Γ_1)

$$\int_\Omega (\bar{\bar{\mathbf{A}}}\nabla u) \cdot \nabla v \, dx + \int_0^1 k(x_1) \frac{\partial u}{\partial x_1}(x_1, 0) \frac{\partial v}{\partial x_1}(x_1, 0) \, dx_1$$
$$= \int_\Omega fv \, dx + \int_{\Gamma_1} g_1 v \, d\Gamma. \tag{4.303}$$

Conversely, it can be proved that if (4.303) holds for every $v \in \mathscr{V}$, where

$$\begin{aligned} \mathscr{V} = \{v \mid v &\in C^\infty(\bar{\Omega}),\ v(0, x_2) = v(1, x_2) \text{ if } 0 \le x_2 \le 1, \\ v &= 0 \text{ in the neighborhood of } \bar{\Gamma}_0\} \end{aligned} \tag{4.304}$$

where $\Gamma_0 = \{x \mid x \in \{x_1, x_2\},\ 0 < x_1 < 1,\ x_2 = 1\}$, then u is a solution of the boundary-value problem (4.297). Relations (4.301), (4.303) suggest the introduction of the following subspaces of $H^1(\Omega)$:

$$\begin{aligned} V &= \{v \mid v \in H^1(\Omega),\ v(0, x_2) \\ &= v(1, x_2) \text{ a.e. } 0 < x_2 < 1,\ (d/dx_1)v(x_1, 0) \in L^2(0, 1)\}, \end{aligned} \tag{4.305}$$

$$V_0 = \{v \mid v \in V,\ v(x_1, 1) = 0 \text{ a.e. } 0 < x_1 < 1\}. \tag{4.306}$$

Suppose that V is endowed with the scalar product

$$(v, w)_V = (v, w)_{H^1(\Omega)} + \int_0^1 \frac{d}{dx_1} v(x_1, 0) \frac{d}{dx_1} w(x_1, 0) \, dx_1 \tag{4.307}$$

and the corresponding norm

$$\|v\|_V = (v, v)_V^{1/2}. \tag{4.308}$$

We then have the following:

Proposition 4.21. *The spaces V and V_0 are Hilbert spaces for the scalar product and norm defined by* (4.307) *and* (4.308), *respectively. Moreover, the seminorm*

$$v \to \left(\int_\Omega |\nabla v|^2 \, dx + \int_0^1 \left| \frac{dv}{dx_1}(x_1, 0) \right|^2 dx_1 \right)^{1/2}$$

defines a norm equivalent to the V-norm (4.308) *over* V_0.

EXERCISE 4.11. Prove Proposition 4.21.

We now define a bilinear form $a: V \times V \to \mathbb{R}$ and a linear functional $L: V \to \mathbb{R}$ by

$$a(v, w) = \int_{\Omega} (\bar{\bar{\mathbf{A}}}\nabla v) \cdot \nabla w \, dx + \int_0^1 k(x_1) \frac{d}{dx_1} v(x_1, 0) \frac{d}{dx_1} w(x_1, 0) \, dx_1, \tag{4.309}$$

$$L(v) = \int_{\Omega} fv \, dx + \int_{\Gamma_1} g_1 v \, d\Gamma, \tag{4.310}$$

respectively. We suppose that the following hypotheses concerning $\bar{\bar{\mathbf{A}}}$, k, f, g_1 hold:

$$f \in L^2(\Omega), \qquad g_1 \in L^2(\Gamma_1), \tag{4.311}$$

$$k \in L^\infty(0, 1), \quad k(x_1) \geq \alpha_0 > 0 \text{ a.e. on }]0, 1[, \tag{4.312}$$

$$\bar{\bar{\mathbf{A}}} \text{ satisfies (4.47).} \tag{4.313}$$

From the above hypotheses, we find that $a(\cdot, \cdot)$ is bilinear continuous over $V \times V$ and V_0-elliptic, and that $L(\cdot)$ is linear continuous over V; we can therefore apply Theorem 2.1 of Sec. 2.3 to prove:

Proposition 4.22. *If the above hypotheses on $\bar{\bar{\mathbf{A}}}$, k, f, g_1 hold, and if g_0 in $(4.297)_1$ satisfies*

$$g_0 = \tilde{g}_{0|\Gamma_0} \quad \text{with } \tilde{g}_0 \in V, \tag{4.314}$$

then the linear variational problem:

Find $u \in V$ such that $u|_{\Gamma_0} = g_0$ and

$$a(u, v) = L(v), \qquad \forall\, v \in V_0 \tag{4.315}$$

has a unique solution; this solution is also the unique solution in V of the boundary-value problem (4.297).

PROOF. Define $\bar{u} \in V_0$ by $\bar{u} = u - \tilde{g}_0$; $\bar{u}$—if it exists—is clearly a solution of the following linear variational problem in V_0:

$$\bar{u} \in V_0, \quad a(\bar{u}, v) = L(v) - a(\tilde{g}_0, v). \tag{4.316}$$

Since $v \to L(v) - a(\tilde{g}_0, v)$ is linear and continuous over V_0, and since $a(\cdot, \cdot)$ is V_0-elliptic, it follows from Theorem 2.1 of Sec. 2.3 that (4.316) has a (unique) solution, in turn implying the existence of u solving problem (4.315). The above u is clearly unique, since if u_1 and u_2 are two solutions of (4.315), then $u_2 - u_1 \in V_0$ and also $a(u_2 - u_1, u_2 - u_1) = 0$; the V_0-ellipticity of $a(\cdot, \cdot)$ then implies $u_1 = u_2$. □

Remark 4.23. Suppose that $\bar{\bar{\mathbf{A}}}$ is symmetric; this in turn implies the symmetry of $a(\cdot, \cdot)$. From Proposition 2.1 of Sec. 2.4 it then follows that (4.315) is equivalent to the minimization problem:

Find $u \in V$, $u = g_0$ on Γ_0, such that

$$J(u) \le J(v), \qquad \forall\, v \in V, \quad v = g_0 \quad \text{on } \Gamma_0,$$

where

$$J(v) = \frac{1}{2}\int_\Omega (\overline{\overline{\mathbf{A}}}\nabla v) \cdot \nabla v \, dx + \frac{1}{2}\int_0^1 k(x_1)\left|\frac{d}{dx_1} v(x_1, 0)\right|^2 dx_1$$

$$- \int_\Omega fv \, dx - \int_{\Gamma_1} g_1 v \, d\Gamma.$$

4.6.3. *Finite element approximation of problem* (4.297)

In this section we consider the approximation of the boundary-value problem (4.297) via the variational formulation (4.315). Actually the finite element approximation discussed in the sequel is closely related to the approximations of the Neumann, Dirichlet, and Fourier problems discussed in Secs. 4.5.4 and 4.5.5. To approximate (4.297), (4.315), we consider a family $\{\mathscr{T}_h\}_h$ of triangulations of Ω satisfying the hypotheses (i)–(iv) of Sec. 4.5.2 and also:

(v) If $Q = \{0, x_2\}$ is a vertex of $\mathscr{T}_h$, then $Q' = \{1, x_2\}$ is also a vertex of $\mathscr{T}_h$, and conversely (i.e., $\mathscr{T}_h$ preserves the periodicity of the functions of the space V (cf. (4.305)).

With H_h^1 still defined by (4.197), we approximate the above spaces V and V_0 (cf. (4.305), (4.306)) by

$$V_h = V \cap H_h^1 = \{v_h \,|\, v_h \in H_h^1,\, v_h(0, x_2) = v_h(1, x_2) \text{ if } 0 \le x_2 \le 1\}, \tag{4.317}$$

$$V_{0h} = V_0 \cap H_h^1 = \{v_h \,|\, v_h \in V_h,\, v_h(x_1, 1) = 0 \text{ if } 0 \le x_1 \le 1\}. \tag{4.318}$$

We suppose that g_0 (in $(4.297)_2$, (4.314)) also satisfies $g_0 \in C^0[0, 1]$ and (with $\overline{\Gamma}_0 = \{x \,|\, x = \{x_1, 1\}, 0 \le x_1 \le 1\}$) we approximate the problem (4.297), (4.315) by:

Find $u_h \in V$ such that $u_h(Q) = g_0(Q)$, $\forall\, Q$ vertex of $\mathscr{T}_h$ located on $\overline{\Gamma}_0$ and

$$a(u_h, v_h) = L(v_h), \qquad \forall\, v_h \in V_{0h}, \tag{4.319}$$

where $a(\cdot, \cdot)$ and L are still defined by (4.209), (4.310), respectively.

We should easily prove that the approximate problem (4.319) has a unique solution if (4.311)–(4.313) hold.

Exercise 4.12. Prove that the approximate problem (4.319) has a unique solution if the above hypotheses hold.

The convergence of the approximate solutions follows from:

Proposition 4.23. *Suppose that the above hypotheses on* $\bar{\bar{\mathbf{A}}}$, k, f, g_0, g_1 *and* $\{\mathcal{T}_h\}_h$ *hold. Also suppose that* $g_0 = \tilde{g}_0|_{\Gamma_0}$, *where* $\tilde{g}_0$ *is Lipschitz continuous over* $\bar{\Omega}$. *If* $\{\mathcal{T}_h\}_h$ *is a regular family of triangulations of* Ω *(in the sense of Sec. 4.5.3.4), we then have*

$$\lim_{h\to 0} \|u_h - u\|_V = 0, \tag{4.320}$$

where u_h *(resp.,* u*) is the solution of* (4.319) *(resp.,* (4.297), (4.315)*).*

PROOF. We sketch the proof in the case where $g_0 = 0$ on Γ_0, implying that we can take $\tilde{g}_0 = 0$ over $\bar{\Omega}$. Problems (4.315), (4.319) reduce to:

Find $u \in V_0$ *such that*

$$a(u, v) = L(v), \qquad \forall\, v \in V_0. \tag{4.321}$$

Find $u_h \in V_{0h}$ *such that*

$$a(u_h, v_h) = L(v_h), \qquad \forall\, v_h \in V_{0h}, \tag{4.322}$$

respectively.

To apply Theorem 3.2 of Sec. 3.3 (with V_h replaced by V_{0h}), it suffices to find $\mathscr{V}$ and r_h obeying (3.1). Define $\mathscr{V}$ by (4.304); it has been proved by H. Beresticky and F. Mignot (personal communications) that

$$\bar{\mathscr{V}}^V = V_0.$$

If $v \in \mathscr{V}$ and if r_h is still defined by (4.255), we have—since condition (v) on $\{\mathcal{T}_h\}_h$ holds—$r_h v \in V_{0h}$; on the other hand, we still have (see the proof of Proposition 4.19 for more details)

$$\lim_{h\to 0} \|r_h v - v\|_{H^1(\Omega)} = 0, \qquad \forall\, v \in \mathscr{V}. \tag{4.323}$$

Since

$$\|w\|_V = \left(\|w\|^2_{H^1(\Omega)} + \left\|\frac{d}{dx_1} w(x_1, 0)\right\|^2_{L^2(0,1)}\right)^{1/2},$$

it follows from (4.323) that to complete the proof of the present proposition, it suffices to prove that

$$\lim_{h\to 0} \left\|\frac{d}{dx_1}(r_h v - v)(x_1, 0)\right\|_{L^2(0,1)} = 0, \qquad \forall\, v \in \mathscr{V}. \tag{4.324}$$

Let us denote by $\gamma_1 v$ the function $x_1 \to v(x_1, 0)$; we clearly have

$$\gamma_1(r_h v - v) = s_h \gamma_1 v - \gamma_1 v,$$

where s_h is the interpolation operator defined as follows:

$s_h \phi \in C^0[0, 1], \qquad \forall\, \phi \in C^0[0, 1]$,

$s_h \phi(\xi_1) = \phi(\xi_1)$ for any vertex $\{\xi_1, 0\}$ of $\mathcal{T}_h$ belonging to $\bar{\Gamma}_1$,

$s_h \phi|_{[\xi_1, \xi'_1]} \in P_1$ for any pair $\{\xi_1, 0\}$, $\{\xi'_1, 0\}$ of consecutive vertices of $\mathcal{T}_h$ belonging to $\bar{\Gamma}_1$

(we recall that $\bar{\Gamma}_1 = \{x | x = \{x_1, 0\}, 0 \le x_1 \le 1\}$).

Since $v \in \mathscr{V}$ implies that $\gamma_1 v \in C^\infty[0, 1]$, it follows from standard approximation results that

$$\lim_{h \to 0} \left\| \frac{d}{dx_1}(s_h \gamma_1 v - \gamma_1 v) \right\|_{L^2(0,1)} = 0, \qquad \forall\, v \in \mathscr{V},$$

i.e., (4.324) holds. □

EXERCISE 4.13. Prove Proposition 4.23 if $g_0 \neq 0$.

Remark 4.24. The approximation by variational methods of problems closely related to (4.297) is considered in Aubin [2].

4.6.4. *Some comments on the practical solution of the approximate problem* (4.319)

To solve the approximate problem (4.319), we can use a penalty method similar to the one used in Sec. 4.4.4 to approximate the Dirichlet problem by a Fourier one; the main advantage of this formulation is that the discrete space under consideration is still H_h^1 defined by (4.197). A possible penalty approximation of (4.319) is (with $\varepsilon > 0$):

Find $u_h^\varepsilon \in H_h^1$ *such that*, $\forall\, v_h \in H_h^1$,

$$a(u_h^\varepsilon, v_h) + \frac{1}{\varepsilon}\int_0^1 (u_h^\varepsilon(1, x_2) - u_h^\varepsilon(0, x_2))(v_h(1, x_2) - v_h(0, x_2))\, dx_2$$
$$+ \frac{1}{\varepsilon}\int_0^1 u_h^\varepsilon(x_1, 1) v_h(x_1, 1)\, dx_1 = L(v_h) + \frac{1}{\varepsilon}\int_0^1 g_{0h}(x_1) v_h(x_1, 1)\, dx_1, \tag{4.325}$$

where $g_{0h} \in C^0[0, 1]$ coincides with g_0 at those vertices of $\mathscr{T}_h$ located on $\bar{\Gamma}_0$ $(=\{x \mid x = \{x_1, 1\}, 0 \le x_1 \le 1\})$ and is linear (i.e., belongs to P_1) between two consecutive vertices of $\mathscr{T}_h$ located on $\bar{\Gamma}_0$.

Such an approximation is justified by

$$\lim_{\varepsilon \to 0} u_h^\varepsilon = u_h, \tag{4.326}$$

where u_h^ε and u_h are the solutions of (4.325) and (4.319), respectively.

EXERCISE 4.14. Prove (4.326).

Another possibility is to work directly with the spaces V_h and V_{0h}, taking into account the periodicity conditions $u_h(0, x_2) = u_h(1, x_2)$, $v_h(0, x_2) = v_h(1, x_2)$ and also the fact that $u_h(x_1, 1) = g_{0h}(x_1)$, $v_h(x_1, 1) = 0$, $\forall\, v_h \in V_{0h}$. This second approach will require an explicit knowledge of vector bases for

V_h and V_{0h}; obtaining such bases from the basis of H_h^1 defined by (4.198) (see Sec. 4.5.2) is not very difficult and is left to the reader as an exercise.

We should again use numerical integration to compute the matrices and right-hand sides of the linear systems equivalent to the approximate problems (4.319) and (4.325).

We shall conclude by mentioning that the methods discussed in Sec. 4.5.7 still apply to the solution of the above linear systems.

4.7. On some problems of the mechanics of continuous media and their variational formulations

4.7.1. *Synopsis*

In this section, which very closely follows Mercier [1, Chapter 2] and Ciarlet [2], we briefly discuss some important problems of the *mechanics of continuous media* which are the three-dimensional *linear elasticity equations* (Sec. 4.7.2), the *plate problem* (Sec. 4.7.3), and *Stokes problem* (Sec. 4.7.4). After describing the partial differential equations modelling the physical phenomena, we discuss the variational formulations and various questions concerning the existence and uniqueness of the solutions of these problems.

4.7.2. *Three-dimensional linear elasticity*

Let $\Omega \subset \mathbb{R}^3$ be a bounded domain. Let Γ be the boundary of Ω and suppose that $\Gamma = \Gamma_0 \cup \Gamma_1$ with Γ_0, Γ_1 such that $\int_{\Gamma_0 \cap \Gamma_1} d\Gamma = 0$ (a typical situation is shown in Fig. 4.6). We suppose that Ω is occupied by an elastic continuous medium and that the resulting elastic body is fixed along Γ_0. Let $\mathbf{f} = \{f_i\}_{i=1}^3$ be a density of body forces acting in Ω and $\mathbf{g} = \{g_i\}_{i=1}^3$ be a density of surface forces acting on Γ_1. We denote by $\mathbf{u}(x) = \{u_i(x)\}_{i=1}^3$ the displacement of the body at x.

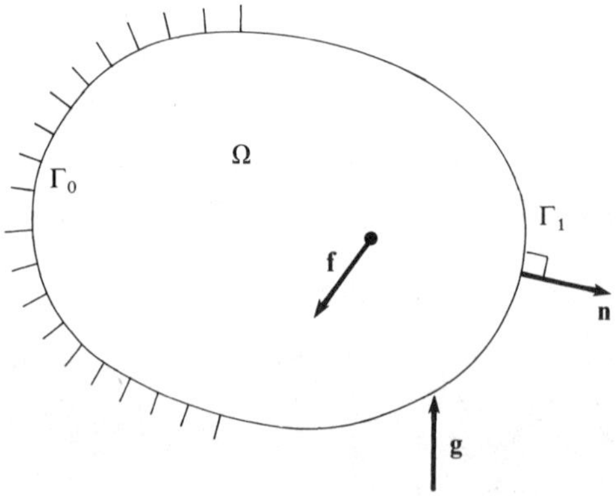

Figure 4.6

In linear elasticity the stress-strain relation is

$$\sigma_{ij}(\mathbf{u}) = \lambda(\nabla \cdot \mathbf{u})\delta_{ij} + 2\mu\varepsilon_{ij}(\mathbf{u}), \qquad \text{with } \varepsilon_{ij}(\mathbf{u}) = \frac{1}{2}\left(\frac{\partial u_i}{\partial x_j} + \frac{\partial u_j}{\partial x_i}\right), \tag{4.327}$$

where σ_{ij} and ε_{ij} denote the components of the stress and strain tensors, respectively; λ and μ are positive constants and are known as the *Lamme coefficients*.

The problem is to find the tensor $\boldsymbol{\sigma} = (\sigma_{ij})$, the displacement $\mathbf{u} = \{u_i\}_{i=1}^3$ if Ω, $\mathbf{f} \in (L^2(\Omega))^3$ and $\mathbf{g} \in (L^2(\Gamma_1))^3$ are given.

The equilibrium equations are

$$\frac{\partial}{\partial x_j}\sigma_{ij} + f_i = 0 \quad \text{in } \Omega, \quad i = 1, 2, 3, \tag{4.328$_1$}$$

$$\sigma_{ij}n_j = g_i \quad \text{on } \Gamma_i, \quad i = 1, 2, 3, \tag{4.328$_2$}$$

$$u_i = 0 \quad \text{on } \Gamma_0, \quad i = 1, 2, 3. \tag{4.328$_3$}$$

We have used the summation convention of repeated indices in the above equations.

To obtain a variational formulation of the linear elasticity problem (4.327), (4.328), we define a space V, a bilinear form $a(\cdot, \cdot)$ and a linear functional L by

$$V = \{\mathbf{v} \mid \mathbf{v} \in (H^1(\Omega))^3, \mathbf{v} = \mathbf{0} \text{ on } \Gamma_0\}, \tag{4.329}$$

$$a(u, v) = \int_\Omega \sigma_{ij}(\mathbf{u})\varepsilon_{ij}(\mathbf{v})\, dx, \tag{4.330}$$

$$L(v) = \int_{\Gamma_1} g_i v_i \, d\Gamma + \int_\Omega f_i v_i \, dx, \tag{4.331}$$

respectively. Using (4.327), $a(\mathbf{u}, \mathbf{v})$ can be written as

$$a(\mathbf{u}, \mathbf{v}) = \int_\Omega \{\lambda\nabla \cdot \mathbf{u}\nabla \cdot \mathbf{v} + 2\mu\varepsilon_{ij}(\mathbf{u})\varepsilon_{ij}(\mathbf{v})\}\, dx, \tag{4.332}$$

from which it is clear that $a(\cdot, \cdot)$ is symmetric. The space V is a Hilbert space for the $(H^1(\Omega))^3$-norm, and the functionals $a(\cdot, \cdot)$ and L are clearly continuous over V. Proving the V-ellipticity of $a(\cdot, \cdot)$ is nontrivial; actually this ellipticity property follows from the so-called *Korn inequality* for which we refer to Duvaut and Lions [1] (see also Ciarlet [3] and Nitsche [1]).

Now consider the variational problem associated with V, a, and L, i.e.:

Find $\mathbf{u} \in V$ *such that*

$$a(\mathbf{u}, \mathbf{v}) = L(\mathbf{v}), \qquad \forall\, \mathbf{v} \in V; \tag{4.333}$$

from the above properties of V, $a(\cdot, \cdot)$, L, the variational problem (4.333) has a unique solution (from Theorem 2.1 of Sec. 2.3). Applying the Green–Ostrogradsky formula, (4.33) shows that the boundary-value problem corresponding to (4.333) is precisely (4.328).

The finite element solution of (4.328), via (4.333), is discussed in great detail in e.g., Ciarlet [1]–[3], Zienkiewicz [1], and Bathe and Wilson [1].

Remark 4.25. The term $a(\mathbf{u}, \mathbf{v})$ can be interpreted as the work of the internal elastic forces and $L(\mathbf{v})$ as the work of the external (body and surface) forces. Thus, the equation

$$a(\mathbf{u}, \mathbf{v}) = L(\mathbf{v}), \qquad \forall\, \mathbf{v} \in V$$

is a reformulation of the *virtual work theorem.*

4.7.3. *A thin plate problem*

We follow the presentation of Ciarlet [3, Chapter 1]. Let Ω be a bounded domain of $\mathbb{R}^2$ and consider V, $a(\cdot, \cdot)$, L defined by

$$V = H_0^2(\Omega) = \overline{\mathscr{D}(\Omega)}^{H^2(\Omega)} = \left\{v \,|\, v \in H^2(\Omega),\ v = \frac{\partial v}{\partial n} = 0 \text{ on } \Gamma\right\}, \tag{4.334}$$

$$\begin{aligned} a(u, v) &= \int_\Omega \left\{\Delta u \Delta v + (1 - \sigma)\left(2 \frac{\partial^2 u}{\partial x_1\, \partial x_2} \frac{\partial^2 v}{\partial x_1\, \partial x_2} \right.\right. \\ &\qquad \left.\left. - \frac{\partial^2 u}{\partial x_1^2} \frac{\partial^2 v}{\partial x_2^2} - \frac{\partial^2 u}{\partial x_2^2} \frac{\partial^2 v}{\partial x_1^2}\right)\right\} dx \\ &= \int_\Omega \left\{\sigma \Delta u \Delta v + (1 - \sigma)\left(\frac{\partial^2 u}{\partial x_1^2} \frac{\partial^2 v}{\partial x_1^2} + \frac{\partial^2 u}{\partial x_2^2} \frac{\partial^2 v}{\partial x_2^2} \right.\right. \\ &\qquad \left.\left. + 2 \frac{\partial^2 u}{\partial x_1\, \partial x_2} \frac{\partial^2 v}{\partial x_1\, \partial x_2}\right)\right\} dx, \end{aligned} \tag{4.335}$$

$$L(v) = \int_\Omega f v \, dx, \qquad f \in L^2(\Omega). \tag{4.336}$$

The associated variational problem:

Find $u \in V$ such that

$$a(u, v) = L(v), \qquad \forall\, v \in V, \tag{4.337}$$

corresponds to the variational formulation of the *clamped plate problem*, which concerns the equilibrium position of a plate of constant (and very small) thickness under the action of a transverse force whose density is proportional to f. The constant σ is the Poisson coefficient of the plate material $(0 < \sigma < \frac{1}{2})$. If $f = 0$, the plate is in the plane of coordinates $\{x_1, x_2\}$. The condition $u \in H_0^2(\Omega)$ takes into account the fact that the plate is clamped.

The derivation of (4.337) from (4.333) is discussed in Ciarlet and Destuynder [1]. The Poisson coefficient σ satisfies $0 < \sigma < \frac{1}{2}$; the bilinear form $a(\cdot, \cdot)$ is $H_0^2(\Omega)$-elliptic, since we have

$$a(v, v) = \sigma \|\Delta v\|_{L^2(\Omega)}^2 + (1 - \sigma)|v|_{2,\Omega}^2 \tag{4.338}$$

(where $|\cdot|_{2,\Omega}$ has been defined in Sec. 4.5.3.2).

Actually, using the fact that $H_0^2(\Omega) = \overline{\mathscr{D}(\Omega)}^{H^2(\Omega)}$, we can easily prove that

$$a(v, w) = \int_\Omega \Delta v \Delta w \, dx, \qquad \forall\, v, w \in H_0^2(\Omega). \tag{4.339}$$

From (4.339) it follows that the solution u of the variational problem (4.337) is also the unique solution in $H^2(\Omega)$ of the biharmonic problem

$$\Delta^2 u = f \text{ in } \Omega, \qquad u = \frac{\partial u}{\partial n} = 0, \tag{4.340}$$

and conversely (here $\Delta^2 = \Delta\Delta$). Problem (4.340) also plays an important role in the analysis of incompressible fluid flows (see, e.g., Girault and Raviart [1], Glowinski and Pironneau [1], and Glowinski, Keller, and Reinhart [1] for further details).

For the finite element solution of (4.337), (4.340), see Strang and Fix [1], Ciarlet and Raviart [1], Ciarlet [3], and Brezzi [1], and the references therein.

4.7.4. *The Stokes problem*

As mentioned in Chapter VI, Sec. 5.2, the motion of an incompressible viscous fluid is modelled by the Navier–Stokes equations; if we neglect the nonlinear term $(\mathbf{u} \cdot \nabla)\mathbf{u}$, the steady case reduces to the steady Stokes problem

$$-\Delta \mathbf{u} + \nabla p = \mathbf{f} \text{ in } \Omega, \qquad \nabla \cdot \mathbf{u} = 0 \text{ in } \Omega, \qquad \mathbf{u}|_\Gamma = \mathbf{g} \quad \text{with} \int_\Gamma \mathbf{g} \cdot \mathbf{n} \, d\Gamma = 0; \tag{4.341}$$

in (4.341), Ω is the flow domain ($\Omega \subset \mathbb{R}^N$, $N = 2$ or 3 in practical applications), Γ is its boundary, $\mathbf{u} = \{u_i\}_{i=1}^N$ is the flow velocity, p is the pressure, and $\mathbf{f}$ is the density of external forces.

For simplicity we suppose that $\mathbf{g} = \mathbf{0}$ on Γ (for the case $\mathbf{g} \neq \mathbf{0}$, see Chapter VII, Sec. 5, and also Appendix III) and also that Ω is bounded.

There are many possible variational formulations of the Stokes problem (4.341), and some of them are described in Chapter VII, Sec. 5; we shall concentrate on one of them, obtained as follows:

Let

$$\boldsymbol{\phi} = \{\phi_i\}_{i=1}^N \in (\mathscr{D}(\Omega))^N \quad (\Rightarrow \boldsymbol{\phi}|_\Gamma = \mathbf{0});$$

taking the $\mathbb{R}^N$-scalar product of $\boldsymbol{\phi}$ with both sides of the first equation (4.341) and integrating over Ω, we obtain

$$-\int_\Omega \Delta \mathbf{u} \cdot \boldsymbol{\phi} \, dx + \int_\Omega \nabla p \cdot \boldsymbol{\phi} \, dx = \int_\Omega \mathbf{f} \cdot \boldsymbol{\phi} \, dx \tag{4.342}$$

Using the Green–Ostrogradsky formula (4.33), it follows from (4.342) that

$$\int_\Omega \nabla \mathbf{u} \cdot \nabla \boldsymbol{\phi}\, dx - \int_\Gamma \frac{\partial \mathbf{u}}{\partial \mathbf{n}} \cdot \boldsymbol{\phi}\, d\Gamma - \int_\Omega p \nabla \cdot \boldsymbol{\phi}\, dx + \int_\Gamma p \boldsymbol{\phi} \cdot \mathbf{n}\, d\Gamma = \int_\Gamma \mathbf{f} \cdot \boldsymbol{\phi}\, dx \tag{4.343}$$

(with $\nabla \mathbf{u} \cdot \nabla \boldsymbol{\phi} = \sum_{i=1}^N \nabla u_i \cdot \nabla \phi_i$). Since $\boldsymbol{\phi} = \mathbf{0}$ on Γ, the above relation (4.343) reduces to

$$\int_\Omega \nabla \mathbf{u} \cdot \nabla \boldsymbol{\phi}\, dx = \int_\Omega \mathbf{f} \cdot \boldsymbol{\phi}\, dx + \int_\Omega p \nabla \cdot \boldsymbol{\phi}\, dx. \tag{4.344}$$

Now suppose that $\boldsymbol{\phi} \in \mathscr{V}$, where

$$\mathscr{V} = \{\boldsymbol{\phi} \mid \boldsymbol{\phi} \in (\mathscr{D}(\Omega))^N, \nabla \cdot \boldsymbol{\phi} = 0\};$$

from (4.344) it follows that

$$\int_\Omega \nabla \mathbf{u} \cdot \nabla \boldsymbol{\phi}\, dx = \int_\Omega \mathbf{f} \cdot \boldsymbol{\phi}\, dx, \qquad \forall\, \boldsymbol{\phi} \in \mathscr{V}. \tag{4.345}$$

Relation (4.345) suggests the introduction of V, $a(\cdot, \cdot)$, L defined by

$$V = \{\mathbf{v} \mid \mathbf{v} \in (H_0^1(\Omega))^N, \nabla \cdot \mathbf{v} = 0\}, \tag{4.346}$$

$$a(\mathbf{v}, \mathbf{w}) = \int_\Omega \nabla \mathbf{v} \cdot \nabla \mathbf{w}\, dx, \qquad \forall\, \mathbf{v}, \mathbf{w} \in (H^1(\Omega))^N, \tag{4.347}$$

$$L(\mathbf{v}) = \int_\Omega \mathbf{f} \cdot \mathbf{v}\, dx, \tag{4.348}$$

respectively, and then, in turn, the following variational problem:

Find $\mathbf{u} \in V$ *such that*

$$a(\mathbf{u}, \mathbf{v}) = L(\mathbf{v}), \qquad \forall\, \mathbf{v} \in V. \tag{4.349}$$

Since the mapping $\mathbf{v} \to \nabla \cdot \mathbf{v}$ is linear and continuous from $(H_0^1(\Omega))^N$ into $L^2(\Omega)$, V is a closed subspace of $(H_0^1(\Omega))^N$ and, since Ω is bounded is therefore a Hilbert space for the scalar product $\{\mathbf{v}, \mathbf{w}\} \to \int_\Omega \nabla \mathbf{v} \cdot \nabla \mathbf{w}\, dx$. The bilinear form $a(\cdot,\cdot)$ is clearly continuous and $(H_0^1(\Omega))^N$-elliptic, and the linear functional L is continuous over V if $\mathbf{f} \in (L^2(\Omega))^N$. From the above properties of V, $a(\cdot, \cdot)$, L, it follows from Theorem 2.1 of Sec. 2.3 that the variational problem (4.349) has a unique solution.

Since we have (cf. Ladyshenskaya [1])

$$\overline{\mathscr{V}}^{(H_0^1(\Omega))^N} = V,$$

it follows from (4.345), (4.349) that if $\{\mathbf{u}, p\}$ is a solution of the Stokes problem (4.341), then $\mathbf{u}$ is also the solution of (4.349). Actually the reciprocal property is true, but proving it is nontrivial, particularly obtaining a pressure $p \in L^2(\Omega)$ from the variational formulation (4.349); for the reciprocal property we refer to, e.g., Ladyshenskaya, *loc. cit.*, Lions [1], Temam [1], Tartar [1], and Girault and Raviart [1].

We refer to Chapter VII, Sec. 5 for finite element approximations of the Stokes problem (4.341) (see also the references therein) and also to Appendix III for some complements.

5. Further Comments: Conclusion

Variational methods provide powerful and flexible tools for solving a large variety of boundary-value problems for partial differential operators. The various examples discussed in this appendix are all classical (or almost classical) boundary-value problems for elliptic operators, but in fact the variational approach can also be used to solve first-order systems as shown in Friedrichs [1] and Lesaint [1].

As a last example showing the flexibility of variational formulations and methods, we would like to discuss the approximate calculation of the *flux* associated with the solution of an elliptic boundary-value problem. For simplicity we consider the Dirichlet problem (with Ω bounded):

$$-\nabla \cdot (\bar{\bar{\mathbf{A}}}\nabla u) = f \text{ in } \Omega, \qquad u = g \text{ on } \Gamma, \tag{5.1}$$

whose variational formulation is given (see Sec. 4.3.2) by:

Find $u \in H^1(\Omega)$ *such that* $u = g$ *on* Γ *and*

$$\int_\Omega (\bar{\bar{\mathbf{A}}}\nabla u) \cdot \nabla v \, dx = \int_\Omega fv \, dx, \qquad \forall\, v \in H_0^1(\Omega). \tag{5.2}$$

If the hypotheses on $\bar{\bar{\mathbf{A}}}$, f, g made in Sec. 4.3 hold, we know (from Sec. 4.3.2) that (5.1), (5.2) has a unique solution in $H^1(\Omega)$.

We call flux the boundary function[29]

$$\lambda = ((\bar{\bar{\mathbf{A}}}\nabla u) \cdot \mathbf{n})|_\Gamma,$$

where $\mathbf{n}$ still denotes the unit vector of the outward normal at Γ. There are many situations in which it is important to know (at least approximately) the flux λ; this can be achieved through a finite element approximation of (5.1), (5.2) as discussed below.

With H_h^1 and H_{0h}^1 as in Sec. 4.5.2, we approximate (5.1), (5.2) by:

Find $u_h \in H_h^1$ *such that* $u_h = g_h$ *on* Γ *and*

$$\int_\Omega (\bar{\bar{\mathbf{A}}}\nabla u_h) \cdot \nabla v_h \, dx = \int_\Omega f v_h \, dx, \qquad \forall\, v_h \in H_{0h}^1, \tag{5.3}$$

---

[29] Usually $\lambda \in H^{-1/2}(\Gamma)$.

where g_h is an approximation of g belonging to the space γH_h^1 defined by

$$\gamma H_h^1 = \{\mu_h \mid \mu_h \in C^0(\Gamma), \exists\, v_h \in H_h^1 \text{ such that } \mu_h = v_h|_\Gamma\}$$

(actually γH_h^1 is also the space of those functions continuous over Γ and piecewise linear on the edges of $\mathcal{T}_h$ supported by Γ).

Concerning the approximation of λ, a naive method would be to define it by

$$\lambda_h = ((\overline{\overline{\mathbf{A}}}\nabla u_h) \cdot \mathbf{n})|_\Gamma, \tag{5.4}$$

which is possible since ∇u_h is piecewise constant over Ω; actually (5.4) yields very inaccurate results. A much better approximation of λ is obtained as follows.

From the Green–Ostrogradsky formula (4.33) we know that λ and u satisfy

$$\int_\Gamma \lambda v \, d\Gamma = \int_\Gamma (\overline{\overline{\mathbf{A}}}\nabla u) \cdot \mathbf{n} v \, d\Gamma = \int_\Omega \nabla \cdot (\overline{\overline{\mathbf{A}}}\nabla u) v \, dx + \int_\Omega (\overline{\overline{\mathbf{A}}}\nabla u) \cdot \nabla v \, dx,$$
$$\forall\, v \in H^1(\Omega).$$

Since $-\nabla \cdot (\overline{\overline{\mathbf{A}}}\nabla u) = f$, we finally have

$$\int_\Gamma \lambda v \, d\Gamma = -\int_\Omega f v \, dx + \int_\Omega (\overline{\overline{\mathbf{A}}}\nabla u) \cdot \nabla v \, dx, \qquad \forall\, v \in H^1(\Omega). \tag{5.5}$$

Starting from (5.5) to approximate λ, we shall define an approximation λ_h of λ as the solution of the linear variational problem (in which u_h is known from a previous computation):

Find $\lambda_h \in \gamma H_h^1$ such that

$$\int_\Gamma \lambda_h v_h \, d\Gamma = -\int_\Omega f v_h \, dx + \int_\Omega (\overline{\overline{\mathbf{A}}}\nabla u_h) \cdot \nabla v_h \, dx, \qquad \forall\, v_h \in H_h^1; \tag{5.6}$$

it is easy to see that (5.6) is equivalent to:

Find $\lambda_h \in \gamma H_h^1$ such that

$$\int_\Gamma \lambda_h \mu_h \, d\Gamma = -\int_\Omega f \tilde{\mu}_h \, dx + \int_\Omega (\overline{\overline{\mathbf{A}}}\nabla u_h) \cdot \nabla \mu_h \, dx, \qquad \forall \mu_h \in \gamma H_h^1, \tag{5.7}$$

where $\tilde{\mu}_h$ is the extension of μ_h over $\bar{\Omega}$ such that $\tilde{\mu}_{h|T} = 0$, $\forall\, T \in \mathcal{T}_h$, such that $\partial T \cap \Gamma = \emptyset$. The variational problem (5.7) is equivalent to a linear system whose matrix is symmetric, positive definite, and sparse.

As a final comment, we would like to mention that the above method, founded on the application of the Green–Ostrogradsky formula, can also be applied to the computation of fluxes through lines (or surfaces if $\Omega \subset \mathbb{R}^3$) inside $\bar{\Omega}$; this is done, for example, in some solution methods for partial differential equations using domain decomposition.

APPENDIX II

A Finite Element Method with Upwinding for Second-Order Problems with Large First-Order Terms

1. Introduction

Upwinding finite element schemes have been a subject of very active research in recent years; in this direction we shall mention, among others, Lesaint [2], Tabata [1], Heinrich, Huyakorn, Zienkiewicz, and Mitchell [1], Christie and Mitchell [1], Ramakrishnan [1], Brooks and Hugues [1] and also Fortin and Thomasset [1], Girault and Raviart [2], Johnson [2], Bredif [1], and Thomasset [1], these last five references being concerned more particularly with the Navier–Stokes equations for incompressible viscous fluids.

In this appendix we would like to describe a method (due to Bristeau and Glowinski) which can be viewed as an extension of the method introduced by Tabata, *loc. cit.*; this method will be described in relation to a particular simple model problem, but generalizations to more complicated problems are quite obvious.

2. The Model Problem

Let Ω be a bounded domain of $\mathbb{R}^2$ and $\Gamma = \partial\Omega$. We consider the problem (with $\varepsilon > 0$)

$$-\varepsilon\Delta u + \boldsymbol{\beta} \cdot \nabla u = f \text{ in } \Omega, \qquad u = 0, \text{ on } \Gamma, \tag{2.1}$$

where $\boldsymbol{\beta} = \{\cos\theta, \sin\theta\}$.

We are mainly interested in solving (2.1) for small values of ε; in the following we shall suppose that $f \in L^2(\Omega)$, and we shall use the notation

$$\frac{\partial v}{\partial \beta} = \boldsymbol{\beta} \cdot \nabla v. \tag{2.2}$$

Problem (2.1) has as variational formulation (see Appendix I, Sec. 4.3):

Find $u \in H_0^1(\Omega)$ such that

$$\varepsilon \int_\Omega \nabla u \cdot \nabla v \, dx + \int_\Omega \frac{\partial u}{\partial \beta} v \, dx = \int_\Omega fv \, dx, \qquad \forall\, v \in H_0^1(\Omega), \tag{2.3}$$

from which we can easily prove, using the Lax–Miligram theorem (see Appendix I, Secs. 2.3 and 4.3), the existence of a unique solution of (2.1) in $H_0^1(\Omega)$.

EXERCISE 2.1. Prove that (2.3) has a unique solution.

Hint: Use the fact that

$$\int_\Omega \frac{\partial v}{\partial \beta} v \, dx = 0, \qquad \forall\, v \in H_0^1(\Omega).$$

3. A Centered Finite Element Approximation

We suppose that Ω is a bounded polygonal domain of $\mathbb{R}^2$. Let $\{\mathcal{T}_h\}_h$ be a family of triangulations of Ω like those in Chapter VII, Sec. 5.3.3.1; to approximate (2.1), (2.3), we use the space

$$H_{0h}^1 = \{v_h \mid v_h \in C^0(\overline{\Omega}),\ v_{h|T} \in P_1,\ \forall\, T \in \mathcal{T}_h,\ v_{h|\Gamma} = 0\}.$$

The obvious approximation of (2.3) using H_{0h}^1 is:

Find $u_h \in H_{0h}^1$ such that, $\forall\, v_h \in H_{0h}^1$,

$$\varepsilon \int_\Omega \nabla u_h \cdot \nabla v_h \, dx + \int_\Omega \frac{\partial u_h}{\partial \beta} v_h \, dx = \int_\Omega f v_h \, dx. \tag{3.1}$$

Problem (3.1) has a unique solution; moreover, if $\{\mathcal{T}_h\}_h$ is such that the angles of $\mathcal{T}_h$ are bounded from below by $\theta_0 > 0$, independent of h, we have

$$\lim_{h \to 0} \|u_h - u\|_{H_0^1(\Omega)} = 0.$$

If ε is "too small," it is well known that u_h obtained from (3.1) is afflicted with spurious oscillations (unless h is very small); it is therefore necessary to use approximations of (2.1), (2.3) more sophisticated than (3.1).

4. A Finite Element Approximation with Upwinding

We have in mind an approximation of (2.1), (2.3) such that

$$\varepsilon \int_\Omega \nabla u_h \cdot \nabla v_h \, dx + \left(\frac{\partial_h u_h}{\partial \beta}, v_h\right)_h = \int_\Omega f v_h \, dx, \qquad \forall\, v_h \in H_{0h}^1, \quad u_h \in H_{0h}^1, \tag{4.1}$$

where the scalar product $(\cdot, \cdot)_h$ is defined by

$$(u_h, v_h)_h = \frac{1}{3} \sum_{T \in \mathcal{T}_h} \operatorname{meas}(T) \sum_{i=1}^{3} u_h(M_{iT}) v_h(M_{iT}), \tag{4.2}$$

where M_{iT}, $i = 1, 2, 3$ are the vertices of T.

Figure 4.1

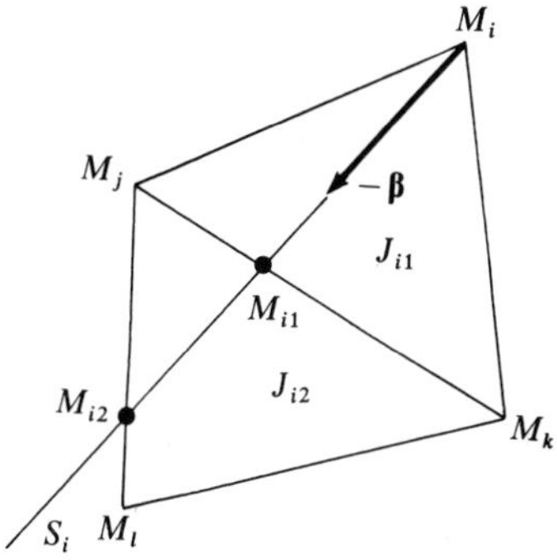

We look for an approximation $\partial_h u_h/\partial\beta$ of $\partial u/\partial\beta$ backward with respect to $\boldsymbol{\beta}$, second-order accurate (unlike the approximation in Tabata [1] which is only first-order accurate), and defined at the vertices of $\mathscr{T}_h$. To define such an approximation, we shall use an interpolation method defined as follows.

Let M_i be a node of $\mathscr{T}_h$ such that

$$M_i \notin \Gamma. \tag{4.3}$$

Let S_i be the half-line starting at M_i and directed by $-\boldsymbol{\beta}$ (see Fig. 4.1). To express $(\partial_h u_h/\partial\beta)(M_i)$, we look for two points of S_i (different from M_i) also belonging to the edges of triangles close to M_i. Let us now describe the choice of these two points.

Step 1. Among the triangles of $\mathscr{T}_h$ having M_i as a vertex, we consider the triangle denoted by J_{i1} which is crossed by S_i. Let M_j, M_k be the two other vertices of J_{i1} (taken with the trigonometric orientation); since (4.3) holds, J_{i1} always exists.

Step 2. Denote by M_{i1} the point at the intersection of S_i and of the side M_jM_k. If M_jM_k is supported by the boundary Γ, then go to Remark 4.1; if the contrary holds, let J_{i2} be the triangle adjacent to J_{i1} along M_jM_k (see Fig. 4.1) and let M_l be the third vertex of J_{i2}; finally, denote by M_{i2} the point where S_i crosses M_jM_l or M_lM_k.

Step 3. We have constructed three points on S_i, which are M_i, M_{i1}, M_{i2}. But M_{i1}, M_{i2} may be very close to each other and even coincide with either M_j or M_k; if this unfortunate situation holds, our second-order approximation degenerates into a first-order one. Thus we may need to define a different M_{i2} if S_i is "close" to one of the eges of J_{i1}. We shall proceed as follows.

We compute the two angles $(\overrightarrow{M_iM_j}, -\boldsymbol{\beta})$ and $(\overrightarrow{M_iM_j}, \overrightarrow{M_iM_k})$ and then consider the following possibilities (other strategies are possible):

(a) If $\frac{1}{4}(\overrightarrow{M_iM_j}, \overrightarrow{M_iM_k}) \le (\overrightarrow{M_iM_j}, -\boldsymbol{\beta}) \le \frac{3}{4}(\overrightarrow{M_iM_j}, \overrightarrow{M_iM_k})$,

then M_{i2} is defined as in Step 2.

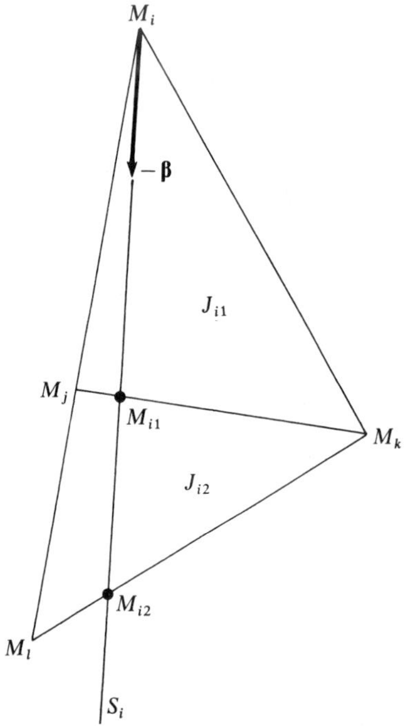

Figure 4.2

(b) If $(\overrightarrow{M_iM_j}, -\boldsymbol{\beta}) < \frac{1}{4}(\overrightarrow{M_iM_j}, \overrightarrow{M_iM_k})$, denote $M_v = M_j$;
if $(\overrightarrow{M_iM_j}, -\boldsymbol{\beta}) > \frac{3}{4}(\overrightarrow{M_iM_j}, \overrightarrow{M_iM_k})$, denote $M_v = M_k$.

If M_{i2} of Step 2 does not belong to M_vM_l, then M_{i1}, M_{i2} will not coincide, and we can use the M_{i2} defined in Step 2 (see Fig. 4.2 (where $M_v = M_j$)).

If M_{i2} belongs to M_vM_l, then $|M_{i1}M_{i2}|$ may be "very small," and we have to modify the definition of M_{i2}; *go to Step* 4 for such a modification.

Step 4. We look for the triangle J_{i3} adjacent to J_{i2} along M_vM_l. If M_vM_l is supported by Γ, such a triangle J_{i3} does not exist; *then go to Remark* 4.1. If J_{i3} exists, let M_m be the third vertex of J_{i3} (see Fig. 4.3); if S_i crosses edge M_lM_m, we denote by M_{i2} the crossing point, and one uses this point (together with M_i, M_{i1}) to approximate $\partial u/\partial\beta$; if S_i crosses M_vM_m, we repeat the process of this step to finally obtain the following situation: either we have to go to Remark 4.1[1] or we obtain M_{i2} as the intersection of S_i and of the polygonal line, the boundary of the polygonal domain consisting of the union of those triangles of $\mathcal{T}_h$ with M_v as a common vertex.

If we have been able to avoid ending at Remark 4.1, we have constructed two points M_{i1}, M_{i2} "close" to vertex M_i. Since u_h is continuous on $\overline{\Omega}$, its values

[1] This cannot happen if $M_v \notin \Gamma$.

Figure 4.3

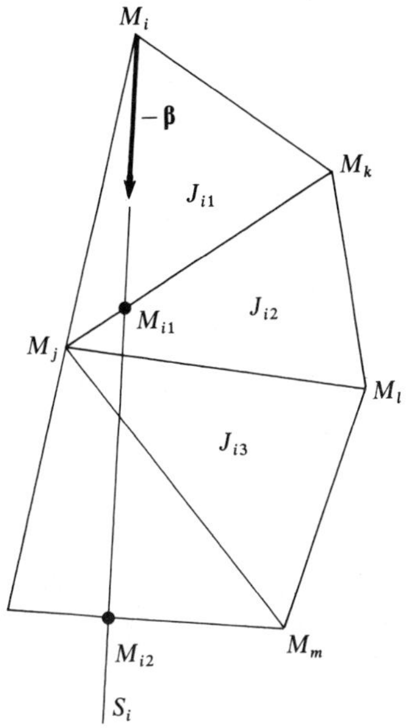

at M_i, M_{i1}, M_{i2} are known; we denote

$$u_{io} = u_h(M_i), \quad u_{i1} = u_h(M_{i1}), \quad i_{i2} = u_h(M_{i2}). \tag{4.4}$$

Let $h_{i1} = |M_i M_{i1}|$, $h_{i2} = |M_i M_{i2}|$; we then define $(\partial_h u_h/\partial\beta)(M_i)$ as the value at M_i of the derivative of a *second-order polynomial*, defined on the half-line S_i and coinciding with u_h at M_i, M_{i1}, M_{i2}. We then obtain

$$\frac{\partial_h u_h}{\partial\beta}(M_i) = \frac{h_{i1}+h_{i2}}{h_{i1}h_{i2}}\,u_{io} - \frac{h_{i2}}{h_{i1}(h_{i2}-h_{i1})}\,u_{i1} + \frac{h_{i1}}{h_{i2}(h_{i2}-h_{i1})}\,u_{i2}. \tag{4.5}$$

Remark 4.1. If M_i is close to Γ and if we have not been able to define M_{i2}, we can proceed as follows:

(i) Either define a centered approximation of $(\partial u/\partial\beta)(M_i)$—$O(h^2)$ accurate—by

$$\frac{\partial_h u_h}{\partial\beta}(M_i) = \frac{1}{A_i}\sum_{T\in\mathscr{T}_i} \operatorname{meas}(T)\,\frac{\partial u_h}{\partial\beta}\bigg|_T, \tag{4.6}$$

where $\mathscr{T}_i$ is the subset of $\mathscr{T}_h$ consisting of those triangles with M_i as a common vertex; $A_i = \operatorname{meas}(\bigcup_{T\in\mathscr{T}_i} T)$; $\partial u_h/\partial\beta|_T = \beta\cdot\nabla u_{h|T}$ (we recall that ∇u_h is piecewise constant on $\overline{\Omega}$).

(ii) Or use an upwinded first-order approximation (as in Tabata [1]) defined by

$$\frac{\partial_h u_h}{\partial \beta}(M_i) = \left.\frac{\partial u_h}{\partial \beta}\right|_{J_{i1}}, \tag{4.7}$$

where the triangle J_{i1} has been defined in Step 1.

5. On the Solution of the Linear System Obtained by Upwinding

Using the upwinded scheme described in Sec. 4, we obtain u_h via the solution of a linear system whose matrix has a bandwidth which is approximately twice the bandwidth of the matrix associated with the centered approximate problem (3.1) of Sec. 3.

The matrix of the above system is definitely nonsymmetric; in the preliminary stage of our numerical experiments, we have chosen to solve the linear system, say

$$\mathbf{A}\mathbf{U}_h = \mathbf{b}_h, \tag{5.1}$$

by the standard method consisting of solving the normal equation

$$\mathbf{A}_h^t \mathbf{A}_h \mathbf{U}_h = \mathbf{A}_h^t \mathbf{b}_h, \tag{5.2}$$

by a Cholesky method, since the matrix in (5.2) is symmetric positive definite. At the moment, we are testing more sophisticated methods like the *Lanczos-type* methods described in, e.g., Widlund [1] (see also Strikwerda [1] for iterative methods for solving finite difference approximations of second-order elliptic equations with large first-order terms).

6. Numerical Experiments

6.1. The test problem

We have chosen to solve

$$-\varepsilon\Delta u + \boldsymbol{\beta} \cdot \nabla u = 1 \text{ in } \Omega, \quad u = 0 \text{ on } \Gamma, \tag{6.1}$$

where Ω is the domain (cavity with a step) of Fig. 6.1.

6.2. Numerical results

The method of Sec. 3, using a centered approximation, produces very poor results as soon as $\varepsilon < 10^{-1}$ (at least with the finite element grids shown in Figs 6.2–6.5); therefore we only present results concerning the upwinding

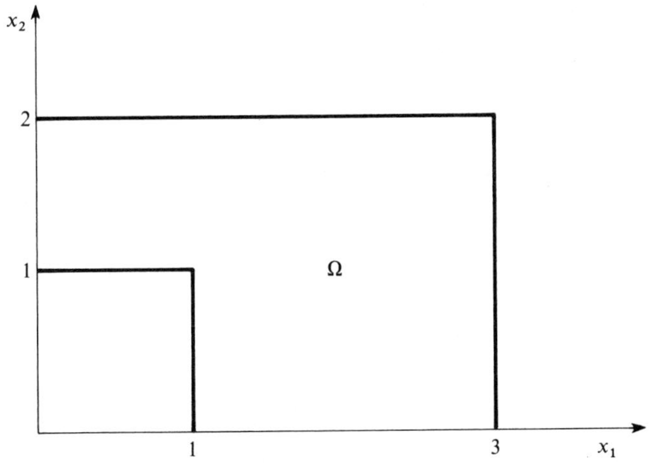

Figure 6.1

method of Sec. 4. For the nodes close to Γ, we have used the first-order approximation defined by (4.7). The domain Ω being the one of Fig. 6.1, we have used the regular triangulations of Figs. 6.4 and 6.5 in our tests. We have then solved (6.1) via the finite element approximate problem (4.1). The above tests are severe since the solution contains strong boundary layer and free layer (generated by the corner) phenomenon, resulting in strong variations of u and/or ∇u.

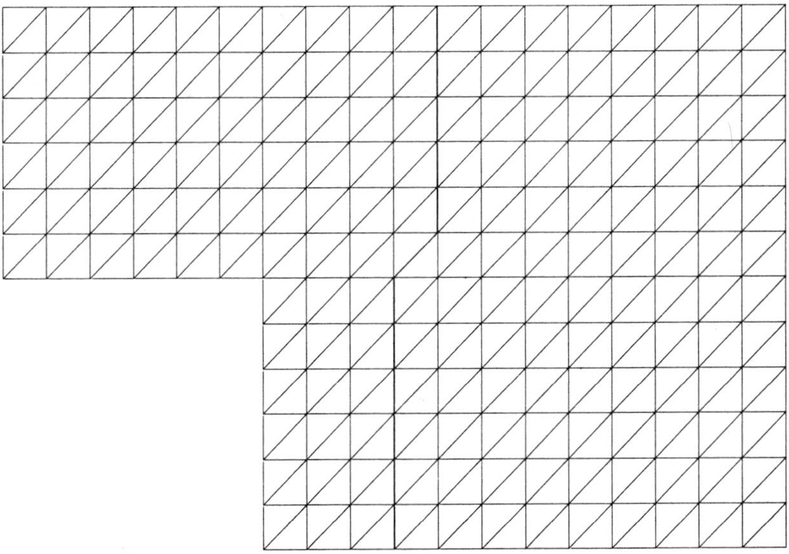

Figure 6.2. Triangulation $\mathscr{T}_h^1$ (360 triangles, 211 nodes).

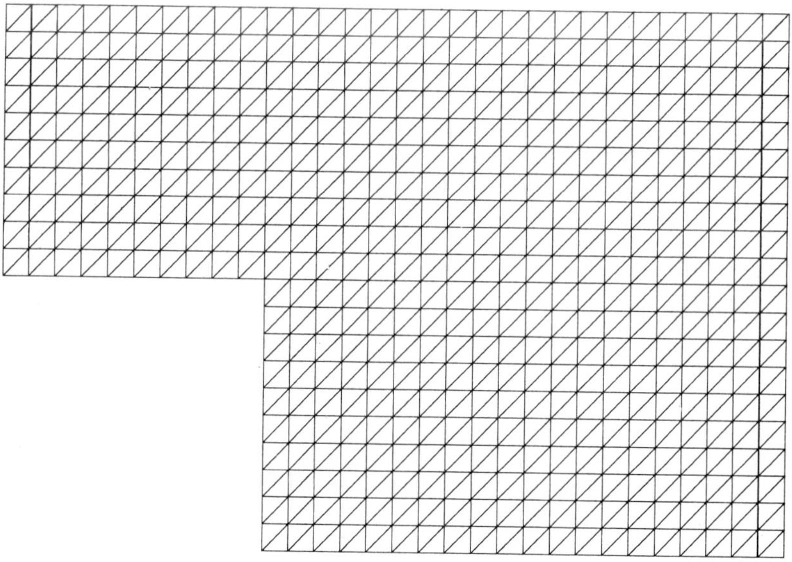

Figure 6.3. Triangulation $\mathscr{T}_h^2$ (1000 triangles, 551 nodes).

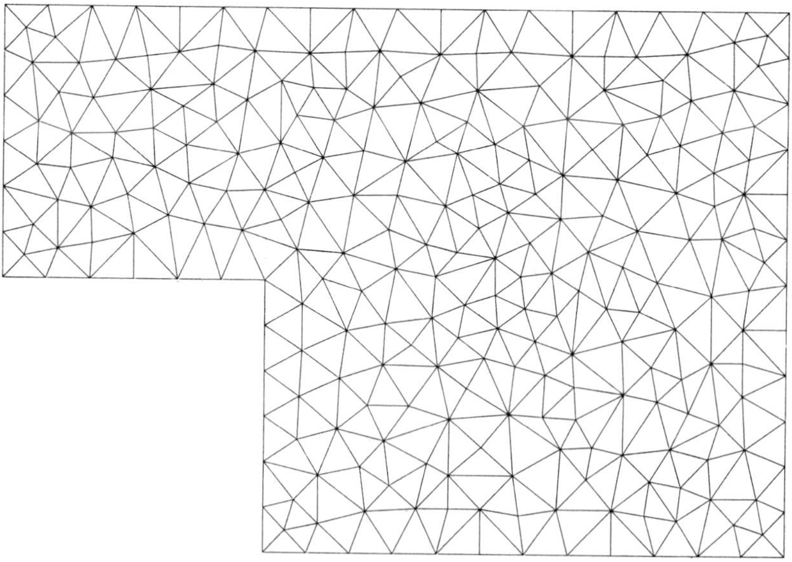

Figure 6.4. Triangulation $\mathscr{T}_h^3$ (472 triangles, 267 nodes).

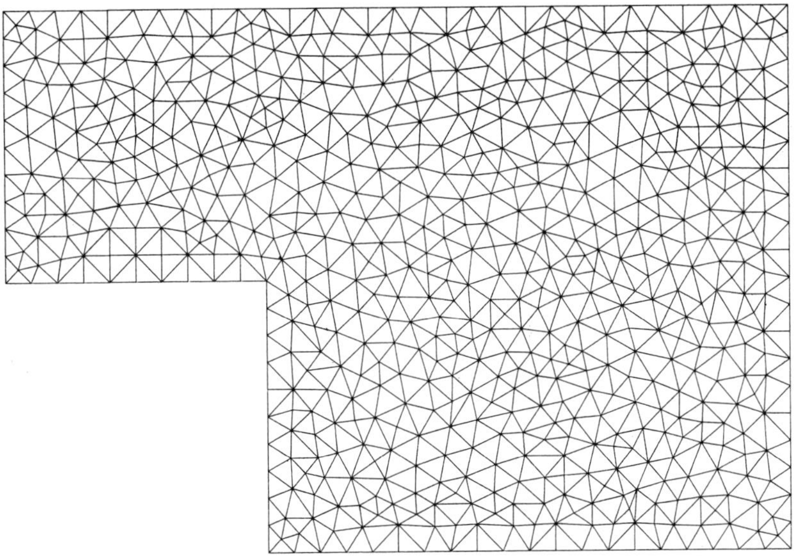

Figure 6.5. Triangulation $\mathscr{T}_h^4$ (1246 triangles, 674 nodes).

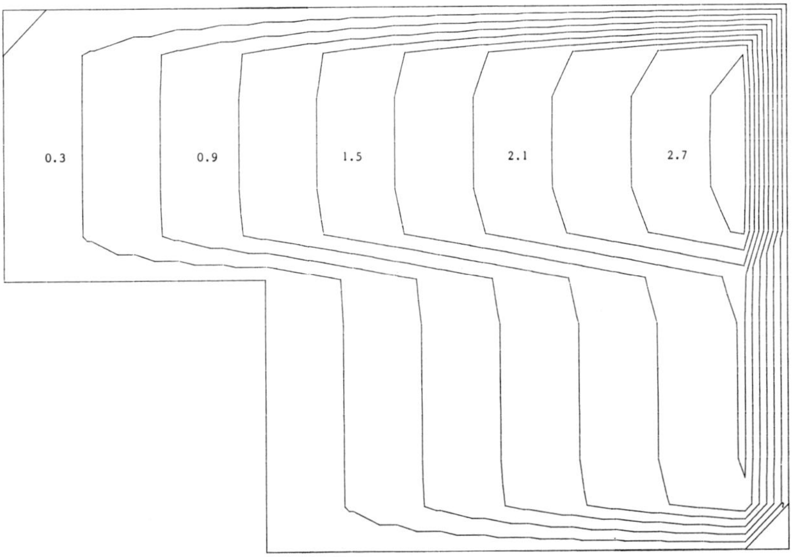

Figure 6.6. Triangulation $\mathscr{T}_h^1$ ($\varepsilon = 10^{-3}$, $\theta = 0$).

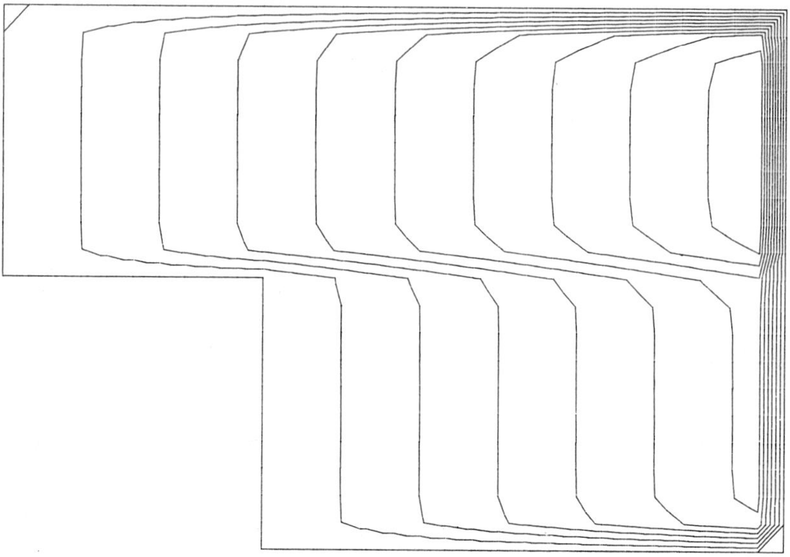

Figure 6.7. Triangulation $\mathscr{T}_h^2$ ($\varepsilon = 10^{-3}$, $\theta = 0$).

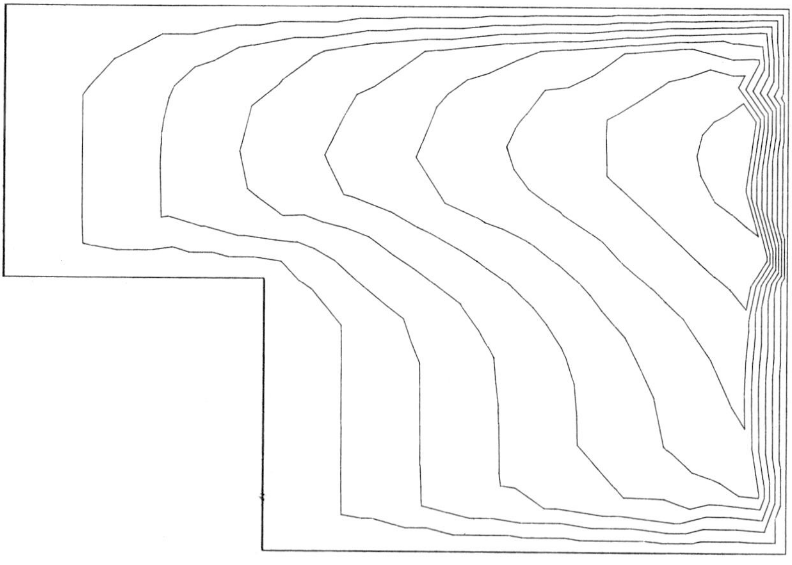

Figure 6.8. Triangulation $\mathscr{T}_h^3$ ($\varepsilon = 10^{-3}$, $\theta = 0$).

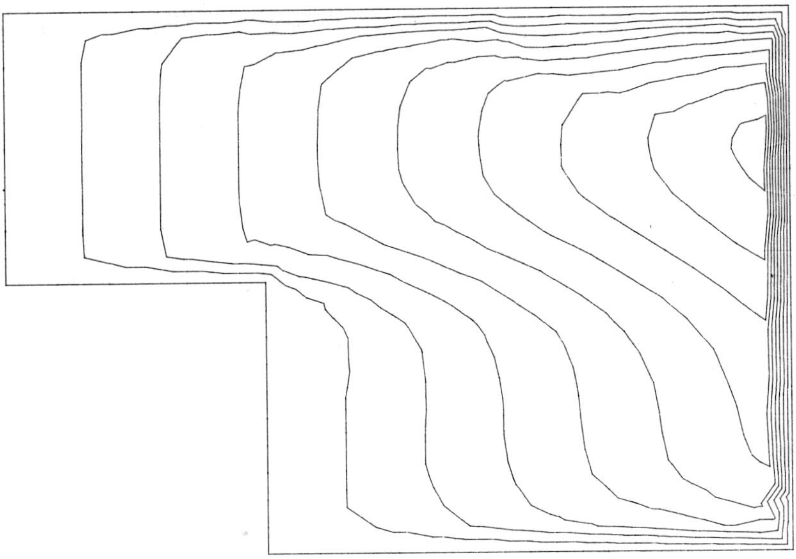

Figure 6.9. Triangulation $\mathscr{T}_h^4$ ($\varepsilon = 10^{-3}, \theta = 0$).

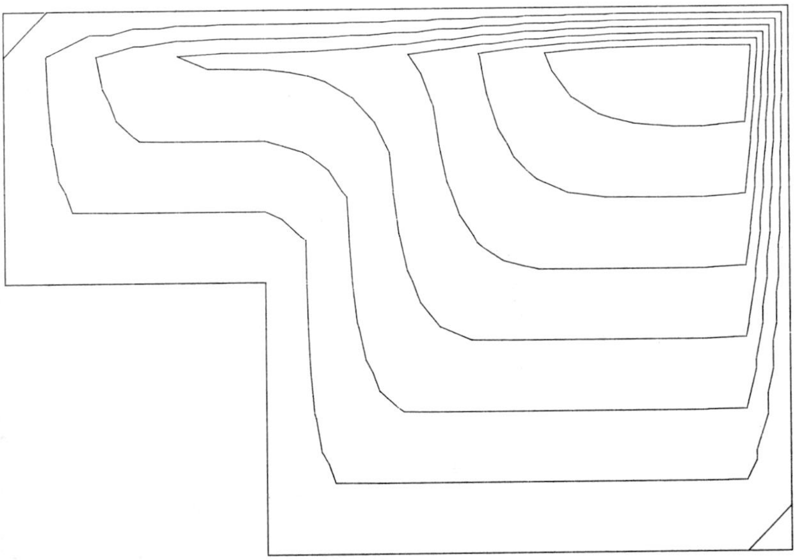

Figure 6.10. Triangulation $\mathscr{T}_h^1$ ($\varepsilon = 10^{-3}, \theta = \pi/3$).

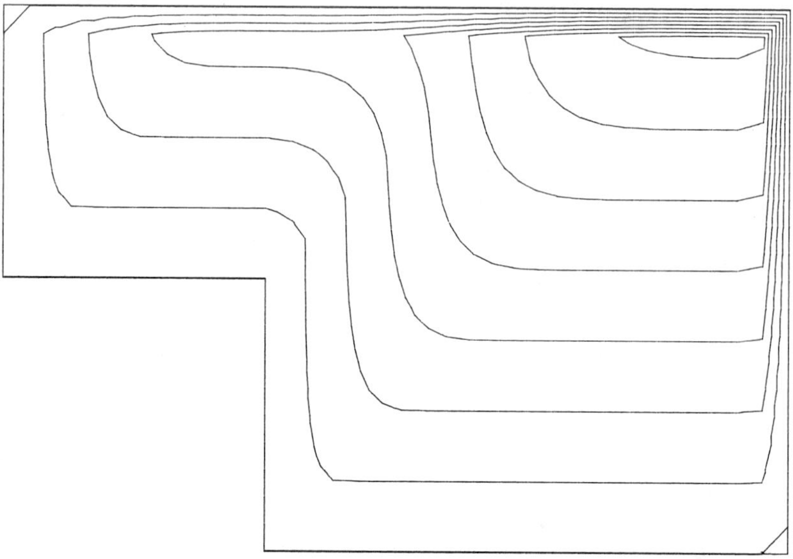

Figure 6.11. Triangulation $\mathscr{T}_h^2$ ($\varepsilon = 10^{-3}, \theta = \pi/3$).

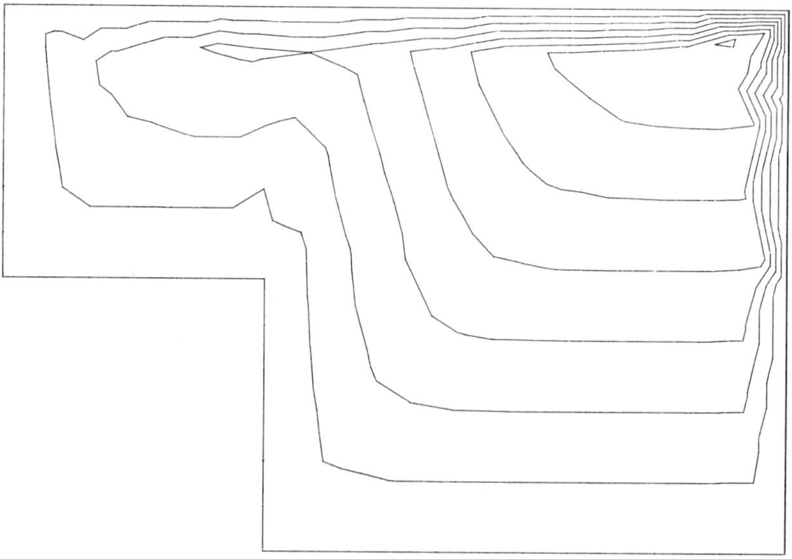

Figure 6.12. Triangulation $\mathscr{T}_h^3$ ($\varepsilon = 10^{-3}, \theta = \pi/3$).

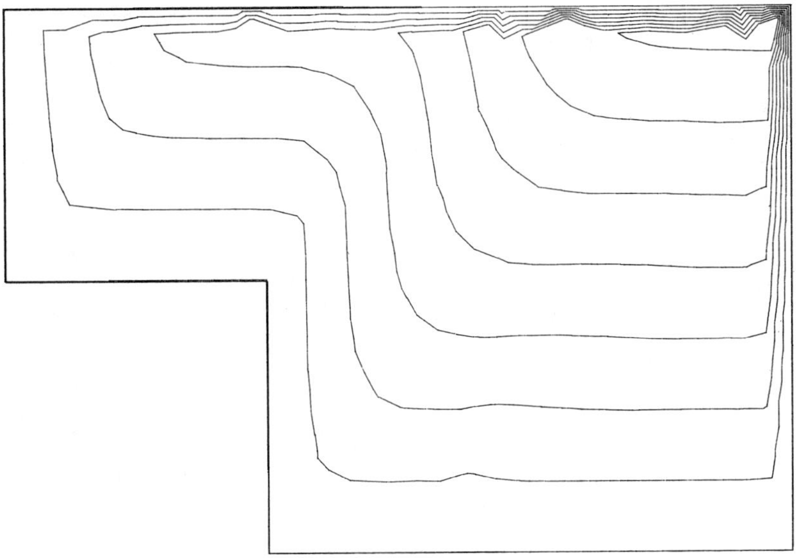

Figure 6.13. Triangulation $\mathscr{T}_h^4$ ($\varepsilon = 10^{-3}, \theta = \pi/3$).

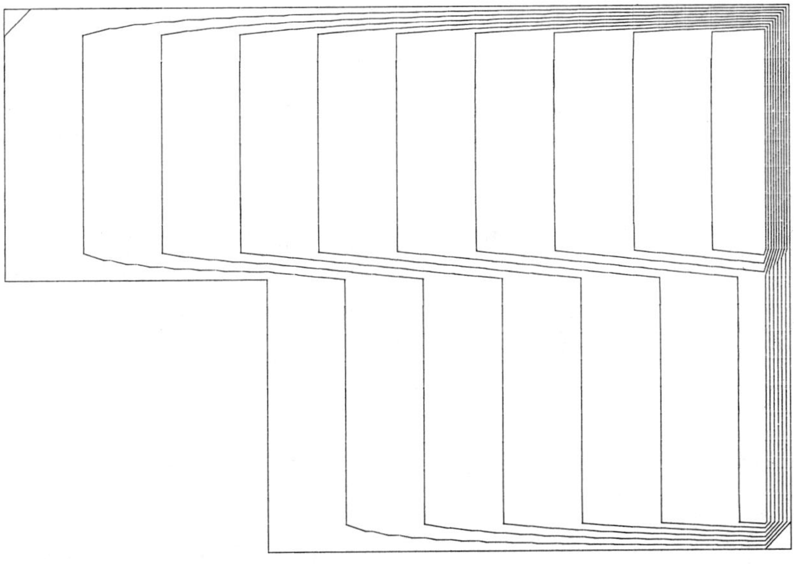

Figure 6.14. Triangulation $\mathscr{T}_h^2$ ($\varepsilon = 10^{-5}, \theta = 0$).

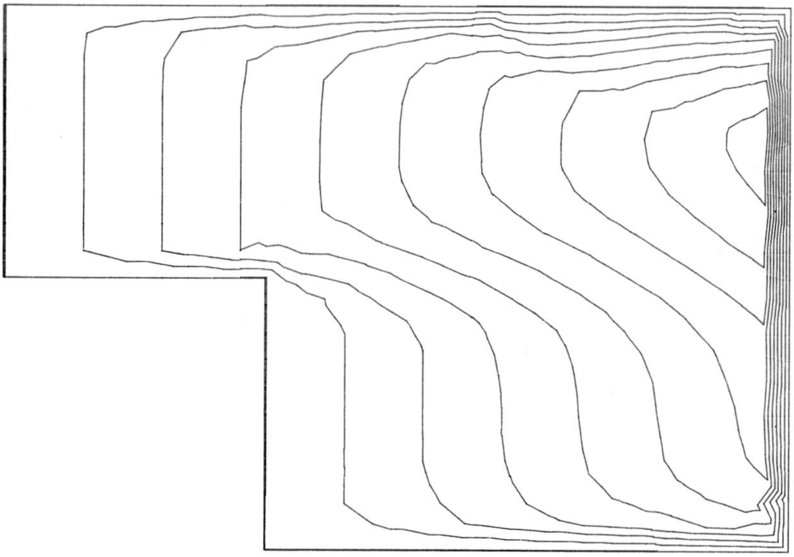

Figure 6.15. Triangulation $\mathscr{T}_h^4$ ($\varepsilon = 10^{-5}, \theta = 0$).

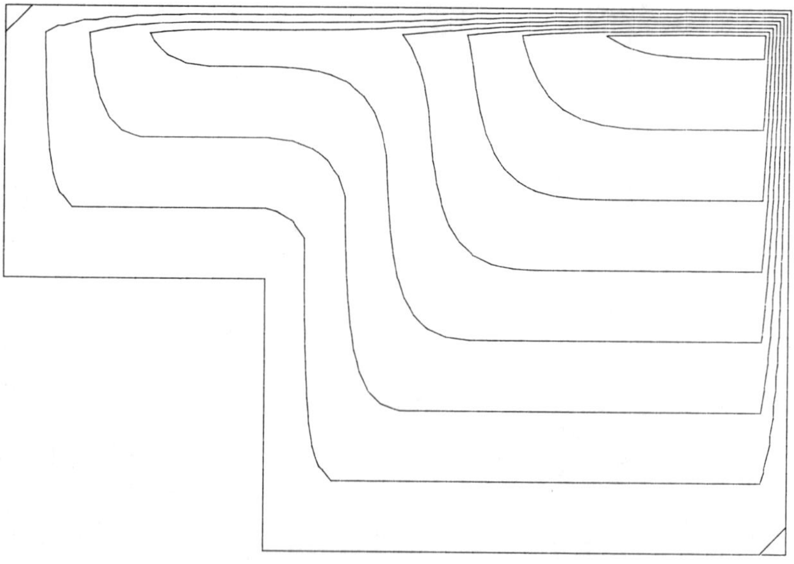

Figure 6.16. Triangulation $\mathscr{T}_h^2$ ($\varepsilon = 10^{-5}, \theta = \pi/3$).

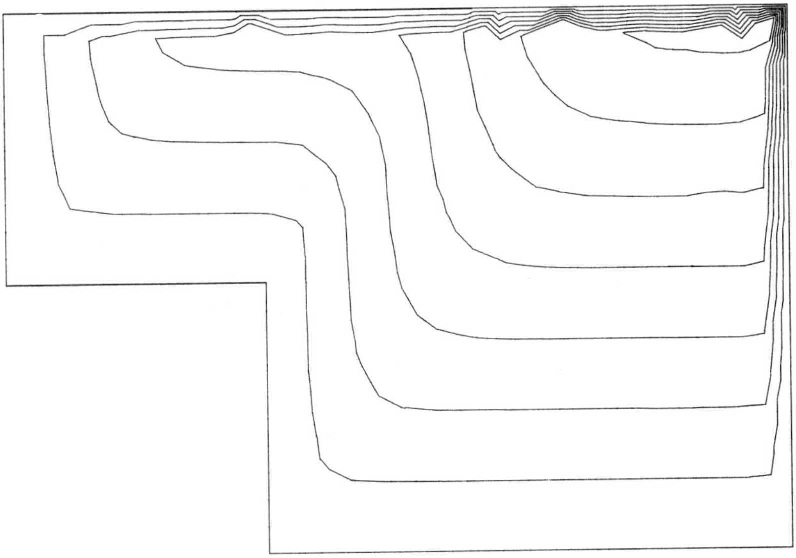

Figure 6.17. Triangulation $\mathscr{T}_h^4$ ($\varepsilon = 10^{-5}, \theta = \pi/3$).

Using the notation $\boldsymbol{\beta} = \{\cos \theta, \sin \theta\}$, we show on Figs. 6.6–6.17 the equipotential lines of the approximate solutions corresponding to various values of ε and θ. The agreement between these results is good, except if $\theta = 0$, for which the results obtained with the regular triangulations $\mathscr{T}_h^1$, $\mathscr{T}_h^2$ are good, unlike those obtained with $\mathscr{T}_h^3$, $\mathscr{T}_h^4$. In fact, the case $\theta = 0$ is quite severe, since it is a case for which the limit, as $\varepsilon \to 0$, of the solution of (6.1) is a discontinuous function; indeed it is easily shown that this limit is

$$\begin{aligned} u(x_1, x_2) &= x_1 \quad \text{if } 1 < x_2 < 2, \\ u(x_1, x_2) &= x_1 - 1 \quad \text{if } 0 < x_2 < 1, \end{aligned} \tag{6.2}$$

and is therefore discontinuous along the line $x_2 = 1$, with a jump of amplitude 1; actually the numerical results obtained with $\mathscr{T}_h^1$, $\mathscr{T}_h^2$ for $\theta = 0$ and very small ε are nearly optimal, since the discontinuities of the limit function (6.2) (including the boundary layers) are supported only by one layer of triangles on which the approximate solution varies continuously without oscillations. The above results can be improved by including special devices to take into account the behavior of the solution in the boundary layers. The poor quality of the results obtained with $\mathscr{T}_h^3$, $\mathscr{T}_h^4$, for $\theta = 0$ and ε very small, seems to be due to the fact, pointed out by Engquist and Kreiss [1], that irregular meshes may introduce very severe phase distorsions. For $\theta \neq 0$ (and $\theta \neq \pi/2$), the limit solution as $\varepsilon \to 0$ is continuous (but not C^1), and therefore the numerical process, is less sensitive to phase distorsion (but exhibits some numerical diffusion (as shown on Figs. 6.10–6.13, 6.16, 6.17) along the line starting from the corner and propagating the discontinuities of the derivatives of the limit function).

7. Concluding Comments

The finite element scheme with upwinding that we have introduced in Sec. 4 of this appendix is quite accurate and extremely robust; it may also handle fairly irregular geometries. The above scheme leads, of course, to a nontrivial coding.

We are presently extending the ideas of Sec. 4 to the solution of more complicated problems (in fluid mechanics, semiconductor simulations, etc.).

APPENDIX III

Some Complements on the Navier–Stokes Equations and Their Numerical Treatment

1. Introduction

In this appendix, wou would like to complete Sec. 5 of Chapter VII, where the numerical solution of the Navier–Stokes equations for incompressible viscous fluids was discussed.

In Sec. 2 we shall discuss the finite element approximation of the Dirichlet boundary condition $\mathbf{u} = \mathbf{g}$ on Γ, with $\int_\Gamma \mathbf{g} \cdot \mathbf{n}\, d\Gamma = 0$, when $\mathbf{g} \neq \mathbf{0}$. In Sec. 3 we shall make comments on the treatment of the nonlinear term, $(\mathbf{u} \cdot \nabla)\mathbf{u}$. In Sec. 4, boundary conditions other than the Dirichlet boundary condition, will be discussed, and their effect on the decomposition properties of the Stokes problem will be considered in Sec. 5.

2. Finite Element Approximation of the Boundary Condition $u = g$ on Γ if $g \neq 0$

Let us consider the steady-state Navier–Stokes equations, i.e.,

$$-\nu\Delta\mathbf{u} + (\mathbf{u} \cdot \nabla)\mathbf{u} + \nabla p = \mathbf{f} \text{ in } \Omega, \quad \nabla \cdot \mathbf{u} = 0 \text{ in } \Omega,$$

$$\mathbf{u} = \mathbf{g} \text{ on } \Gamma, \quad \text{with} \int_\Gamma \mathbf{g} \cdot \mathbf{n}\, d\Gamma = 0 \tag{2.1}$$

($\mathbf{n}$ is the unit vector of the outward normal).

In Chapter VII, Sec. 5.3.3, we have defined the following finite element spaces:

$$V_h = \{\mathbf{v}_h | \mathbf{v}_h \in C^0(\overline{\Omega}) \times C^0(\overline{\Omega}), \mathbf{v}_{h|T} \in P_2 \times P_2, \quad \forall\, T \in \mathcal{T}_h\}, \tag{2.2}$$

$$V_{gh} = \{\mathbf{v}_h | \mathbf{v}_h \in V_h, \mathbf{v}_h = \mathbf{g}_h \text{ on } \Gamma\}, \tag{2.3}$$

where $\mathbf{g}_h$ is a convenient approximation of $\mathbf{g}$ (the following also holds if V_h is defined by (5.21) of Chapter VII, Sec. 5.3.3.1).

A problem of important theoretical and practical interest is to construct $\mathbf{g}_h$ such that

$$\int_\Gamma \mathbf{g}_h \cdot \mathbf{n} \, d\Gamma = 0;$$

for simplicity we suppose that $\mathbf{g}$ is continuous over Γ. We now define the space γV_h as

$$\gamma V_h = \{\boldsymbol{\mu}_h | \boldsymbol{\mu}_h = \mathbf{v}_h|_\Gamma, \quad \mathbf{v}_h \in V_h\}, \tag{2.4}$$

i.e., γV_h is the space of the traces on Γ of those functions $\mathbf{v}_h$ belonging to V_h. Actually, if V_h is defined by (2.2), γV_h is also the space of those functions defined over Γ, taking their values in $\mathbb{R}^2$, continuous over Γ and piecewise quadratic over the edges of $\mathscr{T}_h$ contained into Γ.

Our problem is to construct an approximation $\mathbf{g}_h$ of $\mathbf{g}$ such that

$$\mathbf{g}_h \in \gamma V_h, \qquad \int_\Gamma \mathbf{g}_h \cdot \mathbf{n} \, d\Gamma = 0. \tag{2.5}$$

If $\pi_h \mathbf{g}$ is the unique element of γV_h, obtained from the values taken by $\mathbf{g}$ at those nodes of $\mathscr{T}_h$ belonging to Γ, we usually have $\int_\Gamma \pi_h \mathbf{g} \cdot \mathbf{n} \, d\Gamma \neq 0$. To overcome the above difficulty, we may proceed as follows:

(i) We define an approximation $\mathbf{n}_h$ of $\mathbf{n}$ as the solution of the following linear variational problem in γV_h:

$$\mathbf{n}_h \in \gamma V_h, \quad \int_\Gamma \mathbf{n}_h \cdot \boldsymbol{\mu}_h \, d\Gamma = \int_\Gamma \mathbf{n} \cdot \boldsymbol{\mu}_h \, d\Gamma, \qquad \forall \, \boldsymbol{\mu}_h \in \gamma V_h; \tag{2.6}$$

problem (2.6) is in fact equivalent to a linear system whose matrix is sparse, symmetric, positive definite, and quite easy to compute.

(ii) Then define $\mathbf{g}_h$ by

$$\mathbf{g}_h = \pi_h \mathbf{g} - \left(\frac{\int_\Gamma \pi_h \mathbf{g} \cdot \mathbf{n} \, d\Gamma}{\int_\Gamma \mathbf{n} \cdot \mathbf{n}_h \, d\Gamma} \right) \mathbf{n}_h. \tag{2.7}$$

It is easy to check that (2.6), (2.7) imply (2.5).

3. Some Comments On the Numerical Treatment of the Nonlinear Term $(\boldsymbol{u} \cdot \nabla)\boldsymbol{u}$

Let us denote by $B(\mathbf{u})$ the nonlinear term $(\mathbf{u} \cdot \nabla)\mathbf{u}$ occurring in the Navier–Stokes equations (see Chapter VII, Sec. 5.2). An important property of operator B is

$$\int_\Omega B(\mathbf{v}) \cdot \mathbf{v} \, dx = 0, \qquad \forall \, \mathbf{v} \in (H_0^1(\Omega))^N \text{ such that } \nabla \cdot \mathbf{v} = 0. \tag{3.1}$$

If $\nabla \cdot \mathbf{v} \neq 0$, we have $\int_\Omega B(\mathbf{v}) \cdot \mathbf{v}\, dx \neq 0$, in general. If we now consider the various alternating-direction methods discussed in Chapter VII, Sec. 5, we observe that we do not require the incompressibility condition $\nabla \cdot \mathbf{u} = 0$ to be satisfied for the nonlinear steps. In order to improve the well-posedness properties of the elliptic problems to solve at these nonlinear steps (and also to simplify convergence proofs), one may replace the original nonlinear term $B(\mathbf{u})$ by

$$\tilde{B}(\mathbf{u}) = (\mathbf{u} \cdot \nabla)\mathbf{u} + \tfrac{1}{2}\mathbf{u}(\nabla \cdot \mathbf{u}),$$

following Temam [1]. It is clear that $B(\mathbf{u}) = \tilde{B}(\mathbf{u})$ if $\nabla \cdot \mathbf{u} = 0$. Actually the good property of $\tilde{B}$ is that

$$\int_\Omega \tilde{B}(\mathbf{v}) \cdot \mathbf{v}\, dx = 0, \qquad \forall\, \mathbf{v} \in (H_0^1(\Omega))^N \ \text{(even if } \nabla \cdot \mathbf{v} \neq 0\text{)} \tag{3.2}$$

(see Temam [1] for further details).

EXERCISE 3.1. Prove (3.2)

A similar observation holds for the approximate problems discussed in Chapter VII, Sec. 5. However, we have to mention that the numerical results obtained using either B or $\tilde{B}$ are practically identical, provided that Δt is sufficiently small (Δt is the time step occurring in the alternating-direction methods of Chapter VII, Sec. 5).

4. Further Comments on the Boundary Conditions

4.1. Synopsis

In Chapter VII, Sec. 5, we have considered the solution of the Navier–Stokes equations for boundary conditions of Dirichlet type. Actually, in many problems, these Dirichlet boundary conditions are inappropriate; they have, for example, the property of being *too reflecting* if one approximates flows in an unbounded region of $\mathbb{R}^N$ (like flows around an obstacle or in a channel) by flows in a bounded domain.

The goal of this section is to show that the methods discussed in Chapter VII, Sec. 5 can be generalized to the solution of the following Navier–Stokes problem:

$$\alpha\mathbf{u} - \nu\Delta\mathbf{u} + (\mathbf{u} \cdot \nabla)\mathbf{u} + \nabla p = \mathbf{f} \text{ in } \Omega, \quad \nabla \cdot \mathbf{u} = 0 \text{ in } \Omega,$$

$$\mathbf{u} = \mathbf{g}_0 \text{ on } \Gamma_0, \quad \nu\frac{\partial \mathbf{u}}{\partial n} - p\mathbf{n} = \mathbf{g}_1 \text{ on } \Gamma_1. \tag{4.1}$$

In (4.1), ν, $\mathbf{u}$, p, f are as in Chapter VII, Sec. 5.2, α is a non-negative constant ($\alpha = 0$ for a steady flow, $\alpha > 0$ if (4.1) follows from the time discretization of the time-dependent Navier–Stokes equations), $\mathbf{n}$ is the unit vector of the outward normal at Γ; finally $\mathbf{g}_0$, $\mathbf{g}_1$ are two functions defined over Γ_0, Γ_1, respectively, where Γ_0, Γ_1 are two subsets of the boundary Γ of Ω such that

$$\int_{\Gamma_i} d\Gamma > 0, \qquad \forall\, i = 0, 1, \qquad \Gamma_0 \cup \Gamma_1 = \Gamma, \qquad \int_{\Gamma_0 \cap \Gamma_1} d\Gamma = 0; \tag{4.2}$$

a partition of Γ obeying (4.2) is shown on Fig. 4.1.

Several continuous and discrete variational formulations of (4.1) will be discussed in Secs. 4.2 and 4.3, and their decomposition properties will be discussed in Sec. 5. Some (useful) generalizations of the results of this section may be found in Conca [1].

Remark 4.1. Suppose $\int_{\Gamma_1} d\Gamma > 0$; then the boundary condition

$$\nu \frac{\partial \mathbf{u}}{\partial n} - \mathbf{n}p = \mathbf{g}_1 \text{ on } \Gamma_1$$

adjusts the pressure without ambiguity (unlike the Dirichlet boundary condition $\mathbf{u}|_\Gamma = \mathbf{g}$ for which the pressure is determined only to within an arbitrary constant).

4.2. Variational formulations of (4.1). (I) The continuous case

For simplicity we suppose that $\mathbf{f} \in (L^2(\Omega))^N$ in the sequel, and also that Ω is bounded.

4.2.1. *Standard variational formulations*

Let us define the space $\mathscr{H}_0$ by

$$\mathscr{H}_0 = \{\mathbf{v} \,|\, \mathbf{v} \in (H^1(\Omega))^N,\ \mathbf{v} = \mathbf{0} \text{ on } \Gamma_0\}. \tag{4.3}$$

Now multiplying the first relation (4.1) by $\mathbf{v} \in \mathscr{H}_0$, and integrating by parts (i.e., applying the Green–Ostrogradsky formula (4.33) of Appendix I, Sec. 4), we obtain

$$\alpha \int_\Omega \mathbf{u} \cdot \mathbf{v}\, dx + \nu \int_\Omega \nabla \mathbf{u} \cdot \nabla \mathbf{v}\, dx + \int_\Omega (\mathbf{u} \cdot \nabla)\mathbf{u} \cdot \mathbf{v}\, dx - \int_\Omega p \nabla \cdot \mathbf{v}\, dx$$

$$= \int_\Omega \mathbf{f} \cdot \mathbf{v}\, dx + \int_{\Gamma_1} \left(\nu \frac{\partial \mathbf{u}}{\partial n} - p\mathbf{n} \right) \cdot \mathbf{v}\, d\Gamma,$$

Figure 4.1

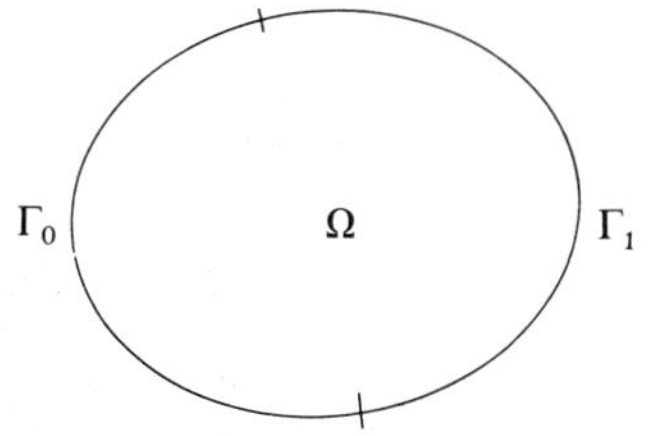

which combined with the last relation (4.1), implies, that any pair $\{\mathbf{u}, p\}$, a solution of (4.1), has to satisfy

$$\alpha \int_\Omega \mathbf{u} \cdot \mathbf{v}\, dx + \nu \int_\Omega \nabla \mathbf{u} \cdot \nabla \mathbf{v}\, dx + \int_\Omega (\mathbf{u} \cdot \nabla)\mathbf{u} \cdot \mathbf{v}\, dx - \int_\Omega p \nabla \cdot \mathbf{v}\, dx$$
$$= \int_\Omega \mathbf{f} \cdot \mathbf{v}\, dx + \int_{\Gamma_1} \mathbf{g}_1 \cdot \mathbf{v}\, d\Gamma, \qquad \forall\, \mathbf{v} \in \mathscr{H}_0. \tag{4.4}$$

We now suppose that $\mathbf{g}_0$ is sufficiently smooth to be the trace on Γ_0 of a function $\tilde{\mathbf{g}}_0 \in (H^1(\Omega))^N$ (if Γ_1 is a "good" subset of Γ—which is always the case in practice—we can choose $\tilde{\mathbf{g}}_0$ to have $\nabla \cdot \tilde{\mathbf{g}}_0 = 0$ on Ω); it is then quite natural—from (4.4)—to introduce the following variational problem:

Find $\{\mathbf{u}, p\} \in (H^1(\Omega))^N \times L^2(\Omega)$ *such that* $\mathbf{u} = \mathbf{g}_0$ *on* Γ_0 *and*

$$\alpha \int_\Omega \mathbf{u} \cdot \mathbf{v}\, dx + \nu \int_\Omega \nabla \mathbf{u} \cdot \nabla \mathbf{v}\, dx + \int_\Omega (\mathbf{u} \cdot \nabla)\mathbf{u} \cdot \mathbf{v}\, dx - \int_\Omega p \nabla \cdot \mathbf{v}\, dx$$
$$= \int_\Omega \mathbf{f} \cdot \mathbf{v}\, dx + \int_{\Gamma_1} \mathbf{g}_1 \cdot \mathbf{v}\, d\Gamma, \qquad \forall\, \mathbf{v} \in \mathscr{H}_0, \quad \nabla \cdot \mathbf{u} = 0 \text{ in } \Omega. \tag{4.5}$$

If we now restrict the test functions $\mathbf{v}$ in (4.5) to the space V_0 defined by

$$V_0 = \{\mathbf{v} \mid \mathbf{v} \in \mathscr{H}_0, \nabla \cdot \mathbf{v} = 0 \text{ in } \Omega\}, \tag{4.6}$$

formulation (4.5) yields the variational problem:

Find $\{\mathbf{u}, p\} \in (H^1(\Omega))^N \times L^2(\Omega)$ *such that* $\mathbf{u} = \mathbf{g}_0$ *on* Γ_0, $\nabla \cdot \mathbf{u} = 0$ *in* Ω, *and*

$$\alpha \int_\Omega \mathbf{u} \cdot \mathbf{v}\, dx + \nu \int_\Omega \nabla \mathbf{u} \cdot \nabla \mathbf{v}\, dx + \int_\Omega (\mathbf{u} \cdot \nabla)\mathbf{u} \cdot \mathbf{v}\, dx$$
$$= \int_\Omega \mathbf{f} \cdot \mathbf{v}\, dx + \int_{\Gamma_1} \mathbf{g}_1 \cdot \mathbf{v}\, d\Gamma, \qquad \forall\, \mathbf{v} \in V_0, \tag{4.7}$$

in which the pressure p is (apparently) eliminated.

Proving the existence of solutions for (4.5), (4.7), and also the equivalence between these problems and (4.1), is beyond the scope of this book, and therefore will not be discussed here (indeed these existence and equivalence results hold if we suppose that $\mathbf{g}_1 \in (L^2(\Gamma_1))^N$, for example). However, proving the existence and uniqueness of solutions for the corresponding Stokes problem (i.e., for the problem obtained by neglecting the nonlinear term $(\mathbf{u} \cdot \nabla)\mathbf{u}$) is not too difficult and left as an exercise to the reader.

4.2.2. *A less standard variational formulation*

Let $\{\mathbf{u}, p\}$ be a solution of (4.1); multiplying the first relation (4.1) by $\mathbf{v} + \nabla\phi$, where $\mathbf{v} \in \mathscr{H}_0$ and $\phi \in \mathscr{D}(\Omega)$, and since $\nabla \cdot \mathbf{u} = 0$ on Ω, we obtain

$$\alpha \int_\Omega \mathbf{u} \cdot \mathbf{v}\, dx + \nu \int_\Omega \nabla\mathbf{u} \cdot \nabla\mathbf{v}\, dx + \int_\Omega (\mathbf{u} \cdot \nabla)\mathbf{u} \cdot (\mathbf{v} + \nabla\phi)\, dx - \int_\Omega p(\Delta\phi + \nabla \cdot \mathbf{v})\, dx = \int_\Omega \mathbf{f} \cdot (\mathbf{v} + \nabla\phi)\, dx + \int_{\Gamma_1} \mathbf{g}_1 \cdot \mathbf{v}\, d\Gamma, \quad \forall \{\mathbf{v}, \phi\} \in \mathscr{H}_0 \times \mathscr{D}(\Omega). \tag{4.8}$$

From the density property $\overline{\mathscr{D}(\Omega)}^{H^2} = H_0^2(\Omega)$,[1] relation (4.8) also holds for any $\{\mathbf{v}, \phi\} \in \mathscr{H}_0 \times H_0^2(\Omega)$. We now define two spaces, W_0 and W_g, by

$$W_0 = \left\{\{\mathbf{v}, \phi\} \mid \{\mathbf{v}, \phi\} \in \mathscr{H}_0 \times H_0^1(\Omega), \int_\Omega \nabla\phi \cdot \nabla w\, dx = \int_\Omega \nabla \cdot \mathbf{v} w\, dx, \forall\, w \in H^1(\Omega)\right\}, \tag{4.9}$$

$$W_g = \left\{\{\mathbf{v}, \phi\} \mid \{\mathbf{v}, \phi\} \in (H^1(\Omega))^N \times H_0^1(\Omega), \mathbf{v} = \mathbf{g}_0 \text{ on } \Gamma_0, \int_\Omega \nabla\phi \cdot \nabla w\, dx = \int_\Omega \nabla \cdot \mathbf{v} w\, dx, \quad \forall\, w \in H^1(\Omega)\right\}, \tag{4.10}$$

respectively. A key result is given by:

Lemma 4.1. *Suppose that the boundary Γ of Ω is sufficiently smooth (or Ω is convex). Then, if $\{\mathbf{v}, \phi\} \in W_0$ or W_g, we have*

$$-\Delta\phi = \nabla \cdot \mathbf{v} \quad \text{in } \Omega, \tag{4.11}$$

$$\phi \in H_0^2(\Omega). \tag{4.12}$$

PROOF. Since W_0 is the particular case of W_g, corresponding to $\mathbf{g}_0 = \mathbf{0}$, we prove the lemma for this latter space. Consider $\{\mathbf{v}, \phi\} \in W_g$; we then have (since $H_0^1(\Omega) \subset H^1(\Omega)$)

$$\phi \in H_0^1(\Omega)$$

[1] $H_0^2(\Omega) = \{\phi \mid \phi \in H^2(\Omega), \phi = \partial\phi/\partial n = 0 \text{ on } \Gamma\}$.

and

$$\int_\Omega \nabla\phi \cdot \nabla w\, dx = \int_\Omega \nabla \cdot \mathbf{v} w\, dx, \qquad \forall\, w \in H_0^1(\Omega),$$

which implies that ϕ is the unique solution in $H_0^1(\Omega)$ of the Dirichlet problem

$$-\Delta\phi = \nabla \cdot \mathbf{v} \text{ in } \Omega, \quad \phi = 0 \text{ on } \Gamma. \tag{4.13}$$

Since $\mathbf{v} \in (H^1(\Omega))^N$, we have $\nabla \cdot \mathbf{v} \in L^2(\Omega)$; it then follows from Chapter I, Sec. 2.4 that the smoothness of Γ (or the convexity of Ω) implies the regularity property

$$\phi \in H^2(\Omega) \cap H_0^1(\Omega). \tag{4.14}$$

Multiplying the two sides of the first relation (4.13) by $w \in H^1(\Omega)$, we obtain, from Green's formula,

$$-\int_\Omega \Delta\phi w\, dx = \int_\Omega \nabla\phi \cdot \nabla w\, dx - \int_\Gamma \frac{\partial\phi}{\partial n} w\, d\Gamma = \int_\Omega \nabla \cdot \mathbf{v} w\, dx, \qquad \forall\, w \in H^1(\Omega). \tag{4.15}$$

But since $\{\mathbf{v}, \phi\} \in W_g$ implies

$$\int_\Omega \nabla\phi \cdot \nabla w\, dx = \int_\Omega \nabla \cdot \mathbf{v} w\, dx, \qquad \forall\, w \in H^1(\Omega),$$

from (4.15) it follows that $\int_\Gamma (\partial\phi/\partial n) w\, d\Gamma = 0, \forall\, w \in H^1(\Omega)$, which in turn implies

$$\frac{\partial\phi}{\partial n} = 0 \text{ on } \Gamma; \tag{4.16}$$

(4.14), (4.16) imply (4.12) and complete the proof of the lemma. □

From Lemma 4.1 it is feasible to take $\{\mathbf{v}, \phi\} \in W_0$ in (4.8), and since (from (4.11)) $\Delta\phi + \nabla \cdot \mathbf{v} = 0$, the integral containing p vanishes; on the other hand, we clearly find that

$$\text{if } \{\mathbf{u}, p\} \in (H^1(\Omega))^N \times L^2(\Omega) \text{ is solution of (4.1), then } \{\mathbf{u}, 0\} \in W_g. \tag{4.17}$$

Collecting the above results and properties, we have finally proved that if $\{\mathbf{u}, p\}$ is solution of (4.1) (and (4.5), (4.7)), then

$$\{\mathbf{u}, 0\} \in W_g \text{ and } \alpha \int_\Omega \mathbf{u} \cdot \mathbf{v}\, dx + \nu \int_\Omega \nabla\mathbf{u} \cdot \nabla\mathbf{v}\, dx + \int_\Omega (\mathbf{u} \cdot \nabla)\mathbf{u} \cdot (\mathbf{v} + \nabla\phi)\, dx$$
$$= \int_\Omega \mathbf{f} \cdot (\mathbf{v} + \nabla\phi)\, dx + \int_{\Gamma_1} \mathbf{g}_1 \cdot \mathbf{v}\, d\Gamma, \qquad \forall\, \{\mathbf{v}, \phi\} \in W_0. \tag{4.18}$$

It is natural to introduce the following variational problem:

Find $\{\mathbf{u}, \psi\} \in W_g$ *such that*

$$\alpha \int_\Omega \mathbf{u} \cdot \mathbf{v}\, dx + \nu \int_\Omega \nabla\mathbf{u} \cdot \nabla\mathbf{v}\, dx + \int_\Omega (\mathbf{u} \cdot \nabla)\mathbf{u} \cdot (\mathbf{v} + \nabla\phi)\, dx$$
$$= \int_\Omega \mathbf{f} \cdot (\mathbf{v} + \nabla\phi)\, dx + \int_{\Gamma_1} \mathbf{g}_1 \cdot \mathbf{v}\, d\Gamma, \qquad \forall\, \{\mathbf{v}, \phi\} \in W_0. \tag{4.19}$$

We have seen that any solution $\{\mathbf{u}, p\} \in (H^1(\Omega))^N \times L^2(\Omega)$ of (4.1) (and (4.5), (4.7)) is such that $\{\mathbf{u}, \psi\}$ is solution of (4.19) with $\psi = 0$. Actually the reciprocal property is true and we have:

Proposition 4.1. *There is equivalence between* (4.19) *and* (4.1); *moreover, if* $\{\mathbf{u}, \psi\}$ *is solution of* (4.19), *we have* $\psi = 0$.

PROOF. We have already proved that if $\{\mathbf{u}, p\} \in (H^1(\Omega))^N \times L^2(\Omega)$ is solution of (4.1) (and therefore of (4.5), (4.7)) then $\{\mathbf{u}, 0\}$ is solution of (4.19). We now prove the reciprocal property.

Let $\{\mathbf{u}, \psi\} \in W_g$ be a solution of (4.19); from Lemma 4.1 we know that $-\Delta\psi = \nabla \cdot \mathbf{u}$ in Ω and that $\psi \in H_0^2(\Omega)$. Since $\psi \in H_0^2(\Omega)$, we have $\nabla\psi \in (H_0^1(\Omega))^N$ and in fact

$$\{-\nabla\psi, \psi\} \in W_0. \tag{4.20}$$

From (4.20) it follows that we can take $\{\mathbf{v}, \phi\} = \{-\nabla\psi, \psi\}$ in (4.19); we then obtain

$$-\alpha \int_\Omega \mathbf{u} \cdot \nabla\psi \, dx - \nu \int_\Omega \nabla\mathbf{u} \cdot \nabla(\nabla\psi) \, dx = \alpha \int_\Omega \nabla \cdot \mathbf{u}\psi \, dx - \nu \int_\Omega \Delta\psi\nabla \cdot \mathbf{u} \, dx = 0. \tag{4.21}$$

Then combining (4.21) and $-\Delta\psi = \nabla \cdot \mathbf{u}$, we in turn obtain

$$\alpha \int_\Omega |\nabla\psi|^2 \, dx + \nu \int_\Omega |\Delta\psi|^2 \, dx = 0, \tag{4.22}$$

which implies that $\psi = 0$ and also $\nabla \cdot \mathbf{u} = -\Delta\psi = 0$. Thus we have proved that if $\{\mathbf{u}, \psi\}$ is solution of (4.19), then $\psi = 0$ and $\mathbf{u} \in V_g$. Now consider $\mathbf{v} \in V_0$; we clearly have $\{\mathbf{v}, 0\} \in W_0$, and by substitution in (4.19) we finally have proved that $\mathbf{u}$ satisfies

$$\mathbf{u} \in V_g, \quad \alpha \int_\Omega \mathbf{u} \cdot \mathbf{v} \, dx + \nu \int_\Omega \nabla\mathbf{u} \cdot \nabla\mathbf{v} \, dx + \int_\Omega (\mathbf{u} \cdot \mathbf{v})\mathbf{u} \cdot \mathbf{v} \, dx$$
$$= \int_\Omega \mathbf{f} \cdot \mathbf{v} \, dx + \int_{\Gamma_1} \mathbf{g}_1 \cdot \mathbf{v} \, d\Gamma, \qquad \forall\, \mathbf{v} \in V_0,$$

i.e., $\mathbf{u}$ is a solution of (4.7), which, by equivalence with (4.1), completes the proof of the proposition.

Remark 4.2. Proposition 4.1 generalizes Theorem 5.1 of Chapter VII, Sec. 5.3.2.2.

4.3. Variational formulations of (4.1). (II) Finite element approximations

We suppose that Ω is a bounded polygonal domain of $\mathbb{R}^2$.

4.3.1. *Triangulation of Ω. Fundamental discrete spaces*

We follow Chapter VII, Sec. 5.3.3.1. Let $\{\mathcal{T}_h\}_h$ be a family of triangulations of Ω such that $\overline{\Omega} = \bigcup_{T \in \mathcal{T}_h} T$. We set $h(T)$ equal to the length of the greatest side of T, $h = \max_{T \in \mathcal{T}_h} h(T)$, and we suppose that

$$\text{the points of } \Gamma \text{ at the interface of } \Gamma_0 \text{ and } \Gamma_1 \text{ are vertices of } \mathcal{T}_h. \tag{4.23}$$

We then define the following finite element spaces:

$$H_h^1 = \{\phi_h \in C^0(\bar{\Omega}), \phi_{h|T} \in P_1, \quad \forall\, T \in \mathcal{T}_h\}, \tag{4.24}$$

$$H_{0h}^1 = \{\phi_h \in H_h^1, \phi_h = 0 \text{ on } \Gamma\} = H_h^1 \cap H_0^1(\Omega), \tag{4.25}$$

$$V_h = \{\mathbf{v}_h \in (C^0(\bar{\Omega}))^2, \mathbf{v}_{h|T} \in P_2 \times P_2, \quad \forall\, T \in \mathcal{T}_h\}, \tag{4.26}$$

$$V_{gh} = \{\mathbf{v}_h \in V_h, \mathbf{v}_h = \mathbf{g}_{0h} \text{ on } \Gamma_0\}, \tag{4.27}$$

where $\mathbf{g}_{0h}$ is a convenient approximation of $\mathbf{g}_0$ (if $\mathbf{g}_0 = \mathbf{0}$, one takes $\mathbf{g}_{0h} = \mathbf{0}$ in (4.27) to obtain V_{0h}),

$$W_{gh} = \left\{\{\mathbf{v}_h, \phi_h\} \in V_{gh} \times H_{0h}^1, \int_\Omega \nabla\phi_h \cdot \nabla w_h \, dx = \int_\Omega \nabla \cdot \mathbf{v}_h w_h \, dx, \quad \forall\, w_h \in H_h^1\right\}. \tag{4.28}$$

As in Chapter VII, Sec. 5.3.3.1, we can also use the variants of V_{gh}, W_{gh} obtained from

$$V_h = \{\mathbf{v}_h \in (C^0(\bar{\Omega}))^2, \mathbf{v}_{h|T} \in P_1 \times P_1, \quad \forall\, T \in \tilde{\mathcal{T}}_h\}, \tag{4.29}$$

where $\tilde{\mathcal{T}}_h$ is the triangulation of Ω obtained from $\mathcal{T}_h$ by subdivision of each $T \in \mathcal{T}_h$ into four subtriangles, by joining the mid-sides (see Fig. 5.1 of Chapter VII, Sec. 5.3.3.1).

4.3.2. *A first approximation of the Navier–Stokes equations* (4.1)

This is the finite-dimensional analogue of formulation (4.5) defined as follows:

Find $\{\mathbf{u}_h, p_h\} \in V_{gh} \times H_h^1$ *such that*

$$\alpha \int_\Omega \mathbf{u}_h \cdot \mathbf{v}_h \, dx + \nu \int_\Omega \nabla\mathbf{u}_h \cdot \nabla\mathbf{v}_h \, dx + \int_\Omega (\mathbf{u}_h \cdot \nabla)\mathbf{u}_h \cdot \mathbf{v}_h \, dx$$

$$- \int_\Omega p_h \nabla \cdot \mathbf{v}_h \, dx = \int_\Omega \mathbf{f}_h \cdot \mathbf{v}_h \, dx + \int_{\Gamma_1} \mathbf{g}_{1h} \cdot \mathbf{v}_h \, d\Gamma, \qquad \forall\, \mathbf{v}_h \in V_{0h},$$

$$\int_\Omega \nabla \cdot \mathbf{u}_h q_h dx = 0, \qquad \forall\, q_h \in H_h^1, \tag{4.30}$$

where $\mathbf{f}_h$ and $\mathbf{g}_{1h}$ are convenient approximations of $\mathbf{f}$ and $\mathbf{g}_1$, respectively. Formulation (4.30) generalizes (5.156) of Chapter VII, Sec. 5.8.3.

4.3.3. *A second approximation of the Navier–Stokes equations* (4.1)

This is the finite-dimensional analogue of formulation (4.19) defined as follows:

Find $\{\mathbf{u}_h, \psi_h\} \in W_{gh}$ *such that*

$$\alpha \int_\Omega \mathbf{u}_h \cdot \mathbf{v}_h \, dx + \nu \int_\Omega \nabla \mathbf{u}_h \cdot \nabla \mathbf{v}_h \, dx + \int_\Omega (\mathbf{u}_h \cdot \nabla)\mathbf{u}_h \cdot (\mathbf{v}_h + \nabla \phi_h) \, dx$$

$$= \int_\Omega \mathbf{f}_h \cdot (\mathbf{v}_h + \nabla \phi_h) \, dx + \int_{\Gamma_1} \mathbf{g}_{1h} \cdot \mathbf{v}_h \, d\Gamma, \qquad \forall \{\mathbf{v}_h, \phi_h\} \in W_{0h}; \tag{4.31}$$

formulation (4.31) generalizes $(5.38)_1$ of Chapter VII, Sec. 5.3.4. Actually any pair $\{\mathbf{u}_h, \psi_h\}$, a solution of (4.31), is characterized by the existence of a discrete pressure $p_h \in H_h^1$ such that

$$\int_\Omega \nabla p_h \cdot \nabla w_h \, dx + \int_\Omega (\mathbf{u}_h \cdot \nabla)\mathbf{u}_h \cdot \nabla w_h \, dx = \int_\Omega \mathbf{f}_h \cdot \nabla w_h \, dx, \; \forall \; w_h \in H_{0h}^1,$$

$$\alpha \int_\Omega \mathbf{u}_h \cdot \mathbf{v}_h \, dx + \nu \int_\Omega \nabla \mathbf{u}_h \cdot \nabla \mathbf{v}_h \, dx + \int_\Omega (\mathbf{u}_h \cdot \nabla)\mathbf{u}_h \cdot \mathbf{v}_h \, dx - \int_\Omega p_h \nabla \cdot \mathbf{v}_h \, dx$$

$$= \int_\Omega \mathbf{f}_h \cdot \mathbf{v}_h \, dx + \int_{\Gamma_1} \mathbf{g}_{1h} \cdot \mathbf{v}_h \, d\Gamma, \qquad \forall \mathbf{v}_h \in V_{0h}, \quad \{\mathbf{u}_h, \psi_h\} \in W_{gh}, \quad p_h \in H_h^1. \tag{4.32}$$

The decomposition properties of the approximate problem (4.31) will be discussed in Sec. 5.

4.4. Further comments

The boundary condition

$$\nu \frac{\partial \mathbf{u}}{\partial n} - p\mathbf{n} = \mathbf{g}_1 \text{ on } \Gamma_1 \tag{4.33}$$

has no obvious physical meaning. It has, however, the advantage—as mentioned before—of being less reflecting than the Dirichlet boundary condition; also it occurs in a natural way when some domain decomposition techniques are applied to the solution of the Stokes and Navier–Stokes equations. A more physical boundary condition would be to specify the normal stress on Γ_1, i.e., to consider

$$\bar{\bar{\boldsymbol{\sigma}}}\mathbf{n} = \mathbf{g}_1 \text{ on } \Gamma_1, \tag{4.34}$$

where $\bar{\bar{\boldsymbol{\sigma}}} = (\sigma_{ij})$ is the stress tensor defined by

$$\sigma_{ij} = -p\delta_{ij} + \nu\left(\frac{\partial u_i}{\partial x_j} + \frac{\partial u_j}{\partial x_i}\right)$$

(δ_{ij}: Kronecker's symbol).

As may be expected, the numerical treatment of (4.34) is more complicated than that of (4.33) and will not be discussed here.

5. Decomposition Properties of the Continuous and Discrete Stokes Problems of Sec. 4. Application to Their Numerical Solution

5.1. Synopsis

The main goal of this section is to show that the decomposition properties of the Stokes problem discussed in Chapter VII, Sec. 5 (when $\Gamma_0 = \Gamma, \Gamma_1 = \emptyset$) still hold for the Stokes problem associated with the various formulations of the Navier–Stokes equations (4.1) discussed in Sec. 4 of this appendix.

5.2. Formulations of the continuous and discrete Stokes problems

The notation are those of Sec. 4 of this appendix. The Stokes problem that we consider is

$$\alpha \mathbf{u} - \nu \Delta \mathbf{u} + \nabla p = \mathbf{f} \text{ in } \Omega, \qquad \nabla \cdot \mathbf{u} = 0 \text{ in } \Omega,$$
$$\mathbf{u} = \mathbf{g}_0 \text{ on } \Gamma_0, \qquad \nu \frac{\partial \mathbf{u}}{\partial n} - p\mathbf{n} = g_1 \text{ on } \Gamma_1. \tag{5.1}$$

A first variational formulation of (5.1) is:

Find $\{\mathbf{u}, p\} \in (H^1(\Omega))^N \times L^2(\Omega)$ *such that* $\mathbf{u} = \mathbf{g}_0$ *on* Γ_0 *and*

$$\alpha \int_\Omega \mathbf{u} \cdot \mathbf{v}\, dx + \nu \int_\Omega \nabla \mathbf{u} \cdot \nabla \mathbf{v}\, dx - \int_\Omega p \nabla \cdot \mathbf{v}\, dx$$
$$= \int_\Omega \mathbf{f} \cdot \mathbf{v}\, dx + \int_{\Gamma_1} \mathbf{g}_1 \cdot \mathbf{v}\, d\Gamma, \qquad \forall\, \mathbf{v} \in (H^1(\Omega))^N, \quad \mathbf{v} = 0 \textit{ on } \Gamma_0,$$
$$\nabla \cdot \mathbf{u} = 0 \textit{ in } \Omega. \tag{5.2}$$

(5.2) is the Stokes version of (4.5) of Sec. 4.2.1.

A second variational formulation of (5.1) is:

Find $\{\mathbf{u}, \psi\} \in W_g$ *such that*

$$\alpha \int_\Omega \mathbf{u} \cdot \mathbf{v}\, dx + \nu \int_\Omega \nabla \mathbf{u} \cdot \nabla \mathbf{v}\, dx$$
$$= \int_\Omega \mathbf{f} \cdot (\mathbf{v} + \nabla \phi)\, dx + \int_{\Gamma_1} \mathbf{g}_1 \cdot \mathbf{v}\, d\Gamma, \qquad \forall\, \{\mathbf{v}, \phi\} \in W_0, \tag{5.3}$$

which is the Stokes version of (4.19) of Sec. 4.2.2.

From Sec. 4.3 it follows that the discrete problems corresponding to (5.2) and (5.3) are:

Find $\{\mathbf{u}_h, p_h\} \in V_{gh} \times H_h^1$ *such that*

$$\alpha \int_\Omega \mathbf{u}_h \cdot \mathbf{v}_h \, dx + \nu \int_\Omega \nabla \mathbf{u}_h \cdot \nabla \mathbf{v}_h \, dx - \int_\Omega p_h \nabla \cdot \mathbf{v}_h \, dx$$

$$= \int_\Omega \mathbf{f}_h \cdot \mathbf{v}_h \, dx + \int_{\Gamma_1} \mathbf{g}_{1h} \cdot \mathbf{v}_h \, d\Gamma, \qquad \forall\, \mathbf{v}_h \in V_{0h},$$

$$\int_\Omega \nabla \cdot \mathbf{u}_h q_h \, dx = 0, \qquad \forall q_h \in H_h^1 \tag{5.4}$$

and *find* $\{\mathbf{u}_h, \psi_h\} \in W_{gh}$ *such that*

$$\alpha \int_\Omega \mathbf{u}_h \cdot \mathbf{v}_h \, dx + \nu \int_\Omega \nabla \mathbf{u}_h \cdot \nabla \mathbf{v}_h \, dx$$

$$= \int_\Omega \mathbf{f}_h \cdot (\mathbf{v}_h + \nabla \phi_h) \, dx + \int_{\Gamma_1} \mathbf{g}_{1h} \cdot \mathbf{v}_h \, d\Gamma, \qquad \forall\, \{\mathbf{v}_h, \phi_h\} \in W_{0h}, \tag{5.5}$$

respectively. Actually (5.5) is equivalent to:

Find $\{\mathbf{u}_h, \psi_h\} \in W_{gh}$ *and* $p_h \in H_h^1$ *such that*

$$\int_\Omega \nabla p_h \cdot \nabla w_h \, dx = \int_\Omega \mathbf{f}_h \cdot \nabla w_h \, dx, \qquad \forall\, w_h \in H_{0h}^1,$$

$$\alpha \int_\Omega \mathbf{u}_h \cdot \mathbf{v}_h \, dx + \nu \int_\Omega \nabla \mathbf{u}_h \cdot \nabla \mathbf{v}_h \, dx - \int_\Omega p_h \nabla \cdot \mathbf{v}_h \, dx$$

$$= \int_\Omega \mathbf{f}_h \cdot \mathbf{v}_h \, dx + \int_{\Gamma_1} \mathbf{g}_{1h} \cdot \mathbf{v}_h \, d\Gamma, \qquad \forall\, \mathbf{v}_h \in V_{0h}. \tag{5.6}$$

5.3. Solution of (5.1) via (5.2), (5.4)

We now consider an algorithm which is a direct generalization of algorithm (5.311), (5.312) of Chapter VII, Sec. 5.8.7.4.3; this algorithm is defined as follows:

$$p^0 \in L^2(\Omega) \textit{ arbitrarily given}; \tag{5.7}$$

then for $n \geq 0$, p^n *being known, we compute* $\mathbf{u}^n$ *and* p^{n+1} *by*

$$\alpha \mathbf{u}^n - \nu \Delta \mathbf{u}^n = \mathbf{f} - \nabla p^n \textit{ in } \Omega, \quad \mathbf{u}^n = \mathbf{g}_0 \textit{ on } \Gamma_0,$$

$$\nu \frac{\partial \mathbf{u}^n}{\partial n} = \mathbf{g}_1 + p^n \mathbf{n} \textit{ on } \Gamma_1, \tag{5.8}$$

$$p^{n+1} = p^n - \rho \nabla \cdot \mathbf{u}^n, \qquad \rho > 0. \tag{5.9}$$

We observe that the boundary condition on Γ_1 is quite formal since p^n, as an element of $L^2(\Omega)$, usually has no trace on Γ_1; to overcome this difficulty, we shall use a variational formulation of (5.8), namely

$\mathbf{u}^n \in (H^1(\Omega))^N, \qquad \mathbf{u}^n = \mathbf{g}_0 \text{ on } \Gamma_0, \quad \text{and}$

$$\alpha \int_\Omega \mathbf{u}^n \cdot \mathbf{v}\, dx + \nu \int_\Omega \nabla \mathbf{u}^n \cdot \nabla \mathbf{v}\, dx = \int_\Omega \mathbf{f} \cdot \mathbf{v}\, dx + \int_\Omega p^n \nabla \cdot \mathbf{v}\, dx + \int_{\Gamma_1} \mathbf{g}_1 \cdot \mathbf{v}\, d\Gamma,$$

$$\forall\, \mathbf{v} \in (H^1(\Omega))^N, \qquad \mathbf{v} = \mathbf{0} \text{ on } \Gamma_0. \quad (5.8)'$$

About the convergence of algorithm (5.7)–(5.9), we have:

Proposition 5.1. *Suppose that*

$$0 < \rho < 2\frac{\nu}{N}. \tag{5.10}$$

We then have, $\forall\, p^0 \in L^2(\Omega)$,

$$\lim_{n \to +\infty} \{\mathbf{u}^n, p^n\} = \{\mathbf{u}, p\} \textit{ strongly in } (H^1(\Omega))^N \times L^2(\Omega), \tag{5.11}$$

where $\{\mathbf{u}, p\}$ *is the solution of* (5.1), (5.2).

PROOF. Define $\bar{\mathbf{u}}^n$ and $\bar{p}^n$ by $\bar{\mathbf{u}}^n = \mathbf{u}^n - \mathbf{u}$ and $\bar{p}^n = p^n - p$. We clearly have
$\bar{\mathbf{u}}^n \in (H^1(\Omega))^N, \qquad \bar{\mathbf{u}}^n = \mathbf{0} \text{ on } \Gamma_0 \quad \text{and}$

$$\alpha \int_\Omega \bar{\mathbf{u}}^n \cdot \mathbf{v}\, dx + \nu \int_\Omega \nabla \bar{\mathbf{u}}^n \cdot \nabla \mathbf{v}\, dx = \int_\Omega \bar{p}^n \nabla \cdot \mathbf{v}\, dx, \qquad \forall\, \mathbf{v} \in (H^1(\Omega))^N, \quad \mathbf{v} = \mathbf{0} \text{ on } \Gamma_0, \tag{5.12}$$

and (since $\nabla \cdot \mathbf{u} = 0$)

$$\bar{p}^{n+1} = \bar{p}^n - \rho \nabla \cdot \bar{\mathbf{u}}^n. \tag{5.13}$$

From (5.13) it follows that

$$\|\bar{p}^n\|^2_{L^2(\Omega)} - \|\bar{p}^{n+1}\|^2_{L^2(\Omega)} = 2\rho \int_\Omega \bar{p}^n \nabla \cdot \bar{\mathbf{u}}^n\, dx - \rho^2 \int_\Omega |\nabla \cdot \bar{\mathbf{u}}^n|^2\, dx. \tag{5.14}$$

Now taking $\mathbf{v} = \bar{\mathbf{u}}^n$ in (5.12), and combining with (5.14), we obtain

$$\|\bar{p}^n\|^2_{L^2(\Omega)} - \|\bar{p}^{n+1}\|^2_{L^2(\Omega)} = 2\rho\left(\alpha \int_\Omega |\bar{\mathbf{u}}^n|^2\, dx + \nu \int_\Omega |\nabla \bar{\mathbf{u}}^n|^2\, dx\right) - \rho^2 \int_\Omega |\nabla \cdot \bar{\mathbf{u}}^n|^2\, dx. \tag{5.15}$$

Combining (5.15) with relation (5.325) of Chapter VII, Sec. 5.8.7.4.3 (i.e.,

$$\frac{\nu}{N}\|\nabla \cdot \mathbf{v}\|^2_{L^2(\Omega)} \le \alpha \int_\Omega |\mathbf{v}|^2\, dx + \nu \int_\Omega |\nabla \mathbf{v}|^2\, dx, \qquad \forall\, \mathbf{v} \in (H^1(\Omega))^N),$$

we finally obtain

$$\|\bar{p}^n\|^2_{L^2(\Omega)} - \|\bar{p}^{n+1}\|^2_{L^2(\Omega)} \geq \rho\left(2 - \frac{N}{\nu}\rho\right)\left\{\alpha \int_\Omega |\bar{\mathbf{u}}^n|^2\, dx + \nu \int_\Omega |\nabla \bar{\mathbf{u}}^n|^2\, dx\right\},$$

which proves the convergence of $\mathbf{u}^n$ to $\mathbf{u}$, $\forall\, p^0 \in L^2(\Omega)$, if (5.10) holds (we have to remember that Ω bounded implies that

$$\mathbf{v} \to \left(\alpha \int_\Omega |\mathbf{v}|^2\, dx + \nu \int_\Omega |\nabla \mathbf{v}|^2\, dx\right)^{1/2}$$

is a norm on $\{\mathbf{v} \mid \mathbf{v} \in (H^1(\Omega))^N, \mathbf{v} = \mathbf{0} \text{ on } \Gamma_0\}$, equivalent to the $(H^1(\Omega))^N$-norm, and this for all $\alpha \geq 0$). The proof of the convergence of p^n to p is left to the reader (actually we should prove that the convergence of $\{\mathbf{u}^n, p^n\}$ to $\{\mathbf{u}, p\}$ is linear).

Remark 5.1. Using the material of Chapter VII, Sec. 5.8.7.4, it is straightforward to obtain conjugate gradient variants of algorithm (5.7)–(5.9) and also variants derived from an augmented Lagrangian functional reinforcing the incompressibility condition. The same observations hold for the solution of the approximate problem (5.4).

Remark 5.2. When using a finite element variant of algorithm (5.7)–(5.9) to solve the approximate problem (5.4), we have to solve, at each iteration, a discrete elliptic system with boundary conditions of the Dirichlet–Neumann type. The solution of such problems has been discussed in Appendix I, Sec. 4. The same observation holds for the conjugate gradient and augmented Lagrangian algorithms mentioned in Remark 5.1 above.

5.4. Solution of (5.1) via (5.3) and (5.5), (5.6)

We follow (and generalize) Chapter VII, Sec. 5.7, where the situation $\Gamma = \Gamma_0$, $\Gamma_1 = \emptyset$ was treated.

In this section we suppose that $\int_{\Gamma_1} d\Gamma > 0$. The decomposition properties of the Stokes problem (5.1) follow directly from:

Proposition 5.2. *Let $\lambda \in H^{-1/2}(\Gamma)$ and let $A: H^{-1/2}(\Gamma) \to H^{1/2}(\Gamma)$ be defined by the following cascade of Dirichlet and Dirichlet–Neumann problems.*

$$\Delta p_\lambda = 0 \text{ in } \Omega, \qquad p_\lambda = \lambda \text{ on } \Gamma, \tag{5.16}$$

$$\alpha \mathbf{u}_\lambda - \nu \Delta \mathbf{u}_\lambda = -\nabla p_\lambda \text{ in } \Omega, \qquad \mathbf{u}_\lambda = \mathbf{0} \text{ on } \Gamma_0, \qquad \nu \frac{\partial \mathbf{u}_\lambda}{\partial n} = p_\lambda \mathbf{n}\ (= \lambda \mathbf{n}) \text{ on } \Gamma_1, \tag{5.17}$$

$$-\Delta \psi_\lambda = \nabla \cdot \mathbf{u}_\lambda \text{ in } \Omega, \qquad \psi_\lambda = 0 \text{ on } \Gamma, \tag{5.18}$$

and then

$$A\lambda = -\left.\frac{\partial\psi_\lambda}{\partial n}\right|_\Gamma. \tag{5.19}$$

Then A is an isomorphism from $H^{-1/2}(\Gamma)$ onto $H^{1/2}(\Gamma)$. Moreover, the bilinear form $a(\cdot, \cdot)$ defined by

$$a(\lambda, \mu) = \langle A\lambda, \mu\rangle, \qquad \forall\, \lambda, \mu \in H^{-1/2}(\Gamma) \tag{5.20}$$

(where $\langle \cdot, \cdot \rangle$ denotes the duality pairing between $H^{1/2}(\Gamma)$ and $H^{-1/2}(\Gamma)$) is continuous, symmetric, and $H^{-1/2}(\Gamma)$- elliptic.

We do not give the proof of Proposition 5.2; let us mention, however, that it is founded on the relation

$$\langle A\lambda_1, \lambda_2\rangle = \alpha \int_\Omega \mathbf{u}_{\lambda_1} \cdot \mathbf{u}_{\lambda_2}\, dx + \nu \int_\Omega \nabla\mathbf{u}_{\lambda_1} \cdot \nabla\mathbf{u}_{\lambda_2}\, dx, \qquad \forall\, \lambda_1, \lambda_2 \in H^{-1/2}(\Gamma),$$

where $\mathbf{u}_{\lambda_1}, \mathbf{u}_{\lambda_2}$ are the solutions of (5.17) corresponding to $\lambda = \lambda_1$ and $\lambda = \lambda_2$, respectively.

Application of Proposition 5.2 to the solution of the Stokes problem (5.1). We define $p_0, \mathbf{u}_0, \psi_0$ as the solutions of, respectively

$$\Delta p_0 = \nabla \cdot \mathbf{f} \text{ in } \Omega, \qquad p_0 = 0 \text{ on } \Gamma, \tag{5.21}$$

$$\alpha\mathbf{u}_0 - \nu\Delta\mathbf{u}_0 = \mathbf{f} - \nabla p_0 \text{ in } \Omega, \qquad \mathbf{u}_0 = \mathbf{g}_0 \text{ on } \Gamma_0, \qquad \nu\frac{\partial \mathbf{u}_0}{\partial n} = \mathbf{g}_1 + p_0\,\mathbf{n} \text{ on } \Gamma_1, \tag{5.22}$$

$$-\Delta\psi_0 = \nabla \cdot \mathbf{u}_0 \text{ in } \Omega, \qquad \psi_0 = 0 \text{ on } \Gamma. \tag{5.23}$$

The fundamental result is given by:

Theorem 5.1. *Let $\{\mathbf{u}, p\}$ be the solution of the Stokes problem* (5.1). *The trace $\lambda = p|_\lambda$ is the unique solution of the linear variational equation*

$$\lambda \in H^{-1/2}(\Gamma), \qquad \langle A\lambda, \mu\rangle = \left\langle \frac{\partial\psi_0}{\partial n}, \mu \right\rangle, \qquad \forall\, \mu \in H^{-1/2}(\Gamma). \tag{5.24}$$

If we compare the above theorem to Theorem 5.7 of Chapter VII, Sec. 5.7.1.2, we observe that this time—due to $\text{meas}(\Gamma_1) > 0$—the trace of the pressure is uniquely defined by (5.24).

The same decomposition principles can be applied to the discrete Stokes problem (5.5), (5.6); since the resulting methods are trivial variants of the methods discussed in Chapter VII, Sec. 5.7.2, they will not be discussed here any further, except to say that, again, $\text{meas}(\Gamma_1) > 0$ implies that the linear system (discrete analogue of (5.24)), providing the trace of the discrete pressure p_h, has a unique solution.

6. Further Comments

The methods, for solving the Navier–Stokes equations, discussed in Chapter VII, Sec. 5, and in this appendix have been generalized by Conca [1], [2], in order to treat a large variety of boundary conditions involving the stress tensor $\bar{\bar{\boldsymbol{\sigma}}} = (\sigma_{ij})_{i,j}$ defined by

$$\sigma_{ij} = -p\delta_{ij} + 2\nu D_{ij}(\mathbf{u}), \tag{6.1}$$

where $\mathbf{u} = \{u_i\}_{i=1}^N$ and $D_{ij}(\mathbf{u}) = \frac{1}{2}(\partial u_i/\partial x_j + \partial u_j/\partial x_i)$. In this direction, it is quite convenient to use the following equivalent formulation of the Navier–Stokes equations:

$$\alpha u_i - 2\nu \sum_{j=1}^N \frac{\partial}{\partial x_j} D_{ij}(\mathbf{u}) + \sum_{j=1}^N u_j \frac{\partial u_i}{\partial x_j} + \frac{\partial p}{\partial x_i} = f_i \text{ in } \Omega, \qquad i = 1, \ldots, N, \tag{6.2}$$

$$\nabla \cdot \mathbf{u} = 0 \quad \text{in } \Omega. \tag{6.3}$$

We refer to Conca, *loc. cit.*, for further details (see also Engelman, Sani, and Gresho [1] for the practical finite element implementation of various boundary conditions associated with the Navier–Stokes equations).

Some Illustrations from an Industrial Application

The methods described in Chapter VII have been used for the numerical simulation of the aerodynamical performances of a tri-jet engine AMD/BA Falcon 50. Figure A shows the trace on the aircraft of the three-dimensional finite element mesh used for the computation, and Fig. B shows the corresponding Mach distribution (dark: low Mach number, light: high Mach number); the flow is mainly supersonic on the upper part of the wings.

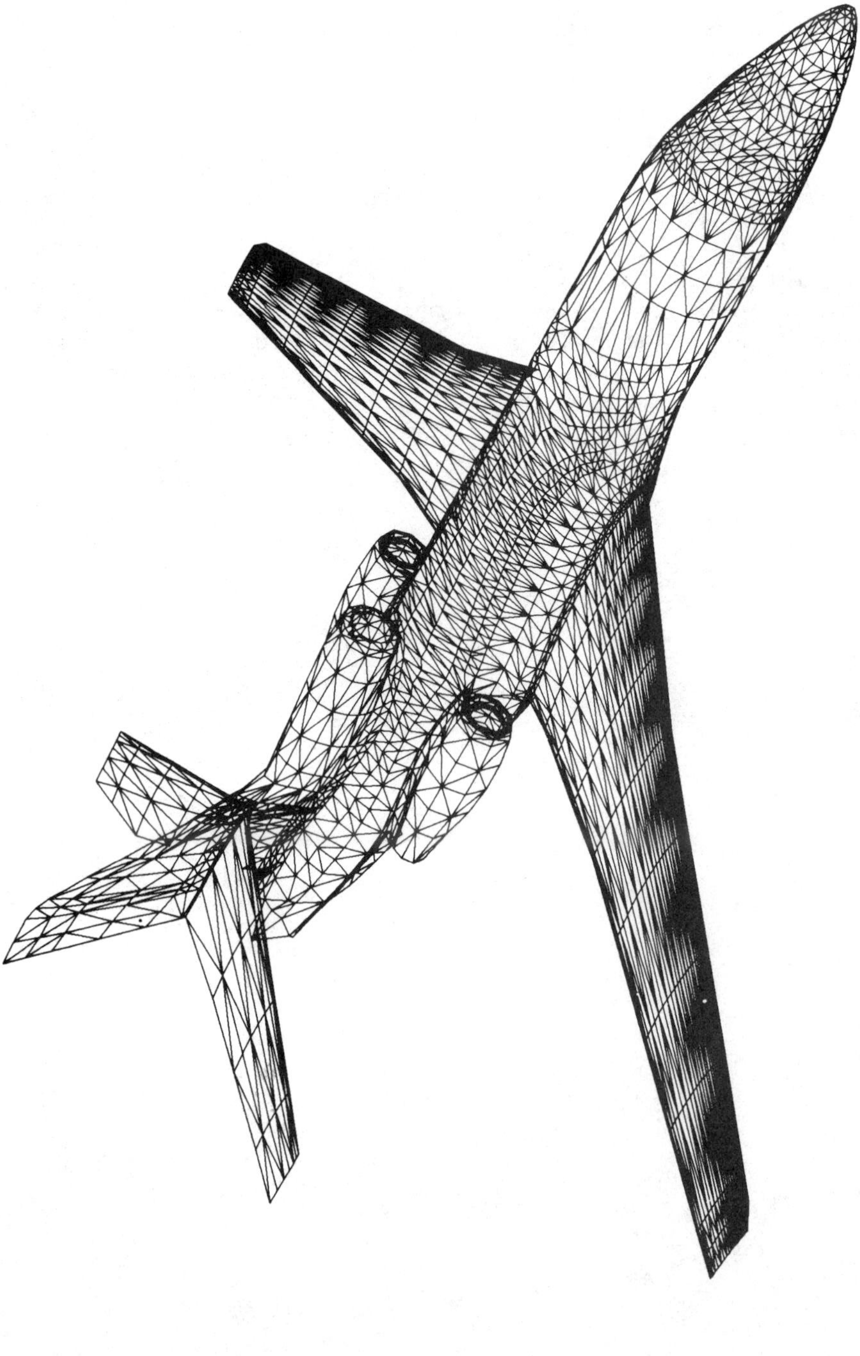

Figure A. Finite element mesh. (Avions Marcel Dassault–Breguet Aviation, Falcon 50).

Figure B. Transonic flow simulation by finite elements: Mach distribution. Mach at infinity: 0.85; angle of attack: 1°. (Avions Marcel Dassault–Breguet Aviation, Falcon 50).

Bibliography

N.B. The following abbreviations for references have been used in the text:

G.L.T. for Glowinski R., Lions J. L., Trémolières R.
B.G.4.P. for Bristeau M. O., Glowinski R., Périaux J., Perrier P., Pironneau O., Poirier G.

AASEN J. O.

[1] On the reduction of a symmetric matrix to tridiagonal form. *BIT* **11**, 233–242 (1971)

ADAMS R. A.

[1] *Sobolev Spaces* (Academic, New York 1975)

AGMON S., DOUGLIS A., NIRENBERG L.

[1] Estimates near the boundary for solution of elliptic partial differential equations satisfying general boundary conditions (I). *Commun. Pure Appl. Math.* **12**, 623–727 (1959)

AMANN H.

[1] Fixed point equations and nonlinear eigenvalue problems in ordered Banach spaces. *SIAM Rev.* **18** (4), 620–709 (1976)

AMARA M., JOLY P., THOMAS J. M.

[1] A mixed finite element method for solving transonic flow equations. *Comput. Methods Appl. Mech. Eng.* **39**, 1–19 (1983)

ARGYRIS J. H., DUNNE P. C.

[1] "The Finite Element Method Applied to Fluid Mechanics", in *Computational Methods and Problems in Aeronautical Fluid Dynamics*, B. L. Hewitt, C. R. Illingworth, R. C. Lock, K. W. Mangler, J. H. McDonnell, C. Richards, F. Walkden (Academic, London 1976) pp. 158–197

AUBIN J. P.

[1] *Mathematical Methods of Game and Economic Theory* (North-Holland, Amsterdam 1979)

[2] *Approximation of Elliptic Boundary Value Problems* (Wiley-Interscience, New York 1972)

AUSLENDER A.

[1] Méthodes numériques pour la décomposition et la minimisation de fonctions non différentiables. *Numer. Math.* **18**, 213–223 (1972)

AXELSSON O.

[1] A class of iterative methods for finite element equations. *Comput. Methods Appl. Mech. Eng.* **19**, 123–138 (1976)

AZIZ A. K., BABUSKA I.

[1] "Survey Lectures on the Mathematical Foundations of the Finite Element Method", *The Mathematical Foundations of the Finite Element Method with Applications to Partial Differential Equations*, ed. by A. K. Aziz (Academic, New York 1972) pp. 3–359

BAIOCCHI C.

[1] Sur un problème à frontière libre traduisant le filtrage de liquides à travers des milieux poreux. *C. R. Acad. Sci. Ser. A* **273**, 1215–1217 (1971)

BAIOCCHI C., CAPELO A.

[1] *Dissequazioni Variazionali e Quasi-variazionali. Applicazioni a problemi di frontiera libera.* Quaderni 4 and 7 dell' Unione Mathematica Italiana. (Pitagora Editrice, Bologna 1978)

BARTELS R., DANIEL J. W.
[1] "A Conjugate Gradient Approach to Nonlinear Boundary Value Problems in Irregular Regions," in *Conference on the Numerical Solution of Differential Equations, Dundee 1973*. Lecture Notes in Mathematics, Vol. 363, ed. by G. A. Watson (Springer, Berlin, Heidelberg, New York 1974) p. 7–77

BARWELL W., GEORGE A.
[1] A comparison of algorithms for solving symmetric indefinite systems of linear equations. *ACM Trans. Math. Software*, **2** (3) 242–251 (1976)

BATHE K. J., WILSON E. L.
[1] *Numerical Methods in Finite Element Analysis* (Prentice-Hall, Englewood Cliffs, NJ 1976)

BAUER F., GARABEDIAN P., KORN D.
[1] *A Theory of Supercritical Wing Sections, with Computer Programs and Examples*, Lectures Notes in Economics and Mathematical Systems, Vol. 66 (Springer, Berlin, Heidelberg, New York 1972)

BAUER F., GARABEDIAN P., KORN D., JAMESON A.
[1] *Supercritical Wing Sections*, Lecture Notes in Economics and Mathematical Systems, Vol. 108 (Springer, Berlin, Heidelberg, New York 1975)
[2] *Supercritical Wing Sections*, Lecture Notes in Economics and Mathematical Systems, Vol. 150 (Springer Berlin, Heidelberg, New York 1977)

BEALE J. T., MAJDA A.
[1] Rates of convergence for viscous splitting of the Navier–Stokes equations. *Math. Comput.* **37**, 243–260 (1981)

BEAM R. M., WARMING R. F.
[1] An implicit factored scheme for the compressible Navier–Stokes equations. *AIAA* **16**, 393–402 (1978)
[2] Alternating Direction implicit methods for parabolic equations with a mixed derivative. *SIAM J. Sci. Stat. Comput.* **1** (1) 131–159 (1980)

BEGIS D.
[1] "Analyse numérique de l'écoulement d'un fluide de Bingham"; Thèse de 3ème cycle, Université Pierre et Marie Curie, Paris (1972)
[2] "Etude numérique de l'écoulement d'un fluide visco-plastique de Bingham par une méthode de lagrangien augmenté." IRIA-Laboria Rpt. 355 (1979)

BEGIS D., GLOWINSKI R.
[1] "Application des méthodes de lagrangien augmenté à la simulation numérique d'écoulements bi-dimensionnels de fluides visco-plastiques incompressibles", in *Méthodes de lagrangien augmenté. Application à la résolution numérique de problèmes aux limites*, ed. by M. Fortin, R. Glowinski (Dunod-Bordas, Paris 1982) pp. 219–240

BENQUE J. P., IBLER B., KERAMSI A., LABADIE G.
[1] "A Finite Element Method for Navier–Stokes Equations," in *Proceedings of the Third International Conference on Finite Elements in Flow Problems, Banff, Alberta, Canada*, 10-13 *June 1980*, ed. by D. H. Norrie, Vol. 1, pp. 110–120

BENSOUSSAN A., LIONS J. L.
[1] *Contrôle Impulsionnel et Inéquations Quasi-Variationnelles* (Dunod-Bordas, Paris 1982)

BERCOVIER M., ENGELMAN M.
[1] A finite element method for the numerical solution of viscous incompressible flows. *J. Comput. Phys.* **30**, 181–201 (1979)

BERCOVIER M., PIRONNEAU O.
[1] Error estimates for finite element method solution of the Stokes problem in the primitive variables. *Numer. Math.* **33**, 211–224 (1979)

BERGER A. E.
[1] The truncation method for the solution of a class of variational inequalities. *Rev. Fr. Autom. Inf. Rech. Opér.* **10** (3) 29–42 (1976)

BERNADOU M., BOISSERIE J. M.
[1] *The Finite Element Method in Thin Shell Theory: Application to Arch Dam Simulations* (Birkhäuser, Boston 1982)

BERS L.
[1] *Mathematical Aspects of Subsonic and Transonic Gas Dynamics* (Chapman and Hall, London 1958)

BLANC M., RICHMOND A. D.
[1] The ionospheric disturbance dynamo. *J. Geophys. Res.* **85** A (4), 1669–1686, (1980)

BOURGAT J. F., DUMAY J. M., GLOWINSKI R.
[1] Large displacement calculations of flexible pipelines by finite elements and nonlinear programming methods. *SIAM J. Sc. Stat. Comput.* **1**, 34–81 (1980)

BOURGAT J. F., DUVAUT G.
[1] Numerical analysis of flows with or without wake past a symmetric two-dimensional profile, with or without incidence. *Int. J. Numer. Methods Eng.* **11**, 975–993 (1977)

BOURGAT J. F., GLOWINSKI R., LE TALLEC P.
[1] "Application des méthodes de lagrangien augmenté à la résolution de problèmes d'élasticité non linéaire finie," in *Méthodes de lagrangien augmenté. Application à la résolution numérique de problèmes aux limites*, ed. by M. Fortin, R. Glowinski (Dunod-Bordas, Paris 1982) pp. 241–278

BREDIF M.
[1] "*Résolution des équations de Navier-Stokes par éléments finis mixtes*"; Thèse de 3ème cycle, Université Pierre et Marie Curie, Paris (1980)

BRENT R.
[1] *Algorithms for Minimization Without Derivatives* (Prentice-Hall, Englewood Cliffs, NJ 1973)

BREZIS H.
[1] "A New Method in the Study of Subsonic Flows," in *Partial Differential Equations and Related Topics*, ed. by J. Goldstein, Lecture Notes in Mathematics, Vol. 446 (Springer, Berlin, Heidelberg, New York 1975) pp. 50–60
[2] Multiplicateur de Lagrange en torsion élasto-plastique. *Arch. Ration. Mech. Anal.* **49**, 30–40 (1972)
[3] Problèmes unilatéraux. *J. Math. Pures Appl.* **9**, (72), 1–168 (1971).
[4] "Monotonicity in Hilbert Spaces and Some Applications to Nonlinear Partial Differential Equations," in *Contributions to Nonlinear Functional Analysis*, ed. by E. Zarantonello (Academic, New York 1971) pp. 101–116.
[5] *Opérateurs maximaux monotones et semi-groupes de contraction dans les espaces de Hilbert* (North-Holland, Amsterdam 1973)

BREZIS H., CRANDALL M., PAZY A.
[1] Perturbation of nonlinear monotone sets in Banach spaces. *Commun. Pure Appl. Math.* **23**, 153–180 (1970)

BREZIS H., SIBONY M.
[1] Equivalence de deux inéquations variationnelles et applications. *Arch. Ration. Mech. Anal.* **41**, 254–265 (1971)
[2] Méthodes d'approximation et d'itération pour les opérateurs monotones. *Arch. Ration. Mech. Anal.* **28**, 59–82 (1968)

BREZIS H., STAMPACCHIA G.
[1] The hodograph method in fluid dynamics in the light of variational inequalities. *Arch. Ration. Mech. Anal.* **61**, 1–18 (1976)
[2] Sur la régularité de la solution d'inéquations elliptiques. *Bull. Soc. Math. Fr.* **96**, 153–180 (1968)

BREZZI F.
[1] "Non-Standard Finite Elements for Fourth Order Elliptic Problems," in *Energy Methods in Finite Element Analysis*, ed. by R. Glowinski, E. Y. Rodin, O. C. Zienkiewicz (Wiley, Chichester 1979) pp. 193–211

BREZZI F., HAGER W. W., RAVIART P.A.
[1] Error estimates for the finite element solution of variational inequalities; Part I: Primal Theory. *Numer. Math.* **28**, 431–443 (1977)
[2] Error estimates for the finite element solution of variational inequalities; Part II: Mixed methods. *Numer. Math.* **31**, 1–16 (1978)

Brezzi F., Rappaz J., Raviart P. A.
[1] Finite dimensional approximation of nonlinear problems, Part I: Branches of nonsingular solutions. *Numer. Math.* **36**, 1–25 (1980)
[2] Finite dimensional approximation of nonlinear problems, Part II: Limit points. *Numer. Math.* **37**, 1–28 (1981)
[3] Finite dimensional approximation of nonlinear problems, Part III: Simple bifurcation points. *Numer. Math.* **38**, 1–30 (1981)

Brezzi F., Sacchi G.
[1] A finite element approximation of variational inequalities related to hydraulics. *Calcolo* **13**, 259–273 (1976)

Bristeau M. O.
[1] "Application de la méthode des éléments finis à la résolution d'inéquations variationnelles d'évolution de type Bingham"; Thèse de 3ème cycle, Université Pierre et Marie Curie, Paris (1975)
[2] "Application of Optimal Control Theory to Transonic Flow Computations by Finite Element Methods," in *Computing Methods in Applied Sciences and Engineering, 1977, II,* ed. by R. Glowinski, J. L. Lions, Lecture Notes in Physics, Vol. 91, (Springer Berlin, Heidelberg, New York 1979) pp. 103–124
[3] "Application of a Finite Element Method to Transonic Flow Problems Using an Optimal Control Approach," in *Computational Fluid Dynamics*, ed. by W. Kollmann (McGraw Hill, New York 1980) pp. 281–328

Bristeau M. O., Glowinski R.
[1] "Finite Element Analysis of the Unsteady Flow of a Viscous-Plastic Fluid in a Cylindrical Pipe," in *Finite Element Methods in Flow Problems*, ed. by J. T. Oden, O. C. Zienkiewicz, R. H. Gallagher, C. Taylor (University of Alabama Press, Huntsville, Alabama 1974) pp.471–488

Bristeau M. O., Glowinski R., Mantel B., Periaux J., Perrier P., Pironneau O.
[1] "A Finite Element Approximation of Navier–Stokes Equations for Incompressible Viscous Fluids. Iterative Methods of Solution," in *Approximation Methods for Navier–Stokes Problems*, ed. by R. Rautmann, Lecture Notes in Mathematics, Vol. 771 (Springer, Berlin Heidelberg, New York 1980) pp.78–128

Bristeau M. O., Glowinski R., Periaux J., Perrier P., Pironneau O.
[1] On the numerical solution of nonlinear problems in fluid dynamics by least squares and finite element methods. (I) Least squares formulations and conjugate gradient solution of the continuous problems. *Comput. Methods. Appl. Mech. Eng.* **17/18**, 619–657 (1979)

Bristeau M. O., Glowinski R., Periaux J., Perrier P., Pironneau O., Poirier G.
[1] "Application of Optimal Control and Finite Element Methods to the Calculation of Transonic Flows and Incompressible Flows," in *Numerical Methods in Applied Fluid Dynamics*, ed. by B. Hunt (Academic, London 1980) pp. 203–312
[2] "Transonic Flow Simulations by Finite Elements and Least Squares Methods," *Finite Elements in Fluids*, Vol. 4, ed. by R. H. Gallagher, D. H. Norrie, J. T. Oden, O. C. Zienkiewicz (Wiley, Chichester, 1982)

Brooks A. N., Hugues T. J. R.
[1] Streamline upwind/Petrov–Galerkin formulations for convection dominated flows with particular emphasis on the incompressible Navier–Stokes equations. *Comput. Methods Appl. Mech. Eng.* **32**, 199–259 (1982)

Broyden, C. G.
[1] A class of methods for solving nonlinear simultaneous equations. *Math. Comput.* **19**, 577–593 (1965).
[2] Quasi-Newton methods and their application to function minimization. *Math. Comput.* **21**, 368–381 (1967).

Bunch J. R., Kaufman L.
[1] Some stable methods for calculating inertia and solving symmetric linear systems. *Math. Comput.* **31**, 163–179 (1977).

BUNCH, J. R., PARLETT, B. N.
[1] Direct methods for solving symmetric indefinite systems of linear equations. *SIAM J. Numer. Anal.* **8**, 639–655 (1971).

CARTAN H.
[1] *Calcul Différentiel*, (Hermann, Paris 1967).

CEA J., GEYMONAT G.
[1] "Une méthode de linearisation via l'optimisation," in *Symposia Mathematica, Vol. X (Convegno di Analesi Numerica, INDAM, Roma, 1972)* (Academic, London 1972) pp. 431–451.

CEA, J.
[1] *Optimisation: Théorie et Algorithmes* (Dunod, Paris 1971)
[2] *Optimization: Theory and Algorithms.* Lecture Notes, Vol. 53 (Tata Institute of Fundamental Research, Bombay, 1978)

CEA J., GLOWINSKI R.
[1] Sur des méthodes d'optimisation par relaxation. *Rev. Fr. Autom. Inf. Rech. Oper.* **R3**, 5–32 (1973).
[2] Méthodes Numériques pour l'écoulement laminaire d'un fluide rigide viscoplastique incompressible. *Intern. J. Comput. Math. Sect. B*, **3**, 225–255 (1972)

CEA J., GLOWINSKI R., NEDELEC J. C.
[1] "Application des méthodes d'optimisation, de différences et d'éléments finis, à l'analyse numérique de la torsion élasto-plastique d'une barre cylindrique," in *Approximations et Méthodes Itératives de Résolution d'Inéquations Variationnelles et de Problèmes Non Linéaires*, Cahier de l'IRIA, No. 12, (1974) pp. 7–138

CHAN T. F., FORTIN M., GLOWINSKI R.
[1] "Résolution numérique de problèmes faiblement non linéaires par des méthodes de lagrangien augmenté," in *Méthodes de lagrangien augmenté. Application à la résolution numérique de problèmes aux limites*, ed. by M. Fortin, R. Glowinski (Dunod-Bordas, Paris 1982) pp. 137–158

CHAN T. F., GLOWINSKI R.
[1] "Finite Element Approximation and Iterative Solution of a Class of Mildly Nonlinear Elliptic Equations." Rpt. STAN-CS-78-674, Computer Science Department, Stanford University, Palo Alto, CA (1978)

CHAN T. F., KELLER H. B.
[1] Arclength continuation and multigrid techniques for nonlinear eigenvalue problems. *SIAM J. Sci. Stat. Comput.* **3**(2), 173–194 (1982)

CHATTOT J. J.
[1] "Méthode variationnelle pour les problèmes hyperboliques et mixtes du premier ordre," in *Computing Methods in Applied Sciences and Engineering*, ed. by R. Glowinski, J. L. Lions (North-Holland, Amsterdam 1980) pp.197–211

CHEUNG T. Y.
[1] Recent developments in the numerical solution of partial differential equations by linear programming. *SIAM Rev.* **20**, 139–167 (1978)

CHORIN A. J.
[1] A numerical method for solving incompressible viscous flow problems. *J. Comput. Phys.* **2**, 12–26 (1967)
[2] "Numerical solution of incompressible flow problems," in *Studies in Numerical Analysis, 2; Numerical Solution of Nonlinear Problems (Symposium, SIAM, Philadelphia, Pa., 1968).* (Soc. Indust. Appli. Math., Philadelphia, Pa., 1970), pp. 64–71.
[3] On the convergence and approximation of discrete approximation to the Navier–Stokes equations. *Math. Comput.* **23**, 341–353 (1968)
[4] Numerical study of slightly viscous flow. *J. Fluid Mech.* **57**, 785–796 (1973)

CHRISTIE I., MITCHELL A. R.
[1] Upwinding of high order Galerkin methods in conduction-convection problems. *Int. J. Numer. Methods. Eng.* **12**, 1764–1771 (1978)

CIARLET P. G.
[1] *The Finite Element Method.* Lecture Notes, Vol. 49 (Tata Institute of Fundamental Research, Bombay, 1975)
[2] *The Finite Element Method for Elliptic Problems* (North-Holland, Amsterdam 1978)
[3] *Numerical Analysis of the Finite Element Method*, Séminaire de Mathématiques Supérieures (Presses de l'Université de Montréal 1976)
[4] *Introduction à l'analyse numérique matricielle et à l'optimisation* (Masson, Paris 1982)

CIARLET P. G., DESTUYNDER P.
[1] Approximation of Three-Dimensional Models by Two-Dimensional Models in Plate Theory," in *Energy Methods in Finite Element Analysis*, ed. by R. Glowinski, E. Y. Rodin, O. C. Zienkiewicz (Wiley, Chichester 1979) pp. 33–45

CIARLET P. G., RABIER P.
[1] *Les équations de Von Karman.* Lecture Notes in Mathematics, Vol.826 (Springer, Berlin, Heidelberg, New York 1980)

CIARLET P. G., RAVIART P. A.
[1] "A Mixed Finite Element Method for the Biharmonic Equation," in *Mathematical Aspects of Finite Elements in Partial Differential Equations*, ed. by C. de Boor (Academic, New York 1974) pp. 125–145

CIARLET P. G., SCHULTZ M. H., VARGA R. S.
[1] Numerical methods of high order accuracy for nonlinear boundary value problems. V. Monotone operator theory. *Numer. Math.* **13**, 51–77 (1969).

CIARLET P. G., WAGSHAL C.
[1] Multipoint Taylor formulas and applications to the finite element method. *Numer. Math.* **17**, 84–100 (1971)

CIAVALDINI J. F., POGU M., TOURNEMINE G.
[1] "Sur l'approximation des écoulements compressibles autour d'un profil régulier placé en atmosphère infinie: estimation asymptotique lorsque l'on borne le domaine extérieur au profil," Rapport de l'Université de Rennes I et de l'Institut National des Sciences Appliquées (March 1979)
[2] Une méthode variationnelle non linéaire pour l'étude dans le plan physique d'écoulements compressibles subcritiques en atmosphère infinie. *C. R. Hebd. Seances Acad. Sci. Ser. A* **281**, 1105–1108 (1975)
[3] Sur la régularité d'écoulements plans, stationnaires et subcritiques autour d'un profil à pointe: ajustement de la circulation pour satisfaire la condition de Kutta–Joukowsky. *C. R. Hebd. Seances Acad. Sci. Ser. A* **285**, 297–300 (1977)

CIAVALDINI J. F., TOURNEMINE G.
[1] A finite element method to compute stationary steady flows in the hodograph plane. *J. Ind. Math. Soc.* **41**, 69–82 (1977)

COMINCIOLI V.
[1] On some oblique derivative problems arising in the fluid flow in porous media. *Appl. Math. Optim.* **1**(4), 313–336 (1975)

CONCA C.
[1] "Approximation de quelques problèmes de type Stokes par une méthode d'éléments finis mixtes," Rpt. 82033, Laboratoire d'Analyse Numérique, Université Pierre et Marie Curie, Paris (1982)
[2] "Etude d'écoulements visqueux incompressibles dans un condenseur," Thèse de 3ème cycle, Université Pierre et Marie Curie, Paris (1982)

CONCUS P.
[1] Numerical solution of the nonlinear magnetostatic field equation in two dimensions. *J. Comput. Phys.* **1**, 330–342 (1967)

CONCUS P., GOLUB G. H.
[1] "A generalized Conjugate Gradient Method for Nonsymmetric Systems of Linear Equations," in *Computing Methods in Applied Sciences and Engineering*, ed. by R. Glowinski, J. L. Lions, Lecture Notes in Economics and Math. Systems, Vol. 134 (Springer, Berlin, Heidelberg, New York 1976) pp. 56–65.

CONCUS P., GOLUB G. H., O'LEARY D. P.
[1] Numerical solution of nonlinear elliptic partial differential equations by a generalized conjugate gradient method. *Computing* **19**, 321–339 (1978)

COURANT R., FRIEDRICHS K. O.
[1] *Supersonic Flows and Shock Waves* (Springer, New York 1976)

CRANDALL M. G., RABINOWITZ P. H.
[1] Bifurcation, perturbation of single eigenvalues and linearized stability. *Arch. Ration. Mech. Anal.* **52**, 161–180 (1973)
[2] Some continuation and variational methods for positive solutions of nonlinear elliptic eigenvalue problems. *Arch. Ration. Mech. Anal.* **58**, 207–218 (1975)

CRISFIELD M. A.
[1] "Finite Element Analysis for Combined Material and Geometric Nonlinearities," in *Nonlinear Finite Element Analysis in Structural Mechanics*, ed. by W. Wunderlich, E. Stein, K. J. Bathe (Springer, Berlin, Heidelberg, New York 1981) pp. 325–338

CROUZEIX M.
[1] "Sur l'approximation des équations différentielles opérationnelles par des méthodes de Runge-Kutta." Thèse d'Etat, Université Pierre et Marie Curie, Paris (1975)
[2] "Etude d'une méthode de linéarisation. Résolution numérique des équations de Stokes stationnaires. Applications aux équations de Navier-Stokes stationnaires;" in *Approximations et Méthodes Itératives de Résolution d'Inéquations Variationnelles et de Problèmes Non Linéaires*, Cah. de l'IRIA, **12**, 139–244 (1974)

CRYER C. W.
[1] The method of Christoferson for solving free boundary problems for infinite journal bearings by means of finite differences. *Math. Comput.* **25**, 435–443 (1971)
[2] The solution of a quadratic programming problem using systematic over relaxation. *SIAM J. Control* **9**(3), 385–392 (1971)

DANIEL J.
[1] *The Approximate Minimization of Functionals* (Prentice Hall, Englewood Cliffs, NJ 1970)

DECONINCK H., HIRSCH C.
[1] "Subsonic and Transonic Computation of Cascade Flows," in *Computing Methods in Applied Sciences and Engineering*, ed. by R. Glowinski, J. L. Lions (North-Holland, Amsterdam, 1980) pp. 175–195.

DENNIS J. E., MORE J. J.
[1] Quasi-Newton methods, motivation and theory. *SIAM Rev.* **19**, 46–89 (1977)

DOSS S., MILLER K.
[1] Dynamic ADI methods for elliptic equations. *SIAM J. Numer. Anal.* **16**(5), 837–856 (1979)

DOUGLAS J., DUPONT T.
[1] "Interior Penalty Procedures for Elliptic and Parabolic Galerkin Methods," in *Computing Methods in Applied Sciences and Engineering*, ed. by R. Glowinski, J. L. Lions, Lecture Notes in Physics, Vol.58, (Springer, Berlin, Heidelberg, New York 1976) pp.207–216.
[2] "Preconditioned Conjugate Gradient Iteration Applied to Galerkin Methods for Mildly Nonlinear Dirichlet Problems," in *Sparse Matrix Computations*, ed. by J. R. Bunch, D. J. Rose (Academic, New York 1976) pp. 333–348.

DOUGLAS J., RACHFORD H. H.
[1] On the numerical solution of the heat conduction problem in 2 and 3 space variables. *Trans. Am. Math. Soc.* **82**, 421–439 (1956)

DUFF I. S.
[1] "Recent Developments in the Solution of Large Sparse Linear Equations," in *Computing Methods in Applied Sciences and Engineering*, ed. by R. Glowinski, J. L. Lions (North-Holland, Amsterdam, 1980) pp. 407–426.
[2] "Sparse Matrix Software for Elliptic PDE's," in *Multigrid Methods*. Lecture Notes in Mathematics, Vol. 960, ed. by W. Hackbusch, U. Trottenberg (Springer, Berlin, Heidelberg, New York 1982) pp. 410–426.

Duff, I. S., Munksgaard, N., Nielsen H. B., Reid J. K.
[1] "Direct Solution of Sets of Linear Equations Whose Matrix is Sparse, Symmetric and Indefinite," Harwell Rpt., C.S.S. Division, A.E.R.E. Harwell, England (1977)

Duvaut G., Lions J. L.
[1] *Les Inéquations en Mécanique et en Physique* (Dunod, Paris 1972)

Eberle A.
[1] "Transonic Potential Flow Computations by Finite Element: Airfoil and Wing Analysis, Airfoil Computation," Lecture at the DGLR/GARTEUR 6 Symposium Transonic Configurations, Bad Harzburg, Germany (June 1978)
[2] "Finite Element Methods for the Solution of the Full Potential Equation in Transonic Steady and Unsteady Flow," in *Finite Elements in Fluids*, Vol. 4, ed. by R. H. Gallagher, D. H. Norrie, J. T. Oden, O. C. Zienkiewicz, (Wiley, Chichester 1982) pp. 483–504

Ekeland I., Temam R.
[1] *Convex Analysis and Variational Problems* (North-Holland, Amsterdam 1976)

Engelman M. S., Sani R. L., Gresho P. M.
[1] The implementation of normal and/or tangential boundary conditions in finite element codes for incompressible fluid flow. *Int. J. Numer. Meth. Fluids*, **2**, 225–238 (1982)

Engquist B., Kreiss H. O.
[1] Difference and finite element methods for hyperbolic differential equations. *Comput. Meth. Appl. Mech. Eng.* **17/18**, 581–596 (1979)

Falk R. S.
[1] "Approximate Solutions of Some Variational Inequalities with Order of Convergence Estimates," Ph.D. Thesis, Cornell University (1971)
[2] Error estimates for the approximation of a class of variational inequalities. *Math. Comput.* **28**, 963–971 (1974)
[3] Approximation of an elliptic boundary value problem with unilateral constraints. *Rev. Fr. Autom. Inf. Rech. Opér.* **R2**, 5–12 (1975)

Falk R. S., Mercier B.
[1] Error estimates for elasto-plastic problems. *Rev. Fr. Autom. Inf. Rech. Opér.* **11**, 135–144 (1977)

Finlayson B. A.
[1] *The Method of Weighted Residuals and Variational Principles* (Academic, New York 1972)

Fletcher R.
[1] *Practical Methods of Optimization*, Vol. 1: *Unconstrained Optimization* (Wiley, Chichester 1980)

Forsythe G. E., Wasow W.
[1] *Finite Difference Methods for Partial Differential Equations* (Wiley, New York 1960)

Fortin M.
[1] "Calcul numérique des écoulements des fluids de Bingham et des fluides visqueux incompressibles par des méthodes d'éléments finis." Thèse d'Etat Université Pierre et Marie Curie, Paris (1972)
[2] Minimization of some non-differentiable functionals by the augmented Lagrangian method of Hestenes and Powell. *Appl. Math. Optim.* **2**, 236–250 (1976)
[3] "Approximation d'un opérateur de projection et application à un schéma de résolution numérique des équations de Navier–Stokes," Thèse de 3ème cycle, Université Paris-Sud, Orsay (1970)

Fortin M., Glowinski R.
[1] *Méthodes de lagrangien augmenté. Application à la résolution numérique de problèmes aux limites*, ed. by M. Fortin, R. Glowinski (Dunod-Bordas, Paris 1982)
[2] "Sur des méthodes de décomposition-coordination par lagrangien augmenté," in *Méthodes de lagrangien augmenté. Application à la résolution numérique de problèmes aux limites*, ed. by M. Fortin, R. Glowinski (Dunod-Bordas, Paris 1982), pp. 91–136
[3] "Méthodes de lagrangien augmenté en programmation quadratique," in *Méthodes de lagrangien augmenté. Application à la résolution numérique de problèmes aux limites*, ed. by M. Fortin, R. Glowinski (Dunod-Bordas, Paris 1982) pp.1–43

FORTIN M., GLOWINSKI R., MARROCCO A.

[1] "Application des méthodes de lagrangien augmenté à la résolution de problèmes aux limites d'ordre deux fortement non linéaires," in *Méthodes de lagrangien augmenté. Application à la résolution numérique de problèmes aux limites*, ed. by M. Fortin, R. Glowinski (Dunod-Bordas, Paris 1982) pp.159–201

FORTIN M., THOMASSET F.

[1] Mixed finite element methods for incompressible flow problems. *J. Comput. Phys.* **31**, 173–215 (1979)

[2] "Application aux équations de Stokes et de Navier–Stokes," in *Méthodes de lagrangien augmenté. Applications à la résolution numérique de problèmes aux limites*, ed. by M. Fortin, R. Glowinski (Dunod-Bordas, Paris 1982) pp. 45–89

FRIEDRICHS K. O.

[1] Symmetric positive linear differential equations. *Commun. Pure Appl. Math.* **11**, 333–418 (1958)

GABAY D.

[1] "Méthodes numériques pour l'optimisation non linéaire." Thèse d'Etat, Université Pierre et Marie Curie, Paris (1979)

[2] "Application de la méthode des multiplicateurs aux inéquations variationnelles," in *Méthodes de lagrangien augmenté. Application à la résolution numérique de problèmes aux limites*, ed. by M. Fortin, R. Glowinski (Dunod-Bordas, Paris, 1982) pp.279–307

GABAY D., MERCIER B.

[1] A dual algorithm for the solution of nonlinear variational problems via finite element approximations. *Comput. Math. Appl.* **2**(1), 17–40 (1976).

GARTLING D. K., BECKER E. B.

[1] Finite element analysis of viscous incompressible fluid flow, 1. *Comput. Meth. Appl. Mech. Eng.* **8**, 51–60 (1976)

[2] Finite element analysis of viscous incompressible fluid flow, 2. *Comput. Meth. Appl. Mech. Eng.* **8**, 127–138 (1976)

GEORGE A.

[1] "Direct Solution of Sparse Positive Definite Systems: Some Basic Ideas and Open Problems," in *Sparse Matrices and Their Uses*, ed. by I. S. Duff (Academic, New York 1981) pp. 283–306

GERMAIN P.

[1] *Mécanique des Milieux Continus*, Vol.1 (Masson, Paris 1973)

GIRAULT V., RAVIART P. A.

[1] *Finite Element Approximation of the Navier-Stokes Equations*, Lecture Notes in Mathematics, Vol. 749, (Springer, Berlin, Heidelberg, New York 1979)

[2] An analysis of upwind schemes for the Navier–Stokes equations. *SIAM J. Numer. Anal.* **9**(2), 312–333 (1982)

GLOWINSKI R.

[1] "Introduction to the Approximation of Elliptic Variational Inequalities," Rpt. 76006, Laboratoire d'Analyse Numérique, Université Pierre et Marie Curie, Paris (1976)

[2] "Analyse Numérique d'Inéquations Variationnelles d'Ordre Quatre," Rpt. 75002, Laboratoire d'Analyse Numérique, Université Pierre et Marie Curie, Paris (1975)

[3] Sur l'approximation d'une inéquation variationnelle elliptique de type Bingham. *Rev. Fr. Autom. Inf. Rech. Opér.* **10**(12), 13–30 (1976)

[4] Sur l'écoulement d'un fluide de Bingham dans une conduite cylindrique. *J. Mec.* **13**(4), 601–621 (1974)

[5] *Introduction à l'Analyse Numérique des Problèmes Non Linéaires.* (Masson, Paris) (to appear)

[6] La méthode de relaxation. *Rend. Mat.* **14** (1971)

GLOWINSKI R., KELLER H. B., REINHART L.

[1] Continuation—Conjugate gradient methods for the least squares solution of nonlinear boundary value problems (to appear).

GLOWINSKI R., LANCHON H.

[1] Torsion élasto-plastique d'une barre cylindrique de section multi-connexe. *J. Méc.* **12**, 151–171 (1973)

Glowinski R., Le Tallec P.

[1] Numerical solution of problems in incompressible finite elasticity by augmented lagrangian methods (I). Two-dimensional and axisymmetric problems. *SIAM J. Appl. Math.* **42**(2), 400–429 (1982)

[2] Numerical solution of problems in incompressible finite elasticity by augmented lagrangian methods (II). Three dimensional problems (to appear).

Glowinski R., Le Tallec P., Ruas V.

[1] "Approximate Solution of Nonlinear Problems in Incompressible Finite Elasticity," in *Nonlinear Finite Element Analysis in Structural Mechanics*, ed. by W. Wunderlich, E. Stein, K. J. Bathe (Springer, Berlin, Heidelberg, New York 1981) pp. 666–695

Glowinski R., Lions J. L., Tremolieres R.

[1] *Analyse Numériques des Inéquations Variationnelles; Vol. 1: Théorie Générale et Premières Applications* (Dunod-Bordas, Paris 1976)

[2] *Analyse Numérique des Inéquations Variationnelles;* Vol. 2: *Applications aux Phénomènes Stationnaires et d'Evolution* (Dunod-Bordas, Paris 1976)

[3] *Numerical Analysis of Variational Inequalities* (North-Holland, Amsterdam, 1981)

Glowinski R., Mantel B., Periaux J., Perrier P., Pironneau O.

[1] "On an Efficient New Preconditioned Conjugate Gradient Method. Application to the in Core Solution of the Navier–Stokes Equations via Nonlinear Least Squares and Finite Element Methods," in *Finite Elements in Fluids*, Vol. 4, ed. by R. H. Gallagher, D. H. Norrie, J. T. Oden, O. C. Zienkiewicz (Wiley, London 1982) pp. 365–401

Glowinski R., Mantel B., Periaux J., Pironneau O.

[1] "A Finite Element Approximation of Navier–Stokes equations for Incompressible Viscous Fluids. Functional Least Squares Methods of Solution," in *Computer Methods in Fluids*, ed. by K. Morgan, C. Taylor, C. A. Brebbia (Pentech, London 1980) pp. 84–133

Glowinski R., Mantel B., Periaux J., Pironneau O., Poirier G.

[1] "An Efficient Preconditioned Conjugate Gradient Method Applied to Nonlinear Problems in Fluid Dynamics," in *Computing Methods in Applied Sciences and Engineering*, ed. by R. Glowinski, J. L. Lions (North-Holland, Amsterdam 1980) pp. 445–487

Glowinski R., Marrocco A.

[1] Analyse numérique du champ magnétique d'un alternateur par éléments finis et sur-relaxation non linéaire. *Comput. Meth. Appl. Mech. Eng.* **3**, 55–85 (1974)

[2] "Etude numérique du champ magnétique d'un alternateur tétrapolaire par la méthode des éléments finis," in *Computing Methods in Applied Sciences and Engineering, Part 1*, ed. by R. Glowinski, J. L. Lions, Lecture Notes in Computer Sciences, Vol. 10 (Springer, Berlin, Heidelberg, New York 1974) pp. 392–409

[3] Sur l'approximation par éléments finis d'ordre un et la résolution par pénalisation-dualité d'une classe de problèmes de Dirichlet non linéaires. *C. R. Acad. Sci. Paris*, **278A**, 1649–1652 (1974)

[4] "On the Solution of a Class of Nonlinear Dirichlet Problems by a Penalty-Duality Method and Finite Elements of Order One," in *Optimization Techniques: IFIP Technical Conference, NOVOSSIBIRSK, U.S.S.R., June 1974*, ed. by G. I. Marchuk, Lecture Notes in Computer Sciences, Vol.27 (Springer, Berlin, Heidelberg, New York 1975), pp.327–333

[5] Sur l'approximation par éléments finis d'ordre un et la résolution par pénalisation-dualité d'une classe de problèmes de Dirichlet non linéaires. *Rev. Fr. Autom. Inf. Rech. Oper. Anal. Numér. R-2*, 41–76 (1975)

[6] Numerical solution of two-dimensional magneto-static problems by augmented lagrangian methods. *Comput. Meth. Appl. Mech. Eng.* **12**, 33–46 (1977)

[7] "Sur l'approximation par éléments finis d'ordre un et la résolution par pénalisation-dualité d'une classe de problèmes de Dirichlet non linéaires," IRIA-Laboria Rpt. 115 (1975) (extended version of Glowinski and Marrocco [5]).

Glowinski R., Periaux J., Pironneau O.

[1] On a mixed finite element approximation of the Stokes problem (II). Solution of the approximate problems (to appear).

[2] An efficient preconditioning scheme for iterative numerical solution of partial differential equations. *Appl. Math. Model.* **4**, 187–192 (1980)

GLOWINSKI R., PIRONNEAU O.
[1] Numerical methods for the first biharmonic equation and for the two-dimensional Stokes problem. *SIAM Rev.* **21**(2), 167–212 (1979)
[2] On a mixed finite element approximation of the Stokes problem (I). Convergence of the approximate solution. *Numer. Math.* **33**, 397–424 (1979)
[3] "On Numerical Methods for the Stokes Problem," in *Energy Methods in Finite Element Analysis*, ed. by R. Glowinski, E. Y. Rodin, O. C. Zienkiewicz (Wiley, Chichester 1979), pp. 243–264
[4] Approximation par éléments finis mixtes du problème de Stokes en formulation vitesse-pression. Convergence des solutions approchées. *C.R. Acad. Sci. Paris* **286A**. 181–183 (1978)
[5] Approximation par élémentes finis mixtes du problème de Stokes en formulation vitesse-pression. Résolution des problèmes approchés. *C.R. Acad. Sci. Paris* **286A**, 225–228 (1978)

GODLEWSKI E.
[1] "Méthodes à pas multiples et de directions alternées pour la discrétisation d'équations d'évolution," Thèse de 3ème cycle, Université Pierre et Marie Curie, Paris (1980)

GOLUB G. H., MEURANT G.
[1] *Résolution Numérique des Grands Systèmes Linéaires*, (Eyrolles, Paris, 1983)

GOLUB G. H., PEREYRA V.
[1] The differentiation of pseudo-inverses and nonlinear least squares problems whose variables separate. *SIAM J. Numer. Anal.* **10**, 413–432 (1973)

GOLUB G. H., PLEMMONS R. J.
[1] Large scale geodetic least squares adjustment by dissection and orthogonal decomposition. *Linear Algebra Appl.* **34**, 3–28 (1980)

GRESHO P., LEE R. L., CHAN S. T., SANI R. L.
[1] "Solution of the Time-Dependent Incompressible Navier–Stokes and Boussinesq Equations Using the Galerkin Finite Element Method," in *Approximation Methods for Navier–Stokes Problems*, ed. by R. Rautmann, Lecture Notes in Mathematics, Vol. 771 (Springer, Berlin, Heidelberg, New York 1980) pp. 203–222

HAYES L. J.
[1] "Generalization of Galerkin Alternating-Direction Methods to Nonrectangular Regions Using Isoparametric Elements," Ph.D. Thesis, University of Texas at Austin, Austin, TX (1977)
[2] Generalization of Galerkin alternating-direction methods to nonrectangular regions using patch approximations. *SIAM J. Numer. Anal.* **18**, 627–643 (1981)

HECHT F.
[1] On weakly divergence free finite element bases (to appear)

HEINRICH J. C., HUYAKORN P. S., ZIENKIEWICZ O. C., MITCHELL A. R.
[1] An "upwind" finite element scheme for two dimensional convective transport equation. *Int. J. Numer. Meth. Eng.* **11**, 131–143 (1977)

HESTENES M.
[1] Multiplier and gradient methods. *J. Optim. Theory Appl.* **4**, 303–320 (1969).
[2] *Conjugate Direction Methods in Optimization* (Springer, Berlin, Heidelberg, New York 1980)

HEWITT B. L., ILLINGWORTH C. R., LOCK R. C., MANGLER K. W., MCDONNELL J. H., RICHARDS C., WALKDEN F. (eds.)
[1] *Computational Methods and Problems in Aeronautical Fluid Dynamics* (Academic, London 1976)

HOLST T.
[1] "An Implicit Algorithm for the Transonic Full-Potential Equation in Conservative Form," in *Computing Methods in Applied Sciences and Engineering*, ed. by R. Glowinski, J. L. Lions (North-Holland, Amsterdam 1980) pp. 157–174

HOOD P., TAYLOR C.
[1] A numerical solution of the Navier–Stokes equations using the finite element technique. *Comput. Fluids* **1**, 73–100 (1973)

HOUSEHOLDER A. S.

[1] *The numerical treatment of a single nonlinear equation* (McGraw-Hill, New York 1970)

HUGHES T. J. R., LIU W. K., BROOKS A.

[1] Finite element analysis of incompressible viscous flows by the penalty function formulation. *J. Comput. Phys.* **30**, 1–40 (1979)

HUNT B.

[1] "The Mathematical Basis and Numerical Principles of the Boundary Integral Method for Incompressible Potential Flow over 3-D Aerodynamic Configurations," in *Numerical Methods in Applied Fluid Dynamics*, ed. by B. Hunt (Academic, London 1980) pp. 47–135

HUTTON A. G.

[1] "A General Finite Element Method for Vorticity and Stream Function Applied to a Laminar, Separated Flow," Central Electricity Generating Board Rpt; Research Department, Berkeley Nuclear Laboratories, Berkeley, CA (August 1975)

ISAACSON E., KELLER H. B.

[1] *Analysis of Numerical Methods* (Wiley, New York 1966)

JAMESON A.

[1] "Numerical Solution of Nonlinear Partial Differential Equations of Mixed Type," in *Numerical Solution of Partial Differential Equations III, Synspade 1975*, ed. by B. O. Hubbard (Academic, New York 1975) pp. 275–320

[2] "Transonic Flow Calculations," in *Numerical Methods in Fluid Dynamics*, ed. by H. J. Wirz, J. J. Smolderen (McGraw-Hill, New York 1978) pp. 1–87

[3] "Numerical Calculation of Transonic Flow Past a Swept Wing by a Finite Volume Method," in *Computing Methods in Applied Sciences and Engineering, 1977, II*, ed. by R. Glowinski, J. L. Lions, Lecture Notes in Physics, Vol. 91 (Springer, Berlin, Heidelberg, New York 1979) pp. 125–148

[4] "Remarks on the Calculation of Transonic Potential Flow by a Finite Volume Method," in *Numerical Methods in Applied Fluid Dynamics*, ed. by B. Hunt (Academic, London 1980) pp.363–386

JOHNSON C.

[1] A convergence estimate for an approximation of a parabolic variational inequality. *SIAM J. Numer. Anal.* **13**(4), 599–606 (1976)

[2] A mixed finite element method for the Navier–Stokes equations. *Rev. Fr. Autom. Inf. Rech. Opér. Ser. Anal. Numer.* **12**(4), 335–348 (1978)

JOURON C.

[1] Résolution numérique du problème des surfaces minima. *Arch. Rat. Mech. Anal.* **59**(4), 311–342 (1975)

KELLER H. B.

[1] "Numerical Solution of Bifurcation and Nonlinear Eigenvalue Problems," in *Applications of Bifurcation Theory*, ed. by P. Rabinowitz (Academic, New York 1977) pp. 359–384

[2] "Global Hometopies and Newton Methods," in *Recent Advances in Numerical Analysis*, ed. by C. de Boor, G. H. Golub (Academic, New York 1978) pp.73–94

KELLOG R. B.

[1] A nonlinear alternating direction method. *Math. Comput.* **23**(195), 23–27 (1969)

KIKUCHI F.

[1] "Finite Element Approximations to Bifurcation Problems of Turning Point Type," in *Computing Methods in Applied Sciences and Engineering, 1977*, Part I, ed. by R. Glowinski, J. L. Lions, Lecture Notes in Mathematics, Vol. 704, (Springer, Berlin, Heidelberg, New York 1979) pp. 252–266

KINDERLHERER D., STAMPACCHIA G.

[1] *An Introduction to Variational Inequalities and Their Applications* (Academic, New York 1980)

KOITER W. I.

[1] "General Theorems for Elastic-Plastic Solids," in *Progress in Solid Mechanics*, Vol. 2, North-Holland, Amsterdam 1960) pp. 165–221

KREISS H. O.
[1] *Numerical Methods for Solving Time-Dependent Problems for Partial Differential Equations*, Séminaire de Mathématiques Supérieures (Presses de l'Université de Montréal, Montréal 1978)

LADYSHENSKAYA O.
[1] *The Mathematical Theory of Viscous Incompressible Flow* (Gordon and Breach, New York 1969)

LANCHON H.
[1] "Torsion elaso-plastique d'un arbre cylindrique de section simplement ou multiplement connexe." Thèse d'Etat, Université Pierre et Marie Curie, Paris (1972)

LANCZOS C.
[1] Solution of systems of linear equations by minimized iterations. *J. Res. Natl. Bur. Stand.* **49**, 33–53 (1952)

LANDAU L., LIFSCHITZ E.
[1] *Mécanique des Fluides* (Mir, Moscow 1953)

LASCAUX P.
[1] *Numerical Methods for Time Dependent Equations. Application to Fluid Flow Problems.* Lecture Notes, Vol. 52, (Tata Institute of Fundamental Research, Bombay 1976)

LAWSON C. L., HANSON R. J.
[1] *Solving Least Squares Problems* (Prentice Hall, Englewood Cliffs, NJ 1974)

LEMARECHAL C.
[1] "An Extension of Davidon Methods to non Differentiable Problems," in *Mathematical Programming Study, 3: Nondifferentiable Optimization*, ed. by M. L. Balinski, P. Wolfe (North-Holland, Amsterdam 1975) pp. 95–109
[2] "Non Differentiable Optimization," in *Nonlinear Optimization: Theory and Algorithms*, ed. by L. C. W. Dixon, E. Spedicato, G. P. Szegö (Birkaüser, Boston 1980) pp.149–199

LESAINT P.
[1] Finite element methods for symmetric hyperbolic equations. *Numer. Math.* **21**, 244–255 (1973)
[2] "Sur la résolution des systèmes hyperboliques du premier ordre par des méthodes d'éléments finis," Thèse d'Etat, Université Pierre et Marie Curie, Paris (1975)

LE TALLEC P.
[1] "Les problèmes d'équilibre d'un corps hyperélastique incompressible en grandes déformations," Thèse d'Etat, Université Pierre et Marie Curie, Paris (1981)
[2] "Simulation numérique d'écoulements visqueux incompressibles par des méthodes d'éléments finis mixtes"; Thèse de 3ème cycle, Université Pierre et Marie Curie, Paris (1978)
[3] A mixed finite element approximation of the Navier–Stokes equations. *Numer. Math.* **35**, 381–404 (1980)

LEVENBERG K.
[1] A method for the solution of certain nonlinear problems in least squares. *Q. J. Mech. Appl. Math.* **2**, 164–168 (1944)

LEVEQUE R.
[1] "Time-Split Methods for Partial Differential Equations," Ph.D. Thesis, Computer Science Dept., Stanford University, Palo Alto, CA (1982)

LEVEQUE R., OLIGER J.
[1] "Numerical Methods Based on Additive Splitting for Hyperbolic Partial Differential Equations," Manuscript NA-81-16, Numerical Analysis Project, Computer Science Dept., Stanford University, Palo Alto, CA (1981)

LIEUTAUD J.
[1] "Approximation d'opérateurs par des méthodes de décomposition," Thèse d'Etat, Université Pierre et Marie Curie, Paris (1968)

LIONS J. L.
[1] *Quelques Méthodes de résolution des problèmes aux limites non linéaires.* (Dunod-Gauthier Villars, Paris 1969)

[2] *Problémes aux limites dans les équations aux dérivées partielles*, Séminaire de Mathématiques Supérieures de l'Université de Montréal (Presses de l'Université de Montreal, Montréal 1962)
[3] *Equations différentielles opérationnelles et problèmes aux limites*, (Springer, Berlin, Göttingen, Heidelberg 1961)
[4] *Contrôle optimal des systèmes gouvernés par des équations aux dérivées partielles* (Dunod-Gauthier Villars, Paris 1968)
[5] *Sur quelques questions d'Analyse, de Mécanique et de Contrôle Optimal* (Presses de l'Université de Montréal, Montréal 1976)
[6] *Elliptic Partial Differential Equations* Lecture Notes, Vol.10 (Tata Institute of Fundamental Research, Bombay 1957)

LIONS J. L., MAGENES E.
[1] *Problèmes aux limites non homogènes*, Vol.1. (Dunod, Paris 1968)

LIONS J. L., STAMPACCHIA G.
[1] Variational Inequalities. *Commun. Pure Appl. Math.* **20**, 493–519 (1967)

LIONS P. L., MERCIER B.
[1] Splitting algorithms for the sum of two nonlinear operators. *SIAM J. Numer. Anal.* **16**(6), 964–979 (1979)

LOZI R.
[1] "Analyse numérique de certains problèmes de bifurcation," Thèse de 3ème cycle, Université de Nice, Nice (1975)

MANTEUFFEL T. A.
[1] "Solving Structure Problems Iteratively with a Shifted Incomplete Cholesky Preconditioning," in *Computing Methods in Applied Sciences and Engineering*, ed. by R. Glowinski, J. L. Lions (North-Holland, Amsterdam 1980) pp. 427–444
[2] An incomplete factorization technique for positive definite linear systems. *Math. Comput.*, **34**(150), 473–497 (1980)

MARCHUK G. I.
[1] *Methods of Numerical Mathematics* (Springer, New York 1975)

MARQUARDT D. W.
[1] An algorith for least squares estimation of nonlinear parameters. *SIAM J. Appl. Math.* **11**, 431–441 (1963)

MATTHIES H., STRANG G.
[1] The solution of nonlinear finite element equations. *Int. J. Numer. Meth. Eng.* **14**, 1613–1626 (1979)

MERCIER B.
[1] *Topics in Finite Element Solution of Elliptic Problems*, Lecture Notes, Vol. 63 (Tata Institute of Fundamental Research, Bombay 1979)

MERCIER B., PIRONNEAU O.
[1] "Some Examples of Implementation and Application of the Finite Element Method," in *Lectures on the Finite Element Method*, ed. by T. W. Seng, K. H. Lee (University of Malaysia Press, Penang, Malaysia 1977)

MIELLOU J. C.
[1] Méthodes de Jacobi, Gauss–Seidel, sur (sous) relaxation par blocs, appliqué à une classe de problèmes non linéaires. *C. R. Acad. Sci. Paris* **273A**, 1257–1260 (1971)
[2] Sur une variante de la méthode de relaxation. *C. R. Acad. Sci. Paris* **275A**, 1107–1110 (1972)

MIGNOT F., MURAT F., PUEL J. P.
[1] Variation d'un point de retournement en fonction du domaine. *Commun. Part. Dif. Equ.* **4**, 1263–1297 (1979)

MIGNOT F., PUEL J. P.
[1] "Sur une classe de problèmes non linéaires avec non linéarité positive, croissante, convexe," in *Comptes-Rendus du Congrès d'Analyse Non Linéaire, Rome*, (*May 1978*) (Pitagora Editrice, Bologna 1979) pp. 45–72

MIKHLIN S. G.
[1] *Variational Methods in Mathematical Physics* (Pergamon, Oxford 1963)
[2] *Numerical Performance of Variational Methods* (Wolters-Noordhoff, Groningen, Holland 1971)

MOORE G., SPENCE A.
[1] The calculation of turning points of nonlinear equations. *SIAM J. Numer. Anal.* **17**, 567–576 (1980)

MORAVETZ C. S.
[1] On the non existence of continuous transonic flows past profiles (I). *Commun. Pure Appl. Math.* **9**, 45–68 (1956)
[2] On the nonexistence of continuous transonic flows past profiles (II). *Commun. Pure Appl. Math.* **10**, 107–131 (1957)
[3] A weak solution for a system of equations of elliptic-hyperbolic type. *Commun. Pure Appl. Math.* **11**, 315–331 (1958)
[4] Mixed equations and transonic flows. *Rend. Mat.* **25**, 1–28 (1960)
[5] The mathematical approach to the sonic barrier. *Bull. Am. Math. Soc.* (New Series) **6**(2), 127–145 (1982)

MORE J. J.
[1] "The Levenberg-Marquardt Algorithm: Implementation and Theory," in *Numerical Analysis Proceedings, Biennal Conference, Dundee 1977*, ed. by G. A. Watson, Lecture Notes in Mathematics, Vol.630 (Springer, Berlin, Heidelberg, New York 1978) pp.105–116

MOSCO U., STRANG G.
[1] One sided approximation and variational inequalities. *Bull. Am. Math. Soc.* **80**, 308–312 (1974)

MOSSOLOV P. P., MIASNIKOV V. P.
[1] Variational methods in the theory of the fluidity of a viscous-plastic medium. *J. Mech. Appl. Math.* **29**, 468–492 (1965)
[2] On stagnant flow regions of a viscous-plastic medium in pipes. *J. Mech. Appl. Math.* **30**, 705–719 (1966)
[3] On qualitative singularities of the flow of a viscous-plastic medium in pipes. *J. Mech. Appl. Math.* **31**, 581–585 (1967)

MURMAN E. M., COLE J. D.
[1] Calculation of plane steady transonic flows. *AIAA J.* **9**, 114–121 (1971)

NECAS J.
[1] *Les Méthodes Directes en Théorie des Equations Elliptiques* (Masson, Paris 1967)

NITSCHE J. A.
[1] On Korn's second inequality. *Rev. Fr. Autom. Inf. Rech. Opér. Anal. Numér.* **15**(3), 237–248 (1981)

NORRIE D. H., DE VRIES G.
[1] *The Finite Element Method. Fundamental and Applications* (Academic, New York 1973)

ODEN J. T.
[1] *Applied Functional Analysis.* (Prentice-Hall, Englewood Cliffs, NJ 1979)

ODEN J. T., KIKUCHI N.
[1] Theory of variational inequalities with application to problems of flow through porous medias. *Int. J. Eng. Sci.*, **18**(10), 1173–1284 (1980)

ODEN J. T., REDDY J. W.
[1] *Mathematical Theory of Finite Elements* (Wiley, New York 1976)

OHTAKE K., ODEN J. T., KIKUCHI N.
[1] Analysis of certain unilateral problems in von Karman plate theory by a penalty method, Part 1. *Comput. Methods Appl. Mech. Eng.* **24**, 187–213 (1980)
[2] Analysis of certain unilateral problems in von Karman plate theory by a penalty method, Part 2. *Comput. Methods Appl. Mech. Eng.* **24**, 317–337 (1980)

O'LEARY D. P.
[1] A discrete Newton algorithm for minimizing a function of many variables. *Math. Program.* **23**, 20–33 (1982)

OPIAL Z.
[1] Weak convergence of the successive approximation for non expansive mappings in Banach spaces. *Bull. Am. Math. Soc.* **73**, 591–597 (1967)

ORTEGA J., RHEINBOLDT W. C.
[1] *Iterative Solution of Nonlinear Equations in Several Variables* (Academic, New York 1970)

OSBORNE M. R., WATSON G. A.
[1] "Nonlinear Approximation Problems in Vector Norms," in *Numerical Analysis Proceedings, Biennial Conference, Dundee 1977*, ed. by G. A. Watson, Lecture Notes in Mathematics, Vol. 630 (Springer, Berlin, Heidelberg, New York 1978) pp. 117–132

OSHER S.
[1] "One-Sided Difference Schemes and Transonic Flow," in *Computing Methods in Applied Sciences and Engineering*, ed. by R. Glowinski, J. L. Lions (North-Holland, Amsterdam 1980) pp. 153–156

PAIGE C. C., SAUNDERS M. A.
[1] Solution of sparse indefinite systems of linear equations. *SIAM J. Numer. Anal.* **12**, 617–629 (1975)

PARLETT B. N.
[1] A new look at the Lanczos algorithm for solving symmetric systems of linear equations. *Linear Algebra Appl.* **29**, 323–346 (1980)

PARLETT B. N., REID J. K.
[1] On the solution of a system of linear equations whose matrix is symmetric but not definite. *BIT* **10**, 386–397 (1970)

PEACEMAN D. H., RACHFORD H. H.
[1] The numerical solution of parabolic and elliptic differential equations. *J. Soc. Ind. Appl. Math.* **3**, (1955), pp. 28–41

PERCELL P.
[1] Note on a global homotopy. *Numer. Funct. Anal. Optim.* **2**(1), 99–106 (1980)

PERIAUX J.
[1] 3-D analysis of compressible potential flows with the finite element method. *Int. J. Numer. Meth. Eng.* **9**, pp. 775–831 (1975)
[2] "Résolution de quelques problèmes non linéaires en aérodynamique par des méthodes d'éléments finis et de moindres carrés." Thèse de 3ème cycle, Université Pierre et Marie Curie, Paris (June 1979)

PERRONNET A.
[1] "MODULEF: A Library of Subroutine for Finite Element Analysis," in *Computing Methods in Applied Sciences and Engineering, 1977, 1*, ed. by R. Glowinski, J. L. Lions, Lecture Notes in Mathematics, Vol. 704 (Springer, Berlin, Heidelberg, New York 1979) pp. 127–153

PIRONNEAU O.
[1] On the transport-diffusion algorithm and its applications to the Navier–Stokes equations. *Numer. Math.* **38**, 309–332 (1982)

POIRIER G.
[1] "Traitment numérique en éléments finis de la condition d'entropie des équations transoniques," Thèse de 3ème Cycle, Université Pierre et Marie Curie, Paris (1981)

POLAK E.
[1] *Computational Methods in Optimization* (Academic, New York 1971)

POWELL M. J. D.
[1] "A Method for Nonlinear Constraints in Minimization Problems," in *Optimization*, ed. by R. Fletcher (Academic, London 1969) pp. 283–298
[2] Restart procedure for the conjugate gradient method. *Math. Program.* **12**, 148–162 (1977)
[3] "A Hybrid Method for Nonlinear Equations," in *Numerical Methods for Nonlinear Algebraic Equations*, ed. by Ph. Rabinowitz (Gordon and Breach, London 1970) pp. 87–114
[4] "A Fortran Subroutine for Solving Systems of Nonlinear Algebraic Equations," in *Numerical Methods for Nonlinear Algebraic Equations*, ed. by Ph. Rabinowitz (Gordon and Breach, London 1970) pp. 115–161

PRAGER W.
[1] *Introduction to Mechanics of Continua* (Ginn and Company, Boston 1961)

PRENTER P. M.
[1] *Splines and Variational Methods* (Wiley, New York 1975)

RAMAKRISHNAN C. V.
[1] An upwind finite element scheme for the unsteady convective diffusive transport equation. *Appl. Math. Model.* **3**, 280–284 (1979)

RANNACHER R.
[1] "On the Finite Element Approximation of the Nonstationary Navier–Stokes Problem," in *Approximation Methods for Navier–Stokes Problems*, ed. by R. Rautmann, Lecture Notes in Mathematics, Vol. 771 (Springer, Berlin, Heidelberg, New York 1980) pp. 408–424

RAVIART P. A.
[1] "The Use of Numerical Integration in Finite Element Method for Solving Parabolic Equations," in *Topics in Numerical Analysis*, Vol. 1, ed. by J. J. H. Miller (Academic, London 1973) pp. 233–264
[2] "Multistep Methods and Parabolic Equations," in *Functional Analysis and Numerical Analysis, Japan-France Seminar, Tokyo and Kyoto, 1976*, ed. by H. Fujita (Japan Society for the Promotion of Science, Tokyo 1978) pp. 429–454
[3] Incompressible finite element methods for the Navier–Stokes equations. *Adv. Water Resources* **5**, 2–8 (1982)

RAVIART P. A., THOMAS J. M.
[1] *Introduction à l'Analyse Numérique des Equations aux Dérivées Partielles* (Masson, Paris 1983)

REINHART L.
[1] "Sur la résolution numérique de problèmes aux limites non linéaires par des méthodes de continuation," Thèse de 3ème cycle, Université Pierre et Marie Curie, Paris (1980)
[2] On the numerical analysis of the Von Karman equations: Mixed finite element approximation and continuation techniques. *Numer. Math.* **39**, 371–404 (1982)

REKTORYS K.
[1] *Variational Methods in Mathematics, Science and Engineering* (Reidel, Dordrecht, Holland 1980)

RICHTMYER R., MORTON K.
[1] *Difference Methods for Initial-Value Problems* (Wiley, New York 1967)

RIESZ F., NAGY B. S.
[1] *Functional Analysis* (Ungar, New York 1955)

ROACHE P. J.
[1] *Computational Fluid Dynamics.* (Hermosa, Albuquerque, NM 1972)

ROCKAFELLAR T. R.
[1] *Convex Analysis* (Princeton University Press, Princeton, NJ 1970)

ROSEN J. B.
[1] "Approximate Computational Solution of Nonlinear Parabolic Equations by Linear Programming," in *Numerical Solution of Nonlinear Differential Equations*, ed. by D. Greenspan (Wiley, New York 1966) pp. 265–296
[2] Approximate solutions and error bounds for quasilinear elliptic boundary value problems. *SIAM J. Numer. Anal.* **7**, 80–103 (1970)

SCHECHTER S.
[1] Iteration methods for nonlinear problems. *Trans. Am. Math. Soc.* **104**, 601–612 (1962)
[2] "Minimization of Convex Functions by Relaxation," in *Integer and Nonlinear Programming*, ed. by J. Abadie (North-Holland, Amsterdam 1970) pp. 177–189
[3] Relaxation methods for convex problems. *SIAM J. Numer. Anal.* **5**, 601–612 (1968).

SCHULTZ M. H.
[1] *Spline Analysis* (Prentice Hall, Englewood Cliffs, NJ 1973)

SCHWARTZ L.
[1] *Théorie des Distributions* (Herman, Paris 1966)

SEGAL A.
[1] On the numerical solution of Stokes equation using the finite element method. *Comput. Meth. Appl. Mech. Eng.* **19**, 165–185 (1979)

SHANNO D. F.
[1] Conjugate gradient methods with inexact line searches. *Math. Opt. Res.* **13**, 244–255 (1978)

SIMPSON R. B.
[1] A method for the numerical determination of bifurcation states of nonlinear systems of equations. *SIAM J. Numer. Anal.* **12**, 439–451 (1975)

STAMPACCHIA G.
[1] "Equations Elliptiques du Second Ordre à Coefficients Discontinus." Séminaire de Mathématiques Supérieures de l'Université de Montréal (Presses de l'Université de Montréal 1965)

STRANG G.
[1] "The Finite Element Method, Linear and Nonlinear Applications," in *Proceedings of the International Congress of Mathematicians* (*Vancouver, B.C., 1974*), Vol. 2 (Canadian Math. Congress, Montreal, Que. 1975), pp. 429–435
[2] On the construction and comparison of difference schemes. *SIAM J. Numer. Anal.* **5**, 506–517 (1968)

STRANG G., FIX G.
[1] *An Analysis of the Finite Element Method* (Prentice-Hall, Englewood Cliffs, NJ 1973)

STRAUSS M. J.
[1] "Variation of Korn's and Sobolev's Inequalities," in *Proceedings of Symposia in Pure Mathematics* Vol.23 (American Mathematical Society, Providence, RI 1973) pp.207–214

STRIKWERDA J.
[1] Iterative methods for the numerical solution of second order elliptic equations with large first order terms. *SIAM J. Sci. Stat. Comput.* **1**(1), 119–130 (1980)

TABATA M.
[1] "A Finite Element Approximation Corresponding to Upwind Finite Differencing." *Mem. Numer. Math. Univ. Kyoto and Tokyo* **4**, 131–143 (1977)

TARTAR L.
[1] "Topics in Nonlinear Analysis," Publications Mathématiques d'Orsay, Université Paris-Sud, Départment de Mathématiques, Paris (1978)

TEMAM R.
[1] *Theory and Numerical Analysis of the Navier-Stokes Equations* (North-Holland, Amsterdam 1977)

THOMASSET F.
[1] *Implementation of Finite Element Methods for Navier–Stokes Equations.* (Springer, New York, Heidelberg, Berlin 1981)

TREMOLIERES R.
[1] "Inéquations Variationnelles: Existence, Approximations, Résolution." Thèse d'Etat, Université Pièrre et Marie Curie, Paris (1972)

VARGA R. S.
[1] *Matrix Iterative Analysis* (Prentice-Hall, Englewood Cliffs, NJ, 1962)

WARMING R. F., BEAM R. M.
[1] On the construction and application of implicit factored schemes for conservation laws. Symposium on Computational Fluid Dynamics, New York, April 16–17, 1977, *SIAM-AMS Proceedings* **11**, 85–129 (1978)
[2] An extension of A-stability to alternating direction implicit methods. *BIT* **19**, 395–417 (1979)
[3] "An Implicit Factored Scheme for the Compressible Navier–Stokes Equations II: The Numerical ODE Connection," in *Proceedings of the AIAA 4th Computational Fluid Dynamics Conference*, Williamsburg, Virginia, July 23-24, 1979, paper 79-1446

WHEELER M. F.
[1] An elliptic collocation-finite element method with interior penalties. *SIAM J. Numer. Anal.* **15**(1), 152–161 (1978)

WIDLUND O.
[1] A Lanczos method for a class of nonsymmetric systems of linear equations. *SIAM J. Numer. Anal.* **15**(4), 801–812 (1978)

WILDE D. J., BEIGHTLER C. S.
[1] *Foundations of Optimization* (Prentice-Hall, Englewood Cliffs, NJ 1967)

WINSLOW A. M.
[1] Numerical solution of a quasi-linear Poisson equation in a non uniform triangle mesh. *J. Comput. Phys.* **2**, 149–172 (1967)

YANENKO N. N.
[1] *The method of fractional steps* (Springer, New York, Heidelberg, Berlin 1971)

YOSIDA K.
[1] *Functional Analysis* (Springer, Berlin 1968)

YOUNG D. M.
[1] *Iterative Solution of Large Linear Systems* (Academic, New York 1971)

ZIENKIEWICZ O. C.
[1] *The Finite Element Method* (McGraw-Hill, London 1977)

Glossary of Symbols

The number on the right indicates the page of first occurrence, the most common symbols are not repeated in the following chapters.

Chapter I

$\mathbb{R}$	real line, i.e. the set of all real numbers	1
$\bar{\mathbb{R}} = \mathbb{R} \cup \{+\infty\} \cup \{-\infty\}$		1
V	a real Hilbert space	1
$(\cdot, \cdot)$ and $\|\cdot\|$	scalar product and norm of V, respectively	1
V^*	the dual space of V	1
$a(\cdot, \cdot)$	a bilinear form, continuous and V-elliptic	1
α	ellipticity constant of $a(\cdot, \cdot)$ $(\alpha > 0)$	1
L	a continuous and linear functional from V to $\mathbb{R}$	1
K	a closed convex and nonempty subset of V	1
j	a convex lower semicontinuous (l.s.c.) and proper functional	1
I_K	indicator functional of K	2
$\mathscr{L}(V, V)$	the space of the operators from V to V, linear and continuous	3
A	the unique operator of $\mathscr{L}(V, V)$ such that $$a(v, w) = (Av, w), \quad \forall\, v, w \in V$$	3
l	the unique element of V such that $L(v) = (l, v)$, $\forall\, v \in V$	3
P_K	the projection operator from V to K in the $\|\cdot\|$ norm	3
$\|A\|$	the norm of operator A, i.e. $$\|A\| = \sup_{v \in V - \{0\}} \frac{\|Av\|}{\|v\|}$$	3
$\|L\|$	the norm of L, i.e. $$\|L\| = \sup_{v \in V - \{0\}} \frac{\lvert L(v)\rvert}{\|v\|}$$	4
$J'(v)$	the Gateaux derivative of J at v; defined by $$(J'(v), w) = \lim_{\substack{t \to 0 \\ t > 0}} \frac{J(v + tw) - J(v)}{t},$$ if such a limit exists	5

$b(\cdot, \cdot)$	a bilinear form, symmetric, continuous and V-elliptic	6
β	ellipticity constant of $b(\cdot, \cdot)$ $(\beta > 0)$	6
λ	a continuous and linear functional from V to $\mathbb{R}$	7
h	a parameter converging to 0	9
$\{V_h\}_h$	a family of closed subspaces of V	9
$\{K_h\}_h$	a family of closed nonempty subsets of V with $K_h \subset V_h, \forall h$	9
χ	a dense subset of K	9
r_h	an operator from χ to V_h (or K_h)	9
U	a dense subspace of V	12
$\{j_h\}$	a family of convex, l.s.c. and proper functionals	12
ε	a positive parameter; to converge to 0	15
j_ε	penalty functional; defined from j by $j_\varepsilon = \dfrac{1}{\varepsilon} j$	15
$\langle \cdot, \cdot \rangle$	the duality pairing between V^* and V	16
$j'_\varepsilon(v)$	the Gateaux differential of j_ε at v defined by $$\langle j'_\varepsilon(v), w \rangle = \lim_{\substack{t \to 0 \\ t > 0}} \frac{j_\varepsilon(v + tw) - j_\varepsilon(v)}{t},$$ if such a limit exists	16
$\mathbb{R}^N$	N-dimensional Euclidean real space	18
$\mathbf{A}$	an $N \times N$ matrix, positive definite	18
$\mathbf{b}$	an element of $\mathbb{R}^N$	18
$\mathbf{A}^t$	the transpose matrix of $\mathbf{A}$	18
$\mathbf{B}$	an $M \times N$ matrix	18
$R(\mathbf{B})$	the range of $\mathbf{B}$	18
$\mathrm{Ker}(\mathbf{B})$	the kernel of $\mathbf{B}$	19
$\mathbf{I}$	the $M \times M$ identity matrix	21
$\rho(\mathbf{B}^t\mathbf{B})$	the spectral radius of $\mathbf{B}^t\mathbf{B}$	23
g_i	a functional from $\mathbb{R}^N$ to $\bar{\mathbb{R}}$	23
g_i^+	the positive part of g_i, i.e. $g_i^+ = \sup(0, g_i)$	24
$\mathbb{R}^N_+$	the positive cone of $\mathbb{R}^N$, i.e. $\mathbb{R}^N_+ = \{\mathbf{v} \in \mathbb{R}^N \mid \mathbf{v} = \{v_i\}_{i=1}^N, v_i \geq 0, \forall i = 1, \ldots, N\}$	25

Chapter II

Ω	a bounded domain (i.e. a bounded connected open set) of $\mathbb{R}^2$	27
$\bar{\Omega}$	closure of Ω	27
$\Gamma = \partial\Omega$	the boundary of Ω	27
$x = \{x_1, x_2\}$	the generic point of $\mathbb{R}^2$	27
$\mathbf{\nabla} = \left\{\dfrac{\partial}{\partial x_1}, \dfrac{\partial}{\partial x_2}\right\}$		27
$C^m(\bar{\Omega})$	space of m-times continuously differentiable real-valued functions for which all the derivatives up to order m are continuous in $\bar{\Omega}$	27
$\mathrm{supp}(v)$	support of the function $v = \{x \mid v(x) \neq 0\}$	27

$C_0^m(\Omega) = \{v \mid v \in C^m(\overline{\Omega}),\ \mathrm{supp}(v)$ is a compact subset of $\Omega\}$ 27
$\|v\|_{m,p,\Omega} = \sum_{|\alpha| \le m} \|D^\alpha v\|_{L^p(\Omega)}$ for $v \in C^m(\overline{\Omega})$ where $\alpha = \{\alpha_1, \alpha_2\}$; α_1, α_2 are non-negative integers. $|\alpha| = \alpha_1 + \alpha_2$ and $D^\alpha = \partial^{|\alpha|}/\partial x_1^{\alpha_1}\, \partial x_2^{\alpha_2}$ 27
$L^p(\Omega) = \{v \mid v$ measurable, $\int_\Omega |v(x)|^p\, dx < +\infty\}$ $(1 \le p < +\infty)$ 27
$W^{m,p}(\Omega)$ completion of $C^m(\overline{\Omega})$ in the above norm

$$= \{v \mid v \in L^p(\Omega),\ D^\alpha v \in L^p(\Omega),\ \forall\, \alpha,\ |\alpha| \le m\}$$ 28

$W_0^{m,p}(\Omega)$ completion of $C_0^m(\overline{\Omega})$ in the above norm 28
$H^m(\Omega) = W^{m,2}(\Omega)$ 28
$H_0^m(\Omega) = W_0^{m,2}(\Omega)$ 28
$\mathscr{D}(\Omega) = C_0^\infty(\Omega)$ 28
$dx = dx_1\, dx_2$ 28
$v|_\Gamma$ the trace of v on Γ 28
$H_0^1(\Omega) = \{v \mid v \in H^1(\Omega),\ v|_\Gamma = 0\} = \overline{\mathscr{D}(\Omega)}^{H^1(\Omega)}$ 28
$H^{-1}(\Omega)$ dual space of $H_0^1(\Omega)$ 28
$\langle \cdot, \cdot \rangle$ duality pairing between $H^{-1}(\Omega)$ and $H_0^1(\Omega)$ 28
$\mathbf{x} \cdot \mathbf{y}$ usual scalar product of $\mathbb{R}^N$; $x \cdot y = \sum_{i=1}^N x_i y_i$ if $\mathbf{x} = \{x_i\}_{i=1}^N$, $\mathbf{y} = \{y_i\}_{i=1}^N$ 28
v^+ and v^- positive and negative parts of v, respectively ($v^+ = \sup(0, v)$, $v^- = \inf(0, v)$, $v = v^+ - v^-$, $|v| = v^+ + v^-$) 29
Δ Laplace operator,

$$\Delta = \frac{\partial^2}{\partial x_1^2} + \frac{\partial^2}{\partial x_2^2} = \nabla^2$$ 30

$\mathscr{T}_h$ a finite element triangulation of Ω 32
T the generic triangle of $\mathscr{T}_h$ 32
$\mathring{T}$ interior of T 32
h length of the largest edge of the triangles of $\mathscr{T}_h$ 33
P_k space of polynomials in x_1, x_2 of degree $\le k$ 33
$v_h|_T$ restriction of v_h to T 33
$V_h^k = \{v_h \mid v_h \in C^0(\overline{\Omega}),\ v_h|_\Gamma = 0$ and $v_h|_T \in P_k,\ \forall\, T \in \mathscr{T}_h\}$ 33
$\mathrm{meas}(T)$ = measure of $T = \int_T dx$ 34
χ_T = characteristic function of T 35
$L^\infty(\Omega)$ space of the measurable functions, bounded on Ω 35
r_h^1 piecewise linear interpolation operator 36
r_h^2 piecewise quadratic interpolation operator 36
$\{\rho_n\}_{n \ge 0}$ a mollifying sequence 37
$d(x, \Gamma)$ distance from x to Γ 38
A symmetric, positive definite $N \times N$ matrix 40
$(\cdot, \cdot)$ usual scalar product of $\mathbb{R}^N$ 40
ω relaxation parameter 40

$$|\nabla v| = \sqrt{\left(\frac{\partial v}{\partial x_1}\right)^2 + \left(\frac{\partial v}{\partial x_2}\right)^2}$$ 41

$\mathrm{meas}(\Omega)$ = measure of Ω 43
$W^{2,\infty}(\Omega) = \{v \mid v \in C^1(\overline{\Omega})$, all first-order derivatives being Lipschitz continuous over $\overline{\Omega}\}$ 44

$v' = \dfrac{dv}{dx}$		44
$r = \sqrt{x_1^2 + x_2^2}$		45
$V_h = \{v_h \mid v_h \in C^0(\overline{\Omega}),\, v_h = 0 \text{ on } \Gamma,\, v_h\vert_T \in P_1,\, \forall\, T \in \mathcal{T}_h\}$		46
$\Pi_T v$	linear interpolate of v at the three vertices of T	52
h_T	diameter of T (i.e. the length of the largest edge of T)	52
Π_h	piecewise linear interpolation operator	52
$\mathscr{L}$	a Lagrangian functional	54
$L_+^\infty(\Omega) = \{q \mid q \in L^\infty(\Omega),\, q \geq 0 \text{ a.e.}\}$		54
$\nabla \cdot \nabla(a\nabla) = \sum_i \dfrac{\partial}{\partial x_i}\left(a \dfrac{\partial}{\partial x_i}\right)$		54
γv	trace of v on Γ	56
$d\Gamma$	superficial measure along the boundary Γ	58
$\dfrac{\partial}{\partial n}$	(outward) normal derivative	58
$L^2(\Gamma) = \{q \mid \int_\Gamma \vert q\vert^2\, d\Gamma < +\infty\}$		64
Λ	positive cone of $L^2(\Gamma) = \{q \mid q \in L^2(\Gamma),\, q \geq 0 \text{ a.e. on } \Gamma\}$	64
P_Λ	projection operator from $L^2(\Gamma)$ to Λ	66
$\Vert\gamma\Vert$	norm of the trace operator $\gamma\colon H^1(\Omega) \to L^2(\Gamma)$	66
$H^{1/2}(\Gamma)$	space of the traces of the functions of $H^1(\Omega)$	70
$L^1(\Gamma) = \{q \mid \int_\Gamma \vert q\vert\, d\Gamma < +\infty\}$		70
$L^\infty(\Gamma) = \{q \mid q \text{ measurable and bounded on } \Gamma\}$		70
$\Lambda = \{\mu \mid \mu \in L^2(\Gamma),\, \vert\mu(x)\vert \leq 1 \text{ a.e. on } \Gamma\}$		70
μ	viscosity of a Bingham fluid	78
g	plasticity yield of a Bingham fluid	78
$\vert\mathbf{q}\vert = \sqrt{q_1^2 + q_2^2}$		82
$\Lambda = \{\mathbf{q} \mid \mathbf{q} \in L^2(\Omega) \times L^2(\Omega),\, \vert\mathbf{q}(x)\vert \leq 1 \text{ a.e.}\}$		82
$\nabla \cdot \mathbf{p} = \operatorname{div} \mathbf{p} = \sum_i \dfrac{\partial p_i}{\partial x_i}$ (if $\mathbf{p} = \{p_i\}_i$)		84
$D_2 v = \begin{pmatrix} \dfrac{\partial^2 v}{\partial x_1^2} & \dfrac{\partial^2 v}{\partial x_1\, \partial x_2} \\ \dfrac{\partial^2 v}{\partial x_1\, \partial x_2} & \dfrac{\partial^2 v}{\partial x_2^2} \end{pmatrix}$		91
$M(T)$	measure of T	96

Chapter III

V, H	two real Hilbert spaces such that $V \subset H$, $\overline{V} = H$	98
V^*, H^*	dual spaces of V and H, respectively	98
$(\cdot,\cdot)$, $((\cdot,\cdot))$	scalar products of H and V, respectively	98
$\vert\cdot\vert$, $\Vert\cdot\Vert$	norms of H and V, respectively	98
$a(\cdot,\cdot)$	a continuous bilinear form from $V \times V \to \mathbb{R}$, elliptic in the following sense: $\exists\, \alpha > 0$ and $\lambda \geq 0$ such that $a(v, v) + \lambda\vert v\vert^2 \geq \alpha\Vert v\Vert^2, \quad \forall\, v \in V$	98

X	a Banach space whose norm is denoted by $\\|\cdot\\|_X$	98
$[0, T]$	a time interval	98
$L^2(0, T; X)$	a space of *time dependent*, X-valued functions, defined as follows: $L^2(0, T; X) = \{x \mid x(t) \in X$ a.e. on $]0, T[$, x is a measurable function of t, $\int_0^T \\|x(t)\\|_X^2\, dt < +\infty\}$	98
f	an element of $L^2(0, T; V^*)$	98
$C^0([0, T], X)$	the space of the X-valued functions continuous on $[0, T]$; also denoted $C^0(0, T; X)$	99
h	a parameter converging to 0	99
$(V_h)_h$	a family of closed subspaces of V, approximating V and H as $h \to 0$	99
$a_h(\cdot, \cdot)$	an approximation of $a(\cdot, \cdot)$, as $h \to 0$	99
f_h	an approximation of f, as $h \to 0$	99
Δt	a time discretization step	99
u_h^n	approximate solution at time $n\Delta t$	99
$(\cdot, \cdot)_h$	an approximation of $(\cdot, \cdot)$, as $h \to 0$	99
$(K_h)_h$	a family of closed convex nonempty subsets of V_h, approximating K as $h \to 0$	101
$(j_h)_h$	a family of convex, proper, l.s.c. functionals defined over V_h, approximating j as $h \to 0$	103

Chapter IV

ϕ	a real-valued, continuous, nondecreasing function defined on $\mathbb{R}$, such that $\phi(0) = 0$	110
Φ	the real-valued, C^1, convex function defined by $\Phi(t) = \int_0^t \phi(\tau)\, d\tau$	111
j	the functional defined by $j(v) = \int_\Omega \Phi(v)\, dx$ if $\Phi(v) \in L^1(\Omega)$	111
τ_n	a truncation operator	114
j_h	an approximation of j	123
$\mathbf{A}$	an $N \times N$ positive-definite matrix	128
$\mathbf{D}$	an $N \times N$ diagonal matrix	128
$\mathbf{u}, \mathbf{f}$	elements of $\mathbb{R}^N$	128
$\mathbf{S}$	a symmetric positive-definite matrix	128
$\mathbf{A}^t$	the transpose matrix of $\mathbf{A}$	129
ω	a relaxation parameter	130
$\nabla \times \mathbf{v} =$ curl of $\mathbf{v}$		134
ϕ	a velocity potential	134
ρ	density of the fluid	134
γ	ratio of specific heats	134
C_*	critical velocity	134

Chapter V

V	a real reflexive Banach space	140
V^*	dual space of V	140

K	a nonempty closed convex subset of V	140
J_0	a convex differentiable functional from V to $\mathbb{R}$	141
J_1	a convex proper l.s.c. functional from V to $\bar{\mathbb{R}}$	141
$\mathrm{Dom}(J) = \{v \mid v \in V, J(v) \in \mathbb{R}\}$		141
$\langle \cdot, \cdot \rangle$	duality pairing between V^* and V	141
A	a monotone operator from V to V^*	141
V_i	a real Hilbert space	152
$V = \prod_{i=1}^{N} V_i$		152
K_i	closed convex nonempty subsets of V_i	153
$K = \prod_{i=1}^{N} K_i$		153
P_i	a projection operator from V_i to K_i	153
ω_i	relaxation parameters	155

Chapter VI

V, H	two real topological vector spaces	166
B	an element of $\mathscr{L}(V, H)$	166
F, G	two convex proper l.s.c. functionals, from H to $\bar{\mathbb{R}}$ and V to $\bar{\mathbb{R}}$, respectively	166
$\mathscr{L}$	a Lagrangian functional	167
r	a nonnegative parameter	168
$\mathscr{L}_r$	an augmented Lagrangian functional	168
J	the functional from V to $\bar{\mathbb{R}}$ defined by $J(v) = F(Bv) + G(v)$	168
$\mathrm{dom}(j) = \{x \mid x \in X, j(x) \in \mathbb{R}\}$		168
$R(B) =$ range of B		168

Chapter VII

$\mathbf{F}$	a mapping from $\mathbb{R}^N$ to $\mathbb{R}^N$ with $\mathbf{F} = \{f_1, \ldots, f_N\}$	195
T	a nonlinear operator from $H_0^1(\Omega)$ to $H^{-1}(\Omega)$	198
k	a time discretization step ($k > 0$)	201
$\\|\cdot\\|_*$	the $H^{-1}(\Omega)$ norm defined by $\\|f\\|_* = \sup_{v \in H_0^1(\Omega) - \{0\}} \frac{\lvert\langle f, v\rangle\rvert}{\\|v\\|_{H^1(\Omega)}}$	203
s	a curvilinear abscissa	207
T.E.	abbreviation for *trailing edge*	213
Σ	a slit of the flow domain	213
p	the pressure	214
Ω_h	an approximation of Ω	217
$H_h^1 = \{v_h \mid v_h \in C^0(\bar{\Omega}_h), v_h\|_T \in P_1, \forall\, T \in \mathscr{T}_h\}$		217
$V_h = \{v_h \mid v_h \in H_h^1, v_h(\text{T.E.}) = 0\}$		217
$N_h = \dim V_h$		218
$B_h = \{w_j\}_{j=1}^{N_h}$		218
X_h	a set of feasible discrete transonic flow solutions	218
$x = \operatorname{Arg\,min}_{y \in C}$	$f(y)$ if $x \in C$ and $f(x) \leq f(y)$, $\forall\, y \in C$	219
M	Mach number	220

$[q]$	jump of a quantity q along a stream line	222
$\{r, \theta\}$	standard polar coordinates	243
$\mathbf{u} = \{u_i\}_{i=1}^N$	flow velocity	245
p	pressure	245
ν	viscosity coefficient	245
Re	Reynold number	245
$\mathbf{f}$	density of external forces	245
$(\mathbf{u} \cdot \nabla)\mathbf{u}$	a symbolic notation for $\left\{\sum_{j=1}^N u_j \frac{\partial u_i}{\partial x_j}\right\}_{i=1}^N$	245
H_h^1, H_{0h}^1, V_h	fundamental finite element spaces	248
V_{gh}, W_{gh}	finite element spaces taking into account the boundary condition $\mathbf{u} = \mathbf{g}$ on Γ	248, 249
$\tilde{\mathscr{T}}_h$	a subtriangulation of $\mathscr{T}_h$	249
$\lvert q \rvert_{1,\Omega} = (\int_\Omega \lvert \nabla q \rvert^2 \, dx)^{1/2}$		249
$\lvert \mathbf{v} \rvert_{0,\Omega} = \lVert \mathbf{v} \rVert_{L^2(\Omega) \times L^2(\Omega)}$		249
$\lVert \cdot \rVert_{X/\mathbb{R}}$	quotient norm defined by $\lVert v \rVert_{X/\mathbb{R}} = \operatorname{Inf}_{c \in \mathbb{R}} \lVert v + c \rVert_X$ where X is a functional Banach space containing the constant functions	251
k	a time discretization step	253
$\Gamma_- = \{x \mid x \in \Gamma, \mathbf{g}(x) \cdot \mathbf{n}(x) < 0\}$		259
$H^{1/2}(\Gamma)$	space of the traces of the functions of $H^1(\Omega)$	266
$\mathscr{H}^{1/2}(\Gamma) = \{\mu \mid \mu \in H^{1/2}(\Gamma), \int_\Gamma \mu \, d\Gamma = 0\}$		266
$H^{-1/2}(\Gamma)$	dual space of $H^{1/2}(\Gamma)$	266
A	a pseudo-differential operator from $H^{-1/2}(\Gamma)$ to $H^{1/2}(\Gamma)$	266
M_h	a complementary subspace of H_{0h}^1 in H_h^1, i.e. $H_h^1 = H_{0h}^1 \oplus M_h$	268
B_h	a basis of M_h	271
$\mathbf{A}_h$	a matrix approximating the operator A	271
$\mathscr{L}$	a Lagrangian functional	293
$\mathscr{L}_r$	an augmented Lagrangian functional	294

Appendix 1

$\mathbb{C}$	field of the complex numbers	325
$\bar{z}$	conjugate of the complex number z	325
$\mathscr{R}z$	real part of z	325
τ	injection operator	325
$\mathscr{V}$	a subspace of V such that $\bar{\mathscr{V}} = V$	326
δ_{x_0} = Dirac measure at x_0		344
$\lVert v \rVert_{m,p,\Omega} = \lVert v \rVert_{W^{m,p}(\Omega)}$		367
$\lvert v \rvert_{m,p,\Omega} = \left(\sum_{\substack{\alpha \\ \lvert\alpha\rvert = m}} \lVert \partial^\alpha v \rVert_{L^p(\Omega)}^p\right)^{1/p}$		367
$\lvert v \rvert_{m,\infty,\Omega} = \max_{\substack{\alpha \\ \lvert\alpha\rvert = m}} \lVert \partial^\alpha v \rVert_{L^\infty(\Omega)}$		367

h_T	length of the largest edge of a triangle T	368
ρ_T	supremum of the diameters of all the disks contained in T	368
$\hookrightarrow$	notation for the inclusion with continuity of the injection	368
π_h	an interpolation operator	369
$\overrightarrow{CA} \times \overrightarrow{CB}$	vector product of $\overrightarrow{CA}$ and $\overrightarrow{CB}$	379
ω	over-relaxation parameter	384
$\mathbf{f}$	a density of body forces	392
$\mathbf{g}$	a density of surface forces	392
$\mathbf{u}$	displacement vector	392
$\boldsymbol{\sigma}$, $\boldsymbol{\varepsilon}$	stress and strain tensors, respectively	393
λ, μ	Lamme coefficients	393
σ	Poisson coefficient	394

Appendix 2 (see previous chapters)

Appendix 3 (see Chapter VII, Sec. 5, for most notation)

$\overline{\overline{\boldsymbol{\sigma}}}$ normal stress tensor; defined by

$$\sigma_{ij} = -p\delta_{ij} + v\left(\frac{\partial u_i}{\partial x_j} + \frac{\partial u_j}{\partial x_i}\right)$$ 424

$$D_{ij}(\mathbf{u}) = \frac{1}{2}\left(\frac{\partial u_i}{\partial x_j} + \frac{\partial u_j}{\partial x_i}\right)$$ 430

Author Index

Italic figures indicate the page numbers, within the Bibliography, in which the author referred to is cited.

Subject Index